CMS/CAIMS Books in Mathematics

Volume 2

CMS/CAIMS Books in Mathematics is a collection of monographs and graduate-level textbooks published in cooperation jointly with the Canadian Mathematical Society- Societé mathématique du Canada and the Canadian Applied and Industrial Mathematics Society-Societé Canadienne de Mathématiques Appliquées et Industrielles. This series offers authors the joint advantage of publishing with two major mathematical societies and with a leading academic publishing company. The series is edited by Karl Dilcher, Frithjof Lutscher, Nilima Nigam, and Keith Taylor. The series publishes high-impact works across the breadth of mathematics and its applications. Books in this series will appeal to all mathematicians, students and established researchers. The series replaces the CMS Books in Mathematics series that successfully published over 45 volumes in 20 years.

More information about this series at http://www.springer.com/series/16627

CMS
SMC

CAIMS
SCMAI

Pinaki Mondal

How Many Zeroes?

Counting Solutions of Systems of Polynomials via Toric Geometry at Infinity

Springer

Pinaki Mondal
Scarborough, ON, Canada

ISSN 2730-650X ISSN 2730-6518 (electronic)
CMS/CAIMS Books in Mathematics
ISBN 978-3-030-75176-0 ISBN 978-3-030-75174-6 (eBook)
https://doi.org/10.1007/978-3-030-75174-6

Mathematics Subject Classification: 14C17, 14M25, 52B20, 14N10

This Springer imprint is published by the registered company Springer Nature Switzerland AG
The registered company address is: Gewerbestrasse 11, 6330 Cham, Switzerland

Contents

Acknowledgements. It was Pierre Milman who wanted me to write a book; it would not have been possible without his constant encouragement and support—with mathematics, and all sorts of things beyond mathematics—throughout these years. Even though the scope of the final version is considerably limited compared to his vision, I offer it as a first step. The encouragement from Eriko Hironaka worked as a catalyst during a critical period when the project was stuck. I sincerely thank Jan Stevens who sent numerous corrections after reading one of the earlier drafts. Najma Ahmad, Kinjal Dasbiswas, Naren Hoovinakatte, and especially, Jonathan Korman read parts of earlier drafts and gave important suggestions. Thanks are also due to the referees and editors, especially Keith Taylor, whose suggestions significantly improved the exposition. Over the last few years the work on this book took a great portion of my time owed to my friends and family, especially my mother Purnima Mondal and brother Protim Mondol. The application of points at infinity to Chickens' Road Crossing problem is due to Shatabdi Sarker; its presentation given in this book is due to Tanzil Rashid.

Scarborough, Canada Pinaki Mondal

Preface

In this book we describe an approach through *toric geometry* to the following problem: "estimate the number (counted with appropriate multiplicity) of isolated solutions of n polynomial equations in n variables over an algebraically closed field $\mathbb{k}$." The outcome of this approach is the number of solutions for "generic" systems in terms of their *Newton polytopes*, and an explicit characterization of what makes a system "generic." The pioneering work in this field was done in the 1970s by Kushnirenko, Bernstein and Khovanskii, who completely solved the problem of counting solutions of generic systems on the "torus" $(\mathbb{k}\setminus\{0\})^n$. In the context of our problem, however, the natural domain of solutions is not the torus, but the affine space $\mathbb{k}^n$. There were a number of works on extending Bernstein's theorem to the case of affine space, and recently it has been completely resolved, the final steps having been carried out by the author.

The aim of this book is to present these results in a coherent way. We start from the beginning, namely, Bernstein's beautiful theorem which expresses the number of solutions of generic systems on the torus in terms of the *mixed volume* of their Newton polytopes. We give complete proofs, over arbitrary algebraically closed fields, of Bernstein's theorem, its recent extension to the affine space, and some other related applications including generalizations of Kushnienko's results on *Milnor numbers* of hypersurface singularities which in 1970s served as a precursor to the development of toric geometry. Our proofs of all these results share several key ideas and are accessible to someone equipped with the knowledge of basic algebraic geometry. This book can serve as a companion to introductory courses on algebraic geometry or toric varieties. While it does *not* provide a comprehensive introduction to algebraic geometry, it does develop the relevant parts of the subject from the beginning (modulo some explicitly stated basic results) with lots of examples and exercises and can be used as a quick introduction to basic algebraic geometry. We hope the readers who take that undertaking will be rewarded by a deep understanding of the affine Bézout problem.

To my parents Purnima Mondal and Monojit Mondal

CHAPTER I

Introduction

1. The problem and the results

This book is about the problem of computing the number of solutions of systems of polynomials, or equivalently, the number of points of intersection of the sets of zeroes of polynomials. In this section we formulate the precise version of the problem we are going to study and give an informal description of the results. One natural observation that simplifies the problem is that *intersection multiplicity* should be taken into account, e.g. even though a tangent line intersects a parabola at only one point, it should be counted with multiplicity two (see fig. 1).

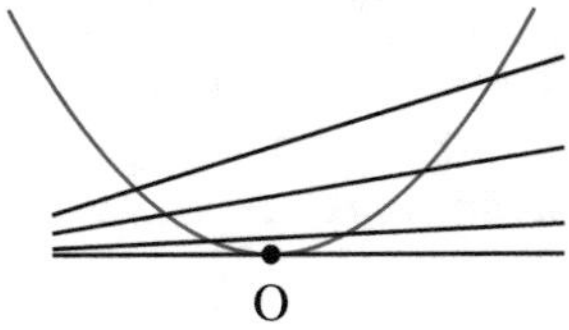

As secants approach the tangent at O more and more closely, both of the two points of intersection move arbitrarily close to O.

FIGURE 1. A tangent line intersects a parabola at a point with multiplicity two.

The geometric intuition for intersection multiplicity is the "principle of continuity," the principle that continuous perturbations of systems result in continuous changes of associated metrics or invariants[1]. Since the number of points of intersection is a *discrete* invariant of a system, it follows that it must not change under a continuous perturbation. However, over real numbers points of intersection may disappear upon an infinitesimal deformation (see fig. 2). On the other hand, this

[1]"Consider an arbitrary figure in general position ... Is it not obvious that if ... one begins to change the initial figure by insensible steps, or applies to some parts of the figure an arbitrary continuous motion, then is it not obvious that the properties and relations established for the initial system remain applicable to subsequent states of this system provided that one is mindful of particular changes, when, say, certain magnitudes vanish, change direction or sign, and so on—changes which one can always anticipate a priori on the basis of reliable rules." – J. V. Poncelet, the foremost exponent of the principle of continuity, in the introduction of *Traité des propriétés projectives des figures* (1822), as cited in [Ros05].

P. Mondal, *How Many Zeroes?*, CMS/CAIMS Books in Mathematics 2,
https://doi.org/10.1007/978-3-030-75174-6_I

problem disappears if one also counts "imaginary" solutions (this is why the intersection theory over complex numbers, or, more generally, an algebraically closed field, is easier than the intersection theory over real numbers). In this book we will consider polynomial systems defined over algebraically closed fields[2].

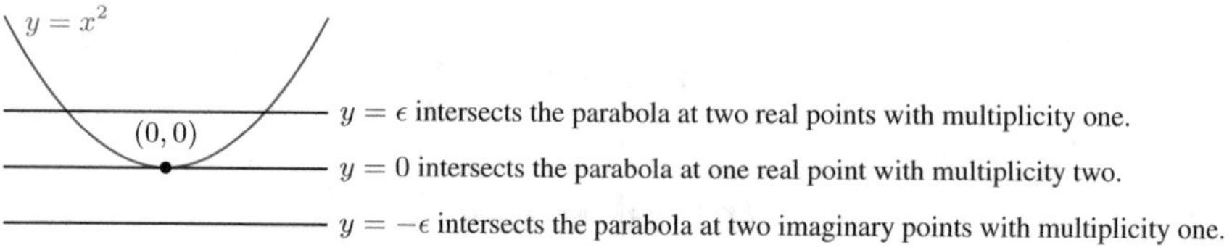
$y = x^2$
$(0, 0)$
$y = \epsilon$ intersects the parabola at two real points with multiplicity one.
$y = 0$ intersects the parabola at one real point with multiplicity two.
$y = -\epsilon$ intersects the parabola at two imaginary points with multiplicity one.

FIGURE 2. Disappearance of real points of intersection.

If there are infinitely many solutions of a system of polynomials, then the solution set has positive dimensional components, and assigning multiplicity to these components is trickier; we bypass this problem in this book and consider only the number of *isolated*[3] solutions. This implies in particular we do not consider "underdetermined systems,"[4] since an underdetermined system over an algebraically closed field can only have either positive dimensional or empty sets of solutions. We also ignore "overdetermined systems"[4] because of the relative difficulty in assigning multiplicities. The final form of the subject of this book is thus the following:

Problem I.1. (Affine Bézout problem). *Given n polynomials in n variables over an algebraically closed field $\mathbb{k}$, give a sharp estimate of the number of its isolated solutions counted with appropriate multiplicity, and determine the conditions under which it is exact.*

For $n = 1$, the fundamental theorem of algebra gives a complete answer: a polynomial of degree d has precisely d zeroes counted with multiplicity. For $n \geq 2$, there is a problem: points of intersection may run off to infinity (see fig. 3).

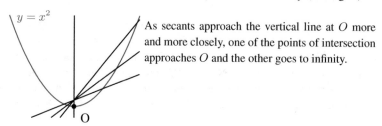

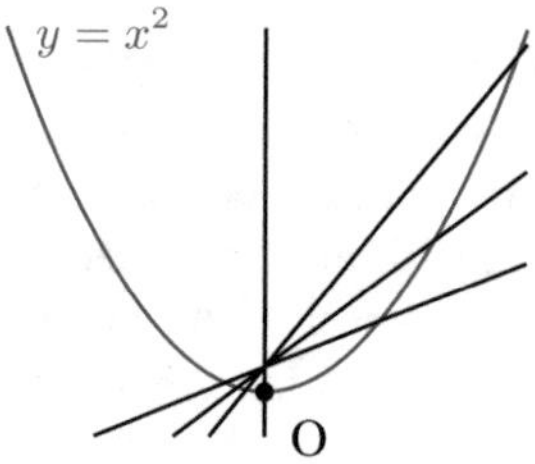
$y = x^2$
O

As secants approach the vertical line at O more and more closely, one of the points of intersection approaches O and the other goes to infinity.

FIGURE 3. A vertical line intersects the parabola at one point with multiplicity one.

[2]... which Poncelet probably would not have approved of, given his attitude toward consideration of complex solutions; see [Gra11, Section 4.2] for a most interesting account of this history.

[3]A point is *isolated* in a set S if it is open in S.

[4]A system is *underdetermined* or *overdetermined* depending on whether the number of equations is smaller or greater than the number of variables.

Any reasonable approach to problem I.1 therefore must take into account "intersections at infinity." A theorem named after E. Bézout (1730–1783) is the most basic result that does it satisfactorily.

THEOREM I.2. (Bézout's theorem, affine version). *The number of isolated solutions in* $\mathbb{k}^n$ *of* n *polynomials in* n *variables is at most the product of their degrees. Moreover, this bound is exact if and only if the only common solution of the* leading forms[5] *of the polynomials is the origin.*

Example I.3. Consider the system in fig. 3 consisting of the parabola $y - x^2 = 0$ and a line $ax + by + c = 0$. The Bézout bound is $2 \times 1 = 2$, and the leading forms are $-x^2$ and $ax + by$. As long as $b \neq 0$, the only solution to $-x^2 = ax + by = 0$ is $(0, 0)$, so that the bound is exact. However, if $b = 0$, i.e. the line is vertical, then any point of the form $(0, k)$, $k \in \mathbb{k}$, is a common solution of the leading forms. Consequently the Bézout bound overestimates the number of solutions in this case, as illustrated in fig. 3.

From the perspective of *projective geometry*, the Bézout bound is the number of intersections of polynomial hypersurfaces in the *projective space* $\mathbb{P}^n$, which is a *compactification* of the affine space $\mathbb{k}^n$ formed by adjoining a "hyperplane at infinity." Therefore the Bézout bound is exact if and only if the hypersurfaces do not intersect at any point at infinity on $\mathbb{P}^n$. However, as Gauss famously remarked,[6] infinity is the limit of some process, and curves which approach arbitrarily close to each other in one process may grow apart in another. A natural class of compactifications of $\mathbb{k}^n$ containing the projective space is that of *weighted projective spaces*. Given an n-tuple $\omega = (\omega_1, \ldots, \omega_n)$ of positive integers, the corresponding *weighted rational curve* C_a^ω through a point $a = (a_1, \ldots, a_n) \in \mathbb{k}^n$ is the curve parametrized by the map $t \mapsto (a_1 t^{\omega_1}, \ldots, a_n t^{\omega_n})$. In the same way that in the projective space straight lines with different slopes are separated at infinity, in the weighted projective space $\mathbb{P}^n(1, \omega)$ the curves C_a^ω corresponding to distinct a are separated at infinity. See fig. 4 for an example with $\omega = (1, 2)$, in which case $\{C_a^\omega\}_a$ is the family of parabolas $\{{a_1}^2 y - a_2 x^2 = 0\}$. The "weight" of a monomial $x_1^{\alpha_1} x_2^{\alpha_2} \cdots x_n^{\alpha_n}$ corresponding to ω is $\omega_1\alpha_1 + \cdots + \omega_n\alpha_n$. If f is a polynomial, then the corresponding *weighted degree* $\omega(f)$ of f is the maximum of the weights of all the monomials appearing in f. The *leading weighted homogeneous form* of f is the sum of all monomials (with respective coefficients) of f with the highest weight. Computing intersection numbers on $\mathbb{P}^n(1, \omega)$ leads to the "weighted Bézout theorem," of which the original theorem of Bézout (theorem I.2) is a special case (corresponding to $\omega = (1, \ldots, 1)$).

---

[5]The *leading form* of a polynomial is the sum of its monomial terms with the highest degree; e.g. if $f = 2x^3 + 7x^2y - 9y^2 + 7xy - x + 1$, then its degree is 3 and the leading form is $2x^3 + 7x^2y$.

[6]Discussing his friend H. Schumacher's purported proof of the parallel postulate, Gauss wrote to him (as cited in [Wat79]), "I protest first of all against the use of an infinite quantity as a completed one, which is never permissible in mathematics. The infinite is only a façon de parler, where one is really speaking of limits to which certain ratios come as close as one likes while others are allowed to grow without restriction."

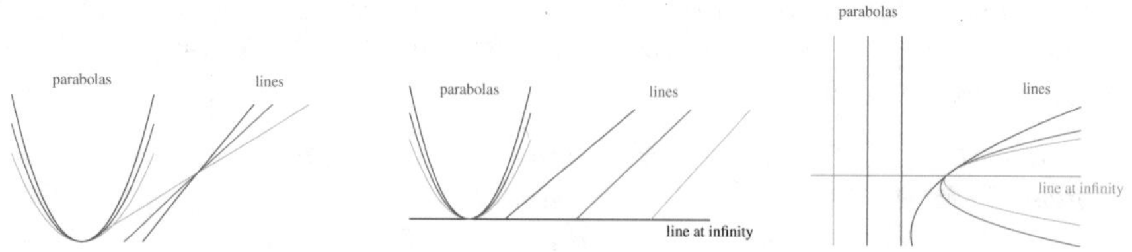

(A) $\Bbbk^2$ (B) $\mathbb{P}^2$ (C) $\mathbb{P}^2(1,1,2)$

FIGURE 4. $\mathbb{P}^2$ separates lines, but not parabolas, at infinity, whereas $\mathbb{P}^2(1,1,2)$ separates parabolas, but not lines, at infinity.

THEOREM I.4. (Weighted Bézout theorem for positive weights). *Let ω be a weighted degree on the ring of polynomials with positive weights ω_i for x_i, $i = 1,\dots,n$. Then the number of isolated solutions of polynomials $f_1,\dots,f_n$ on $\Bbbk^n$ is bounded above by $(\prod_j \omega(f_j))/(\prod_j \omega_j)$. This bound is exact if and only if the leading weighted homogeneous forms of $f_1,\dots,f_n$ have no common solution other than the origin.*

Example I.5. Let $\omega = (1,2)$, $f = y - x^2$ and $g = ax + c$, $a \neq 0$. Then $\omega(f) = 2$, $\omega(g) = 1$, and the leading weighted homogeneous forms of f and g are respectively $y - x^2$ and ax . The only solution to the leading weighted homogeneous forms of f and g with respect to ω is $(0,0)$, so theorem I.4 implies that the number of solutions of $f = g = 0$ is precisely the weighted Bézout bound $(\omega(f)\omega(g)/(\omega(x)\omega(y)) = (2 \times 1)/(1 \times 2) = 1$, as we saw in fig. 3.

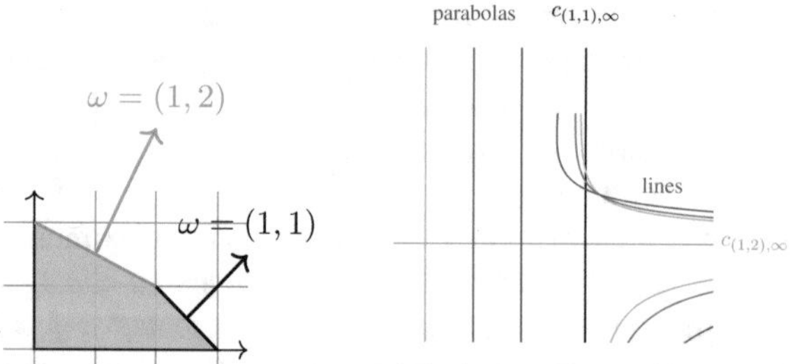

(A) $\mathcal{P}$

(B) A coordinate chart near infinity on $X_{\mathcal{P}}$

FIGURE 5. Parabolas and lines near curves at infinity on $X_{\mathcal{P}}$.

The main class of compactifications considered in this book are *toric varieties* associated to *convex integral polytopes*[7]. If $\mathcal{P}$ is an n-dimensional convex integral

[7]A *convex integral polytope* in $\mathbb{R}^n$ is the convex hull of finitely many points in $\mathbb{R}^n$ with integer coordinates.

polytope in $\mathbb{R}^n$, then the outer normal to each of its $(n-1)$-dimensional faces determines (up to a constant of proportionality) a weighted degree, and in the corresponding toric variety $X_{\mathcal{P}}$, weighted rational curves corresponding to each of these weights are separated. See fig. 5 for an example of a toric variety in which *both* parabolas and lines are separated at infinity. It has two curves at infinity (with respect to $\mathbb{k}^2$) corresponding to the two edges of $\mathcal{P}$ which are not along the axes; we denote these curves by $c_{\omega,\infty}$, where ω is the corresponding weight. Each $c_{\omega,\infty}$ separates the family of weighted rational curves corresponding to ω. Computing intersection numbers of hypersurfaces on toric varieties yields a beautiful result of D. Bernstein, which we now describe.

NP(f) $f = 1 - x + 3x^3 + 4x^2y - 7y^2$

NP(g) $g = 2 + x - y + xy + x^2y + xy^2$

FIGURE 6. Some Newton polytopes in dimension 2.

The *Newton polytope* of a polynomial is the convex hull of all the exponents that appear in its expression, see fig. 6. V. I. Arnold noticed sometime in 1960s or 1970s that invariants of "generic" systems of polynomials tend not to depend on precise values of the coefficients of their monomials, but only on the combinatorial relations of the exponents of these monomials. The study of this phenomenon was a recurring topic at his seminars at Moscow University. While working on Arnold's question on determination of the *Milnor number*[8] at the origin of a generic polynomial, A. Kushnirenko discovered that if all polynomials have the same Newton polytope, then for generic systems the number of isolated solutions which do not belong to any coordinate hyperplane has a strikingly simple expression: it is simply $n!$ times the volume of this polytope! D. Bernstein soon figured out how to remove the restriction on Newton polytopes (about 130 years before this F. Minding [Min41] discovered a special case of Bernstein's theorem in dimension two[9]).

THEOREM I.6. *Let N be the number (counted with appropriate multiplicities) of the isolated zeroes of polynomials $f_1, \ldots, f_n$ on $(\mathbb{k}^*)^n := \mathbb{k}^n \setminus \bigcup_i \{x_i = 0\}$.*

(1) Kushnirenko [Kou76]: *If each f_j has the same Newton polytope $\mathcal{P}$, then $N \leq n! \operatorname{Vol}(\mathcal{P})$. If $\operatorname{Vol}(\mathcal{P})$ is nonzero, then the bound is exact if and only if the following condition holds:*

[8]The *Milnor number* is an invariant of a singularity, see section XI.2.

[9]A. Khovanskii gives a summary of Minding's approach in [BZ88, Section 27.3]; an English translation of [Min41] by D. Cox and J. M. Rojas appears in [GK03].

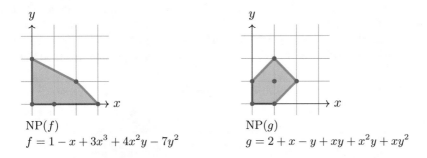

(*) *for each nontrivial weighted degree ω, the corresponding leading forms of $f_1, \dots, f_n$ do not have any common zero on $(\Bbbk^*)^n$.*

(2) Bernstein [Ber75]: *In general N is bounded above by the* mixed volume[10] *of the Newton polytopes of f_j. If the mixed volume is nonzero, then the bound is exact if and only if* (*) *holds.*

Example I.7. If the Newton polytope of each polynomial contains the origin, then theorem I.6 in fact gives an upper bound on the number of isolated solutions on $\Bbbk^n$ and it is in general better than the bounds from theorems I.2 and I.4. For example, using the fact that mixed volume of two planar bodies $\mathcal{P}$ and $\mathcal{Q}$ is simply $\text{Area}(\mathcal{P} + \mathcal{Q}) - \text{Area}(\mathcal{P}) - \text{Area}(\mathcal{Q})$ (example VII.3), we see that Bernstein's bound for the number of solutions of $f = g = 0$ (where f, g are as in fig. 6) is the area of the region shaded in blue in fig. 7, which is equal to 8. Bézout bound, on the other hand is $3 \times 3 = 9$; it is not hard to show that the 9 is also the best possible weighed Bézout bound.

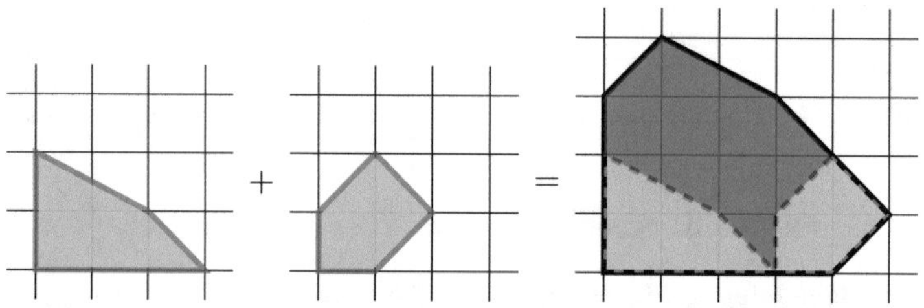

FIGURE 7. Minkowski sum of Newton polytopes of f and g.

The natural domain of solutions of systems of polynomials over a field $\Bbbk$ is however not the torus $(\Bbbk^*)^n$, but the affine space $\Bbbk^n$. There are at least two different ways to extend Bernstein's formula to $\Bbbk^n$. The approach motivated by the *polynomial homotopy* method for solving polynomial systems is as follows: given polynomials $f_1, \dots, f_n$, one starts with a deformed system $f_1 = c_1, \dots, f_n = c_n$ with nonzero c_j. For generic $f_1, \dots, f_n$ all solutions of the deformed system are in fact on the torus, and their number is given by Bernstein's theorem. Then one counts how many of these solutions approach isolated solutions of $f_1, \dots, f_n$ as each $c_j \to 0$. This approach is taken in [Kho78, HS95, LW96, RW96, Roj99]. In particular, B. Huber and B. Sturmfels [HS95] found the general formula through this approach; however they proved it in a special case, and only in characteristic zero. J. M. Rojas [Roj99] observed that Huber and Sturmfels' formula works over all characteristics. The other approach is closer to Bernstein's original proof of his theorem: here one computes the number of "branches" of the curve defined by $f_2 = \cdots = f_n = 0$ and

[10]The *mixed volume* is the canonical multilinear extension (as a functional on convex bodies) of the volume to n-tuples of convex bodies in $\mathbb{R}^n$, see section VII.2 for a precise description.

then the sum of the order of f_1 along these branches. General formulae through this approach were obtained by A. Khovanskii [unpublished][11] and the author [Mon16]. This formula requires knowing the *intersection multiplicity* at the origin of generic systems of polynomials. As an illustration we now state the weighted Bézout formula for weighted degrees with possibly *negative* weights[12]. Let ω be a weighted degree with nonzero weights $\omega_1, \dots, \omega_n$. If $I_- := \{i : \omega_i < 0\}$, then the "general weighted Bézout bound" for the number of isolated zeroes of $f_1, \dots, f_n$ is

$$\sum_{I \subseteq I_-} (-1)^{|I_-|-|I|} \frac{\prod_j \left(\max\{\omega(f_j), 0\} + \sum_{i \in I_-} |\omega_i| \deg_{x_i}(f_j)\right)}{\prod_i |\omega_i|} \tag{1}$$

(theorem X.36). Note that this reduces to the weighted Bézout bound from theorem I.4 in the case that each ω_i is positive, i.e. $I_- = \emptyset$. This bound is exact for generic $f_1, \dots, f_n$, provided $\omega(f_j)$ is *nonnegative* for each j. In the general case, define

$$\mathcal{P}_\omega(f) := \{\alpha = (\alpha_1, \dots, \alpha_n) \in \mathbb{R}^n : \alpha_i \geq 0 \text{ for each } i,\ \langle \omega, \alpha \rangle \leq \omega(f),\ \alpha_k \leq \deg_{x_k}(f) \text{ for each } j \in I_-\}$$

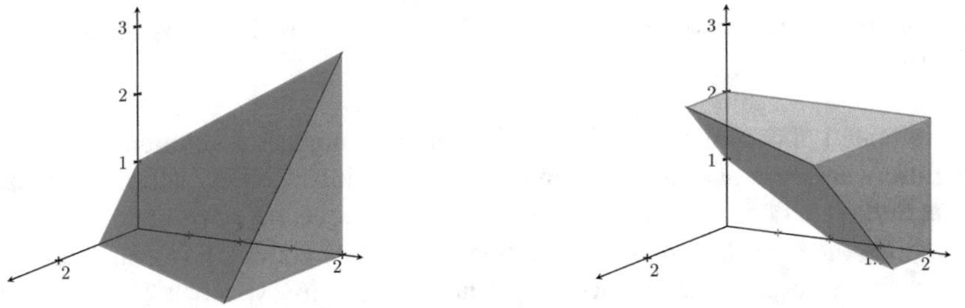

(A) $\omega = (1, -1, 1), \omega(f) = 1$ (B) $\omega = (1, -1, -1), \omega(f) = -1$

FIGURE 8. $\mathcal{P}_\omega(f)$ for $f = x_2^2 + x_2 x_3^2 + x_1 x_2 x_3$.

(see fig. 8). Given $I \subseteq \{1, \dots, n\}$, let $\mathbb{R}^I$ be the $|I|$-dimensional coordinate subspace of $\mathbb{R}^n$ spanned by all x_i, $i \in I$. Then there is a collection $\mathscr{T}$ of subsets of $\{1, \dots, n\}$ such that for each $I \in \mathscr{T}$, the number of distinct j such that $\mathcal{P}_\omega(f_j)$ touches $\mathbb{R}^I$ is precisely $|I|$, and the number of isolated zeroes of $f_1, \dots, f_n$ is bounded by

(2)
$$\sum_{I \in \mathscr{T}} \mathrm{MV}\left(\mathcal{P}_\omega(f_{j_1}) \cap \mathbb{R}^I, \dots, \mathcal{P}_\omega(f_{j_{|I|}}) \cap \mathbb{R}^I\right) \times [\pi_{I'}(\mathcal{P}_\omega(f_{j'_1}), \dots, \pi_{I'}(\mathcal{P}_\omega(f_{j'_{n-|I|}}))]_0$$

[11]Khovanskii described his result to the author at the *Askoldfest* in 2017.

[12]For simplicity here we do not allow zero weights; see theorem X.36 for the statement without this restriction.

(see theorem X.38 for the precise statement), where

- $j_1, \dots, j_{|I|}$ (respectively, $j'_1, \dots, j'_{n-|I|}$) is the collection of indices j such that $\mathcal{P}_\omega(f_j)$ touches (respectively, does not touch) $\mathbb{R}^I$;
- $\mathrm{MV}(\cdot, \dots, \cdot)$ is the mixed volume;
- $I' := \{1, \dots, n\} \setminus I$ is the complement of I, and $\pi_{I'}$ is the natural projection onto the coordinate subspace of $\mathbb{R}^n$ spanned by all $x_{i'}$, $i' \in I'$, and
- $[\cdot, \dots, \cdot]_0$ is the intersection multiplicity at the origin of systems of generic polynomials with given Newton polytopes.

The general formula for generic number of solutions on the affine space is no more difficult; it is of the same type as (2), i.e. it is a sum of products of mixed volumes and generic intersection multiplicities at the origin (see theorem X.4). However, to use it one needs to compute the generic intersection multiplicity at the origin (i.e. the second factor in the summands of (2)). In the special case that each polynomial is "convenient,"[13] a formula for generic intersection multiplicity was given by L. Ajzenberg and A. Yuzhakov [AY83]; a Bernstein-Kushnirenko type "non-degeneracy" condition, i.e. the condition for the bound being exact, was also known for convenient systems (see e.g. [Est12, Theorem 5]). In the general case Rojas [Roj99] gave a formula via Huber and Sturmfels' polynomial homotopy method. The non-degeneracy condition for the general case was established by the author in [Mon16].

As hinted above, the formula for the generic number of solutions on $\mathbb{k}^n$ is straightforward once one has the formula for generic intersection multiplicity at the origin. Sufficient criteria under which the bound is exact can also be obtained easily by adapting the Bernstein-Kushnirenko non-degeneracy condition (*); such criteria were given by several authors including Khovanskii [Kho78], Rojas [Roj99]. Precise non-degeneracy conditions, i.e. which are *necessary and sufficient* for the bound to be exact on $\mathbb{k}^n$, are however more subtle than (*); consider e.g. the problem of characterizing non-degenerate systems on $\mathbb{k}^3$ of the form

$$\begin{aligned} f_1 &= a_1 + b_1x_1x_2 + c_1x_2x_3 + d_1x_3x_1 \\ f_2 &= a_2 + b_2x_1x_2 + c_2x_2x_3 + d_2x_3x_1 \\ f_3 &= x_3(a_3 + b_3x_1x_2 + c_3x_2x_3 + d_3x_3x_1), \end{aligned}$$

where $a_j, b_j, c_j, d_j \in \mathbb{k}^*$ (this system is discussed in example X.16). If all a_j, b_j, c_j, d_j are generic, then it is straightforward to check directly that all common zeroes of f_1, f_2, f_3 on $\mathbb{k}^3$ are isolated and they appear on $(\mathbb{k}^*)^3$. Consequently, Bernstein's theorem implies that the number of solutions is the mixed volume of the Newton polytopes of the f_j, which equals 2. Now if $a_1 = a_2$, $b_1 = b_2$, and the remaining coefficients are generic, then (*) continues to be true, so that Bernstein's theorem applies and number of solutions on $(\mathbb{k}^*)^3$ is still 2; in particular,

[13] A polynomial or power series is *convenient* if for each j, there is $m_j \geq 0$ such that the coefficient of $x_j^{m_j}$ is nonzero.

the system continues to be non-degenerate on $\mathbb{k}^3$. However, in this case the set of common zeroes of f_1, f_2, f_3 on $\mathbb{k}^3$ also has a *positive* dimensional component, namely, the curve $\{x_3 = a_1 + b_1x_1x_2 = 0\}$. This situation never arises in the case of Bernstein's theorem; indeed, existence of a positive dimensional component makes a system violate $(*)$ and its straightforward adaptations. Unlike the Bernstein-Kushnirenko non-degeneracy criterion, the correct non-degeneracy criterion for $\mathbb{k}^n$ needs to accommodate existence of positive dimensional components—it has to be able to differentiate between the cases when such a component leads to a loss of isolated solutions and when it does not; such a criterion was given by the author in [Mon16].

We mentioned above that the pioneering work of Kushnirenko on counting solutions of polynomial systems was motivated by his work on Milnor numbers of hypersurface singularities. In [Kou76] he gave a beautiful formula for a lower bound on the Milnor number and showed that the bound is achieved by *Newton non-degenerate* singularities if either the characteristic is zero or if the polynomial is convenient. It was however clear from the beginning that Newton non-degeneracy is not necessary for the formula to hold, and it also does not imply "finite determinacy."[14] C. T. C. Wall [Wal99] introduced another notion of non-degeneracy which implies finite determinacy and which also guarantees that the Milnor number can be computed by Kushnirenko's formula. S. Brzostowski and G. Oleksik [BO16] found the combinatorial condition which under Newton non-degeneracy is equivalent to finite determinacy. The Milnor number of a hypersurface at the origin is same as the intersection multiplicity at the origin of the partial derivatives of the defining polynomial (or power series). The non-degeneracy condition for intersection multiplicity therefore gives a natural starting point to study Milnor numbers. This condition generalizes both Newton non-degeneracy (for isolated singularities) and Wall's non-degeneracy condition; the author showed in [Mon16] that in positive characteristic this condition is sufficient, and in zero characteristic it is both necessary and sufficient, for the Milnor number to be generic.

The purpose of this book is to give a unified exposition of the results described above. In addition to Bernstein's theorem (over arbitrary algebraically closed fields), classical results proved in this book include weighted homogeneous and multi-homogeneous versions of Bézout's theorem; complete proofs (or even, statements) of these results are otherwise hard to find. We followed Bernstein's original proof for establishing the non-degeneracy conditions of his theorem; in particular we present his simple and ingenious trick to construct a curve of solutions that runs off to infinity in the case that the non-degeneracy condition $(*)$ is not satisfied[15]. This book is the first part of a series of works on a constructive approach to compactifications

[14] i.e. it does not ensure that the singularity at the origin is isolated.

[15] The bound from Bernstein's theorem and the sufficiency of $(*)$ for the bound can be established without much difficulty (and in a very elegant way) using the general machinery of intersection theory (see e.g. [Ful93, Section 5.4]). However, we do not know of any proof of the *necessity* of $(*)$ using this approach which does not involve an adaptation of Bernstein's trick; in all probability it would be much

of affine varieties started in the author's PhD thesis [Mon10], for which the affine Bézout problem served as a motivation. Based on the results of this book, in the next part we give a solution to the general version of the affine Bézout problem, i.e. give a recipe to compute the precise number (counted with multiplicity) of solutions of any given system of n polynomials in n variables. The algorithm is inductive; it consists of finitely many steps, and at each step a non-degeneracy criterion determines if the correct number has been computed. The estimate and non-degeneracy criterion for the number of solutions on $\mathbb{k}^n$ from chapter X of this book serve as the initial step of that algorithm.

2. Prerequisites

We tried to ensure that this book is accessible to someone with the mathematical maturity and algebra background of a second year mathematics graduate student. In the ideal case a reader would be familiar with the properties of algebraic varieties discussed in chapter III, so that (s)he could start with toric varieties in part 2 and only refer to results from part 1 if necessary. However, part 1 is self contained (modulo the dependencies explicitly stated in appendix A and section IV.3.1 and some commutative algebra results stated in appendices B and C)—with proper guidance it can be used as the material for a first course in algebraic geometry. One possible strategy for such a course would be to cover the chapters on algebraic varieties (chapter III), toric varieties (chapter VI), Bernstein-Kushnirenko theorem (chapter VII) and (weighted) Bézout's theorem (chapter VIII). The chapters on intersection multiplicity (chapter IV) and polytopes (chapter V) are included for completion—in a first course the required results from these chapters can simply be explained, perhaps via examples and/or pictures, instead of working out the details of the proofs. In particular, the proofs (and exercises) given in chapter V (polytopes) are elementary and a student should not have much difficulty in following them. The most sophisticated part of chapter IV (intersection multiplicity) is the concept of a "closed subscheme" of a variety and the fact that it can be locally defined by ideals determined by regular functions; the other results are basic facts about intersection multiplicity of n regular functions at a nonsingular point a of an n-dimensional variety (e.g. that they can be defined via the "order" at a of one of the functions along the curve defined by the other functions) and relevant properties of the "order" function at a point on a (possibly non-reduced) curve. While the proofs use somewhat complicated algebra, the statements are intuitive, at least if one has some familiarity with basic properties of (complex) analytic functions.

3. Organization

Part 1 and the first chapter of part 2 have been designed as parts of a textbook, with many exercises and examples. The goal was to develop efficiently (and in an elementary way) the theory needed to prove the results in the subsequent chapters.

---

more difficult otherwise, since establishing positivity of excess intersections is in general a hard problem. Bernstein's trick is a nontrivial example of an elementary argument faring better than a formidable machinery.

These latter chapters are more like those of a monograph; there are no exercises, but they do contain a number of examples. We now give a short description of each chapter.

In chapter III we develop the required theory of algebraic varieties. We tried to stress the geometric point of view where possible. A number of results have been developed through exercises; ample hints have been provided to ensure that no single step of any exercise is very difficult. In chapter IV we describe basic properties of intersection multiplicity (of n regular functions at a nonsingular point of an n-dimensional variety), in particular how it can be computed using curves. After giving simple examples to illustrate that a satisfactory treatment of intersection multiplicity would need to incorporate non-reduced rings, we give a short introduction to "closed subschemes of a variety"[16]. A number of examples presented in chapter III and IV were taken from answers to the question *Algebraic geometry examples* [hba] posed by R. Borcherds on *MathOverflow*. Chapter V is a compilation (with complete proof) of the properties of convex polyhedra which, together with the results of chapter III, constitute the foundation on which we introduce toric varieties in chapter VI. In chapter VI we mainly discuss those properties of toric varieties which are required for the results in the subsequent chapters. In chapter VII we prove Bernstein's theorem and present some of its basic applications to convex geometry. In chapter VIII we apply Bernstein's theorem to prove the weighted homogeneous and multi-homogeneous versions of Bézout's theorem. Chapter IX contains the results on the generic bound and non-degeneracy conditions for intersection multiplicity at the origin, which we use in chapter X to compute the generic bounds and non-degeneracy conditions for the number of solutions of polynomial systems on $\mathbb{k}^n$. It turns out that one can as easily replace $\mathbb{k}^n$ by an arbitrary Zariski open subset of $\mathbb{k}^n$—the results of chapter X are derived in this greater generality. In chapter X we also use the main results to derive generalizations of weighted homogeneous and multi-homogeneous versions of Bézout's theorem applicable to weighted degrees with possibly zero or negative weights. In chapter XI we apply the results from chapter IX to the study of Milnor numbers; in particular, we derive and generalize classical results of Kushnirenko on Milnor numbers. Chapter VII, X and XI end with selections of open problems (mostly combinatorial in nature).

[16]We decided to omit definitions of general sheaves and schemes since we do not use these notions anywhere in this book. On the other hand, once one really understands the special cases of "sheaves of ideals" and "closed subschemes of a variety," which are discussed in chapter IV, the leap to the general notions will be natural.

On the projective space chickens have more than one way of crossing roads

CHAPTER II

A brief history of points at infinity in geometry

In this chapter we give a brief historical overview of the concept of points at infinity in geometry and the subsequent introduction of homogeneous coordinates on projective spaces.

1. Points at infinity

Points at infinity seem to have first cropped up in Johannes Kepler's work on conics in *Ad Vitellionem paralipomena quibus astronomiae pars optica traditur*[1] (1604). It is in this text that Kepler introduces the term *focus*[2] to denote each of the (unique) pair of points inside a conic such that the rays from any point on the conic make equal angles to the tangent at that point. For a circle the foci coincide at the center, and they separate as the circle deforms into an ellipse. As one continues to deform the ellipse so that in the end it turns into a parabola, Kepler concludes that "In the Parabola one focus ... is inside the conic section, the other to be imagined either inside or outside, lying on the axis at an infinite distance from the first, so that if we draw the straight line ... from this blind focus to any point ... on the conic section, the line will be parallel to the axis ..." [FG87, pp. 186–187], see fig. 1.

[1]"Literally 'Things omitted by' (or 'Supplements to') 'Witelo with which the optical part of astronomy is concerned'. Witelo's *Perspectiva*, probably written in the 1270s, appeared in several new editions in the sixteenth century, and seems to have been the standard textbook on Optics" [FG87, pp. 221–222].

[2]*Focus* is the Latin word for hearth. "Since light was reflected to the focus, ... the focus of the mirror was the position in which one would place the material one wished to burn" [FG87, p. 222].

P. Mondal, *How Many Zeroes?*, CMS/CAIMS Books in Mathematics 2,
https://doi.org/10.1007/978-3-030-75174-6_II

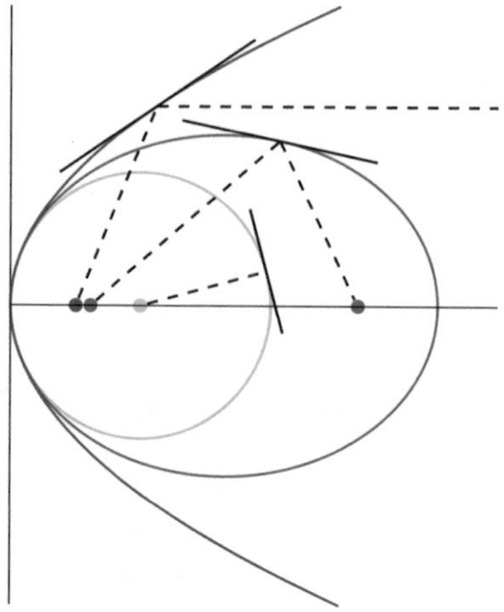

FIGURE 1. Foci of families of conics.

Another inspiration, albeit indirect, of points at infinity is the theory of linear perspective. Application of perspectives was already present in early fourteenth century paintings from Italy [And07, Chapter I], the earliest surviving written account of geometric construction of perspective being Leon Battista Alberti's *De Pictura* (1435). By the seventeenth century there were numerous treatises on perspective. In 1639 Girard Desargues, who had worked as a military engineer and written on perspective, circulated fifty copies of his *Brouillon project d'une atteinte aux evenmens des rencontres du cone avec un plan* ("Rough draft of an essay on the results of taking plane sections of a cone"). At the very beginning of *Brouillon project* Desargues introduced the notion that parallel lines intersect at a point at infinity and parallel planes intersect at a line at infinity; constructing essentially the projective plane $\mathbb{P}^2(\mathbb{R})$ and the three-dimensional projective space $\mathbb{P}^3(\mathbb{R})$ over $\mathbb{R}$. He made extensive use of the lines and planes at infinity to give a unified treatment of families of parallel lines and families of lines through a common point. The subject of projective geometry was born in *Brouillon project*.

2. Homogeneous coordinates

The birth, however, went practically unnoticed. Desargues's manuscript was thought to have been lost and it did not inspire much new work (other than Blaise Pascal's *Essay pour les coniques* (1640) which contains Pascal's theorem on conics). Projective geometry was revived in the nineteenth century largely due to Jean-Victor Poncelet, who fought in Napoleon's army in the battle of Krasnoi in November 1812, and then was a prisoner of war in Saratov till Napoleon's defeat in mid 1814. In the prison "he occupied himself summarizing all he knew of the mathematical sciences in notebooks that he then distributed to his fellow prisoners who wanted to finish an education disrupted by the incessant military campaigns"

[Gra11, p. 13]. In the process he discovered, and upon his return to France, championed, the unifying aspect of projective geometry (as opposed to the "analytic geometry" of René Descartes). A fundamental tool of this new geometry was the duality between points and lines on the plane. Initially applied by Charles Julien Brianchon and Poncelet to conics, the duality principle was extended to all planar curves by Joseph Diaz Gergonne[3]. All the details of the duality principle however were not clear, e.g. even though the principle suggests that dualising twice one should get back the original curve, it was soon discovered that dualising a curve of degree higher than two results in a curve of degree higher than that of the original curve. Poncelet had some ideas about resolving this paradox by taking into account the effects of cusps and double points on a curve, but his ideas were not very precise. The resolution came through the algebraic treatment of projective geometry by August Möbius in *Der Barycentrische Calcül* (1827). Möbius observed that weights w_0, w_1 placed at the ends of a (weightless) rod uniquely determines a point P on the rod, namely, their *Barycenter*, i.e. the center of gravity; the ordered pair $[w_0 : w_1]$ (we write it in this way to distinguish from the Cartesian coordinates of P) are the *Barycentric coordinates* of P. It is straightforward to work out the relation between the Cartesian and barycentric coordinates, e.g. if we identify the rod with the closed interval $[a, b]$ on the real line, then the barycentric coordinates $[w_0 : w_1]$ of $x \in [a, b]$ satisfies

$$x = \frac{aw_0 + bw_1}{w_0 + w_1}.$$

This formula can be readily extended to allow for w_0 and w_1 to be zero or negative. It follows that each point on the real line has barycentric coordinates, see fig. 2. It is also clear that the barycentric coordinates are *homogeneous*, i.e. $[w_0\lambda : w_1\lambda]$ denote the same point as $[w_0 : w_1]$ for every nonzero $\lambda \in \mathbb{R}$. Finally, note that if $w_0 + w_1 = 0$, then $[w_0 : w_1]$ does not correspond to any point on the line; Möbius defined it as a point lying at infinity.

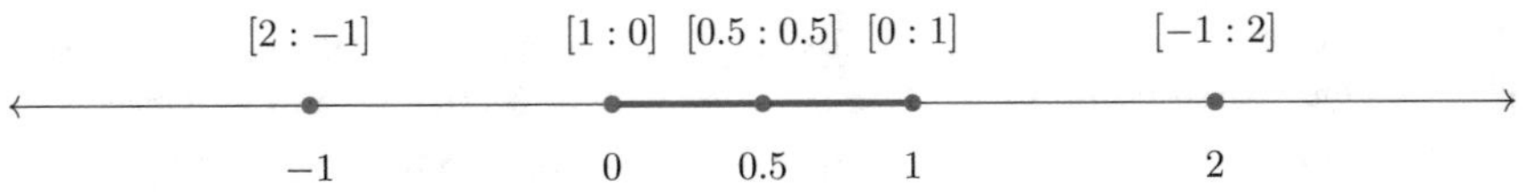

FIGURE 2. Barycentric coordinates in dimension one with respect to the interval $[0, 1]$.

In dimension two one starts with a triangle Δ; assume for convenience that the vertices of Δ are the points with Cartesian coordinates $(0, 0)$, $(1, 0)$ and $(0, 1)$. Then

[3] "... it is one thing to realize that dualising a figure is a good way to obtain new theorems, which is what Poncelet did, and quite another thing to claim that points and lines are interchangeable concepts which must logically be treated on a par. This was the view that Gergonne put forward in 1825. Interpreted in such generality, Gergonne's principle of duality is one of the most profound and simple ideas to have enriched geometry since the time of the Greeks ..." [Gra11, p. 55].

the Cartesian coordinates (x, y) and the barycentric coordinates $[w_0 : w_1 : w_2]$ of a point P on the plane with respect to Δ are related as follows (see Figure 3):

$$x = \frac{w_1}{w_0 + w_1 + w_2}, \quad y = \frac{w_2}{w_0 + w_1 + w_2}.$$

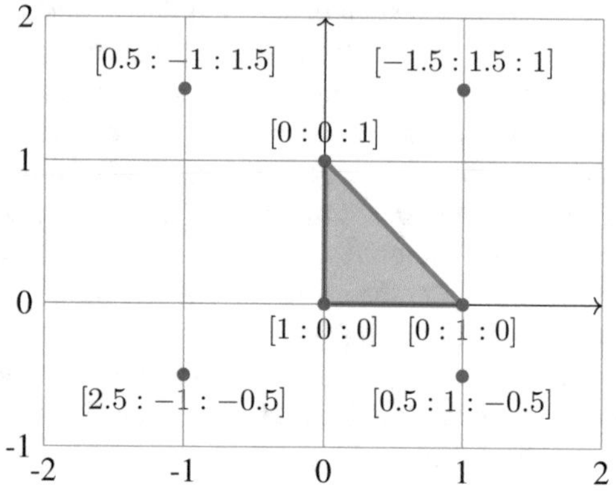

FIGURE 3. Barycentric coordinates in dimension two with respect to Δ.

As in the case of the real line, the barycentric coordinates of the points at infinity are $[w_0 : w_1 : w_2]$ with $w_0+w_1+w_2 = 0$. Möbius observed that many computations with barycentric coordinates become simpler upon a change of coordinates of the form

$$[w_0 : w_1 : w_2] \mapsto [w_0 + w_1 + w_2 : w_1 : w_2].$$

These new coordinates are nowadays usually denoted as *homogeneous coordinates*. In particular, the equation of the line $ax + by + c = 0$ changes in the barycentric coordinates to $aw_1 + bw_2 + c(w_0 + w_1 + w_2) = 0$, and this in turn becomes $aw_1 + bw_2 + cw_0 = 0$ in homogeneous coordinates. And in homogeneous coordinates the points at infinity are described by $w_0 = 0$. As he was finishing *Der Barycentrische Calcül*, Möbius heard of the duality between points and lines studied by the French geometers, and noticed that the homogeneous coordinates give a natural algebraic approach to duality, namely, the line $aw_1+bw_2+cw_0 = 0$ corresponds simply to the point with homogeneous coordinates $[a : b : c]$, and vice versa. This automatically ensured that concurrent lines go to collinear points under duality and that dualizing twice one gets back to the original curve. Julius Plücker, possibly independently of Möbius, gave an analogous theory of homogeneous coordinates in 1830, and later used it to completely resolve the duality paradox. The homogeneous coordinates were soon extended to higher dimensions, which opened the door to algebraic study of higher dimensional projective spaces.

3. Projective space

Take an arbitrary field $\Bbbk$ and fix coordinates $(x_0, \dots, x_n)$ on $\Bbbk^{n+1}$, $n \geq 0$. The *n-dimensional projective space* $\mathbb{P}^n$ *over* $\Bbbk$ is the set of lines (with respect to $(x_0, \dots, x_n)$) in $\Bbbk^{n+1}$ through the origin. Every point $(a_0, \dots, a_n) \in \Bbbk^{n+1} \setminus \{0\}$ determines a unique line through the origin which passes through it; the *homogeneous coordinate* of this line is $[a_0 : \cdots : a_n]$. For each $j = 0, \dots, n$, let $U_j := \{[a_0 : \cdots : a_n] : a_j \neq 0\}$. The map

$$(a_1, \dots, a_n) \mapsto [a_1 : \cdots : a_{j-1} : 1 : a_j : \cdots : a_n]$$

gives a one-to-one correspondence between $\Bbbk^n$ and U_j. In the case that $\Bbbk = \mathbb{R}$ or $\mathbb{C}$, one can use this correspondence to induce a topology on U_j (by declaring a subset of U_j to be open if and only if its pre-image in $\mathbb{R}^n$ or $\mathbb{C}^n$ is open). It is straightforward to check that these topologies are compatible (i.e. they induce the same topology on their intersections), and accordingly turns $\mathbb{P}^n = \bigcup_{j=0}^n U_j$ into a manifold. For a general $\Bbbk$, the usual topology put on $\Bbbk^n$ is the *Zariski topology*, in which the closed subsets are zero-sets of systems of polynomials, and the identification of the U_j with $\Bbbk^n$ is used to give $\mathbb{P}^n$ the structure of an *algebraic variety over* $\Bbbk$. We review algebraic varieties and Zariski topology in chapter III. The complement H_j of U_j in $\mathbb{P}^n$ is the set of lines (through the origin) which lie on the j-th coordinate hyperplane, so that if we identify $\Bbbk^n$ with U_0, then the set of points at infinity, i.e. the complement of $\Bbbk^n$ in $\mathbb{P}^n$, is precisely $H_0 := \{[0 : a_1 : \cdots : a_n]\}$, and the homogeneous coordinates on $\mathbb{P}^n$ are precisely those introduced by Möbius. Note that H_0 is naturally isomorphic to $\mathbb{P}^{n-1}$.

Proposition II.1. *The closure in $\mathbb{P}^n$ of each straight line on $\Bbbk^n$ has a unique point at infinity. Two coplanar lines intersect at a common point at infinity if and only if they are parallel.*

PROOF. Here we treat the case that $\Bbbk = \mathbb{C}$ and the topology on $\mathbb{P}^n$ is that induced from the Euclidean topology on $\mathbb{C}^n \cong \mathbb{R}^{2n}$; see exercise III.61 for the case of general $\Bbbk$ and Zariski topology on $\mathbb{P}^n$. Let $L = \{(a_1, \dots, a_n) + t(b_1, \dots, b_n) : t \in \mathbb{C}\}$ be a line on $\mathbb{C}^n \cong U_0$ (note that this means $(b_1, \dots, b_n) \neq (0, \dots, 0)$). In homogeneous coordinates

$$L = \{[1 : a_1 + tb_1 : \cdots : a_n + tb_n] : t \in \mathbb{C}\}.$$

Therefore the set of all points on $\mathbb{C}^{n+1}$ which correspond to points on L is

$$L' = \{(s, sa_1 + stb_1, \dots, sa_n + stb_n) : s, t \in \mathbb{C}, s \neq 0\}.$$

Since $(b_1, \dots, b_n) \neq 0$, it is straightforward to check that the closure $\bar{L}'$ of L' in $\mathbb{C}^{n+1}$ is the plane spanned by $(1, a_1, \dots, a_n)$ and $(0, b_1, \dots, b_n)$ (see fig. 4). The points at infinity on the closure of L in $\mathbb{P}^n$ correspond to the points $(x_0, \dots, x_n)$ on $\bar{L}' \setminus \{0\}$ with $x_0 = 0$, i.e. the set of points $(0, \lambda b_1, \dots, \lambda b_n)$, $\lambda \in \mathbb{C} \setminus \{0\}$. Since all these points correspond to the single point $[0 : b_1 : \cdots : b_n]$ on $\mathbb{P}^n$, this proves both assertions of the proposition. □

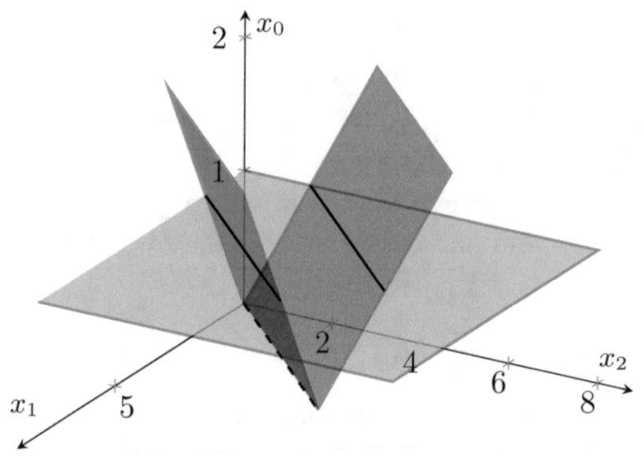

FIGURE 4. U_0 can be identified with the hyperplane $x_0 = 1$ on $\mathbb{k}^{n+1}$. Every line on U_0 corresponds to a plane on $\mathbb{k}^{n+1}$. Parallel lines on U_0 intersect at a point at infinity on $\mathbb{P}^n$ since the corresponding planes on $\mathbb{k}^{n+1}$ intersect along a line through the origin on $x_0 = 0$.

Proposition II.1 shows that the projective space incorporates the intuition from the theory of perspectives that two parallel lines intersect at a point at infinity. The connection of projective spaces with the affine Bézout problem comes from the following property (which you will see in example III.40): if $\mathbb{k}$ is algebraically closed, then for each polynomial $f \in \mathbb{k}[x_1, \ldots, x_n]$, there is a correspondence between the following sets:

(1) points at infinity on the closure in $\mathbb{P}^n$ of $\{x \in \mathbb{k}^n : f(x) = 0\}$ $\longleftrightarrow$ $\{x \in \mathbb{k}^n \setminus \{0\} : \mathrm{ld}(f)(x) = 0\}$

where $\mathrm{ld}(f)$ is the leading form of f. This correspondence provides the geometric explanation for the condition from Bézout's theorem (theorem I.2) under which the Bézout estimate for the number of solutions is exact. A proof of Bézout's theorem usually consists of showing that

- given n polynomials $f_1, \ldots, f_n \in \mathbb{k}[x_1, \ldots, x_n]$, the number of isolated points (counted with appropriate multiplicity) on the intersection of the closures in $\mathbb{P}^n$ of $\{f_i = 0\}$ is at most the product of the degrees of the f_i, and
- if $\mathbb{k}$ is algebraically closed, then this bound is attained with points inside $\mathbb{k}^n$ if and only if there is no "intersection at infinity" on $\mathbb{P}^n$, i.e.

$$\bigcap_i \overline{\{f_i = 0\}} = \bigcap_i \{f_i = 0\}.$$

Most of the mathematics of this book takes place on *toric varieties* (introduced in chapter VI), a class of algebraic varieties of which the projective space is a special case. We will study natural analogues of Bézout's theorem on different classes of toric varieties, and Bézout's theorem will fall out as a special case (corollary VIII.3).

Part 1

Preliminaries

CHAPTER III

Quasiprojective varieties over algebraically closed fields

In this chapter we give a quick introduction to algebraic varieties, focusing mainly on the properties used in parts 2 and 3. Appendix B includes a discussion of the relevant concepts and results from commutative algebra including Hilbert's basis theorem (theorem B.4) and Nullstellensatz (theorem B.8) which are part of the foundation of modern algebraic geometry. There are a few results including Krull's "principal ideal theorem" (theorem A.1) that we use without proof—these are listed in appendix A. Throughout this book $\mathbb{k}$ denotes an algebraically closed field. We start the discussion with *affine varieties* over $\mathbb{k}$. General *quasiprojective varieties* are studied from section III.4 onward. Unless explicitly stated otherwise, in this book a *variety* will mean a quasiprojective variety over $\mathbb{k}$.

1. Affine varieties

An *affine space* over $\mathbb{k}$ is simply the set $\mathbb{k}^n$ of n-tuples of elements from $\mathbb{k}$ for some $n \geq 0$. Fix a system of coordinates $(x_1, \ldots, x_n)$ on $\mathbb{k}^n$, i.e. $\mathbb{k}^n = \{(x_1, \ldots, x_n) : x_1, \ldots, x_n \in \mathbb{k}\}$. A *subvariety* of $\mathbb{k}^n$ is the set of zeroes of a collection of polynomials in $(x_1, \ldots, x_n)$[1]. An *affine variety* is simply the subvariety of an affine space. A *hypersurface* of $\mathbb{k}^n$ is the set of zeroes of a single polynomial.

Example III.1. Since every nonzero polynomial in a single variable has only finitely many zeroes, every proper subvariety of $\mathbb{k}$ consists of finitely many points. Hypersurfaces of $\mathbb{k}^2$ are special cases of (affine) *algebraic curves*[2]. See fig. 1 for pictures of real points of some curves on $\mathbb{C}^2$. The curve in fig. 1b is an "elliptic curve" and the one in fig. 1c is a "deltoid" which was investigated by L. Euler in 1745 in relation to a problem in optics; we refer to [BK86, Chapter I] for many pictures of algebraic curves along with their history.

[1]Traditionally in the definition of a subvariety it was common to include also the requirement that a subvariety V should be *irreducible*, i.e. whenever $V = V_1 \cup V_2$, where V_1, V_2 are sets of zeroes of systems of polynomials, then either $V = V_1$ or $V = V_2$; however, our definition is also widely used now.

[2]A *curve* is a variety of *dimension* one. We discuss dimensions in section III.11.

P. Mondal, *How Many Zeroes?*, CMS/CAIMS Books in Mathematics 2,
https://doi.org/10.1007/978-3-030-75174-6_III

(A) $x^2 + 4y^2 = 9$ (B) $x^2 = y^3 - 4y + 4$ (C) $(x^2+y^2+9)^2 - 8(x^3 - 3xy^2) = 108$

FIGURE 1. Real points of some curves on $\mathbb{C}^2$.

Given a subset $\mathfrak{q}$ of $\Bbbk[x_1, \dots, x_n]$, we write $V(\mathfrak{q})$ for the subvariety determined by $\mathfrak{q}$, i.e. $V(\mathfrak{q}) = \{a \in \Bbbk^n : f(a) = 0 \text{ for all } f \in \mathfrak{q}\}$. Hilbert's Basis Theorem (theorem B.4) implies that the ideal $\langle \mathfrak{q} \rangle$ of $\Bbbk[x_1, \dots, x_n]$ generated by $\mathfrak{q}$ is in fact generated by *finitely many* polynomials $f_1, \dots, f_k$. But then it is immediate to check that $V(\mathfrak{q})$ is precisely the set of common zeroes of $f_1, \dots, f_k$, i.e. $V(\mathfrak{q}) = V(f_1, \dots, f_k)$. This proves the fundamental fact that every affine variety is determined by *finitely many* polynomials:

Proposition III.2. *Every subvariety of $\Bbbk^n$ is the set of common zeroes of finitely many polynomials.* □

It is straightforward to check that there is a unique topology on $\Bbbk^n$ for which the closed sets are precisely the subvarieties on $\Bbbk^n$ (exercise III.3); it is called the *Zariski topology*. In this book $\Bbbk^n$ (and its subsets) will always be assumed to be equipped with the Zariski topology; in particular, by "closed" or "open" subsets we will mean Zariski closed or Zariski open subsets. If X is a subset of $\Bbbk^n$, we write $I(X)$ for the set of all polynomials $f \in \Bbbk[x_1, \dots, x_n]$ such that $f(a) = 0$ for all $a \in X$. It is straightforward to check that $I(X)$ is a *radical ideal* of $\Bbbk[x_1, \dots, x_n]$ (exercise III.7).

Example III.3. If X is a singleton consisting of a single point $a = (a_1, \dots, a_n) \in \Bbbk^n$, then $I(X)$ is the ideal $\mathfrak{m}_a$ of $\Bbbk[x_1, \dots, x_n]$ generated by $x_1 - a_1, \dots, x_n - a_n$. Indeed, it is clear that $\mathfrak{m}_a \subseteq I(X)$. Since $I(X)$ is a proper ideal of $\Bbbk[x_1, \dots, x_n]$ and $\mathfrak{m}_a$ is a *maximal* ideal of $\Bbbk[x_1, \dots, x_n]$ (since $\Bbbk[x_1, \dots, x_n]/\mathfrak{m}_a \cong \Bbbk$), it follows that $I(X) = \mathfrak{m}_a$.

Example III.4. If $X = \Bbbk^n$, then $I(X) = 0$. Indeed, since an algebraically closed field is infinite, no nonzero polynomial in $(x_1, \dots, x_n)$ vanishes at all points of $\Bbbk^n$ (exercise III.11), so that $I(\Bbbk^n) = 0$.

Some basic properties of the operators $V(\cdot)$ and $I(\cdot)$ are presented in exercises III.1 to III.10—the reader is urged to go over them. Given an ideal $\mathfrak{q}$ of $\Bbbk[x_1, \dots, x_n]$, it can be seen directly from the definitions that $V(\sqrt{\mathfrak{q}}) = V(\mathfrak{q})$ and $I(V(\mathfrak{q})) \supseteq \sqrt{\mathfrak{q}}$ (e.g. see exercises III.5 and III.9); Hilbert's Nullstellensatz (theorem B.8) implies that the latter containment is in fact an equality:

$$I(V(\mathfrak{q})) = \sqrt{\mathfrak{q}}. \tag{4}$$

The Nullstellensatz sets up the basic correspondence between algebra and geometry underlining algebraic geometry over algebraically closed fields:

THEOREM III.5. *There is a one-to-one correspondence between subvarieties of $\Bbbk^n$ and radical ideals of $\Bbbk[x_1,\dots,x_n]$ given by $I(\cdot)$ and $V(\cdot)$. Given a subvariety X of $\Bbbk^n$ and a radical ideal $\mathfrak{q}$ of $\Bbbk[x_1,\dots,x_n]$, one has $V(I(X)) = X$ and $I(V(\mathfrak{q})) = \mathfrak{q}$.*

PROOF. This follows immediately from identity (4) and exercises III.7 and III.10. □

In many ways the Zariski topology is distinctly different from the Euclidean topology on $\mathbb{R}$ and $\mathbb{C}$, in part due to "fewer" closed sets. We have already seen that the only proper closed subsets of $\Bbbk$ are finite subsets. See exercises III.12 and III.13 for some other quirks of Zariski topology.

1.1. Exercises.

EXERCISE III.1. If $\mathfrak{q}_1 \subseteq \mathfrak{q}_2 \subseteq \Bbbk[x_1,\dots,x_n]$, show that $V(\mathfrak{q}_1) \supseteq V(\mathfrak{q}_2)$.

EXERCISE III.2. Given subsets $\mathfrak{q}_1, \mathfrak{q}_2$ of $\Bbbk[x_1,\dots,x_n]$, show that

(1) $V(\mathfrak{q}_1) \cap V(\mathfrak{q}_2) = V(\mathfrak{q}_1 \cup \mathfrak{q}_2) = V(\mathfrak{q}_1 + \mathfrak{q}_2)$.

(2) $V(\mathfrak{q}_1) \cup V(\mathfrak{q}_2) = V(\mathfrak{q}_1 \cap \mathfrak{q}_2) = V(\mathfrak{q}_1\mathfrak{q}_2)$, where $\mathfrak{q}_1\mathfrak{q}_2 = \{f_1f_2 : f_j \in \mathfrak{q}_j,\ j = 1,2\}$. [Hint: the inclusions $V(\mathfrak{q}_1) \cup V(\mathfrak{q}_2) \subseteq V(\mathfrak{q}_1 \cap \mathfrak{q}_2) \subseteq V(\mathfrak{q}_1\mathfrak{q}_2)$ follows from exercise III.1, so that it suffices to show $V(\mathfrak{q}_1\mathfrak{q}_2) \subseteq V(\mathfrak{q}_1) \cup V(\mathfrak{q}_2)$. For the latter containment pick $a \in V(\mathfrak{q}_1\mathfrak{q}_2) \setminus V(\mathfrak{q}_1)$. There is $f \in \mathfrak{q}_1$ such that $f(a) \neq 0$. Consider the products fg with $g \in \mathfrak{q}_2$ to show that $a \in V(\mathfrak{q}_2)$.]

EXERCISE III.3. Show that the collection of subvarieties on an affine space satisfies the axioms of a topology, namely, that it contains the empty set and the affine space itself, and it is closed under finite unions and arbitrary intersections. [Hint: use exercise III.2.]

EXERCISE III.4. Given finitely many distinct points $a_1,\dots,a_N \in \Bbbk^n$, construct a polynomial $f \in \Bbbk[x_1,\dots,x_n]$ which is nonzero at a_1 but zero at a_j for each $j \neq 1$. Conclude that there is a Zariski open neighborhood of a_1 in $\Bbbk^n$ which does not contain any a_j for $j \neq 1$.

EXERCISE III.5. Given an ideal $\mathfrak{q}$ of $\Bbbk[x_1,\dots,x_n]$ and $m \in \mathbb{Z}_{\geq 0}$, let $\mathfrak{q}^m$ be the ideal generated by all $f_1\cdots f_m$ for $f_1,\dots,f_m \in \mathfrak{q}$. Show that $V(\mathfrak{q}^m) = V(\mathfrak{q})$ for each $m \geq 1$. Conversely, show that $V(\mathfrak{q}) = V(\sqrt{\mathfrak{q}})$, where $\sqrt{\mathfrak{q}}$ is the *radical* of $\mathfrak{q}$.

EXERCISE III.6. Given any $f \in \Bbbk[x_1,\dots,x_n]$, show that the set $\Bbbk^n \setminus V(f)$ is *Zariski open* (i.e. open with respect to the Zariski topology). Deduce that every Zariski open subset of $\Bbbk^n$ has a open covering by finitely many subsets of the form $\Bbbk^n \setminus V(f)$. [Hint: for finiteness you need to use Hilbert's basis theorem (theorem B.4).]

EXERCISE III.7. Show that $I(X)$ is a radical ideal of $\Bbbk[x_1,\dots,x_n]$ for each $X \subseteq \Bbbk^n$.

EXERCISE III.8. Compute $I(X)$ for $X \subseteq \Bbbk^n$ in the following cases:

(1) $n = 1$, X consists of two distinct points in $\Bbbk$.
(2) $n = 2$, X is the x-axis.
(3) $n = 2$, X is the union of x and y-axes.
(4) $n = 2$, X is the union of x-axis and the point $(1, 0)$.
(5) $n = 1$, $\Bbbk = \mathbb{C}$, $X = \mathbb{Z} = \{0, \pm 1, \pm 2, \ldots\}$.

EXERCISE III.9. If $X_1 \subseteq X_2 \subseteq \Bbbk^n$, show that $I(X_1) \supseteq I(X_2)$.

EXERCISE III.10. Given $X \subseteq \Bbbk^n$, the closure of X in $\Bbbk^n$ under the Zariski topology is the subvariety $\bar{X} := V(I(X))$ (we say that $\bar{X}$ is the *Zariski closure* of X in $\Bbbk^n$). [Hint: it suffices to show that every subvariety V of $\Bbbk^n$ containing X also contains $V(I(X))$. If $V = V(\mathfrak{q})$, then show that $\mathfrak{q} \subseteq I(X)$.]

EXERCISE III.11. Let k be a field and $x_1, \ldots, x_n$, $n \geq 1$, be indeterminates over k.

(1) If k is infinite, then show that for each nonzero polynomial $f \in k[x_1, \ldots, x_n]$, there is $a \in k^n$ such that $f(a) \neq 0$. [Hint: prove it for $n = 1$. In the general case, after renumbering the x_j if necessary, f can be expressed as a nonzero polynomial in x_n with coefficients in $k[x_1, \ldots, x_{n-1}]$. Apply induction and reduce to case $n = 1$.]
(2) If k is finite, then show that there is a nonzero polynomial $f \in k[x_1, \ldots, x_n]$ such that $f(a) = 0$ for all $a \in k^n$.

EXERCISE III.12. (1) Show that any pair of nonempty open subsets of $\Bbbk^n$ has a nonempty intersection. [Hint: due to exercise III.6 it suffices to show that $V(f_1 f_2) \neq \emptyset$ for non-constant polynomials f_1, f_2. Now use exercise III.11.]
(2) Deduce that $\Bbbk^n$ is *not* Hausdorff for any $n \geq 1$.

EXERCISE III.13. Recall that a topological space is *compact* if each of its open covers has a finite subcover. Show that every affine variety is compact[3]. [Hint: due to exercise III.2 it suffices to prove that "if $\mathfrak{q}$ is the sum of a collection $\{\mathfrak{q}_i\}_{i \in \mathcal{I}}$ of ideals of $\Bbbk[x_1, \ldots, x_n]$, then $\mathfrak{q}$ is actually the sum of finitely many of the $\mathfrak{q}_i$." Now use the fact that $\Bbbk[x_1, \ldots, x_n]$ is Noetherian.]

EXERCISE III.14. This exercise illustrates a fundamental property of algebraically closed fields.

(1) Show that every non-constant polynomial over $\Bbbk$ in $(x_1, \ldots, x_n)$ vanishes at some point of $\Bbbk^n$. [Hint: proceed by induction on n. Treat polynomials in $(x_1, \ldots, x_n)$ over $\Bbbk$ as polynomials in x_n over $\Bbbk[x_1, \ldots, x_{n-1}]$, and use the inductive hypothesis to reduce to the case of $n = 1$.]
(2) Show by examples that the preceding statement may be false if $\Bbbk$ is not algebraically closed.

[3]In many algebraic geometry and commutative algebra texts this property is defined as *quasicompactness*, and "compactness" is reserved for spaces which are both quasicompact and Hausdorff.

2. (Ir)reducibility

A topological space X is called *reducible* (respectively, *irreducible*) if it can (respectively, cannot) be represented as the union of two proper closed subsets. The following is a compilation of a few basic properties of irreducible sets—all these follow directly from the definition of irreducible sets; their verification is left as an exercise.

Proposition III.6. *Let X be a topological space.*

(1) X is irreducible if and only if any two of its nonempty open subsets has a nonempty intersection. In particular, if X is irreducible, then no proper subset of X can be both open and closed in X.

(2) Let $V_1, \ldots, V_k$ be closed subsets of X. If V is a closed irreducible subset of X such that $V \subseteq \bigcup_j V_j$, then $V \subseteq V_j$ for some j.

(3) If W is a dense subset of X, then W is irreducible if and only if X is irreducible.

(4) If X is irreducible, then every nonempty open subset of X is irreducible and dense in X.

(5) If X is irreducible, then the image of X under a continuous map is irreducible. □

The relevance of irreducibility in algebraic geometry comes from the observation that under the basic correspondence between subvarieties of $\mathbb{k}^n$ and radical ideals of $\mathbb{k}[x_1, \ldots, x_n]$ given in theorem III.5, irreducible subvarieties correspond to *prime* ideals:

Proposition III.7. *A subvariety X of $\mathbb{k}^n$ is irreducible if and only if $I(X)$ is a prime ideal of $\mathbb{k}[x_1, \ldots, x_n]$.*

PROOF. If $I(X)$ is not prime, then there are $f_1, f_2 \notin I(X)$ such that $f_1 f_2 \in I(X)$. Then $X \subseteq V(f_1 f_2) = V(f_1) \cup V(f_2)$ (exercise III.2), but $X \not\subseteq V(f_j)$ for any j. It follows that X is not irreducible (proposition III.6). On the other hand, if X is not irreducible, then there are $\mathfrak{q}_1, \mathfrak{q}_2 \subseteq \mathbb{k}[x_1, \ldots, x_n]$ such that $X \subseteq V(\mathfrak{q}_1) \cup V(\mathfrak{q}_2)$, but $X \not\subseteq V(\mathfrak{q}_j)$ for any j. Then we can pick $f_j \in \mathfrak{q}_j$ which does not vanish everywhere on X. But $f_1 f_2$ vanishes on $V(\mathfrak{q}_1) \cup V(\mathfrak{q}_2) \supseteq X$, so that $f_1 f_2 \in I(X)$ even though neither f_1 nor f_2 is in $I(X)$. It follows that $I(X)$ is not prime, as required. □

Example III.8. A finite set of points in $\mathbb{k}^n$ is irreducible if and only if it consists of only one point. Since the intersection of any pair of nonempty open subsets of $\mathbb{k}^n$ is nonempty (exercise III.12), proposition III.6 implies that $\mathbb{k}^n$ is irreducible. Note that $I(\mathbb{k}^n)$, being the zero ideal (example III.4), is prime in the polynomial ring. The concept of irreducibility is simple but powerful; see exercise III.18 for a simple proof of the *Cayley-Hamilton theorem* on matrices using the irreducibility of $\mathbb{k}^n$.

An *irreducible component* of X is a closed irreducible subset which is not properly contained in any other irreducible subset of X. Each point of a finite subset S of $\mathbb{k}^n$ is an irreducible component of S. Consider e.g. the case that $X = \{a_1, \ldots, a_k\} \subset \mathbb{k}$. Then $I(X)$ is the ideal of $\mathbb{k}[x]$ generated by $(x - a_1)(x - a_2) \cdots (x - a_k)$. Note that

$$I(X) = \bigcap_j \langle x - a_j \rangle$$

and the ideals $\langle x - a_j \rangle$ are precisely the ideals of the irreducible components of X; this is a manifestation of the following general property of affine varieties:

THEOREM III.9. *Let X be a subvariety of $\Bbbk^n$. Then there is a unique* minimal representation[4]

$$I(X) = \bigcap_j \mathfrak{p}_j \tag{5}$$

of $I(X)$ as the intersection of finitely many prime ideals. The operators $I(\cdot)$ and $V(\cdot)$ induce a one-to-one correspondence between irreducible components of X and the prime ideals $\mathfrak{p}_j$ that appear in (5). *In particular, X has* finitely many *irreducible components.*

PROOF. At first we prove the following: every subvariety of $\Bbbk^n$ is the union of finitely many closed irreducible subsets. Indeed, otherwise there is a subvariety X_0 of $\Bbbk^n$ which is not the union of finitely many closed irreducible subsets. In particular X_0 is reducible, and it can be expressed as the union of proper closed subsets Y_1, Y_2. At least one of the Y_j must also have the property that it is not the union of finitely many closed irreducible subsets; denote it by X_1. By the same arguments X_1 has a proper closed subset X_2 which is not the union of finitely many closed irreducible subsets, and continuing in this way we can construct an infinite chain $X \supsetneq X_1 \supsetneq X_2 \supsetneq \cdots$ of subvarieties of $\Bbbk^n$. But then there is an infinite strictly ascending chain of ideals $I(X_0) \subsetneq I(X_1) \subsetneq I(X_2) \subsetneq \cdots$ of ideals of $\Bbbk[x_1, \ldots, x_n]$, which violates the Noetherianity of polynomial rings (theorem B.4). This proves the claim. Now pick closed irreducible subsets $X_1, \ldots, X_k$ of X such that

$$X = \bigcup_j X_j. \tag{6}$$

Discarding some X_j from the union if necessary, we may assume that the presentation in (6) is *minimal*, i.e. $X_j \not\subseteq \bigcup_{i \neq j} X_i$ for any j. It is then straightforward to see (e.g. using assertion (2) of proposition III.6) that X_j are precisely the irreducible components of X. This in particular proves the last assertion of theorem III.9.

Let $\mathfrak{p}_j := I(X_j)$, $j = 1, \ldots, k$. Then $\mathfrak{p}_j$ are prime (proposition III.7) and $X = V(\bigcap_j \mathfrak{p}_j)$ (exercise III.2). Since $\bigcap \mathfrak{p}_j$ is radical (exercise III.16), the Nullstellensatz (theorem B.8) implies that $I(X) = \bigcap_j \mathfrak{p}_j$. The minimality of this representation of $I(X)$ follows from the minimality of the representation in (6). It is straightforward to see that any such minimal representation is unique; it is left as an exercise (exercise III.17). □

[4]A representation $\mathfrak{q} = \bigcap_{i=1}^k \mathfrak{q}_i$ is *minimal* if $\mathfrak{q}_j \not\supseteq \bigcup_{i \neq j} \mathfrak{q}_i$ for any j.

Example III.10. Consider the hypersurface $V(f)$ of $\mathbb{k}^n$ determined by $f \in \mathbb{k}[x_1, \dots, x_n]$. Recall that $\mathbb{k}[x_1, \dots, x_n]$ is a *unique factorization domain.* If $f_1, \dots, f_k$ are the distinct *irreducible factors* of f, then the irreducible components of $V(f)$ are precisely $V(f_j)$, $j = 1, \dots, k$. Moreover, the ideals generated by the f_j are prime and the ideal generated by $\prod_j f_j$ is radical, so that $I(V(f)) = \langle \prod_j f_j \rangle = \bigcap_j \langle f_j \rangle = \bigcap_j I(V(f_j))$.

2.1. Exercises.

EXERCISE III.15. Prove proposition III.6.

EXERCISE III.16. Show that intersections of any collection of radical ideals in a ring is also a radical ideal.

EXERCISE III.17. If $\mathfrak{p}$ is a prime ideal of a ring R containing the intersection of finitely many ideals $\mathfrak{q}_1, \dots, \mathfrak{q}_k$, then show that $\mathfrak{p} \supseteq \mathfrak{q}_j$ for some j. Conclude that an ideal of R can have (up to reordering) at most one minimal presentation as the intersection of finitely many prime ideals.

EXERCISE III.18. Given an $n \times n$ matrix A over a field F, its *characteristic polynomial* $p(\lambda)$ is the polynomial $\det(A - \lambda \mathbb{1}_n) \in F[\lambda]$, where det denotes the determinant, $\mathbb{1}_n$ is the $n \times n$ identity matrix and λ is an indeterminate over F. The *Cayley-Hamilton theorem* states that $p(A)$ is the zero matrix. In this exercise following a suggestion on *MathOverflow* [hh] we outline a proof of the Cayley-Hamilton theorem using the following fact from linear algebra: if $p(\lambda)$ has n distinct roots in F, then A is *diagonalizable*[5].

(1) Show that to prove Cayley-Hamilton theorem it suffices to assume that F is algebraically closed. In all steps below assume F is algebraically closed.
(2) Show that Cayley-Hamilton theorem is true for diagonalizable matrices.
(3) Identify the space of $n \times n$ matrices over F with the affine space F^{n^2}. Show that to prove Cayley-Hamilton theorem it suffices to show that the set of diagonalizable matrices is Zariski dense in F^{n^2}. [Hint: $I(F^{n^2})$ is the zero ideal.]
(4) In example III.121 we will see that the following is true:

(7) Given an algebraically closed field $\mathbb{k}$ and a positive integer n, there is a nonempty Zariski open subset U of $\mathbb{k}^{n+1}$ such that for each $(c_0, \dots, c_n) \in U$, the polynomial $c_0\lambda^n + c_1\lambda^{n-1} + \cdots + c_n$ has n distinct roots in $\mathbb{k}$.

[5]Recall that a matrix A is *diagonalizable* if $A = PDP^{-1}$ for an invertible matrix P and a diagonal matrix D.

Use this fact to show that the set of diagonalizable matrices contains a dense Zariski open subset of F^{n^2}. [Hint: use the irreducibility of the affine space, exercise III.6 and assertion (III.6) of proposition III.6.]

(5) It turns out that there is a unique irreducible polynomial Δ in $\Bbbk[c_0, \ldots, c_n]$ such that $c_0\lambda^n + c_1\lambda^{n-1} + \cdots + c_n$ has multiple roots if and only if $\Delta(c_0, \ldots, c_n) = 0$; it is called the *discriminant*. For $n = 2$ use the quadratic formula to explicitly compute the discriminant. Discriminants are discussed in many introductory algebra or algebraic geometry books, e.g. [Gri96, Section II.2].

3. Regular functions, coordinate rings and morphisms of affine varieties

Let X be a subvariety of $\Bbbk^n$ with coordinates $(x_1, \ldots, x_n)$. A *regular function* on X is a function $\phi : X \to \Bbbk$ which is the restriction of a polynomial in $(x_1, \ldots, x_n)$ over $\Bbbk$. The set of regular functions on X, equipped with the natural $\Bbbk$-algebra structure, is called the *coordinate ring* of X, and denoted as $\Bbbk[X]$. The restriction map induces a natural surjective homomorphism $\Bbbk[x_1, \ldots, x_n] \to \Bbbk[X]$. Since $f|_X \equiv 0$ if and only if $f \in I(X)$, it follows that

$$\Bbbk[X] \cong \Bbbk[x_1, \ldots, x_n]/I(X). \tag{8}$$

Example III.11. If X is a singleton, then $\Bbbk[X] \cong \Bbbk$ (this follows from example III.3 and identity (8)). On the other extreme, if $X = \Bbbk^n$, then $\Bbbk[X] \cong \Bbbk[x_1, \ldots, x_n]$. If H is the *hyperbola* $V(xy - 1) \subseteq \Bbbk^2$, then $\Bbbk[H] \cong \Bbbk[x, y]/\langle xy - 1\rangle \cong \Bbbk[x, x^{-1}]$.

Example III.12. All the coordinate rings computed in example III.11 are *integral domains*. In fact it turns out that $\Bbbk[X]$ is an integral domain if and only if X is irreducible (exercise III.19). If X is the union of the x and y-axes in $\Bbbk^2$, then $I(X) = \langle xy\rangle$ (example III.10), so that $\Bbbk[X] \cong \Bbbk[x, y]/\langle xy\rangle$, which in particular is not an integral domain. If X is the union of the hyperbola $H := V(xy - 1) \subseteq \Bbbk^2$ from example III.11 and the x-axis (which we denote by A) on $\Bbbk^2$, then

$$\Bbbk[X] \cong \Bbbk[x, y]/\langle y(xy - 1)\rangle \cong \Bbbk[x, y]/\langle y\rangle \times \Bbbk[x, y]/\langle xy - 1\rangle \cong \Bbbk[A] \times \Bbbk[H] \tag{9}$$

(exercise III.20). More generally, you will prove in exercise III.21 that the coordinate ring of a pairwise disjoint union of affine varieties is isomorphic to the product of their coordinate rings.

A map $\phi : X \to Y$ between affine varieties is called a *morphism* if for every regular function h on Y, the pullback $h \circ \phi$ is a regular function on X. An *isomorphism* is a bijective morphism whose inverse is also a morphism; the notation $X \cong Y$ is a shorthand for the statement that X and Y are *isomorphic*, i.e. there is an isomorphism between X and Y. It is straightforward to check that a morphism $\phi : X \to Y$ induces a $\Bbbk$-algebra homomorphism $\phi^* : \Bbbk[Y] \to \Bbbk[X]$ given by $h \mapsto h \circ \phi$, and that ϕ is an isomorphism if and only if ϕ^* is a $\Bbbk$-algebra isomorphism between $\Bbbk[X]$ and $\Bbbk[Y]$ (exercise III.22).

Example III.13. Projections constitute a basic source of morphisms. Let X be the parabola $y = x^2$ in $\mathbb{k}^2$. Note that $\mathbb{k}[X] = \mathbb{k}[x,y]/\langle y - x^2\rangle \cong \mathbb{k}[x]$, so that $X \cong \mathbb{k}$. Indeed, the projection onto x-axis realizes this isomorphism, but the projection onto y-axis induces a two-to-one morphism $X \to \mathbb{k}$ (exercise III.24). Note that the map $(x,y) \mapsto (x, y - x^2)$ is an *automorphism*[6] of $\mathbb{k}^2$ which maps X onto the x-axis. This is a special case of the celebrated *line embedding theorem* of S. S. Abhyankar and T. T. Moh [AM75] which states that "if X is a subvariety of $\mathbb{k}^2$ isomorphic to $\mathbb{k}$ then there is an automorphism of $\mathbb{k}^2$ which maps X onto the x-axis." The following analogous statement remains a conjecture (named after Abhyankar and A. Sathaye, who was a student of Abhyankar) for $n \geq 2$: "if X is a subvariety of $\mathbb{k}^{n+1}$ isomorphic to $\mathbb{k}^n$ then there is an automorphism of $\mathbb{k}^{n+1}$ which maps X onto the hyperplane $x_n = 0$."

Example III.14. The image of the morphism $\mathbb{k} \to \mathbb{k}^2$ given by $t \mapsto (t^2, t^3)$ is the variety $X = V(x^3 - y^2)$. Considered as a morphism from $\mathbb{k}$ to X, this induces a bijection between the points of $\mathbb{k}$ and X, but it is *not* an isomorphism (exercise III.25).

Example III.15. Let f be a nonzero polynomial in $(x_1, \ldots, x_n)$ and X be the hypersurface of $\mathbb{k}^{n+1}$ defined by $x_{n+1}f = 1$. Then $\mathbb{k}[X] \cong \mathbb{k}[x_1, \ldots, x_n, 1/f]$. Let $\pi : X \to \mathbb{k}^n$ be the projection onto the first n-coordinates. The image of π is the proper open subset $U_f := \mathbb{k}^n \setminus V(f)$ of $\mathbb{k}^n$. It is clear that π is one-to-one; in fact π induces a *homeomorphism* with respect to Zariski topology (exercise III.29). We will see in example III.37 that π is actually an *isomorphism of quasiprojective varieties.*

Example III.16. The morphism $\sigma : \mathbb{k}^2 \to \mathbb{k}^2$ given by $(x,y) \mapsto (x, xy)$ maps the whole y-axis to the origin, but it is one-to-one at every point not on the y-axis. The image of σ, which is the union of the origin and all points *not* on the y-axis, is neither open nor closed in $\mathbb{k}^2$ (exercise III.30). However, it is the union of a closed subset and an open subset, which is a "constructible set"; in section III.12 we discuss constructible sets and prove Chevalley's theorem that the image of every morphism is a constructible set.

3.1. Exercises.

EXERCISE III.19. Show that an affine variety is irreducible if and only if its coordinate ring is an integral domain. [Hint: use proposition III.7 and identity (8).]

EXERCISE III.20. Consider the "diagonal" map $\delta : \mathbb{k}[x,y] \mapsto \mathbb{k}[x,y]/\langle y\rangle \times \mathbb{k}[x,y]/\langle xy - 1\rangle$ which maps $f \mapsto (f + \langle y\rangle, f + \langle xy - 1\rangle)$.

(1) Show that δ is a surjective $\mathbb{k}$-algebra homomorphism and $\ker(\delta) = \langle y(xy-1)\rangle$.
(2) Verify the $\mathbb{k}$-algebra isomorphisms presented in (9). [Hint: use example III.10 and the preceding assertion.]

---

[6]An *automorphism* of $\mathbb{k}^n$ is an isomorphism from $\mathbb{k}^n$ to itself.

(3) Find a polynomial $f \in \Bbbk[x, y]$ such that the $f|_A = x|_A$ and $f|_H = y|_H$ (where A is the x-axis and H is the hyperbola $V(xy - 1)$ in $\Bbbk^2$).

EXERCISE III.21. Let $X := \bigcup_{j=1}^k X_j$, where X_j are subvarieties of $\Bbbk^n$ with pairwise empty intersection. Write $\mathfrak{q}_j := I(X_j)$ and let $\delta : \Bbbk[x_1, \ldots, x_n] \to \prod_j \Bbbk[x_1, \ldots, x_n]/\mathfrak{q}_j$ be the "diagonal" map that sends $f \mapsto \prod_j (f + \mathfrak{q}_j)$. Show that

(1) $I(X) = \cap_j \mathfrak{q}_j$.
(2) $\ker(\delta) = \cap_j \mathfrak{q}_j$.
(3) $\mathfrak{q}_i$ and $\mathfrak{q}_j$ are *coprime*[7] if $i \neq j$. [Hint: you need to use that $X_i \cap X_j = \emptyset$.]
(4) Deduce that δ is surjective. [Hint: it suffices to show there is $f \in \Bbbk[x_1, \ldots, x_n]$ such that $\delta(f) = (1, 0, \ldots, 0)$. Due to the preceding assertion for each $i > 1$, there is $f_i \in \mathfrak{q}_i$ and $g_i \in \mathfrak{q}_1$ such that $f_i + g_i = 1$. Take $f := \prod_{i>1} f_i$.]
(5) Deduce that $\Bbbk[X] \cong \prod_j \Bbbk[X_j]$.

EXERCISE III.22. Show that a morphism $\phi : X \to Y$ between affine varieties induces a $\Bbbk$-algebra homomorphism $\phi^* : \Bbbk[Y] \to \Bbbk[X]$ via pullback. Show that ϕ is an isomorphism if and only if ϕ^* is a $\Bbbk$-algebra isomorphism.

EXERCISE III.23. Show that an isomorphism of affine varieties induces a homeomorphism between them.

EXERCISE III.24. Let X be the parabola $y = x^2$ in $\Bbbk^2$. Let π_x (respectively, π_x) be the morphism from X to $\Bbbk$ induced by the projection onto x-axis (respectively, y-axis). Compute the pullback maps $\Bbbk[t] \to \Bbbk[X]$ (where t is the coordinate on $\Bbbk$) induced by π_x and π_y. Deduce that π_x is an isomorphism, but π_y is not. Compute $\pi_x^{-1} : \Bbbk \to X$.

EXERCISE III.25. Prove the statements from example III.14. [Hint: compute $\Bbbk[X]$.]

EXERCISE III.26. Let R be a finitely generated $\Bbbk$-algebra which is *reduced*[8]. Show that there is an affine variety X such that $\Bbbk[X] \cong R$. [Hint: Construct a surjective $\Bbbk$-algebra homomorphism from a polynomial ring over $\Bbbk$ to R.] Exercise III.22 shows that X is unique up to an isomorphism.

EXERCISE III.27. The correspondence between $\Bbbk$-algebra homomorphisms of finitely generated $\Bbbk$-algebras and morphisms of affine varieties can be pushed a little bit further than exercise III.22. Given affine varieties X, Y and any $\Bbbk$-algebra homomorphism $\Phi : \Bbbk[Y] \to \Bbbk[X]$, show that there is a morphism $\phi : X \to Y$ such that $\Phi = \phi^*$. This, together with exercises III.22 and III.26, shows that the categories of affine varieties and finitely generated reduced $\Bbbk$-algebras are equivalent.

[7] Two ideals $\mathfrak{p}, \mathfrak{q}$ of a ring R are *coprime* if $\mathfrak{p} + \mathfrak{q} = R$.

[8] A ring is *reduced* if it does not have any nonzero nilpotent elements.

EXERCISE III.28. (1) Show that every open subset of an affine variety X has a finite covering by open subsets of the form $X \setminus V(f)$ for regular functions f on X. [Hint: use exercise III.6.]

(2) Show that a morphism $\phi : Y \to X$ of affine varieties is a continuous map with respect to the Zariski topology.

EXERCISE III.29. Let $f \in \mathbb{k}[x_1, \ldots, x_n]$. Show that the projection onto the first n-coordinates of $\mathbb{k}^{n+1}$ induces a homeomorphism between the subvariety $V(x_{n+1}f - 1)$ of $\mathbb{k}^{n+1}$ and the open subset $\mathbb{k}^n \setminus V(f)$ of $\mathbb{k}^n$.

EXERCISE III.30. Consider the morphism $\sigma : \mathbb{k}^2 \to \mathbb{k}^2$ given by $(x, y) \mapsto (x, xy)$. Show that

(1) σ induces a homeomorphism from $\mathbb{k}^2 \setminus V(x)$ to itself.
(2) Image of σ is $\{(0, 0)\} \cup (\mathbb{k}^2 \setminus V(x))$.
(3) Image of σ is not closed in $\mathbb{k}^2$. [Hint: $\mathbb{k}^2$ is irreducible. Use proposition III.6.]
(4) Image of σ is not open in $\mathbb{k}^2$. [Hint: the y-axis is isomorphic to $\mathbb{k}$, so any proper open set of the y-axis is infinite.]

4. Quasiprojective varieties

Let $(x_0, \ldots, x_n)$ be a system of coordinates on $\mathbb{k}^{n+1}$, $n \geq 0$. Consider the relation $\sim$ on $\mathbb{k}^{n+1} \setminus \{0\}$, $n \geq 0$, defined as follows: if $a, b \in \mathbb{k}^{n+1} \setminus \{0\}$ with coordinates (with respect to $(x_0, \ldots, x_n)$), respectively, $(a_0, \ldots, a_n)$ and $(b_0, \ldots, b_n)$, then $a \sim b$ if and only if there is $\lambda \in \mathbb{k} \setminus \{0\}$ such that $a_j = \lambda b_j$ for each $j = 0, \ldots, n$. Note that $a \sim b$ if and only if a and b belong to the same line through the origin on $\mathbb{k}^{n+1}$. This immediately implies that $\sim$ is an equivalence relation; the *n-dimensional projective space* $\mathbb{P}^n(\mathbb{k})$, or simply $\mathbb{P}^n$, is the set of the equivalence classes of $\sim$; in other words $\mathbb{P}^n$ is the set of lines through the origin on $\mathbb{k}^{n+1}$. We denote the equivalence class of $\sim$ containing a by $[a_0 : \cdots : a_n]$ and say that $[a_0 : \cdots : a_n]$ is the *homogeneous coordinate* of the point of $\mathbb{P}^n$ determined by a. Let f be a *homogeneous*[9] polynomial of degree d in $(x_0, \ldots, x_n)$, then it is straightforward to check that

$$f(\lambda a_0, \ldots, \lambda a_n) = \lambda^d f(a_0, \ldots, a_n) \tag{10}$$

for each $\lambda, a_0, \ldots, a_n \in \mathbb{k}$. It follows that given $a = (a_0, \cdots, a_n) \in \mathbb{k}^{n+1}$, $f(a_0, \ldots, a_n) = 0$ if and only if $f(b_0, \ldots, b_n) = 0$ for all $(b_0, \ldots, b_n)$ in the equivalence class of a. Consequently the set $V(f) := \{[a_0 : \cdots : a_n] : f(a_0, \ldots, a_n) = 0\}$ of zeroes of f is a well-defined subset of $\mathbb{P}^n$; we say that $V(f)$ is the *hypersurface* of $\mathbb{P}^n$ determined by f. There is a unique topology on $\mathbb{P}^n$ in which the basic closed subsets are intersections of hypersurfaces of $\mathbb{P}^n$ (exercise III.31); it is called the *Zariski topology*. A *projective variety* is a Zariski closed subset of the projective space (equipped with the topology induced from the Zariski topology of $\mathbb{P}^n$). A *quasiprojective variety* is a Zariski open subset of a projective variety (also equipped

[9]Recall that a polynomial is *homogeneous* if each of its monomials has the same degree.

with the Zariski topology induced from $\mathbb{P}^n$). Given a quasiprojective variety X, a *quasiprojective subset* of X is a subset which is also a quasiprojective variety, and a *subvariety* of X is a Zariski closed subset of X.

Example III.17. For each $j = 0, \ldots, n$, $V(x_j) \subseteq \mathbb{P}^n$ is the set of all $[a_0 : \cdots : a_n]$ such that $a_j = 0$. It is straightforward to check that the projection onto the coordinates excluding the j-th coordinate induces a homeomorphism between $V(x_j)$ and $\mathbb{P}^{n-1}$ (exercise III.33). The complementary Zariski open subset $U_j := \mathbb{P}^n \setminus V(x_j)$ has a one-to-one correspondence with $\mathbb{k}^n$ via the map

$$\mathbb{k}^n \ni (x_1, \ldots, x_n) \mapsto [x_1 : \cdots : x_j : 1 : x_{j+1} : \cdots : x_n] \in U_j. \tag{11}$$

We leave it as an exercise (exercise III.34) to check that the inverse to the above map is given by

$$U_j \ni [x_0 : \cdots : x_n] \mapsto (\frac{x_0}{x_j}, \ldots, \frac{x_{j-1}}{x_j}, \frac{x_{j+1}}{x_j}, \ldots, \frac{x_n}{x_j}) \in \mathbb{k}^n. \tag{12}$$

It is clear that $\mathbb{P}^n = \bigcup_j U_j$. We say that the U_j are *basic open subsets* of $\mathbb{P}^n$.

Example III.18. If f is a homogeneous polynomial of degree d in (x_0, x_1) then f can be expressed as $\prod_{i=1}^d (a_i x_0 - b_i x_1)$ for $(a_i, b_i) \in \mathbb{k}^2 \setminus \{0\}$ (exercise III.35), so that $V(f) = \{[b_i : a_i] : i = 1, \ldots, d\} \subseteq \mathbb{P}^1$. It follows that all proper subvarieties of $\mathbb{P}^1$ are finite sets. This is a special case of the general fact that all proper subvarieties of a "curve" are finite sets (example III.79).

Example III.19. The subset $(\mathbb{P}^n \setminus V(x_0)) \cup V(x_0, x_1)$ of $\mathbb{P}^n$ is neither-closed-nor-open for $n \geq 2$ (exercise III.40). This is the projective analogue of the neither closed nor open subset of $\mathbb{k}^2$ from example III.16.

Let $X = V(f_i : i \in \mathcal{I})$ be the subvariety of $\mathbb{P}^n$ defined by a collection $\{f_i\}_{i \in \mathcal{I}}$ of homogeneous polynomials in $(x_0, \ldots, x_n)$. The *cone* $C(X)$ *over* X is the affine subvariety of $\mathbb{k}^{n+1}$ determined by $\{f_i\}_{i \in \mathcal{I}}$ Proposition III.21 below shows that $C(X)$ is a union of lines through the origin (i.e. $C(X)$ is an actual "cone"), and these lines are in one-to-one correspondence with points on X.

Example III.20. The real points of the cone over $X := V(x_0^2 - x_1^2 - x_2^2) \subseteq \mathbb{P}^2(\mathbb{C})$ is the "circular double cone" pictured in fig. 2. Different cross sections of $C(X)$ yields different representations of points on X, e.g. the real points of X are in one-to-one correspondence with

- a circle (intersection with $x_0 = 1$),
- a hyperbola (intersection with $x_1 = 1$) and two "points at infinity" (namely, $[1 : 0 : 1], [1 : 0 : -1]$),
- a parabola (intersection with $x_0 + x_1 = 1$) and one "point at infinity" (namely, $[1 : -1 : 0]$).

This is a manifestation of the equivalence of "compactifications" of conic sections under "projective transformations."

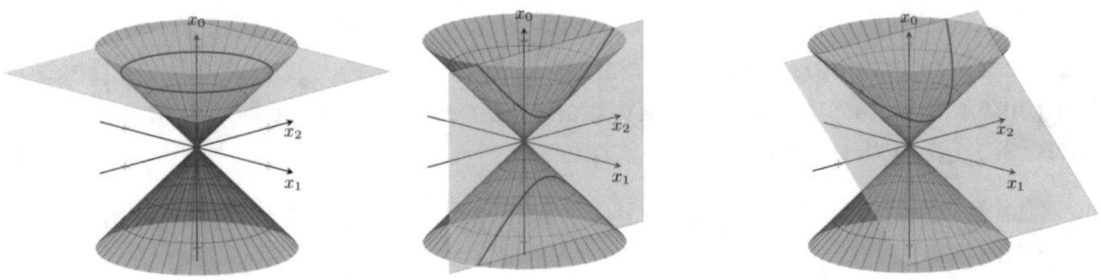

FIGURE 2. Real points and cross sections of $C(X)$ for $X := V(x_0^2 - x_1^2 - x_2^2) \subset \mathbb{P}^2$.

Proposition III.21. *For every $a = [a_0 : \cdots : a_n] \in X$, every point of the line in $\Bbbk^{n+1}$ determined by a is in $C(X)$. Conversely, for every $(a_0, \ldots, a_n) \in C(X)\setminus\{0\}$, the corresponding point $[a_0 : \cdots : a_n]$ in $\mathbb{P}^n$ is in X.*

PROOF. This is a straightforward consequence of the homogeneity of the f_i and identity (10) above. □

Recall that the affine variety $C(X)$ determines an ideal $I(C(X))$ of $\Bbbk[x_0, \ldots, x_n]$ consisting of all polynomials that vanish on $C(X)$. The same construction can be used to define an ideal associated to X, provided we consider only *homogeneous* generators. A *homogeneous ideal* of $\Bbbk[x_0, \ldots, x_n]$ is an ideal generated by homogeneous polynomials. The *homogeneous ideal of* X, denoted by $I(X)$, is the ideal of $\Bbbk[x_0, \ldots, x_n]$ generated by *homogeneous* polynomials that vanish on all points of X.

Proposition III.22. $I(X) = I(C(X))$. *In particular, there are* finitely many *homogeneous polynomials $f_1, \ldots, f_k$ in $(x_0, \ldots, x_n)$ such that $X = V(f_1, \ldots, f_k) \subseteq \mathbb{P}^n$.*

PROOF. Recall that a *homogeneous component* of $f = \sum_\alpha c_\alpha x^\alpha \in \Bbbk[x_0, \ldots, x_n]$ (where x^α is a shorthand for $x_0^{\alpha_0} \cdots x_n^{\alpha_n}$) is a polynomial of the form $\sum_{|\alpha|=d} c_\alpha x^\alpha$ for some $d \geq 0$, where $|\alpha| := \alpha_0 + \cdots + \alpha_n$. Since $I(X)$ is generated by homogeneous polynomials, it has the following property (exercise III.36):

(13) a polynomial f in $(x_0, \ldots, x_n)$ is in $I(X)$ if and only if all homogeneous components of f are in $I(X)$.

Property (13) coupled with Hilbert's Basis theorem (theorem B.4) implies that $I(X)$ is generated by finitely many homogeneous polynomials $f_1, \ldots, f_k$; it is straightforward to check that $X = V(f_1, \ldots, f_k)$. It remains to prove that $I(X) = I(C(X))$. The containment $I(X) \subseteq I(C(X))$ follows immediately from proposition III.21 and identity (10). For the opposite containment pick $a = [a_0 : \ldots : a_n] \in \mathbb{P}^n$. Let $L(a) := \{(ta_0, \ldots, ta_n) : t \in \Bbbk\}$ be the "line in $\Bbbk^{n+1}$ represented by a."

Claim III.22.1. *A polynomial vanishes on $L(a)$ if and only if each of its homogeneous components vanishes at a.*

PROOF. Pick $f \in \Bbbk[x_0, \dots, x_n]$ such that $f|_{L(a)} \equiv 0$. Write $f = \sum_{d=0}^{e} f_d$, where each f_d is homogeneous of degree d. Then $f(ta_0, \dots, ta_n) = \sum_{d=0}^{e} t^d f_d(a)$ is a polynomial in t with infinitely many zeroes, so it must be identically zero. Therefore $f_d(a) = 0$ for each d, as required. □

Claim III.22.1 and proposition III.21 imply that $I(C(X)) \subseteq I(X)$, which completes the proof. □

There is a homogeneous ideal of $\Bbbk[x_0, \dots, x_n]$ which is *not* the homogeneous ideal of any nonempty subvariety of $\mathbb{P}^n$: it is the ideal $\mathfrak{m}_+$ generated by $x_0, \dots, x_n$ [why?]. It is sometimes called the *irrelevant* ideal. Note that *all* proper homogeneous ideals of $\Bbbk[x_0, \dots, x_n]$ are contained in $\mathfrak{m}_+$. The correspondence between affine varieties and radical ideals described in theorem III.5 has a projective counterpart, also due to the Nullstellensatz, provided one excludes $\mathfrak{m}_+$—this is described in exercise III.37.

4.1. Exercises.

EXERCISE III.31. Show that the collection of Zariski closed sets of $\mathbb{P}^n$ satisfies the axioms of a topology. [Hint: mimic the solution to exercise III.3.]

EXERCISE III.32. Show that every Zariski open subset of $\mathbb{P}^n$ has an open cover by subsets of the form $\mathbb{P}^n \setminus V(f)$ for homogeneous polynomials $f \in \Bbbk[x_0, \dots, x_n]$.

EXERCISE III.33. Show that the map from the subvariety $V(x_0)$ of $\mathbb{P}^n$ to $\mathbb{P}^{n-1}$ defined by $[0 : x_1 : \cdots : x_n] \mapsto [x_1 : \cdots : x_n])$ induces a homeomorphism with respect to the Zariski topology.

EXERCISE III.34. Show that the map from U_j to $\Bbbk^n$ from (12) is well defined and it is the inverse to the map from $\Bbbk^n$ to U_j given in (11).

EXERCISE III.35. Let f be a homogeneous polynomial in (x_0, x_1) of degree d.

(1) Show that f/x_0^d is a polynomial in x_1/x_0 over $\Bbbk$.
(2) Since $\Bbbk$ is algebraically closed, deduce that f can be expressed as a product of *linear* homogeneous polynomials.

EXERCISE III.36. Let I be a homogeneous ideal of $\Bbbk[x_0, \dots, x_n]$.

(1) Show that for every $f \in I$, all homogeneous components of f are also in I. [Hint: Given $f \in I$, express it as a sum of products of polynomials and homogeneous generators of I. Equate homogeneous components of both sides of the equation.]
(2) Deduce that I is prime if and only if for all *homogeneous* polynomials $f_1, f_2 \in \Bbbk[x_0, \dots, x_n]$, $f_1 f_2 \in I$ if and only if either f_1 or f_2 is in I.
(3) Show that the radical $\sqrt{I}$ of I is also homogeneous. [Hint: assume by contradiction that $\sqrt{I}$ is not homogeneous. Then there is $f \in \sqrt{I}$ such that *no* homogeneous component of f is in $\sqrt{I}$. Pick $k \geq 1$ such that $f^k \in I$. If f_d is the homogeneous component of f of degree $d = \deg(f)$, then $f_d^k \in I$, so that $f_d \in \sqrt{I}$, which is a contradiction.]

EXERCISE III.37. Let $\mathfrak{m}_+$ be the maximal ideal of $\Bbbk[x_0, \ldots, x_n]$ generated by $x_0, \ldots, x_n$. Use the Nullstellensatz (theorem B.8) to prove that there is a one-to-one correspondence between nonempty subvarieties of $\mathbb{P}^n$ and radical homogeneous ideals properly contained in $\mathfrak{m}_+$. [Hint: use exercise III.36.]

EXERCISE III.38. Let X be a subvariety of $\mathbb{P}^n$. Show that X is irreducible if and only if $I(X)$ is a prime ideal of $\Bbbk[x_0, \ldots, x_n]$. [Hint: mimic the proof of proposition III.7. Use exercise III.36.]

EXERCISE III.39. Show that $I(\mathbb{P}^n) = 0$. Deduce that $\mathbb{P}^n$ is irreducible.

EXERCISE III.40. Show that $(\mathbb{P}^n \setminus V(x_0)) \cup V(x_0, x_1)$ is neither open nor closed in $\mathbb{P}^n$ if $n \geq 2$. [Hint: $\mathbb{P}^n$ and $V(x_0)$ are irreducible. Use proposition III.6.]

EXERCISE III.41. Show that every Zariski open subset of a quasiprojective variety $X \subseteq \mathbb{P}^n$ has a *finite* open cover by subsets of the form $X \setminus V(f)$ for homogeneous polynomials $f \in \Bbbk[x_0, \ldots, x_n]$. [Hint: use proposition III.22.]

EXERCISE III.42. Given $X \subseteq \mathbb{P}^n$, show that the closure of X in $\mathbb{P}^n$ under the Zariski topology is $\bar{X} := V(I(X)) = \{[x_0 : \cdots : x_n] : f(x_0, \ldots, x_n) = 0$ for each homogeneous $f \in I(X)\}$; we say that $\bar{X}$ is the *Zariski closure* of X in $\mathbb{P}^n$.

EXERCISE III.43. let f be a homogeneous polynomial in $(x_0, \ldots, x_n)$, $n \geq 1$, and let $U_j = \mathbb{P}^n \setminus V(x_j)$ be a basic open subset of $\mathbb{P}^n$ from example III.17.

(1) Show that $V(f) \cap U_j = \emptyset$ if and only if $f = cx_j^m$ for some $m \geq 0$ and $c \in \Bbbk \setminus \{0\}$. [Hint: let $m = \deg(f)$. Then f/x_j^m induces a well-defined map from U_j to $\Bbbk$ which can be expressed as a polynomial in $(x_i/x_j)_{i \neq j}$. Use exercise III.14 and the one-to-one correspondence between U_j and $\Bbbk^n$ given by (12).]
(2) Deduce that if f is not a constant, then the hypersurface $V(f) \subset \mathbb{P}^n$ is nonempty.
(3) Show by examples that the preceding statement may be false if $\Bbbk$ is not algebraically closed.

EXERCISE III.44. Let S_1, S_2 be finite subsets of $\mathbb{P}^n$ such that $S_1 \cap S_2 = \emptyset$. Show that there is a hypersurface X of $\mathbb{P}^n$ containing S_1 but not containing any point of S_2. [Hint: for each $a \in S_1$, one can choose $b_0, \ldots, b_n \in \Bbbk$ such that $b_0x_0 + \cdots + b_nx_n$ vanishes at a but not at any point of S_2.]

5. Regular functions

If f, g are homogeneous polynomials in $(x_0, \ldots, x_n)$ of the same degree, then f/g is a well-defined function on the Zariski open subset $\mathbb{P}^n \setminus V(g)$ of $\mathbb{P}^n$; we say that f/g is a *rational function* on $\mathbb{P}^n$ which is *regular* on $\mathbb{P}^n \setminus V(g)$. In general, a *regular function* on a quasiprojective variety X is a function $\phi : X \to \Bbbk$ which can be "locally represented by rational functions," i.e. for each $x \in X$, there is an open neighborhood U of x in X such that $\phi|_U$ is the restriction to U of a rational function f/g such that g does *not* vanish at any point of U. The set of regular functions on X has the natural structure of a $\Bbbk$-algebra; we denote it by $\Bbbk[X]$.

Example III.23. For each i, j, x_i/x_j is a regular function on the basic open set $U_j := \mathbb{P}^n \setminus V(x_j)$ of $\mathbb{P}^n$. It follows that all polynomials in $(x_0/x_j, \ldots, x_n/x_j)$ are regular functions on U_j. We will shortly see (in proposition III.28 below) that these are in fact *all* regular functions on U_j.

Example III.24. Let $X := V(x_0x_3 - x_1x_2) \setminus V(x_1, x_3) \subseteq \mathbb{P}^3$. Note that X is the union of open subsets $X_1 := X \setminus V(x_1)$ and $X_3 := X \setminus V(x_3)$. Since $x_0/x_1 = x_2/x_3$ on $X_1 \cap X_3$, it follows that the function $f : X \to \mathbb{k}$ defined by x_0/x_1 on X_1 and by x_2/x_3 on X_3 is a regular function on X. In exercise III.55 you will prove that $\mathbb{k}[X] = \mathbb{k}[f]$.

Given $J \subseteq \mathbb{k}[X]$, we denote by $V(J)$ the "subvariety of X determined by J," i.e. the set of points on X on which each $f \in J$ vanishes; proposition III.25 below implies that $V(J)$ is indeed a subvariety of X. Proposition III.25 can be verified directly from the definitions; we leave its proof as an exercise.

Proposition III.25. *If f is a regular function on a quasiprojective variety X, then $V(f) := \{x \in X : f(x) = 0\}$ is Zariski closed in X. If in addition X is irreducible and f is zero on a nonempty open subset of X, then it is zero everywhere on X.* □

Consider the basic open subsets U_j of $\mathbb{P}^n$, $j = 0, \ldots, n$. Recall (from example III.17) that the map

$$U_j \ni [x_0 : \cdots : x_n] \mapsto (\frac{x_0}{x_j}, \ldots, \frac{x_{j-1}}{x_j}, \frac{x_{j+1}}{x_j}, \ldots, \frac{x_n}{x_j}) \in \mathbb{k}^n$$

induces a bijection between U_j and $\mathbb{k}^n$. Denote the coordinates on $\mathbb{k}^n$ by $(u_1, \ldots, u_n)$ so that the above map from U_j to $\mathbb{k}^n$ is given by

$$u_i = \begin{cases} x_{i-1}/x_j & \text{if } 1 \leq i \leq j \\ x_{i+1}/x_j & \text{if } j < i \leq n. \end{cases} \tag{14}$$

Recall that $\mathbb{k}^n$ already comes with a Zariski topology and a set of regular functions; we next show that these objects are compatible with the corresponding objects on U_j arising from its structure as a quasiprojective variety.

Proposition III.26. *The Zariski topology on U_j induced from $\mathbb{P}^n$ is the same as the Zariski topology on U_j induced from its identification with the affine space $\mathbb{k}^n$ with coordinates $(u_1, \ldots, u_n)$ via the map given by* (14).

PROOF. A subset V of U_j which is closed with respect to the Zariski topology induced from $\mathbb{k}^n$ is the set of zeroes of a collection of polynomials in $(x_i/x_j)_{i \neq j}$. Since x_i/x_j are regular functions on U_j, proposition III.25 implies V is closed with respect to the Zariski topology induced from $\mathbb{P}^n$. Conversely, if a subset V' of U_j is closed with respect to the Zariski topology induced from $\mathbb{P}^n$, exercise III.46 implies that V' is also closed with respect to the Zariski topology induced from $\mathbb{k}^n$, as required. □

Proposition III.26 in particular implies that "subvarieties" of U_j are the same regardless of whether we identify U_j with the affine space $\mathbb{k}^n$ or an open subset of $\mathbb{P}^n$. In proposition III.28 we will show "regular functions" also remain the same regardless of the consideration. Prior to that we give a description of regular functions on quasiprojective subsets of U_j in terms of rational functions[10] in $(u_1, \dots, u_n)$.

Proposition III.27. *Let X be a quasiprojective subset of U_j. Given a map $\phi : X \to \mathbb{k}$, the following are equivalent:*

(1) ϕ is a regular function on X,

(2) for each $x \in X$, there is an open neighborhood U of x in X such that $\phi|_U$ can be represented as f/g for some $f, g \in \mathbb{k}[u_1, \dots, u_n]$ such that g does not *vanish at any point of U.*

PROOF. This follows immediately from exercise III.46 and the observation that if f, g are homogeneous polynomials of same degree d in $(x_0, \dots, x_n)$, then $f/g = (f/x_j^d)/(g/x_j^d)$. □

Proposition III.28. *Let X be a subvariety of U_j. The set of regular functions on X (when X is regarded as a quasiprojective variety) are precisely the restrictions of polynomials in $(u_1, \dots, u_n)$. In particular,*

$$\mathbb{k}[X] \cong \mathbb{k}[u_1, \dots, u_n]/I(X), \tag{15}$$

where[11] *$I(X)$ is the set of polynomials in $(u_1, \dots, u_n)$ that are identically zero on X.*

PROOF. [12]Note that identity (15) is an immediate consequence of the first assertion and identity (8) from section III.3. Consequently we only prove the first assertion. Without loss of generality we may assume that $j = 0$. We will also identify a point on U_0 with the corresponding point in $\mathbb{k}^n$ via (14); in other words, we identify $a := (a_1, \dots, a_n) \in \mathbb{k}^n$ with the point $\hat{a} := [1 : a_1 : \cdots : a_n] \in U_0$. Since each u_i is a regular function on X, we only need to show that every regular function on X is the restriction of a polynomial in $(u_1, \dots, u_n)$. Given a regular function ρ on X, let $\mathfrak{q}$ be the ideal of $\mathbb{k}[u_1, \dots, u_n]$ consisting of all polynomials g such that $g\rho$ is the restriction of a polynomial in $(u_1, \dots, n)$ on some nonempty open subset of X. It is clear that $I(X) \subseteq \mathfrak{q}$, so that $V(\mathfrak{q}) \subseteq X$ (where $V(\mathfrak{q})$ is the subvariety of U_0 determined by $\mathfrak{q}$). On the other hand, proposition III.27 implies that for each $a = (a_1, \dots, a_n) \in X$, there is $g \in \mathfrak{q}$ such that $g(a) \neq 0$. Taken together, these observations imply that $V(\mathfrak{q}) = \emptyset$. The Nullstellensatz (theorem B.8) then implies that $1 \in \mathfrak{q}$, i.e. ρ agrees with a polynomial on some nonempty open subset of X. In the case that X is *irreducible*, it then follows that ρ is a polynomial on all of X (proposition III.25) and the proposition is true. In the general case, since X has

[10]Recall that classically a "rational function" means quotients of polynomials.

[11]Note that $I(X)$ denotes somewhat different objects depending on whether X is a projective variety or an affine variety. We hope that the intended meaning will always be clear from context.

[12]This proof is inspired by [Mum95, Proof of Proposition 1.11].

finitely many *irreducible components* (due to theorem III.9 and proposition III.26), we can proceed by induction on the number of irreducible components of X. So assume that proposition is true for all subvarieties of U_0 with at most k irreducible components, where $k \geq 1$, and that X has $k+1$ irreducible components. Let ρ be a regular function on X. Let $\mathfrak{r}$ be the ideal of all polynomials $h \in \Bbbk[u_1, \ldots, u_n]$ such that $h\rho$ is the restriction of a polynomial on *all of* X.

Claim III.28.1. *For each* $a = (a_1, \ldots, a_n) \in X$*, there is* $g \in \mathfrak{r}$ *such that* $g(a) \neq 0$*.*

PROOF. Let X_1 be an irreducible component of X, and X_2 be the union of the other irreducible components of X. Due to the induction hypothesis there are polynomials $f_1, f_2 \in \Bbbk[u_1, \ldots, u_n]$ such that ρ agrees with f_i on X_i, $i = 1, 2$. Proposition III.27 implies that there are polynomials $h', g' \in \Bbbk[u_1, \ldots, u_n]$ and an open neighborhood U' of a in X such that g' does not vanish at any point on U' and $g'\rho = h'$ on U'. Pick $g'' \in \Bbbk[u_1, \ldots, u_n]$ such that $g''(a) \neq 0$ and $U \supseteq X \setminus V(g'')$ (this is possible, e.g. due to exercise III.28). Then $g'f_i - h' \equiv 0$ on $X_i \setminus V(g'')$, which implies that $g''(g'f_i - h') \in I(X_i)$. Consequently $(g''g'\rho)|_{X_i} = (g''h')|_{X_i}$ for each i, and the claim holds with $g := g'g''$. □

Since $I(X) \subseteq \mathfrak{r}$, it follows (as in the case of $\mathfrak{q}$) that $V(\mathfrak{r}) = \emptyset$, and therefore $1 \in \mathfrak{r}$. But then ρ is the restriction of a polynomial in $(u_1, \ldots, u_n)$, as required. □

Example III.29. Proposition III.28 in particular implies that $\Bbbk[U_j] = \Bbbk[x_0/x_j, \ldots, x_n/x_j]$ for each j. We now show that $\Bbbk[\mathbb{P}^n] = \Bbbk$, i.e. the only regular functions on $\mathbb{P}^n$ are the constants (in fact, we will see in section III.9 that this is true for *all* irreducible projective varieties). Write $y_i := x_i/x_0$, $i = 1, \ldots, n$. A regular function ρ on $\mathbb{P}^n$ restricts to $f \in \Bbbk[x_1/x_0, \ldots, x_n/x_0] = \Bbbk[y_1, \ldots, y_n]$ on U_0, and to $g \in \Bbbk[x_0/x_1, x_2/x_1, \ldots, x_n/x_1] = \Bbbk[1/y_1, y_2/y_1, \ldots, y_n/y_1]$ on U_1. Since $U_0 \setminus V(y_1) \subseteq U_1$, it follows that $f(y_1, \ldots, y_n) = g(1/y_1, y_2/y_1, \ldots, y_n/y_1)$ in the field of fractions of $\Bbbk[y_1, \ldots, y_n]$ (exercise III.49). Equating the degrees[13] of both sides shows that $\deg(f) = 0$, i.e. f is a *constant*. Since $\mathbb{P}^n$ is irreducible (exercise III.39), it follows that ρ is constant on all of $\mathbb{P}^n$.

5.1. Exercises.

EXERCISE III.45. Prove proposition III.25.

EXERCISE III.46. Let f be a homogeneous polynomial of degree d in $(x_0, \ldots, x_n)$ and $U_j = \mathbb{P}^n \setminus V(x_j)$, $0 \leq j \leq n$, be a basic open subset of $\mathbb{P}^n$. Show that

(1) f/x_j^d can be expressed as a polynomial in $(x_i/x_j)_{i \neq j}$.
(2) $V(f) \cap U_j = \{x \in U_j : f/x_j^d = 0\}$.
(3) Deduce that every Zariski closed subset of U_j is the set of zeroes on U_j of a collection of polynomials in $(x_i/x_j)_{i \neq j}$.

EXERCISE III.47. Due to proposition III.26 we can treat $\Bbbk$ as a quasiprojective variety with Zariski topology induced from $\mathbb{P}^1$. Prove that every regular function

[13] Recall that the *degree* of f_1/f_2, where f_1, f_2 are polynomials, is defined as $\deg(f_1) - \deg(f_2)$.

$f : X \to \mathbb{k}$ on a quasiprojective variety X is *Zariski continuous*, i.e. continuous with respect to Zariski topology.

EXERCISE III.48. Let X be a quasiprojective subset of $\mathbb{k}^n$ (when $\mathbb{k}^n$ is treated as a quasiprojective variety)[14]. Show that X is an open subset of a subvariety of $\mathbb{k}^n$.

EXERCISE III.49. Let $f_1, g_1, f_2, g_2 \in k[x_1, \ldots, x_n]$, where k is an infinite field and $x_1, \ldots, x_n$ are indeterminates over k. Assume $f_1(a)/g_1(a) = f_2(a)/g_2(a)$ for all $a \in k^n \setminus \{x : g_1(x)g_2(x) = 0\}$. Show that $f_1/g_1 = f_2/g_2$ as *rational functions* in $(x_1, \ldots, x_n)$, i.e. as elements of the field of fractions of $k[x_1, \ldots, x_n]$. [Hint: use exercise III.11.]

6. Morphisms of quasiprojective varieties; affine varieties as quasiprojective varieties

A map $\phi : Y \to X$ of quasiprojective varieties is called a *morphism* if for every Zariski open subset U of X and every regular function h on U, the pullback $h \circ \phi$ is a regular function on $\phi^{-1}(U)$. An *isomorphism* is a bijective morphism whose inverse is also a morphism; we write $X \cong Y$ to denote that X and Y are *isomorphic*, i.e. there is an isomorphism between X and Y. We say that a morphism $\phi : Y \to X$ is a *closed embedding* if $\phi(Y)$ is a (closed) subvariety of X and ϕ is an isomorphism between Y and $\phi(Y)$. In section III.5 we have seen that the identification of $\mathbb{k}^n$ with a basic open set of $\mathbb{P}^n$ imbues every subvariety of $\mathbb{k}^n$ with the structure of a quasiprojective variety. From now on we treat every affine variety as a quasiprojective variety via this identification. We saw in section III.5 that this identification is compatible with the topology or regular functions on affine varieties. In proposition III.35 below we show that this is also compatible with morphisms, i.e. a map between affine varieties is a morphism (respectively, isomorphism) of affine varieties if and only if it is a morphism (respectivel, isomorphism) of quasiprojective varieties. First we go over some examples and basic properties of morphisms of quasiprojective varieties.

Example III.30. Morphisms from a quasiprojective variety Y to $\mathbb{k}$ are precisely regular functions on Y.

Example III.31. If $\phi : \mathbb{P}^n \to \mathbb{k}^m$ is a morphism for some m, n, then the pullback of each coordinate on $\mathbb{k}^m$ is a regular function on $\mathbb{P}^n$, and therefore is a constant (example III.29). It follows that the only possible morphism from a projective space to an affine space is a trivial map which maps everything to a point. In fact we will see that this property holds for *all* irreducible projective varieties (exercise III.78).

The following result is immediate from the definitions; we leave its proof as an exercise.

Proposition III.32. *A morphism $\phi : Y \to X$ of quasiprojective varieties induces via pullback a $\mathbb{k}$-algebra homomorphism $\phi^* : \mathbb{k}[X] \to \mathbb{k}[Y]$. If ϕ is an isomorphism of quasiprojective varieties, then ϕ^* is an isomorphism of $\mathbb{k}$-algebras.*

[14] These sets are sometimes referred to as *quasiaffine* varieties.

Example III.33. A constant morphism from $\mathbb{P}^n$ to a point for $n \geq 1$ shows that the converse of the last assertion of proposition III.32 is *not* true in general. Contrast this with the case of morphisms between *affine* varieties (exercise III.22).

Our next result shows that for quasiprojective subsets of affine spaces, morphisms can be characterized in terms of the affine coordinates. It is a counterpart of proposition III.27 which characterizes regular functions on these spaces in terms of the affine coordinates. The proof is straightforward using proposition III.27 and the definitions of rational functions and morphisms on quasiprojective varieties—we leave it as an exercise.

Proposition III.34. *Let X be a quasiprojective subset of $\mathbb{k}^n$ with coordinates $(x_1, \ldots, x_n)$, and Y be a quasiprojective variety. Given a map $\phi : Y \to X$, the following are equivalent:*

(1) ϕ is a morphism of quasiprojective varieties;
(2) for each i, $x_i \circ \phi$ is a regular function on Y.

If Y is a quasiprojective subset of an affine space $\mathbb{k}^m$ with coordinates $(y_1, \ldots, y_m)$, then the preceding properties are equivalent to the following:

(3) for each i, $x_i \circ \phi$ can be locally represented on Y by rational functions in $(y_1, \ldots, y_m)$. □

Corollary III.35. *A map between affine varieties is a morphism (respectively, isomorphism) of quasiprojective varieties if and only if it is a morphism (respectively, isomorphism) of affine varieties.*

PROOF. Since regular functions on affine varieties are precisely polynomials in the coordinate functions (proposition III.28), proposition III.34 implies that a map $Y \to X$ between affine varieties is a morphism of quasiprojective varieties if and only if the pullback of all regular functions on X is a regular function on Y. The latter property is precisely what defines a morphism of affine varieties. □

Example III.36. Corollary III.35 in particular implies that all the examples of morphisms of affine varieties from section III.3 are also examples of morphisms between quasiprojective varieties. For example, the map $\mathbb{k} \to \mathbb{k}^2$ given by $x \mapsto (x, x^2)$ defines an isomorphism between $\mathbb{k}$ and the *parabola* $V(y - x^2) \subseteq \mathbb{k}^2$. Since the parabola is closed in $\mathbb{k}^2$, this is actually a closed embedding.

Example III.37. (Continuation of example III.15). Let $X = V(\mathfrak{q}) \subseteq \mathbb{k}^n$, where $\mathfrak{q}$ is an ideal in $\mathbb{k}[x_1, \ldots, x_n]$. Given a polynomial f in $(x_1, \ldots, x_n)$, define $Y := V(\mathfrak{q}, x_{n+1}f - 1) \subseteq \mathbb{k}^{n+1}$. The projection onto $(x_1, \ldots, x_n)$-coordinates maps Y bijectively onto the Zariski open subset $X \setminus V(f)$ of X, with its inverse given by $(x_1, \ldots, x_n) \mapsto (x_1, \ldots, x_n, 1/f)$. Proposition III.34 implies that both these maps are morphisms, so that $X \setminus V(f) \cong Y$ as quasiprojective varieties; in particular, $X \setminus V(f)$ is isomorphic to an *affine variety*. It follows (due to proposition III.32) that

$$\mathbb{k}[X \setminus V(f)] \cong \mathbb{k}[Y] \cong \mathbb{k}[x_1, \ldots, x_n]/\langle I(X), x_{n+1}f - 1\rangle \cong \mathbb{k}[X]_f, \tag{16}$$

where $\mathbb{k}[X]_f$ is the *localization* (see section B.7) of $\mathbb{k}[X]$ at f. Taking $n = 1$, $\mathfrak{q} =$ the zero ideal, and $f = x_1$ yields that $\mathbb{k} \setminus \{0\} \cong V(x_1x_2 - 1)$ and $\mathbb{k}[\mathbb{k} \setminus \{0\}] \cong \mathbb{k}[x_1, 1/x_1]$ (fig. 3).

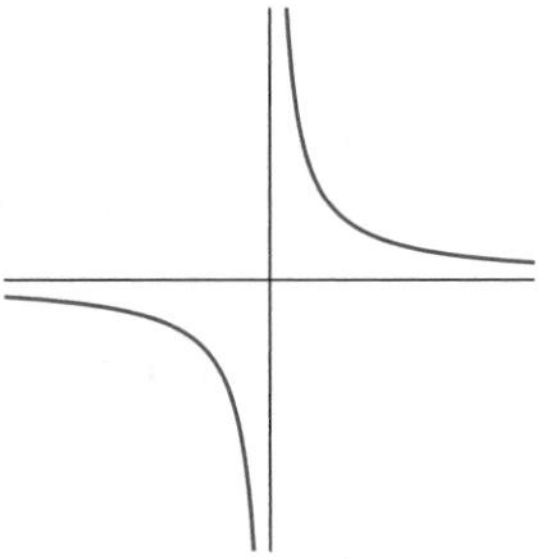

FIGURE 3. The projection onto x_1-axis induces an isomorphism between the hyperbola $x_1x_2 - 1 = 0$ and $\mathbb{k} \setminus \{0\}$.

Using example III.37 we can show that all quasiprojective varieties are built by "gluing" finitely many affine varieties:

Proposition III.38. *Every quasiprojective variety has a finite open cover by open subsets isomorphic to affine varieties.*

PROOF. Let W be the open subset of a subvariety of $\mathbb{P}^n$. Then W is covered by $W \cap U_j$, where U_j, $j = 0, \ldots, n$, are the basic open subsets of $\mathbb{P}^n$. The correspondence between U_j and $\mathbb{k}^n$ identifies $W \cap U_j$ with an open subset of an affine variety, which has a finite open covering by subsets isomorphic to $X \setminus V(f)$ for an affine variety X and a regular function f on X (exercise III.28). Since $X \setminus V(f)$ is isomorphic to an affine variety (example III.37), the proof is complete. □

The identification of $\mathbb{k}^n$ with the basic open subset U_0 of $\mathbb{P}^n$ identifies all subvarieties of $\mathbb{k}^n$ with quasiprojective subsets of $\mathbb{P}^n$. We next describe the closure of these sets in $\mathbb{P}^n$. For each $f \in \mathbb{k}[x_1, \ldots, x_n]$, the *homogenization* of f in $(x_0, \ldots, x_n)$ is the homogeneous polynomial $\tilde{f} := x_0^{\deg(f)} f(x_1/x_0, \ldots, x_n/x_0)$. The homogenization of an ideal $\mathfrak{q}$ of $\mathbb{k}[x_1, \ldots, x_n]$ in $(x_0, \ldots, x_n)$ is the ideal of $\mathbb{k}[x_0, \ldots, x_n]$ generated by the homogenization of all $f \in \mathfrak{q}$.

Proposition III.39. *Let $\mathfrak{q}$ be the any ideal of $\mathbb{k}[x_1, \ldots, x_n]$. The Zariski closure of $V(\mathfrak{q}) \subseteq \mathbb{k}^n$ in $\mathbb{P}^n$ is the subvariety $V(\tilde{\mathfrak{q}})$ of $\mathbb{P}^n$ defined by the homomgenization of $\mathfrak{q}$.*

PROOF. Recall that the identification between $\mathbb{k}^n$ and U_0 is given by the map $(x_1, \ldots, x_n) \mapsto [1 : x_1 : \cdots : x_n]$. It follows that the closure of $X := V(\mathfrak{q})$ in $\mathbb{P}^n$ is

$$\begin{aligned}\bar{X} &= \bigcap \{V(h) : h \text{ homogeneous in } (x_0, \ldots, x_n),\ h(1, a_1, \ldots, a_n) = 0 \text{ for all } (a_1, \ldots, a_n) \in X\} \\ &= \bigcap \{V(h) : h \text{ homogeneous in } (x_0, \ldots, x_n),\ h(1, x_1, \ldots, x_n) \in \sqrt{\mathfrak{q}}\}\end{aligned}$$

$$= \bigcap\{V(h) : h \text{ homogeneous in } (x_0, \ldots, x_n),\ h^k(1, x_1, \ldots, x_n) \in \mathfrak{q} \text{ for some } k \geq 0\}$$
$$= \bigcap\{V(h) : h \text{ homogeneous in } (x_0, \ldots, x_n),\ h^k = x_0^m \tilde{f} \text{ for some } f \in \mathfrak{q},\ k, m \geq 0\}$$
$$= \bigcap\{V(h) : h \text{ homogeneous in } (x_0, \ldots, x_n),\ h^k \in \tilde{\mathfrak{q}} \text{ for some } k \geq 0\}$$
$$= V(\tilde{\mathfrak{q}})$$

where the last equality follows from the correspondence between subvarieties of projective varieties and radical homogeneous ideals of $\mathbb{k}[x_0, \ldots, x_n]$ (exercise III.37). □

Example III.40. If $\mathfrak{q}$ is a principal ideal generated by $f \in \mathbb{k}[x_1, \ldots, x_n]$, then $\tilde{\mathfrak{q}}$ is also a principal ideal generated by $\tilde{f}$, which has the same degree as f. It follows that if X is the hypersurface (respectively, *hyperplane*[15]) of $\mathbb{k}^n$ determined by f, then $\bar{X}$ is the hypersurface (respectively, hyperplane) of $\mathbb{P}^n$ determined by $\tilde{f}$. Note that the set of "points at infinity" on $\bar{X}$ is $V(\tilde{f}) \setminus U_0 = V(\tilde{f}) \cap V(x_0) = \{[0 : x_1 : \cdots : x_n] \in \mathbb{P}^n : \mathrm{ld}(f)(x_1, \ldots, x_n) = 0\}$ where $\mathrm{ld}(f)$ is the *leading form*[16] of f; this is the relation which yields the correspondence (1) from chapter II.

Example III.41. In general, if $X = V(f_1, \ldots, f_k)$, then $\bar{X} \subsetneq V(\tilde{f}_1, \ldots, \tilde{f}_k)$. E.g. if $f_1 = x_1$ and $f_2 = x_2 - x_1^2$, then $X := V(f_1, f_2) \subset \mathbb{k}^2$ is the singleton consisting of the origin, so that $\bar{X} = X$, whereas $V(\tilde{f}_1, \tilde{f}_2) = V(x_1, x_0x_2 - x_1^2) = \{[1 : 0 : 0], [0 : 0 : 1]\} \subset \mathbb{P}^2$ (cf. example I.3).

6.1. Exercises.

EXERCISE III.50. (1) Show that every morphism is *Zariski continuous*, i.e. continuous with respect to the Zariski topology.
(2) Show that every bijection of $\mathbb{P}^1$ is Zariski continuous.
(3) Deduce that there are Zariski continuous maps from $\mathbb{P}^1$ to $\mathbb{P}^1$ which are not morphisms.

EXERCISE III.51. Prove proposition III.32.

EXERCISE III.52. Prove proposition III.34. [Hint: use proposition III.27.]

EXERCISE III.53. Verify the isomorphisms in (16) from example III.37.

EXERCISE III.54. (Converse to proposition III.39) Let V be a subvariety of $\mathbb{P}^n$ defined by a homogeneous ideal $\mathfrak{q}$ of $\mathbb{k}[x_0, \ldots, x_n]$. The basic open subset U_0 of $\mathbb{P}^n$ is isomorphic to $\mathbb{k}^n$ with coordinates $(x_1/x_0, \ldots, x_n/x_0)$. Let

$$\mathfrak{q}_0 := \{f/x_0^{\deg(f)} : f \text{ is a homogeneous polynomial in } \mathfrak{q}\}.$$

Show that

(1) $\mathfrak{q}_0$ is an ideal of $\mathbb{k}[x_1/x_0, \ldots, x_n/x_0]$.

[15] A *hyperplane* is a hypersurface defined by a linear polynomial.

[16] The leading form of a polynomial $f = \sum c_\alpha x^\alpha$ of degree d is the sum of all monomial terms $c_\alpha x^\alpha$ of degree d.

(2) $V \cap U_0$ is the subvariety of U_0 defined by q_0. In particular, the restriction of a projective hypersurface to U_0 is an affine hypersurface.

EXERCISE III.55. Let $X := V(x_0x_3 - x_1x_2) \setminus V(x_1, x_3) \subseteq \mathbb{P}^3$.

(1) Identify $U_3 := \mathbb{P}^3 \setminus V(x_3)$ with $\mathbb{k}^3$ with coordinates $(x, y, z) := (x_0/x_3, x_1/x_3, x_2/x_3)$. Show that $X \cap U_3 = V(x - yz)$. [Hint: use exercise III.54.]
(2) Deduce that $\mathbb{k}[X \cap U_3] \cong \mathbb{k}[y, z]$.
(3) Similarly show that $\mathbb{k}[X \cap U_1] \cong \mathbb{k}[u, v]$, where $(u, v) = (x_0/x_1, x_3/x_1)$.
(4) Deduce that $\mathbb{k}[X] \cong \mathbb{k}[z] \cong \mathbb{k}[u]$.

EXERCISE III.56. [Local nature of morphisms] Given a map $\phi : Y \to X$ between quasiprojective varieties, show that the following are equivalent:

(1) ϕ is a morphism.
(2) for every $a \in X$ and for each open neighborhood U of a in X, there is an open neighborhood U' of a in X such that $U' \subseteq U$ and $h \circ \phi$ is a regular function on $\phi^{-1}(U')$ for all $h \in \mathbb{k}[U']$.

EXERCISE III.57. Let $[x_0 : \cdots : x_n]$ be homogeneous coordinates on $\mathbb{P}^n$ and $U_j := \mathbb{P}^n \setminus V(x_j)$, $j = 0, \ldots, n$, be the basic open subsets of $\mathbb{P}^n$. Let X be a quasiprojective subset of $\mathbb{P}^n$ and Y be a quasiprojective variety. Given a map $\phi : Y \to X$, show that the following are equivalent:

(1) ϕ is a morphism.
(2) $(x_i/x_j) \circ \phi$ is a regular function on $\phi^{-1}(U_j \cap X)$ for each i, j.

[Hint: use exercise III.56 and proposition III.34.]

EXERCISE III.58. Let $\phi_j : X \to Y$ be morphisms of varieties, $j = 1, 2$. Assume $\phi_1(x) = \phi_2(x)$ for all x on a dense subset of X. Then show that $\phi_1 = \phi_2$. [Hint: use exercise III.56 and proposition III.25.]

EXERCISE III.59. Consider an $(n+1) \times (n+1)$-matrix A over $\mathbb{k}$ with entries $a_{i,j}$, $i, j = 0, \ldots, n$. If $\det(A) \neq 0$, then show that the linear isomorphism of $\mathbb{k}^{n+1}$ given by $x \mapsto Ax$ (where x is regarded as a column vector with $(n+1)$-rows) induces an *automorphism*[17] of $\mathbb{P}^n$. [Hint: use exercise III.57.]

EXERCISE III.60. Show that the map $\phi : \mathbb{P}^1 \to \mathbb{P}^2$ given by $[x_0 : x_1] \to [x_0^2 : x_0x_1 : x_2^2]$ induces an isomorphism between $\mathbb{P}^1$ and the subvariety $X := V(x_0x_2 - x_1^2)$ of $\mathbb{P}^2$. Since X is closed in $\mathbb{P}^3$, ϕ is in fact a *closed embedding* of $\mathbb{P}^1$ into $\mathbb{P}^2$.

EXERCISE III.61. Let $a = (a_1, \ldots, a_n), b = (b_1, \ldots, b_n) \in \mathbb{k}^n$, $b \neq 0$, and L be the line $\{a + bt : t \in \mathbb{k}\} \subset \mathbb{k}^n$. We consider L as a subset of $\mathbb{P}^n$ by identifying $\mathbb{k}^n$ with the basic open subset $U_0 = \mathbb{P}^n \setminus V(x_0)$ of $\mathbb{P}^n$. In this exercise you will calculate the closure $\bar{L}$ of L in $\mathbb{P}^n$.

---

[17]Recall that an *autmorphism* of X is an isomorphism from X to itself. In fact it turns out that *all* automorphisms of $\mathbb{P}^n$ are of this form (see, e.g. [Sha94, Exercise III.1.17]).

(1) Show that there is an automorphism ϕ of $\mathbb{P}^n$ such that $\phi(L) = \{[1 : t : 0 : \cdots : 0] : t \in \Bbbk\}$. [Hint: use exercise III.59.]
(2) Show that the homogeneous ideal generated by all polynomials vanishing on $\phi(L)$ is generated by $x_2, \ldots, x_n$. Compute the closure of $\phi(L)$ in $\mathbb{P}^n$.
(3) Show that the closure $\bar{L}$ of L in $\mathbb{P}^n$ is $L \cup \{[0 : b_1 : \cdots : b_n]\}$.
(4) Conclude that two lines in $\Bbbk^n$ intersect at infinity in $\mathbb{P}^n$ if and only if they are parallel, i.e. have the same "direction vector."

7. Rational functions and rational maps on irreducible varieties

Let X be an irreducible quasiprojective variety. A *rational function* on X is a regular function on a nonempty Zariski open subset of X. Formally, define a binary relation $\sim$ on the collection of pairs (f, U), where U is a nonempty Zariski open subset of X and f is a regular function on U, as follows:

$$(f, U) \sim (f', U') \text{ if and only if } f \text{ and } f' \text{ agree on } U \cap U',$$

we will see in proposition III.42 below that $\sim$ is an *equivalence relation*; a *rational function* is an equivalence class of $\sim$. We write $\Bbbk(X)$ for the set of rational functions on X.

Proposition III.42. *Let X be an irreducible variety.*

(1) $\sim$ is an equivalence relation. In particular, rational functions on X are well defined.
(2) If U is a nonempty open subset of X, then $\Bbbk(X) \cong \Bbbk(U)$.
(3) $\Bbbk(X)$ is a field.
(4) If X is also affine, then $\Bbbk(X)$ is isomorphic to the field of fractions of $\Bbbk[X]$.

PROOF. It is clear that $(f, U) \sim (f, U)$, and if $(f, U) \sim (f', U')$ then $(f', U') \sim (f, U)$. Now, given $(f, U) \sim (f', U')$ and $(f', U') \sim (f'', U'')$, it is clear that $f = f''$ on $U \cap U' \cap U''$. Since X is irreducible, it follows that $U \cap U' \cap U''$ is nonempty, and it is dense in $U \cap U''$ (proposition III.6). Consequently $f = f''$ on $U \cap U''$ (proposition III.25) and $(f, U) \sim (f'', U'')$. This proves the first assertion. Similarly, given open subsets U, U' of X, $U \cap U' \neq \emptyset$, and for every regular function f' on U', $(f', U') \sim (f'|_{U \cap U'}, U \cap U')$. This immediately implies that $\Bbbk(X) \cong \Bbbk(U)$. If $(f, U) \in \Bbbk(X)$ and f is not identically zero on U, then it is nonzero on a nonempty open subset U'' of X (proposition III.25) and $(1/f, U'')$ is the multiplicative inverse of (f, U) in $\Bbbk(X)$. This implies that $\Bbbk(X)$ is a field. The last assertion then follows from the fact that regular functions on a nonempty open subset of an affine variety X can be represented as quotients of regular functions on X (proposition III.27). □

Example III.43. Assertion (4) of proposition III.42 implies that the rational functions on $\Bbbk^n$ are the quotients of polynomials in $(x_1, \ldots, x_n)$, i.e. the notion of "rational functions" is compatible with its classical usage.

Example III.44. Assertion (2) of proposition III.42 implies that $\mathbb{k}(\mathbb{P}^n) \cong \mathbb{k}(\mathbb{k}^n) \cong \mathbb{k}(x_1, \ldots, x_n)$. Since $\mathbb{k}[\mathbb{P}^n] = \mathbb{k}$ (example III.29), assertion (4) of proposition III.42 in general does not hold when X is not affine.

A *rational map* ϕ from an irreducible variety X to a variety Y is a morphism from a nonempty open subset U of X to Y; usually a rational map is denoted by a broken arrow

$$\phi : X \dashrightarrow Y$$

It is called *dominant* if the image of U is dense in Y, and *birational* if Y is irreducible and there is a rational map $\psi : Y \dashrightarrow X$ such that $\psi \circ \phi$ (respectively, $\phi \circ \psi$) restricts to an automorphism of a nonempty open subset of X (respectively, Y). We say that *X and Y are birational* or *X is birational to Y* if there is a birational map from X to Y. A *rational variety* is a variety birational to $\mathbb{k}^n$ for some $n \geq 0$.

Example III.45. Every morphism is trivially a rational map. The mapping $x \mapsto (x, 1/x)$ defines a rational map $\mathbb{k} \dashrightarrow \mathbb{k}^2$ which is *not* a morphism, and *not* dominant. This map induces an isomorphism between $\mathbb{k} \setminus \{0\}$ and the hyperbola $H := V(xy - 1) \subseteq \mathbb{k}^2$ (example III.37). It follows that $\mathbb{k}$ and H are birational, and H is a rational variety.

Example III.46. An irreducible variety is trivially birational to all of its nonempty open subsets. In particular, $\mathbb{P}^n$ is a rational variety.

Example III.47. A basic source of rational maps which are not morphisms is the projection from $\mathbb{P}^n$ to a coordinate subspace. For example, given $m < n$, the map $[x_0 : \cdots : x_n] \mapsto [x_0 : \cdots : x_m]$ defines a rational map $\mathbb{P}^n \dashrightarrow \mathbb{P}^m$ which is not defined on the subvariety $Z := V(x_0, \ldots, x_m)$ of $\mathbb{P}^n$. Note that $Z \cong \mathbb{P}^{n-m-1}$.

We outline the proof of the following properties of birational maps in exercises III.64 to III.66 below.

Proposition III.48. *Two varieties are birational if and only if their fields of rational functions are isomorphic. Every irreducible variety is birational to a hypersurface.*

Example III.49. For $X := V(x^3 - y^2) \subseteq \mathbb{k}^2$, we have that $\mathbb{k}[X] \cong \mathbb{k}[x, y]/\langle x^3 - y^2 \rangle$. Since $x = (y/x)^2$ in $\mathbb{k}(X)$, it follows that $\mathbb{k}(X) \cong \mathbb{k}(y/x)$. Proposition III.48 therefore implies that X is birational to $\mathbb{k}$; in particular X is rational. Indeed, we saw in example III.14 that the map $\phi : t \mapsto (t^2, t^3)$ is a bijective morphism from $\mathbb{k}$ to X. It is straightforward to check that ϕ induces an isomorphism between $\mathbb{k} \setminus \{0\}$ and $X \setminus \{(0, 0)\}$ with the inverse given by $t = y/x$.

Cubic curves $V(y^2 - x(x-1)(x-\lambda)) \subseteq \mathbb{k}^2$ are *not* birational to $\mathbb{k}$ for $\lambda \neq 0, 1$ and $\mathrm{char}(\mathbb{k}) \neq 2$. Usually this is proven using "differential forms," which we do not develop in this book; see [Rei88, Section 2.2] for an elementary proof without using differential forms.

7.1. Exercises.

EXERCISE III.62. Let X be a reducible affine variety. Show that

(1) There are nonzero regular functions f, g on X, such that $fg = 0$ on X.
(2) The relation $\sim$ from the definition of rational functions is *not* an equivalence relation when applied to X. [Hint: $(f, X \setminus V(f)) \sim (g, X \setminus V(g)) \sim (\alpha f, X \setminus V(f))$, where $\alpha \neq 0, 1$.]
(3) The above problem can be rectified if in the definition of rational functions we only allow open subsets which have nonempty inersection with *every* irreducible component of X.
(4) The resulting set of "rational functions" on X is the localization of $\Bbbk[X]$ at the set of non zero-divisors; in other words, $\Bbbk(X) \cong \{f/g : f, g \in \Bbbk[X],\ g \text{ is } \textit{not} \text{ a zero-divisor in } \Bbbk[X]\}$.

EXERCISE III.63. Let $\phi : X \dashrightarrow Y$ be a dominant rational map. Show that if X is irreducible, then so is Y. [Hint: morphisms are continuous maps (exercise III.50); use proposition III.6.]

EXERCISE III.64. Let ϕ be a rational map from X to Y, both irreducible varieties. Show that

(1) If ϕ is dominant, then ϕ induces an injection $\phi^* : \Bbbk(Y) \hookrightarrow \Bbbk(X)$. [Hint: if f is a nonzero regular function on Y, then show that f can not be identically zero on the image of X.]
(2) Conversely, if ϕ induces a well-defined map from $\Bbbk(Y) \to \Bbbk(X)$, then ϕ must be dominant.

EXERCISE III.65. Let X and Y be irreducible varieties. In this exercise you will show that X is birational to Y if and only if $\Bbbk(X) \cong \Bbbk(Y)$.

(1) Show that if X is birational to Y then $\Bbbk(X) \cong \Bbbk(Y)$. [Hint: use exercise III.64.]
(2) Assume there is an isomorphism $F : \Bbbk(Y) \to \Bbbk(X)$. Choose an affine open subset of Y isomorphic to a subvariety Y' of $\Bbbk^m$ with coordinates $(y_1, \ldots, y_m)$. Show that
    (a) The correspondence $x \mapsto (F(y_1)(x), \ldots, F(y_m)(x))$ defines a well-defined rational map $\phi : X \dashrightarrow \Bbbk^m$.
    (b) The image of ϕ is a dense subset of Y'. [Hint: a polynomial g in $(y_1, \ldots, y_m)$ vanishes on the image of ϕ if and only if $g \in I(Y')$.]
    (c) ϕ induces a birational map from X to Y.

EXERCISE III.66. Let X be an irreducible variety. In this exercise you will show that there is $n \geq 1$ and a polynomial f in $(x_1, \ldots, x_n)$ such that X is birational to $V(f) \subset \Bbbk^n$.

(1) Show that $\Bbbk(X)$ is a finitely generated field extension of $\Bbbk$. [Hint: it suffices to consider the case that X is affine. Use identity (8) and proposition III.42.]
(2) Use corollary B.37 to show that there are $x_1, \ldots, x_n \in \Bbbk(X)$ such that
    (a) $x_1, \ldots, x_{n-1}$ are algebraically independent over $\Bbbk$,

(b) x_n is algebraic over $\mathbb{k}(x_1, \ldots, x_{n-1})$,
(c) $f(x_n) = 0$ for some irreducible $f \in \mathbb{k}[x_1, \ldots, x_n]$, and
(d) $\mathbb{k}(X) = \mathbb{k}(x_1, \ldots, x_n)$.

(3) Compute the field of rational functions of $Y := V(f) \subset \mathbb{k}^n$ and use exercise III.65 to conclude that X is birational to Y.

EXERCISE III.67. Given a rational map $\phi : X \to Y$ between irreducible varieties, show that the following are equivalent:

(1) ϕ is a birational map;
(2) there are open subsets U' of U and V of Y such that $\phi|_{U'} : U' \to V$ is an isomorphism.

8. Product spaces, Segre map, Veronese embedding

8.1. Product spaces, Segre map. Usually the topology considered on products of topological spaces is the "product topology," whose open sets are unions of products of open subsets of each factor. The product topology on algebraic varieties however is very restrictive:

Example III.50. Since the only proper closed subsets of $\mathbb{k}$ are finite sets of points (example III.1), the proper closed subsets of $\mathbb{k}^2 \cong \mathbb{k} \times \mathbb{k}$ under the product topology are finite unions of sets of the form $S_1 \times S_2$, where at least one of the S_i is finite. Given a polynomial f in (x, y), it follows that $V(f)$ on $\mathbb{k}^2$ is closed in the product topology if and only if f can be expressed as $g(x)h(y)$ for polynomials g, h in one variable (exercise III.68). In particular the product topology on $\mathbb{k}^2$ is *different* from the Zariski topology.

A more natural topology on products of varieties is constructed as follows: consider projective spaces $\mathbb{P}^m, \mathbb{P}^n$ with homogeneous coordinates, respectively, $[x_0 : \cdots : x_m]$ and $[y_0 : \cdots : y_n]$. A polynomial f in $(x_0, \ldots, x_m, y_0, \ldots, y_n)$ is called *bi-homogeneous* of *bi-degree* (d, e) in the x_i and y_j if each monomial that appears in f has degree d in the x-variables and e in the y-variables. The set of zeroes of a bi-homogeneous polynomial on $\mathbb{P}^m \times \mathbb{P}^n$ is well defined, and the Zariski topology on $\mathbb{P}^m \times \mathbb{P}^n$ is by definition the (unique) topology whose basic closed subsets are intersections of zero-sets of bi-homogeneous polynomials. If V, W are quasiprojective varieties with their Zariski closures being subvarieties respectively of $\mathbb{P}^m, \mathbb{P}^n$, $j = 1, 2$, then the Zariski topology on $V \times W$ is the topology induced from the Zariski topology on $\mathbb{P}^m \times \mathbb{P}^n$.

Example III.51. If X, Y are Zariski closed subsets of respectively $\mathbb{P}^m$ and $\mathbb{P}^n$, then $X \times Y$ is Zariski closed in $\mathbb{P}^m \times \mathbb{P}^n$, since every homogeneous polynomial in $(x_0, \ldots, x_m)$ is trivially bi-homogeneous in $(x_0, \ldots, x_m, y_0, \ldots, y_n)$. In particular, the Zariski topology on $\mathbb{P}^m \times \mathbb{P}^n$ is at least as fine as the product topology.

Example III.52. The Zariski topology on $\mathbb{k}^m \times \mathbb{k}^n$ comes from identifying $\mathbb{k}^m$ (respectively, $\mathbb{k}^n$) with the basic open subset $\mathbb{P}^m \setminus V(x_0)$ (respectively, $\mathbb{P}^n \setminus V(y_0)$), so that $\mathbb{k}^m \times \mathbb{k}^n$ is identified with the open subset $U_{00} := (\mathbb{P}^m \times \mathbb{P}^n) \setminus V(x_0 y_0)$ of $\mathbb{P}^m \times \mathbb{P}^n$. Write $x'_i := x_i/x_0$ and $y'_j := y_j/y_0$. It is straightforward to check (exercise III.69) that there is a one-to-one correspondence between the following collections of sets:

(17) intersections of U_{00} and sets of zeroes on $\mathbb{P}^m \times \mathbb{P}^n$ of bi-homogeneous polynomials in $(x_0, \ldots, x_m, y_0, \ldots, y_n)$ $\longleftrightarrow$ sets of zeroes on $\Bbbk^{m+n}$ of polynomials in $(x'_1, \ldots, x'_m, y'_1, \ldots, y'_n)$

It follows that the Zariski topology on $\Bbbk^m \times \Bbbk^n$ is the same as the Zariski topology on $\Bbbk^{m+n}$. Combined with the preceding examples, we see that the Zariski topology on $\Bbbk^m \times \Bbbk^n$ is finer than the product topology.

An equivalent formulation of the Zariski topology on product spaces can be given via the *Segre map* $s : \mathbb{P}^m \times \mathbb{P}^n \to \mathbb{P}^N$, where $N := (m+1)(n+1) - 1 = mn + m + n$. It is defined as follows: let z_{ij}, $0 \leq i \leq m$, $0 \leq j \leq n$, be a system of homogeneous coordinates on $\mathbb{P}^N$; then

$$s : ([x_0 : \cdots : x_m], [y_0 : \cdots : y_n]) \mapsto [x_i y_j]_{i,j}, \tag{18}$$

where the right-hand side denotes the point on $\mathbb{P}^N$ with homogeneous coordinates $z_{ij} = x_i y_j$. The fundamental result in this section is the following result; we outline its proof in exercise III.71.

Proposition III.53. *The image of $\mathbb{P}^m \times \mathbb{P}^n$ under s is the subvariety Z of $\mathbb{P}^N$ defined by (homogeneous) quadrics of the form $z_{ij}z_{kl} - z_{il}z_{kj}$, and s induces a homeomorphism between $\mathbb{P}^m \times \mathbb{P}^n$ and Z.*

Example III.54. When $m = n = 1$, $N = mn + m + n = 3$, and the image of $\mathbb{P}^1 \times \mathbb{P}^1 \hookrightarrow \mathbb{P}^3$ is the hypersurface $V(x_0x_3 - x_1x_2)$ (we already encountered this variety in example III.24). It is called a *ruled surface*, since it comes with "rulings" given by lines of the form $\{a\} \times \mathbb{P}^1$ and $\mathbb{P}^1 \times \{b\}$ for $a, b \in \mathbb{P}^1$. Under the usual identification of basic open subsets with affine spaces, the intersection of the ruled surface with $\Bbbk^3$ is the hypersurface $V(z - xy)$ (exercise III.54) and the rulings are given by the lines $\{(a, t, at) : t \in \Bbbk\}$ and $\{(t, b, bt) : t \in \Bbbk\}$ for $a, b \in \Bbbk$ (see fig. 4).

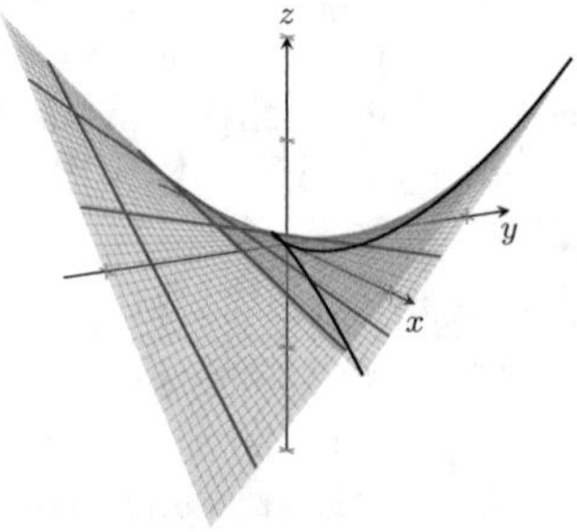

FIGURE 4. The rational normal curve and some rulings on the ruled surface $z = xy \subseteq \mathbb{R}^3$.

Proposition III.53 implies that the Zariski topology on a product space agrees with the topology induced by the Zariski topology on the projective space upon identification of the product space with its image under the Segre map. From now on we view $\mathbb{P}^m \times \mathbb{P}^n$ as a projective variety by identifying it with the subvariety $s(\mathbb{P}^m \times \mathbb{P}^n)$ of $\mathbb{P}^{(m+1)(n+1)-1}$.

8.2. Veronese embedding. Let d be a positive integer and $\mathcal{V}_d := \{\alpha = (\alpha_0, \ldots, \alpha_n) \in \mathbb{Z}_{\geq 0}^{n+1} : \sum_{j=0}^{n} \alpha_j = d\}$ be the set of exponents of monomials of degree d in $(x_0, \ldots, x_n)$. The *degree-d Veronese map* $\nu_d : \mathbb{P}^n \to \mathbb{P}^{|\mathcal{V}_d|-1}$ is given by

$$\nu_d : [x_0 : \cdots : x_n] \mapsto [x^\alpha : \alpha \in \mathcal{V}_d], \tag{19}$$

where x^α is a shorthand for $x_0^{\alpha_0} \cdots x_n^{\alpha_n}$.

Proposition III.55. *ν_d is a* closed embedding, *i.e. $\nu_d(\mathbb{P}^n)$ is a subvariety of $\mathbb{P}^{|\mathcal{V}_d|-1}$, and ν_d is an isomorphism between $\mathbb{P}^n$ and $\nu_d(\mathbb{P}^n)$.*

PROOF. It is straightforward to see that ν_d can be expressed as a composition

$$\mathbb{P}^n \xrightarrow{\delta} \mathbb{P}^n \times \cdots \times \mathbb{P}^n \xrightarrow{s_d} \mathbb{P}^{(n+1)^d-1} \overset{\pi}{\dashrightarrow} \mathbb{P}^{|\mathcal{V}_d|-1},$$

where $\delta : x \mapsto (x, \ldots, x)$ is the diagonal map to the product of d copies of $\mathbb{P}^n$, s_d is the Segre map on the d factors, and π is a projection which omits "redundant" coordinates of $s_d \circ \delta$ (i.e. for each $\alpha \in \mathcal{V}_d$, π retains only one of the coordinates of $s_d \circ \delta$ equalling x^α). Since s_d and δ are closed embeddings (proposition III.53 and exercise III.74), so is $s_d \circ \delta$. It is then straightforward to check that $\pi \circ s_d \circ \delta$ is a closed embedding as well. □

Example III.56. For $n = 1$, ν_d maps $[x_0 : x_1] \mapsto [x_0^d : x_0^{d-1}x_1 : \cdots : x_1^d] \in \mathbb{P}^d$ and its image is called the *rational normal curve of degree d.* Under the identifications of basic open subsets of projective spaces with affine spaces, this becomes the map $\Bbbk \to \Bbbk^d$ given by $t \mapsto (t, t^2, \ldots, t^d)$ so that the rational normal curve of degree d in $\Bbbk^d$ is the subvariety defined by $x_2 = x_1^2, x_3 = x_1^3, \ldots, x_d = x_1^d$. Note that for $d = 3$, the rational curve lies on the ruled surface from example III.54 (see fig. 4).

We use proposition III.55 to show that the complement of a projective hypersurface in a projective variety is affine. It is the projective analogue of the fact that the complement of a hypersurface in an affine variety is also affine (example III.37).

Proposition III.57. *Let X be a subvariety of $\mathbb{P}^n$ and f be a non-constant homogeneous polynomial in $(x_0, \ldots, x_n)$. Then $X \setminus V(f)$ is isomorphic to an affine variety.*

PROOF. Let $d := \deg(f) \geq 1$. If $d = 1$, then after a suitable linear automorphism of $\mathbb{P}^n$ from exercise III.59, we may assume that $f = x_0$. But then $X \setminus V(f)$ is a Zariski closed subset of the basic open set $U_0 := \mathbb{P}^n \setminus V(x_0)$. Since $U_0 \cong \Bbbk^n$, it follows that $X \setminus V(f)$ is isomorphic to an affine variety, as required. In the case that $d > 1$, let $Y := \nu_d(X) \subset \mathbb{P}^{|\mathcal{V}_d|-1}$, then ν_d maps $X \setminus V(f)$ isomorphically onto

$Y \setminus V(h)$ for a linear polynomial h on ν_d, and the result follows from the $d = 1$ case. □

8.3. Exercises.

EXERCISE III.68. Given a polynomial f in (x, y), show that the set $f(x, y) = 0$ on $\mathbb{k} \times \mathbb{k}$ is closed in the product topology if and only if f can be expressed as $g(x)h(y)$ for polynomials g, h in one variable.

EXERCISE III.69. Show that the correspondence between the collections of sets in (17) is bijective.

EXERCISE III.70. Show that the Zariski topology on $\mathbb{k}^m \times \mathbb{P}^n$ can be described as follows: let $(x_1, \ldots, x_m)$ be a system of coordinates polynomial coordinates of $\mathbb{k}^m$ and $[y_0 : \cdots : y_n]$ be homogeneous coordinates on $\mathbb{P}^n$. Then the closed sets on $\mathbb{k}^m \times \mathbb{P}^n$ are (finite) intersections of zero-sets of polynomials $f \in \mathbb{k}[x_1, \ldots, x_m, y_0, \ldots, y_n]$ which are *homogeneous in the y-variables*, i.e. of the form

$$f = \sum_{\beta_0 + \cdots + \beta_n = d} (\text{coefficient}) x_1^{\alpha_1} \cdots x_m^{\alpha_m} y_0^{\beta_0} \cdots y_n^{\beta_n}$$

for some $d \geq 0$ (in this case d is said to be the *degree* of f in $(y_0, \ldots, y_n)$).

EXERCISE III.71. In this exercise you will prove proposition III.53.

(1) Show that the image of s is contained in Z.
(2) Show that to prove the first assertion of proposition III.53 it suffices to prove that for each i, j, s induces a homeomorphism between $s^{-1}(Z \setminus V(z_{ij}))$ and $Z \setminus V(z_{ij})$.
(3) It suffices to consider the case $i = j = 0$. Show that $s^{-1}(Z \setminus V(z_{00})) = (\mathbb{P}^m \setminus V(x_0)) \times (\mathbb{P}^n \setminus V(y_0))$.
(4) Write

$$\begin{aligned} u_i &:= z_{i0}/z_{00}, \quad i = 1, \ldots, m, \\ v_j &:= z_{0j}/z_{00}, \quad j = 1, \ldots, n, \\ w_{ij} &:= z_{ij}/z_{00}, \quad i = 1, \ldots, m,\ j = 1, \ldots, n. \end{aligned}$$

Show that $Z \setminus V(z_{00})$ is isomorphic to the subvariety of $\mathbb{k}^{m+n+mn}$ with coordinates $(u_1, \ldots, u_m, v_1, \ldots, v_n, w_{11}, \ldots, w_{mn})$ determined the equations $w_{ij} = u_i v_j$, $i = 1, \ldots, m$, $j = 1, \ldots, n$ [Hint: use exercise III.54.]. Deduce that $Z \setminus V(z_{00}) \cong \mathbb{k}^{m+n}$.
(5) Write $x'_i := x_i/x_0$, $i = 1, \ldots, n$, $y'_j := y_j/y_0$, $j = 1, \ldots, n$. Show that the restriction of s to $(\mathbb{P}^m \setminus V(x_0)) \times (\mathbb{P}^n \setminus V(y_0))$ can be described as the map from $\mathbb{k}^m \times \mathbb{k}^n \to \mathbb{k}^{m+n+mn}$ given by

$$\begin{aligned} u_i &= x'_i, \quad i = 1, \ldots, m, \\ v_j &= y'_j, \quad j = 1, \ldots, n, \end{aligned}$$

$$w_{ij} = x'_i y'_j, \quad i = 1, \dots, m,\ j = 1, \dots, n.$$

(6) Deduce that s restricts to a homemorphism between $\mathbb{k}^m \times \mathbb{k}^n$ and $Z \setminus V(z_{00}) \cong \mathbb{k}^{m+n}$. [Hint: use example III.52.]

EXERCISE III.72. If X is a subvariety of $\mathbb{k}^m$ with coordinates $(x_1, \dots, x_m)$ and Y is a subvariety of $\mathbb{k}^n$ with coordinates $(y_1, \dots, y_n)$, show that $X \times Y$ is isomorphic to a subvariety of $\mathbb{k}^{n+m}$ with coordinates $(x_1, \dots, x_m, y_1, \dots, y_n)$, and $I(X \times Y)$ in $\mathbb{k}[x_1, \dots, x_m, y_1, \dots, y_n]$ is generated by $I(X) \subseteq \mathbb{k}[x_1, \dots, x_m]$ and $I(Y) \subseteq \mathbb{k}[y_1, \dots, y_n]$. [Hint: identify $\mathbb{k}^m$ and $\mathbb{k}^n$ with basic open subsets, respectively, of $\mathbb{P}^m$ and $\mathbb{P}^n$. Follow the steps of the isomorphism in exercise III.71.]

EXERCISE III.73. Let X, Y be quasiprojective varieties and $\pi_X : X \times Y \to X$ be the natural projection.

(1) In the case that X is a quasiprojective subset of $\mathbb{k}^m$ with coordinates $(x_1, \dots, x_m)$, show that $x_i \circ \pi_X$ is a regular function on $X \times Y$. [Hint: reduce to the case that Y is affine. Use exercise III.72.]
(2) Deduce that π_X is a morphism. [Hint: morphisms can be checked locally (exercise III.57). Use the preceding assertion and proposition III.34].

EXERCISE III.74. Let $\phi : X \to Y$ be a morphism of varieties.

(1) Show that the *graph* of ϕ, i.e. the set $\mathrm{gr}(\phi) := \{(x, \phi(x)) : x \in X\}$ is a subvariety (i.e. a Zariski closed subset) of $X \times Y$ and it is isomorphic to X. [Hint: use exercise III.73 to express $\mathrm{gr}(\phi)$ as the set of zeroes of regular functions on $X \times Y$.]
(2) If $\psi : X \to Z$ is another morphism of varieties, then show that the map $X \to Y \times Z$ given by $x \mapsto (\phi(x), \psi(x))$ is a morphism. [Hint: reduce to the case that Y, Z are affine. Use exercise III.72.]
(3) Deduce that if $\phi' : X' \to Y'$ is another morphism of varieties, then the map $X \times X' \to Y \times Y'$ given by $(x, x') \mapsto (\phi(x), \phi'(x'))$ is a morphism. [Hint: $\phi \circ \pi_X$ and $\phi' \circ \pi_{X'}$ are morphisms from $X \times X'$ to respectively Y and Y'.]

9. Completeness and compactification

While studying manifolds, it is often necessary to compactify them in order that the intersection theory of submanifolds is well behaved. However, the Noetherianness of finitely generated $\mathbb{k}$-algebras (Hilbert's basis theorem) implies that *all* varieties satisfy the usual definitions of compactness (exercise III.79). The property which plays in the case of varieties the role similar to that of compactness in the case of manifolds is *completeness*. A subset of a variety X is called *complete* if for every variety Y, the projection map $\pi : X \times Y \to Y$ is closed, i.e. it maps closed sets on to closed sets.

Example III.58. If $n \geq 1$, the projection onto the x_{n+1}-coordinate maps the hypersurface $V(x_1 x_{n+1} - 1)$ of $\mathbb{k}^{n+1} = \mathbb{k}^n \times \mathbb{k}$ to $\mathbb{k} \setminus \{0\}$, which is not closed in $\mathbb{k}$. It follows that $\mathbb{k}^n$ is *not* complete for any $n \geq 1$.

Example III.59. We can push the reasoning in the preceding example a bit further. If f is a regular function on a complete variety X, consider the subset $Z := V(ft - 1) \subseteq X \times \Bbbk$, where t is the coordinate on $\Bbbk$. Since Z closed in $X \times \Bbbk$, its projection $\pi(Z)$ is closed in $\Bbbk$. Since $\pi(Z) \subsetneq \Bbbk$, it must be finite, i.e. f takes *finitely many* values. This implies, e.g. that the only complete subsets of affine varieties are finite sets of points. See exercise III.75 for some other immediate consequences.

For many "natural" topological spaces (e.g. the Euclidean topology), compactness is equivalent to completeness—this is discussed in exercises III.80 to III.81. The following properties of complete sets are straightforward to verify and left as exercises (exercise III.76):

Proposition III.60. *(1) A complete subset of a variety is also Zariski closed.*
(2) Every subvariety of a complete variety is complete.
(3) The image of a complete variety under a morphism is also complete. □

The following is the main result of this section:

THEOREM III.61. *The projective space $\mathbb{P}^n$ is complete for all n.*

We give a proof of theorem III.61 following [Mum95, Proof of Theorem 2.23]. In fact we prove theorem III.62 below; the equivalence of this result with theorem III.61 is left as exercise III.77.

Proposition III.62. *The projection map $\pi : \mathbb{P}^n \times \Bbbk^m \to \Bbbk^m$ is closed for each m.*

PROOF. Choose affine coordinates $(y_1, \ldots, y_m)$ on $\Bbbk^m$ and homogeneous coordinates $[x_0 : \cdots : x_n]$ on $\mathbb{P}^n$. Let Z be a closed subset of $\mathbb{P}^n \times \Bbbk^m$. Then $Z = V(f_1, \ldots, f_k)$ for polynomials $f_j(x, y) \in \Bbbk[x_0, \ldots, x_n, , y_1, \ldots, y_m]$ which are homogeneous of degree $d_j \geq 0$ in the x-variables (exercise III.70). For each $b \in \Bbbk^m$, it follows from the correspondence between subvarieties of $\mathbb{P}^n$ and homogeneous ideals of $\Bbbk[x_0, \ldots, x_n]$ (exercise III.37) that $y \notin \pi(Z)$ if and only if there is $d \geq 1$ such that $\mathfrak{m}_+^d$ is contained in the ideal of $\Bbbk[x_0, \ldots, x_n]$ generated by $f_1(x, b), \ldots, f_k(x, b)$, where $\mathfrak{m}_+$ is the *irrelevant* maximal ideal of $\Bbbk[x_0, \ldots, x_n]$ generated by $x_0, \ldots, x_n$. Therefore it suffices to show that for each $d \geq 1$, the set $\{b \in \Bbbk^m : \mathfrak{m}_+^d$ is contained in the ideal of $\Bbbk[x_0, \ldots, x_n]$ generated by $f_1(x, b), \ldots, f_k(x, b)\}$ is Zariski open in $\Bbbk^m$. For each integer e, let V_e be the vector space of homogeneous polynomials of degree e in $(x_0, \ldots, x_n)$ (with $V_e = 0$ if $e < 0$) and let $m_e := \dim_{\Bbbk}(V_e)$. For each $b \in \Bbbk^m$, and $d \geq 1$, consider the linear map

$$T^d(b) : V_{d-d_1} \bigoplus \cdots \bigoplus V_{d-d_k} \to V_d \qquad (g_1(x), \ldots, g_k(x)) \mapsto \sum_{j=1}^{k} f_j(x, b) g_j(x)$$

Let $[T^d(b)]$ be the matrix of $T^d(b)$ with respect to fixed bases of $\bigoplus_j V_{d-d_j}$ and V_d. It is straightforward to check that $T^d(b)$ is surjective if and only if there is an $m_d \times m_d$-minor of $[T^d(b)]$ with nonzero determinant. It follows that the set $\{b \in \Bbbk^m : T^d(b)$ is surjective$\}$ is open in $\Bbbk^m$, as required. □

A *compactification*[18] of a variety X is a complete variety $\bar{X}$ containing an dense open subset isomorphic to X. Theorem III.61 implies that every quasiprojective variety has a compactification:

Proposition III.63. *(1) Every projective variety is complete. In particular, if X is a quasiprojective subset of $\mathbb{P}^n$, then the closure $\bar{X}$ of X in $\mathbb{P}^n$ is a compactification of X.*

(2) Let $\phi : X \to Y$ be a morphism of varieties such that $\phi(X)$ is dense in Y. Then ϕ can be extended to a surjective *morphism $\phi' : X' \to Y$ where X' is a variety which contains (an isomorphic copy of) X as a dense open subset.*

PROOF. The first assertion is immediate from proposition III.60 and theorem III.61. For the second assertion, let $\bar{X}$ be a compactification of X, and take X' to be the closure in $\bar{X} \times Y$ of the graph $\mathrm{gr}(\phi) := \{(x, \phi(x)) : x \in X\}$ of ϕ, and ϕ' to be the natural projection from X' to Y. The completeness of $\bar{X}$ implies that ϕ' is surjective. It remains to prove that $\mathrm{gr}(\phi)$ is open in X'. Indeed, since $\mathrm{gr}(\phi)$ is already closed in $X \times Y$, it follows that $\mathrm{gr}(\phi) = X' \cap (X \times Y)$. Since $X \times Y$ is open in $\bar{X} \times Y$, it follows that $\mathrm{gr}(\phi)$ is open in X', as required. □

9.1. Exercises.

EXERCISE III.75. Use example III.59 to prove that

(1) Every regular function on a complete irreducible variety is constant, i.e. $\Bbbk[X] = \Bbbk$.
(2) If f is a morphism from an irreducible complete variety to the affine space, then the image of f is a point.
(3) If $\bar{X}$ is a compactification of an affine variety X and C is a complete subset of $\bar{X}$ consisting of infinitely many points, then $C \cap (\bar{X} \setminus X) \neq \emptyset$.

EXERCISE III.76. Show that

(1) A complete subset of a variety is also (Zariski) closed. [Hint: Given $Z \subseteq X$, the diagonal subset $Z' := \{(z, z) : z \in Z\}$ is closed in $Z \times X$ (exercise III.74).]
(2) Every subvariety of a complete variety is complete. [Hint: if Z is a subvariety (i.e. Zariski closed subset) of X, then $Z \times Y$ is closed in $X \times Y$.]
(3) The image of a complete variety under a morphism is also complete. [Hint: for every variety Y, every morphism $\phi : Z \to Z'$ of varieties lifts to a morphism $\phi' : Z \times Y \to Z' \times Y$ (exercise III.74).]

EXERCISE III.77. Show that in order to prove that $\mathbb{P}^n$ is complete, it suffices to prove that the projection map $\pi : \mathbb{P}^n \times \Bbbk^m \to \Bbbk^m$ is closed for each m. [Hint: every variety has a finite open covering by closed subsets of the affine space (proposition III.38).]

---

[18] We would have liked to use "completion" instead of "compactification"; however, it might have suggested a (misleading) connection with the notion of completion of local rings (discussed in section III.14).

EXERCISE III.78. Show that a morphism from an irreducible projective variety to an affine variety must be a constant map. [Hint: use propositions III.60 and III.63.]

EXERCISE III.79. Recall that a topological space is called *compact* if each of its open covers has a finite subcover, and it is called *sequentially compact* if every infinite sequence of points has a convergent subsequence. Show that

(1) Every variety is compact. [Hint: use exercise III.13 and proposition III.38.]
(2) Every variety is sequentially compact. [Hint (the following strategy requires the notion of *dimension* covered in section III.11): start with a sequence of points on a quasiprojective variety X and take its closure Z in X. Reduce to the case that Z is irreducible and affine. Then consider two cases: (i) there is a nonzero regular function on Z that vanishes on an infinite subsequence, and (ii) every nonzero regular function vanishes on only finitely many points from that sequence.]

EXERCISE III.80. We say that a topological space X is *first countable* if for every point $x \in X$, there is a sequence of open neighborhoods $\{U_j\}_j$ of x in X such that every open neighborhood U of x in X contains some U_j. Show that the following are equivalent:

(1) X is first countable.
(2) For every point $x \in X$, there is a sequence of open neighborhoods $\{U_j\}_j$ of x in X such that for every open neighborhood U of x in X, there is N such that U contains U_j for each $j \geq N$.

EXERCISE III.81. Let X be a first-countable topological space. Consider the following properties:

(1) X is sequentially compact.
(2) For every first-countable topological space Y, the projection map $X \times Y \to Y$ is closed with respect to the product topology on $X \times Y$.

Show that (1) implies (2). If X is a first countable T_1-space[19], show that (2) implies (1).

10. Image of a morphism: Part I

Since projective varieties are complete (proposition III.63), the image of a morphism from a projective variety must be a subvariety of the target space, and in addition must be a *finite set* if the target space is an affine variety. The image of a morphism from an affine variety can be wilder—we have seen that it does not have to be closed or open (example III.16); the following example[20] shows that it can be a nontrivial projective variety:

[19]A topological space is T_1 if for every pair of distinct points, each has an open neighborhood not containing the other.

[20]It was motivated by a comment by R. Borcherds to the answer [hbb] by A. Bayer on *MathOverflow*.

Example III.64. Consider the morphism $\phi : \mathbb{k} \to \mathbb{P}^1$ given by $x \mapsto [(x - a_1)(x - a_2) : (x - b_1)(x - b_2)]$, where $a_i, b_j \in \mathbb{k}$ such that $a_i \neq b_j$ for any $i, j = 1, 2$. Then ϕ is surjective if (and only if) $a_1 + a_2 \neq b_1 + b_2$ (exercise III.82).

A morphism $\phi : X \to Y$ of varieties is said to be *dominant* or *dominating* if $\phi(X)$ is Zariski dense in Y, or equivalently, if $\phi(X)$ is not contained in any proper subvariety of Y. Any morphism $\phi : X \to Y$ turns into a dominant map if the target space is changed from Y to the Zariski closure of $\phi(X)$ in Y.

Proposition III.65. *If a morphism $\phi : X \to Y$ of quasiprojective varieties is dominant (in particular, if it is surjective), then the pullback $\phi^* : \mathbb{k}[X] \to \mathbb{k}[Y]$ is injective.*

PROOF. Given a nonzero regular function f on Y, the set $Y \setminus V(f)$ is nonempty and, due to proposition III.25, Zariski open in Y. Since ϕ is dominant, $\phi(X)$ intersects $Y \setminus V(f)$, so that $U := \phi^{-1}(Y \setminus V(f))$ is a nonempty Zariski open subset of X. Since $\phi^*(f)$ is nonzero on U, the proposition follows. □

Example III.66. The converse of proposition III.65 is *not* true. Indeed, since $\mathbb{k}[\mathbb{P}^n] = \mathbb{k}$, if $\phi : \mathbb{P}^n \to \mathbb{P}^n$ is a constant morphism that maps every point to a fixed point on $\mathbb{P}^n$, then ϕ^* is injective even though ϕ is far from being dominant if $n \geq 1$.

Let X be an affine variety and $\mathfrak{q}$ be an ideal of $\mathbb{k}[X]$. Given generators $f_1, \ldots, f_N$ of $\mathfrak{q}$, the *blow up* $\mathrm{Bl}_{\mathfrak{q}}(X)$ *of X at* $\mathfrak{q}$ is the closure in $X \times \mathbb{P}^{N-1}$ of the graph of the map $X \setminus V(\mathfrak{q}) \to \mathbb{P}^{N-1}$ given by $x \mapsto [f_1(x) : \cdots : f_N(x)]$. The following result shows that $\mathrm{Bl}_{\mathfrak{q}}(X)$ depends only on $\mathfrak{q}$, not on the choice of the f_j.

Proposition III.67. *Let $g_1, \ldots, g_q$ (respectively, $h_1, \ldots, h_r$) be generators of $\mathfrak{q}$ in $\mathbb{k}[X]$. Write Y (respectively, Z) for the closure in $X \times \mathbb{P}^{q-1}$ (respectively, $X \times \mathbb{P}^{r-1}$) of the graph of the map $\phi_g : x \mapsto [g_1(x) : \cdots : g_q(x)]$ (respectively, $\phi_h : x \mapsto [h_1(x) : \cdots : h_r(x)]$). Then $Y \cong Z$.*

PROOF. Here we only consider the special case that $r = q + 1$ and $h_j = g_j$, $j = 1, \ldots, q$; the general case is left as exercise III.83. There are $g'_1, \ldots, g'_q \in \mathbb{k}[X]$ such that $h_{q+1} = g'_1 g_1 + \cdots + g'_q g_q$. Let $\psi : X \times \mathbb{P}^{q-1} \to X \times \mathbb{P}^q$ be the map that sends

$$(x, [z_1 : \cdots : z_q]) \mapsto (x, [z_1 : \cdots : z_q : g'_1(x) z_1 + \cdots + g'_q(x) z_q]).$$

It is straightforward to check that ψ is a morphism and, if $\pi : X \times \mathbb{P}^q \dashrightarrow X \times \mathbb{P}^{q-1}$ is the natural projection given by $(x, [z_1 : \cdots : z_{q+1}]) \mapsto (x, [z_1 : \cdots : z_q])$, then $\pi \circ \psi$ is identity on $\mathrm{gr}(\phi_g)$, and therefore it is identity everywhere on Y (exercise III.58). On the other hand, $\mathrm{gr}(\phi_h) \subseteq V(z_{q+1} - \sum_{j=1}^q g'_j(x) z_j)$, since the latter is closed in $X \times \mathbb{P}^q$, it follows that $Z \subseteq V(z_{q+1} - \sum_{j=1}^q g'_j(x) z_j)$. This implies that π is well defined on Z. Since $\psi \circ \pi$ is identity on $\mathrm{gr}(\phi_h)$, it then follows that $\psi \circ \pi$ is identity on Z. Consequently, ϕ induces an isomorphism $Y \cong Z$. □

We next study the restriction to $\mathrm{Bl}_{\mathfrak{q}}(X)$ of the natural projection from $X \times \mathbb{P}^{N-1}$ to X; often this map is what one means by the "blow up of X at $\mathfrak{q}$." The next result shows that σ induces an isomorphism on $X' := X \setminus V(\mathfrak{q})$ and its image is the closure $\bar{X}'$ of X' in X. (Note that if X is irreducible and $\mathfrak{q}$ is not the zero ideal, then $\bar{X}' = X$.)

Proposition III.68. *Let $\mathfrak{q}$ be an ideal of the coordinate ring of an affine variety X and $\sigma : \mathrm{Bl}_{\mathfrak{q}}(X) \to X$ be the blow up of X at $\mathfrak{q}$. Let $\bar{X}'$ be the closure of $X' := X \setminus V(\mathfrak{q})$ in X. Then*

(1) σ maps $\mathrm{Bl}_{\mathfrak{q}}(X)$ onto $\bar{X}'$, and

(2) σ induces an isomorphism $\mathrm{Bl}_{\mathfrak{q}}(X) \setminus \sigma^{-1}(V(\mathfrak{q})) \cong X \setminus V(\mathfrak{q}) = \bar{X}' \setminus V(\mathfrak{q}) = X'$. In particular, if X is irreducible and $\mathfrak{q}$ is a nonzero ideal, then $\mathrm{Bl}_{\mathfrak{q}}(X)$ is irreducible and birational to X.

PROOF. Since $\mathbb{P}^{N-1}$ is complete and $\mathrm{Bl}_{\mathfrak{q}}(X)$ is closed in $\bar{X}' \times \mathbb{P}^{N-1}$, it follows that its projection to $\bar{X}'$ is closed. Since $\sigma(\mathrm{Bl}_{\mathfrak{q}}(X))$ clearly contains X', it follows that $\sigma(\mathrm{Bl}_{\mathfrak{q}}(X)) = \bar{X}'$, as required for the first assertion. The second assertion follows directly from assertion (1) of exercise III.74. □

If V is an irreducible subvariety of X not contained in $V(\mathfrak{q}) \subset X$, then the *strict transform* of V on $\mathrm{Bl}_{\mathfrak{q}}(X)$ is the closure in $\mathrm{Bl}_{\mathfrak{q}}(X)$ of $\sigma^{-1}(V \setminus V(\mathfrak{q}))$. In the case that $\mathfrak{q}$ is the ideal $I(Y)$ of $\mathbb{k}[X]$ consisting of regular functions vanishing on a subvariety Y of X, we also write $\mathrm{Bl}_Y(X)$ for $\mathrm{Bl}_{\mathfrak{q}}(X)$, and call it the "blow up of Y in X."

Example III.69. If Y is a point in $X := \mathbb{k}^n$, then $\sigma : \mathrm{Bl}_Y(X) \to X$ is an isomophism away from Y, and $\sigma^{-1}(Y) \cong \mathbb{P}^{n-1}$ (exercise III.84). In particular, for $n \geq 2$, the point is "blown up" to a hyperplane in $\mathbb{P}^n$. This is probably the origin of the name "blow up" (see fig. 5). In section VI.11 we revisit the blow up at a point of $\mathbb{k}^n$ from the perspective of "toric varieties" as a special case of "weighted blow ups" at a point.

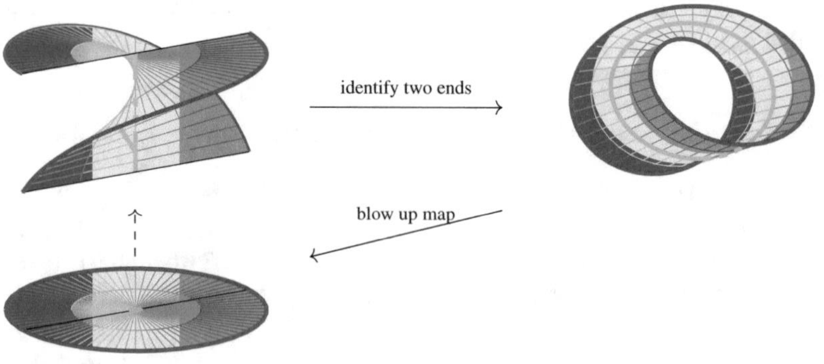

FIGURE 5. Blow up of a point on a disk in $\mathbb{R}^2$ is a *Möbius strip*.

Example III.70. More generally, if Y is a "linear subspace of dimension m" in $X := \mathbb{k}^n$, then $\sigma : \mathrm{Bl}_Y(X) \to X$ is an isomorphism away from Y and $\sigma^{-1}(y) \cong \mathbb{P}^{n-m-1}$ for each $y \in Y$ (exercise III.85).

Example III.71. If X is an open affine subset of $\bar{X}$ and $Z := V(\mathfrak{q}) \subseteq X$ consists of finitely many points, then the blow up map $\sigma : \mathrm{Bl}_{\mathfrak{q}}(X) \to X$ canonically extends to a map $\bar{\sigma} : \bar{Y} \to \bar{X}$ such that

(i) $\bar{Y}$ contains $Y := \mathrm{Bl}_{\mathfrak{q}}(X)$ as an open subset,

(ii) $\bar{\sigma}|_Y = \sigma|_Y$,

(iii) $\bar{\sigma}$ induces an isomorphism $\bar{Y} \setminus E \cong \bar{X} \setminus Z$, where $E := \sigma^{-1}(Z)$ is the "exceptional divisor" of σ.

Indeed, since $\mathrm{Bl}_{\mathfrak{q}}(X)$ is a closed subset of $X \times \mathbb{P}^N$ for some N, and σ is the projection onto X, it is straightforward to check that we can take $\bar{Y}$ to be the closure of $\mathrm{Bl}_{\mathfrak{q}}(X)$ in $\bar{X} \times \mathbb{P}^N$ and $\bar{\sigma}$ to be the projection onto $\bar{X}$. In particular, this gives a construction of the blow up of $\mathbb{P}^n$ at a point from the blow up of $\mathbb{k}^n$ at a point.

The next two results describe coordinate rings of closures of morphisms from affine varieties. These results are used in chapter VI to compute coordinate rings of "toric varieties."

Proposition III.72. *Let $\phi : X \to \mathbb{k}^N$ be the morphism from an affine variety X given by $x \mapsto (f_1(x), \ldots, f_N(x))$, where $f_1, \ldots, f_N$ are regular functions X. Let Z be the closure of $\phi(X)$ in $\mathbb{k}^N$ and R be the $\mathbb{k}$-subalgebra of $\mathbb{k}[X]$ generated by $f_1, \ldots, f_N$.*

(1) $\mathbb{k}[Z] \cong R$.

(2) Let $\pi : \mathbb{k}^N \to \mathbb{k}^M$ be the natural projection onto the first M coordinates, where $M \leq N$, and let H be the coordinate subspace of $\mathbb{k}^N$ spanned by the first M coordinates. Then $Z \cap H$ is contained in the closure of $\pi \circ \phi(X)$.

(3) If $\phi(X) \subseteq (\mathbb{k}^*)^N$ (where $\mathbb{k}^* := \mathbb{k} \setminus \{0\}$), then the coordinate ring of the closure of $\phi(X)$ in $(\mathbb{k}^*)^N$ is isomorphic to $R_{f_1 \cdots f_N} := \mathbb{k}[f_1, f_1^{-1}, \ldots, f_N, f_N^{-1}]$.

PROOF. Let $\phi^* : \mathbb{k}[y_1, \ldots, y_N] \to R$ be the ring homomorphism which maps each $y_i \mapsto f_i$. Let $I(Z)$ be the ideal of $\mathbb{k}[y_1, \ldots, y_N]$ consisting of polynomials that vanish on Z. Then $g \in I(Z)$ if and only if g vanishes on $\phi(X)$ if and only if $g(f_1, \ldots, f_N)(x)$ is zero for all $x \in X$ if and only if $\phi^*(g) = 0$ in $\mathbb{k}[X]$. Therefore $I(Z) = \ker \phi^*$, and $R \cong \mathbb{k}[y_1, \ldots, y_N]/I(Z) \cong \mathbb{k}[Z]$ (see eq. (8) for the last isomorphism). The second assertion follows from a general property of retractions—see exercise III.86. Finally, since $(\mathbb{k}^*)^N = \mathbb{k}^n \setminus V(y_1 \cdots y_N)$, the closure of $\phi(X)$ in $(\mathbb{k}^*)^N$ is $Z \setminus V(y_1 \cdots y_N)$ (exercise III.87), so that the coordinate ring of $Z \setminus V(y_1 \cdots y_N)$ is $\mathbb{k}[Z]_{y_1 \cdots y_N}$ (example III.37), which by the first assertion is isomorphic to $R_{f_1 \cdots f_N}$, as required for the third assertion. □

Corollary III.73. *Let $f_0, \ldots, f_N$ be regular functions on an* irreducible *affine variety X such that no f_i is identically zero on X. Let Z be the closure in $\mathbb{P}^N$ of the image of the map $\phi : X \setminus V(f_0, \ldots, f_N) \to \mathbb{P}^N$ defined by $x \mapsto [f_0(x) :$*

$\cdots : f_N(x)]$. *Let* $[z_0 : \cdots : z_N]$ *be homogeneous coordinates on* $\mathbb{P}^N$ *and* $U_i := \mathbb{P}^N \setminus V(z_i)$, $0 \le i \le N$. *Then*

(1) $\mathbb{k}[Z \cap U_i] \cong \mathbb{k}[f_0/f_i, \ldots, f_N/f_i]$,
(2) $\mathbb{k}[Z \cap U_i \cap U_j] \cong \mathbb{k}[f_0/f_i, \ldots, f_N/f_i, f_i/f_j]$.
(3) *Assume there exists* $b \in \mathbb{Z}_{>0}$ *and* $b_1, \ldots, b_N \in \mathbb{Z}_{\ge 0}$ *such that* $b = \sum_{i=1}^n b_i$ *and* $f_0^b = \prod_{i\ge1} f_i^{b_i}$. *Then* $Z \cap U_0 \subseteq \bigcup_{i\ge1}(Z \cap U_i)$.
(4) *Let* $H = V(z_{M+1}, \ldots, z_N)$, $H' := V(z_0, \ldots, z_M)$ *for some* $M \le N$, *and* $\pi : \mathbb{P}^N \setminus H' \to H$ *be the natural projection onto the first* M *coordinates. Then* $Z \cap H$ *is contained in the closure of* $\pi \circ \phi(X \setminus V(f_0, \ldots, f_M))$.

PROOF. Since each $U_i \cong \mathbb{k}^N$, assertion (1) follows from proposition III.72. Write $Z_i := Z \cap U_i$ and $Z_{ij} := Z \cap U_i \cap U_j$. Since Z_i is an affine variety, f_j/f_i is a regular function on Z_i, and $Z_{ij} = Z_i \setminus V(f_j/f_i)$, it follows that $\mathbb{k}[Z_{ij}] = \mathbb{k}[Z_i]_{f_j/f_i} = \mathbb{k}[f_0/f_i, \ldots, f_N/f_i, f_i/f_j]$ (see example III.37 for the first equality), which proves assertion (2). For assertion (3), note that $z_0^b - \prod_{i\ge1} z_i^{b_i}$ is identically zero on Z. It follows that $Z \cap U_0 = Z \setminus V(z_0^b) = Z \setminus V(\prod_{i\ge1} z_i^{b_i}) \subseteq \bigcup_{i\ge1}(Z \cap U_i)$, as required. For the last assertion it suffices to show that $Z \cap H \cap U_i$ is contained in the closure of $\pi \circ \phi(X \setminus V(f_i))$ for each $i = 0, \ldots, M$. This follows from assertion (2) of proposition III.72. □

10.1. Exercises.

EXERCISE III.82. Prove the assertion from example III.64.

EXERCISE III.83. Show that proposition III.67 is true in general if it holds for the special case that $r = q + 1$ and $h_j = g_j$, $j = 1, \ldots, q$.

EXERCISE III.84. Let $a = (a_1, \ldots, a_n) \in \mathbb{k}^n$. Recall that the ideal $I(a)$ of all polynomials vanishing at a is generated by $x_1 - a_1, \ldots, x_n - a_n$, so that $\mathrm{Bl}_a(\mathbb{k}^n)$ is the closure in $\mathbb{k}^n \times \mathbb{P}^{n-1}$ of the graph $\mathrm{gr}(\phi)$ of the map $\phi : \mathbb{k}^n \setminus \{a\} \to \mathbb{P}^{n-1}$ given by

$$(x_1, \ldots, x_n) \mapsto [x_1 - a_1 : \cdots : x_n - a_n]$$

(1) Given $b := (b_1, \ldots, b_n) \in \mathbb{k}^n \setminus \{0\}$, let $L_b := \{(a_1 + tb_1, \ldots, a_n + tb_n) : t \in \mathbb{k}\}$ be the line through a in the "direction" of b. Show that the closure of $\mathrm{gr}(\phi|_{L_b\setminus\{a\}})$ is $\mathrm{gr}(\phi|_{L_b\setminus\{a\}}) \cup \{[b_1 : \cdots : b_n]\}$.
(2) Deduce that if $\sigma : \mathrm{Bl}_a(\mathbb{k}^n) \to \mathbb{k}^n$ is the blow up map, then $\sigma^{-1}(a) = \mathbb{P}^{n-1}$.

EXERCISE III.85. Let $Y = V(x_1, \ldots, x_k) \subseteq X := \mathbb{k}^n$. Show that $\mathrm{Bl}_Y(X)$ is the closure in $\mathbb{k}^n \times \mathbb{P}^{k-1}$ of the graph $\mathrm{gr}(\phi)$ of the map $\phi : \mathbb{k}^n \setminus Y \to \mathbb{P}^{k-1}$ given by

$$(x_1, \ldots, x_n) \mapsto [x_1 : \cdots : x_k].$$

If $\sigma : \mathrm{Bl}_Y(X) \to X$ is the blow up map, then show that $\sigma^{-1}(y) = \{y\} \times \mathbb{P}^{k-1}$ for each $y \in Y$. [Hint: follow the steps of exercise III.84. In particular, for each

$a = (0, \dots, 0, a_{k+1}, \dots, a_n) \in Y$ and each $b := (b_1, \dots, b_k) \in \Bbbk^k \setminus \{0\}$, consider the restriction of ϕ to the line $L_b := \{(tb_1, \dots, tb_k, , a_{k+1}, \dots, a_n) : t \in \Bbbk\}$.]

EXERCISE III.86. Let Y be a topological space, and $r : Y \to H$ be a *retract* onto a subset $H \subset Y$ (which means r restricts to the identity map on H). Given $W \subset Y$, show that $\overline{W} \cap H \subset \overline{r(W)}$.

EXERCISE III.87. Let U be an open subset of a topological space Y and $W \subseteq U$. Show that the closure of W in U is the intersection of U and the closure of W in Y.

11. Dimension

The *dimension* $\dim(X)$ of an irreducible quasiprojective variety X is the transcendence degree over $\Bbbk$ of the field $\Bbbk(X)$ of rational functions on X. In general $\dim(X)$ is the maximum of the dimensions of all its irreducible components. We say that X has *pure dimension* d if each of its irreducible components has dimension d. If $a \in X$ and there is a Zariski open neighborhood U of a in X which has pure dimension d, we say that X has pure dimension d *near* a. The *codimension* of a subvariety Y of X is the integer $\dim(X) - \dim(Y)$.

Example III.74. Since $\Bbbk(\Bbbk^n) \cong \Bbbk(x_1, \dots, x_n)$, it follows that $\dim(\Bbbk^n) = n$, as expected. Since for irreducible varieties rational functions are determined by any nonempty open subset, it follows that $\dim(\mathbb{P}^n) = n$ as well. A variety is zero dimension if and only if it is a finite set (exercise III.88).

Example III.75. Let $X := V(xyz) \subseteq \Bbbk^3$ be the union of the three coordinate planes. Each of these planes is isomorphic to $\Bbbk^2$, and therefore has dimension two. Cosequently X has pure dimension two. Let Y be the union of X with the "diagonal" line $L = \{(t, t, t) : t \in \Bbbk\}$. It is straightforward to check that $t \mapsto (t, t, t)$ induces an isomorphism $L \cong \Bbbk$, so that $\dim(L) = 1$. It follows that $\dim(Y) = 2$ and Y is *not* pure dimensional. However, since X and L intersects only at the origin, Y is pure dimensional near every point except for the origin.

The following fact underlies all essential properties of dimension of varieties. It is a special case of Krull's Principal Ideal Theorem (theorem A.1).

Proposition III.76. *If f is a non-constant polynomial in $(x_1, \dots, x_n)$, then the hypersurface $V(f)$ of $\Bbbk^n$ has pure dimension $n - 1$.*

PROOF. Recall that the irreducible components of $X := V(f)$ are hypersurfaces $V(f_i)$, where f_i are irreducible factors of f (example III.10). Therefore without loss of generality we may assume f is irreducible, and moreover, that x_n appears in f with a nonzero coefficient. There are nonnegative integers $d_1 > d_2 > \cdots$ with $d_1 \geq 1$ and nonzero polynomials c_j in $(x_1, \dots, x_{n-1})$ such that $f = \sum_j c_j(x_1, \dots, x_{n-1}) x_n^{d_j}$. The irreducibility of f implies that $I(X) = \langle f \rangle$ (example III.10), so that no polynomial in $(x_1, \dots, x_{n-1})$ identically vanishes on X. This implies that

(1) $x_1, \ldots, x_{n-1}$ are algebraically independent over $\mathbb{k}$ in $\mathbb{k}(X)$, i.e. $\operatorname{tr.deg.}_{\mathbb{k}}(\mathbb{k}(X)) \geq n-1$;
(2) $c_j \neq 0 \in \mathbb{k}(X)$ for any j, so that x_n is algebraic over $\mathbb{k}(x_1, \ldots, x_n)$ in $\mathbb{k}(X)$ via the relation

$$\sum_j c_j(x_1, \ldots, x_{n-1}) x_n^{d_j} = 0$$

It follows that $\operatorname{tr.deg.}_{\mathbb{k}}(\mathbb{k}(X)) = n-1$, as required. □

Example III.77. A *curve* is a variety of pure dimension one and a *surface* is a variety of pure dimension two. Proposition III.76 implies that for all non-constant polynomials, $V(f) \subseteq \mathbb{k}^n$ is a curve if $n = 2$ and a surface if $n = 3$.

The dimension of a variety is an analogue of the dimension of smooth manifolds. In particular, when $\mathbb{k} = \mathbb{C}$ and X is irreducible, it follows from the results in section III.13 that there is an open Zariski dense subset of X (consisting of *nonsingular* points of X) which is a smooth manifold of (complex) dimension $\dim(X)$. Dimensions of subvarieties of a variety, however, in some ways behave rather differently from dimensions of submanifolds of a manifold:

THEOREM III.78. *Every proper subvariety of an irreducible variety X has smaller dimension than that of X.*

PROOF. Let Y be a proper subvariety of X. Since the field of rational functions does not change after restricting to a (nonempty) open subset, taking an open neighborhood of a point on Y if necessary we may assume without loss of generality that X, Y are affine. Since X, Y are also irreducible, it follows that $X = V(\mathfrak{p})$ and $Y = V(\mathfrak{q})$ for prime ideals $\mathfrak{p} \subsetneq \mathfrak{q}$ of $\mathbb{k}[x_1, \ldots, x_n]$, $n \geq 1$. Let $d := \dim(X)$. Without loss of generality we may assume that $x_1, \ldots, x_d$ are algebraically independent over $\mathbb{k}$ in $\mathbb{k}(X)$. It suffices to show [why?] that $x_1, \ldots, x_d$ are *not* algebraically independent over $\mathbb{k}$ in $\mathbb{k}(Y)$. Pick $q \in \mathfrak{q} \setminus \mathfrak{p}$. Since $x_1, \ldots, x_d, q$ are algebraically dependent over $\mathbb{k}$ in $\mathbb{k}(X)$, there is an irreducible polynomial $F(x_1, \ldots, x_d, y) \in \mathbb{k}[x_1, \ldots, x_d, y]$ such that $F(x_1, \ldots, x_d, q) = 0$ in $\mathbb{k}[X]$. Since $q \neq 0$ in $\mathbb{k}[X]$, it follows that F is *not* a multiple of y. Therefore $F(x_1, \ldots, x_d, 0)$ is a nonzero polynomial in $\mathbb{k}[x_1, \ldots, x_d]$ which represents an algebraic relation of $x_1, \ldots, x_d$ in $\mathbb{k}[Y]$, as required. □

Example III.79. Let Z be a curve and $Z' \subseteq Z$ be such that Z' contains a dense open subset of Z. Then theorem III.78 implies that $Z \setminus Z'$ has finitely many points. Consequently Z' is also open in Z and therefore Z' is a quasiprojective variety. This in particular implies that the image of a morphism from a curve is a quasiprojective variety. Note that this is *not* true for varieties of dimension greater than one, e.g. the image of the morphism $\mathbb{k}^2 \to \mathbb{k}^2$ given by $(x, y) \mapsto (x, xy)$ is not a quasiprojective variety (example III.16).

We presented theorem III.78 above due to the simple proof (which follows [Mum95, Proof of Proposition 1.14]). However, if one is willing to use more elaborate machinery, namely, Krull's principal ideal theorem (theorem A.1), then we can obtain the following more precise version of theorem III.78 in a straightforward way—its proof is left as exercise III.90.

THEOREM III.80. *Let $f_1, \dots, f_k$ be regular functions on a variety X of pure dimension n.*

(1) *If f_1 does not identically vanish on any irreducible component of X, then $V(f_1) \subseteq X$ is either empty or has pure dimension $n-1$.*
(2) *The dimension of every irreducible component of $V := V(f_1, \dots, f_k) \subseteq X$ is at least $n-k$. In other words, to define a subvariety of codimension m one needs at least m equations.*
(3) *If $\dim(V) = n-k$, then for all $I \subseteq [k]$, $V(f_i : i \in I)$ has pure dimension $n - |I|$ near every point of V.* □

Example III.81. Proposition III.76 is a special case of theorem III.80. In general the dimension of $V(f_1, \dots, f_k)$ may be greater than $n-k$, e.g. $V(x(x-y), y(x-y))$ is the diagonal line $x = y$ on $\mathbb{k}^2$, and therefore has dimension $1 > 2-2$.

Example III.82. Consider $\phi : \mathbb{k} \to \mathbb{k}^3$ given by $t \mapsto (t^3, t^4, t^5)$. In exercise III.91 you will show that

(1) $X := \phi(\mathbb{k})$ is a codimension two subvareity of $\mathbb{k}^3$ which is the set of zeroes of two polynomials (in other words, X is a "set theoretic complete intersection" in $\mathbb{k}^3$),
(2) but the ideal of polynomials vanishing on X can not be generated by two polynomials (in other words, X is *not* a "complete intersection" in $\mathbb{k}^3$).

Corollary III.83. (Curve selection lemma I). *Let U be a nonempty Zariski open subset of an irreducible variety X and $a \in X$. If $\dim(X) > 0$, then there is an irreducible curve C on X containing a such that $C \cap U$ is nonempty.*

PROOF. We prove this by induction on $n := \dim(X)$. If $n = 1$, then it is trivially true, so assume $n > 1$. Without loss of generality we may assume that X is affine.

Claim III.83.1. *There is a regular function f on X such that $f(a) = 0$ and U intersects every irreducible component of $V(f) \subseteq X$.*

PROOF. By theorem III.78 the dimension of $Z := X \setminus U$ is less than n. If $\dim(Z) < n-1$, then due to theorem III.80 the claim is satisfied by any non-constant $f \in \mathbb{k}[X]$ with $f(a) = 0$. So assume $\dim(Z) = n-1 > 0 = \dim(\{a\})$. Then for each irreducible $(n-1)$-dimensional component Z_i of Z, the ideal $I(Z_i)$ of regular functions vanishing on Z_i cannot contain the ideal $I(a)$ of regular functions vanishing at a. It then follows from the "prime avoidance" phenomenon (assertion (2) of theorem B.2) that there is $f \in I(a)$ such that f does not identically vanish on any $(n-1)$-dimensional component of Z. Then f satisfies the claim. □

Let X' be an irreducible component of $V(f)$ containing a. Then $U' := U \cap X \neq \emptyset$. Since $\dim(X') = n - 1$ (theorem III.80), we are done by induction. □

Intuitively, the image of a map cannot have bigger dimension than the source. The following result makes it precise:

Proposition III.84. *Let $\phi : X \to Y$ be a dominant morphism of varieties. Then* $\dim(X) \geq \dim(Y)$.

PROOF. This is immediate from the observation that if X and Y are irreducible, then $\phi^* : \mathbb{k}(Y) \to \mathbb{k}(X)$ is an injection (exercise III.64). □

Now we study the dimension of the fibers of a morphism. Consider e.g. the map $\phi : \mathbb{k}^2 \to \mathbb{k}^2$ from example III.16 given by $(x, y) \mapsto (x, xy)$. The image of ϕ is $(\mathbb{k}^2 \setminus V(x)) \cup \{(0,0)\}$. For most points on the image, in fact for all $(u, v) \in \phi(\mathbb{k}^2) \setminus \{(0,0)\}$, the fiber $\phi^{-1}(u, v)$ over (u, v) is zero dimension, and consists of a single point $(u, v/u)$. On the other hand, $\phi^{-1}(0,0)$ is all of the y-axis, and hence has dimension one. In particular, there is a nonempty Zariski open subset of the image over which the dimension of fibers is constant, and it is possible that some fibers have higher dimension. Our next result shows that this is true in general.

THEOREM III.85. *Let X be an irreducible variety and $\phi : X \to Y$ be a surjective morphism. Then*

(1) For every $y \in Y$ and every irreducible component V of $\phi^{-1}(y)$, $\dim(V) \geq \dim(X) - \dim(Y)$.
(2) Moreover, there is a nonempty Zariski open subset U of Y such that $\phi^{-1}(y)$ is of pure dimension $\dim(X) - \dim(Y)$ *for each $y \in U$.*

PROOF. [Sha94, Proof of Theorem I.7]. We may assume without loss of generality that Y is affine. Let $m := \dim(Y)$. Take $y \in Y$. The following can be proved via an induction on m, we leave its proof as an exercise (exercise III.94).

Claim III.85.1. *There are regular functions $h_1, \ldots, h_m$ on Y such that $V(h_1, \ldots, h_m) \cap U = \{y\}$ for some open neighborhood U of y in Y.* □

Pick $h_1, \ldots, h_m, U$ as in claim III.85.1. Since X is irreducible, it follows that $\dim(\phi^{-1}(U)) = \dim(X)$ (exercise III.89). Since $\phi^{-1}(y) \cap \phi^{-1}(U)$ is the zero-set of $h_j \circ \phi$, $j = 1, \ldots, m$, assertion (1) follows from theorem III.80. For the second assertion we proceed by induction on $n := \dim(X)$. Clearly it is true for $n = 0$. Let X' be an arbitrary nonempty open *affine* subset of $\phi^{-1}(Y)$. Due to the first assertion it suffices [why?] to prove that

(20) there is a nonempty Zariski open subset Y' of Y such that $\dim(\phi^{-1}(y) \cap X') \leq n - m$ for each $y \in Y' \cap \phi(X')$.

Since X' is dense in X and ϕ is continuous (exercise III.47) and surjective, it follows that $\phi(X')$ is dense in Y. Therefore ϕ induces an injection $\mathbb{k}[Y] \to \mathbb{k}[X']$ (proposition III.65); we will consider $\mathbb{k}[Y]$ as a subring of $\mathbb{k}[X']$. Let $g_1, \ldots, g_M$ (respectively $f_1, \ldots, f_N$) be $\mathbb{k}$-algebra generators of $\mathbb{k}[Y]$ (respectively $\mathbb{k}[X']$). Since $\dim(X') = \dim(X) = n$, the transcendence degree of $\mathbb{k}(X')$ over $\mathbb{k}(Y)$

is $n - m$. Therefore we may assume without loss of generality that $f_1, \dots, f_{n-m}$ are algebraically independent over $\Bbbk(Y)$ and f_j is algebraically dependent over $\Bbbk(Y)(f_1, \dots, f_{n-m})$ for each $j > n - m$. Fix j, $n - m + 1 \leq j \leq N$, and pick a nonzero polynomial $F_j(u_1, \dots, u_{n-m}, v)$ in variables $u_1, \dots, u_{n-m}, v$ with coefficients in $\Bbbk[Y]$ such that $F_j(f_1, \dots, f_{n-m}, f_j) = 0$. Write

$$F_j(u_1, \dots, u_{n-m}, v) = \sum_{i=0}^{d_j} F_{j,i}(u_1, \dots, u_{n-m}) v^i,$$

where d_j is the degree of F_j in v, and write

$$F_{j,d_j}(u_1, \dots, u_{n-m}) = \sum_{\alpha=(\alpha_1,\dots,\alpha_{n-m})} g_{j,\alpha} u_1^{\alpha_1} \cdots u_{n-m}^{\alpha_{n-m}},$$

where each $g_{j,\alpha} \in \Bbbk[Y]$. Fix $\alpha^j = (\alpha_1^j, \dots, \alpha_{n-m}^j)$ such that g_{j,α^j} is a *nonzero* regular function on Y. Let $\tilde{Y} := Y \setminus V(\prod_j g_{j,\alpha^j})$. Theorem III.78 implies that $\tilde{Y}$ is a nonempty open subset of Y.

Claim III.85.2. *There is a nonempty open subset Y' of $\tilde{Y}$ such that for each $y \in Y'$, no irreducible component of $\phi^{-1}(y)$ is contained in $V := \bigcup_{j=n-m+1}^{N} V(F_{j,d_j}) \cap X'$.*

PROOF. If $\phi(V)$ is not dense in Y, then we can simply take Y' to be the complement in $\tilde{Y}$ of the closure of $\phi(V)$. So assume $\phi(V)$ is dense in Y. Then $\phi|_V$ can be extended to a surjective morphism $\phi' : V' \to Y$, where V' is a variety containing V as a dense subset (proposition III.63). Since $\dim(V') = \dim(V)$ (exercise III.89), theorem III.78 implies that $\dim(V') < n$. The inductive hypothesis then implies that there is a nonempty open subset Y' of $\tilde{Y}$ such that $\dim(\phi'^{-1}(y)) = \dim(V') - m$ for each $y \in Y'$. The first assertion of theorem III.85 therefore implies that for each $y \in Y'$, no component of $\phi^{-1}(y)$ can be contained in V', as required. □

Let Y' be as in claim III.85.2. Fix $y \in Y' \cap \phi(X')$ and an irreducible component W of $\phi^{-1}(y) \cap X'$. Let $\bar{f}_j := f_j|_W$, $j = 1, \dots, N$. Now fix j, $n-m < j \leq N$. Let $\bar{F}_j(u_1, \dots, u_{n-m}, v)$ be the polynomial in $\Bbbk[u_1, \dots, u_{n-m}, v]$ constructed from F_j by evaluating each coefficient from $\Bbbk[Y]$ at y. Since $\bar{F}_j(\bar{f}_1, \dots, \bar{f}_{n-m}, \bar{f}_j) = 0$, claim III.85.2 implies that $\bar{f}_j$ is algebraically dependent on $\Bbbk(\bar{f}_1, \dots, \bar{f}_{n-m})$ [why?]. Since $\Bbbk[W] = \Bbbk[\bar{f}_1, \dots, \bar{f}_N]$, and since $\Bbbk(W)$ is the quotient field of $\Bbbk[W]$ (proposition III.42), it follows that $\dim(W) \leq n - m$, as required. □

We next show that given a surjective morphism $\phi : X \to Y$ of irreducible varieties, the set $Y_0 := \{y \in Y : \dim(\phi^{-1}(y)) = \dim(X) - \dim(Y)\}$ does *not* have to be open in Y (however, we will see in the next section that it is a *constructible* subset of Y).

Example III.86. ([rshs]). The map $\psi : \Bbbk^3 \to \Bbbk^3$ given by $(x, y, z) \mapsto (x, y, z^2)$ is surjective. For each $\xi \in \Bbbk \setminus \{0\}$, let L_ξ be the line $V(y, z - \xi)$ on $\Bbbk^3$. Assume $\mathrm{char}(\Bbbk) \neq 2$. Then $\psi^{-1}(L_{\xi^2})$ is the disjoint union of L_ξ and $L_{-\xi}$. Consider the blow up $\sigma : \mathrm{Bl}_{L_\xi}(\Bbbk^3) \to \Bbbk^3$ of $\Bbbk^3$ at L_ξ. Recall that σ is isomorphism on $\sigma^{-1}(\Bbbk^3 \setminus L_\xi)$ and $\sigma^{-1}(c) \cong \mathbb{P}^1$ for each $c \in L_\xi$ (example III.70). Now fix $c = (\rho, 0, \xi) \in L_\xi$. Let $X := \mathrm{Bl}_{L_\xi}(\Bbbk^3) \setminus \sigma^{-1}(c)$ and ϕ be the restriction of $\psi \circ \sigma$ on X. Then it is straightforward to check that $\phi : X \to \Bbbk^3$ is surjective, and the set $Y_0 := \{a \in \Bbbk^3 : \dim(\phi^{-1}(a)) = 0\}$ is $(\Bbbk^3 \setminus L_{\xi^2}) \cup \{(\rho, 0, \xi^2)\}$, and $Y_1 := \{a \in \Bbbk^3 : \dim(\phi^{-1}(a)) = 0\}$ is $L_{\xi^2} \setminus \{(\rho, 0, \xi^2)\}$. In particular, neither Y_0 nor Y_1 is open or closed in $Y := \Bbbk^3$.

11.1. Exercises.

EXERCISE III.88. Show that a zero-dimensional variety is the union of finitely many points, in other words, a positive dimensional variety contains infinitely many points.

EXERCISE III.89. Let $X \subseteq Y$ be irreducible quasiprojective varieties, and let X' be the closure of X in X'. Show that $\dim(X') = \dim(X)$.

EXERCISE III.90. Prove theorem III.80 [Hint: use Krull's Principal Ideal Theorem (theorem A.1)].

EXERCISE III.91. Let $\phi : \Bbbk \to \Bbbk^3$ be the morphism given by $t \mapsto (t^3, t^4, t^5)$. Show that

(1) $\phi(\Bbbk)$ is defined by equations $y^3 - x^4 = 0$ and $z^3 - x^5 = 0$ on $\Bbbk^3$. [Hint: if (a, b, c) is a solution to these equations, and α is a third root of a, then $b = \zeta_1 \alpha^4$ and $c = \zeta_2 \alpha^5$ for third roots ζ_1, ζ_2 of 1. One can choose a third root ζ of 1 such that $\phi(\zeta\alpha) = (a, b, c)$.]
(2) X is irreducible and $\dim(X) = 1$. [Hint: $\Bbbk$ is birational to $\phi(\Bbbk)$.]
(3) The ideal $I(X)$ of polynomials vanishing on X can not be generated by less than three polynomials. [Hint: for all monomials of degree ≤ 3 in (x, y, z), compute its pullback by ϕ. Show that the only degree 2 polynomial in $I(X)$ is $g_1 := y^2 - xz$, and modulo multiples of g_1, the only degree 3 polynomials in $I(X)$ are $g_2 := x^3 - yz$ and $g_3 := z^2 - x^2y$. Conclude that there can not be two polynomials $f_1, f_2 \in I(X)$ such that g_1, g_2, g_3 are in the ideal generated by f_1, f_2.]
(4) $\dim(X) = 1$.

EXERCISE III.92. Let X be a subvariety of $\mathbb{P}^n$ of dimension d.

(1) If $d > 0$, show that X intersects every hypersurface on $\mathbb{P}^n$. [Hint: the complement of a hypersurface in $\mathbb{P}^n$ is affine, and every projective variety is complete.]
(2) Deduce that $\dim(X \cap V(f)) \geq d - 1$ for every homogeneous polynomial f (in the homogeneous coordinates on $\mathbb{P}^n$). [Hint: theorem III.80.]
(3) Deduce that $X \cap V(f_1, \ldots, f_d) \neq \emptyset$ for all homogeneous $f_1, \ldots, f_d$.
(4) Show that there are homogeneous polynomials $f_1, \ldots, f_d$ such that $X \cap V(f_1, \ldots, f_d)$ has finitely many elements. [Hint: given a subvariety Y of $\mathbb{P}^n$, fix a point a_j on each irreducible component of Y, and choose

a homogeneous polynomial f which does not vanish at any a_j. Then $\dim(Y \cap V(f)) = \dim(Y) - 1$.]

EXERCISE III.93. Let X be the "ruled surface" $\mathbb{P}^1 \times \mathbb{P}^1$ and Y be a ruling on X (i.e. Y is of the form $\{a\} \times \mathbb{P}^1$ or $\mathbb{P}^1 \times \{a\}$ for some $a \in \mathbb{P}^1$). We treat X as a subvariety of some projective space $\mathbb{P}^n$ (recall from example III.54 that we may take $n = 3$) with homogeneous coordinates $[x_0 : \cdots : x_n]$. Show that

(1) Y is a subvariety of pure codimension one in X.
(2) There is no homogeneous polynomial f in $(x_0, \ldots, x_n)$ such that $Y = X \cap V(f)$. [Hint: $X \setminus V(f)$ is affine (proposition III.57). There are rulings on $X \setminus Y$ which are isomorphic to $\mathbb{P}^1$ and therefore complete.]

EXERCISE III.94. Let X be an affine variety in $\mathbb{k}^n$ of dimension $m \geq 1$ and $a \in X$. Fix $d \geq 1$. Show that it is possible to find m polynomials $h_1, \ldots, h_m$ in $(x_1, \ldots, x_n)$ of degree d such that a is an *isolated point* of $V(h_1, \ldots, h_m) \cap X$, i.e. there is a Zariski open subset U of a in $\mathbb{k}^n$ such that $U \cap X \cap V(h_1, \ldots, h_m) = \{a\}$. [Hint: use induction on dimension; at every step choose a polynomial of degree d which vanishes at a, but does not identically vanish on any of the irreducible components of the variety.]

EXERCISE III.95. Verify the assertions from example III.86.

EXERCISE III.96. Let $\phi : X \to Y$ be a morphism between curves. Assume ϕ does not map any irreducible component of X to a point. Show that ϕ is a finite-to-one map.

EXERCISE III.97. Let $C \subset \mathbb{P}^n$ be a projective curve, and $\phi : C \to \mathbb{P}^1$ be a surjective morphism which is not constant on any of the irreducible components of C. Let $T \subset \mathbb{P}^1$ be a finite set. In this exercise you will show that

(21) there is a finite set $S \subset \mathbb{P}^1 \setminus T$ such that $\phi^{-1}(\mathbb{P}^1 \setminus S)$ is an affine curve.

(1) Pick a point on $\mathbb{P}^1 \setminus T$; denote it by ∞. Show that there is a hypersurface X of $\mathbb{P}^n$ containing $\phi^{-1}(\infty)$ such that $X \cap \phi^{-1}(T) = \emptyset$ and $|X \cap C| < \infty$. [Hint: use exercises III.44 and III.96.]
(2) Show that $C' := C \setminus X$ is an affine curve. [Hint: use proposition III.57.]
(3) Identifying $\mathbb{P}^1 \setminus \{\infty\}$ with $\mathbb{k}$, show that $\phi|_{C'}$ is induced by a regular function f on C'. [Hint: use proposition III.34.]
(4) Let $T' := \phi(X \cap C) \setminus \{\infty\} = \phi(C \setminus C') \cap \mathbb{k}$. Show that $C'' := C' \setminus V(\prod_{t \in T'}(f - t))$ is an affine curve. [Hint: use example III.37.]
(5) Conclude that (21) holds with $S := T' \cup \{\infty\}$.

12. Image of a morphism: Part II - Constructible sets

A *constructible* subset of a topological space is a finite union of open subsets of its closed subsets. In particular, constructible subsets of quasiprojective varieties are simply *finite unions of quasiprojective subsets*. The relevance of constructible sets in algebraic geometry stems from the fundamental result of C. Chevalley that images of morphisms of varieties are constructible sets in Zariski topology (see example III.16). Before we prove this result, we state a "constructible version" of the "curve selection lemma"; it is a straightforward consequence of curve selection lemma I (corollary III.83) and its proof is left as exercise III.100.

Proposition III.87. (Curve selection lemma II). *Let W be a constructible subset of a variety X and $\bar{W}$ be the Zariski closure of W in X. Pick an irreducible component Z of $\bar{W}$ and a point $a \in Z$. If $\dim(Z) \geq 1$, then there is an irreducible curve C on Z containing a such that $C \cap W$ is nonempty and (Zariski) open in C.* □

The main result of this section is the following result of C. Chevalley:

THEOREM III.88. (Chevalley's theorem). *The image $\phi(X)$ of a morphism $\phi : X \to Y$ of varieties is a constructible subset of Y.*

We give a proof of theorem III.88 following [Mum95, Section 2C]. In fact we only prove theorem III.89 below, and leave it as exercise III.104 to show that this is equivalent to Chevalley's theorem. Note that the statement of theorem III.89 is precisely what you get from substituting "closed subsets" by "constructible subsets" in the definition of complete varieties.

THEOREM III.89. *Let X, Y be varieties. Then the projection map $X \times Y \to Y$ maps constructible sets to constructible sets.*

PROOF. It suffices [why?] to consider the case that $X = \mathbb{k}^n$ and $Y = \mathbb{k}^m$. Then taking compositions we can further reduce it to the case of the projection $\pi : \mathbb{k} \times \mathbb{k}^m \to \mathbb{k}^m$. It is straightforward to check, and we leave it as an exercise (exercise III.105) to show, that it suffices to prove the following statement:

(22) If V is an irreducible subvariety of $\mathbb{k} \times \mathbb{k}^m \cong \mathbb{k}^{m+1}$ and U is a nonempty open subset of V, then $\pi(U)$ contains a nonempty open subset of the closure of $\pi(V)$ in $\mathbb{k}^m$.

We now prove (22). Let W be the closure of $\pi(V)$ in $\mathbb{k}^m$. Then π induces an injective map $\mathbb{k}[W] \hookrightarrow \mathbb{k}[V]$ (proposition III.65), so that $\mathbb{k}[V] \cong \mathbb{k}[W][x_1]/\mathfrak{p}$ for some ideal $\mathfrak{p}$ of the polynomial ring $\mathbb{k}[W][x_1]$ in one variable over $\mathbb{k}[W]$. At first consider the case that $\mathfrak{p} = 0$. Then $V = \mathbb{k} \times W$. Let $(a_1, \ldots, a_{m+1})$ be any point of U. Then $W' := \{a_1\} \times W$ is a subvariety of V and therefore $U \cap W'$ is a nonempty open subset of W' whose projection is open in W. Now consider the remaining case that $\mathfrak{p} \neq 0$. Then theorem III.78 and proposition III.84 imply that $\dim(V) = \dim(W)$. Let $\bar{V}$ be the closure of V in $\mathbb{P}^1 \times \mathbb{k}^m$ and $V' := \bar{V} \setminus U$. Since $\mathbb{P}^1$ is complete, it follows that $\pi(\bar{V})$ and $\pi(V')$ are closed in $\mathbb{k}^m$. Since $\pi(\bar{V})$ is closed, it follows that $\pi(\bar{V}) \supset W$. On the other hand $\dim(V') < \dim(\bar{V}) = \dim(V) = \dim(W)$ (theorem III.78 and exercise III.89), and therefore $\pi(V')$ cannot contain W (proposition III.84). Since $\pi(U) \supset \pi(\bar{V}) \setminus \pi(V') \supset W \setminus \pi(V')$, the claim follows. □

We now extend Chevalley's theorem and show that the set of all points y in the target space of a morphism ϕ such that $\phi^{-1}(y)$ has a given dimension is constructible (see example III.86).

Corollary III.90. *Let $\phi : X \to Y$ be a morphism of varieties and k be a nonnegative integer. Then $Y_k := \{y \in Y : \dim(\phi^{-1}(y)) < k\}$ is a constructible subset of Y.*

PROOF. We proceed by double induction on $m_\phi := \dim(\phi(X))$ and k. Due to theorems III.85 and III.88 the corollary is true whenever $m_\phi = 0$ or $k \leq \dim(X) - m_\phi$. Now assume it is true for all ϕ such that $m_\phi < m$. Pick ϕ with $m_\phi = m$ and k' such that the corollary holds for ϕ and k'. We will show that it holds for ϕ and $k'+1$. By the inductive hypothesis $Y \setminus Y_{k'}$ is constructible, and therefore is a union of quasiprojective varieties. Let $Y^0_{k'}$ be an irreducible component of $Y \setminus Y_{k'}$, and $X'^0_{k'}$ be an irreducible component of $\phi^{-1}(Y^0_{k'})$. Note that both $Y^0_{k'}$ and $X'^0_{k'}$ are quasiprojective varieties, so that $Y'^0_{k'} := \phi(X'^0_{k'})$ is constructible due to Chevalley's theorem (theorem III.88). It suffices to show that $Y'^0_{k'} \cap Y_{k'+1}$ is constructible. Let $\phi'^0_{k'}$ be the restriction of ϕ to $X'^0_{k'}$. Then $\phi'^0_{k'} : X'^0_{k'} \to Y'^0_{k'}$ is surjective and by construction $\dim((\phi'^0_{k'})^{-1}(y)) \geq k'$ for each $y \in Y'^0_{k'}$. If $k' < \dim(X'^0_{k'}) - \dim(Y'^0_{k'})$, then theorem III.85 implies that $Y_{k'+1} \cap Y'^0_{k'} = \emptyset$, which is trivially constructible. Otherwise theorem III.85 implies that $k' = \dim(X'^0_{k'}) - \dim(Y'^0_{k'})$ and there is a nonempty Zariski open subset $U'^0_{k'}$ of $Y'^0_{k'}$ such that $\dim((\phi'^0_{k'})^{-1}(y)) = k'$ for each $y \in U'^0_{k'}$. Let $Y''^0_{k'} := Y'^0_{k'} \setminus U'^0_{k'}$ and $\phi''^0_{k'}$ be the restriction of ϕ to $\phi^{-1}(Y''^0_{k'})$. Since $\dim(Y''^0_{k'}) < \dim(Y) = m$, the corollary is true for $\phi''^0_{k'}$ (and *all* values of k), so that $Y_{k'+1} \cap Y''^0_{k'}$ is constructible. Therefore $Y_{k'+1} \cap Y'^0_{k'} = U'^0_{k'} \cup (Y_{k'+1} \cap Y''^0_{k'})$ is constructible as well. □

12.1. Exercises.

EXERCISE III.98. If Y is a constructible subset of X, show that $X \setminus Y$ is also a constructible subset of X.

EXERCISE III.99. If Y is a constructible subset of X and Z is a constructible subset of Y, then show that Z is a constructible subset of X.

EXERCISE III.100. Prove proposition III.87.

EXERCISE III.101. Let $\phi : X \to Y$ be a dominant morphism of varieties and U be a constructible subset of Y. Show that the following are equivalent:

(1) U contains a nonempty Zariski open subset of Y.
(2) $\phi^{-1}(U)$ contains a nonempty Zariski open subset of X.

EXERCISE III.102. Let $\phi : X \to Y$ be a morphism of varieties and U be a constructible subset of X. Show that for each constructible subset V of Y, $\phi^{-1}(V) \cap U$ is a constructible subset of X.

EXERCISE III.103. The *dimension* of a constructible subset U of a variety X is simply the dimension of the Zariski closure of U in X. Let $\phi : X \to Y$ be a dominant morphism of varieties. If X is irreducible, then show that the following are equivalent:

(1) U contains a nonempty Zariski open subset of X.
(2) there is a nonempty Zariski open subset Y' of Y such that for each $y \in Y'$, $\dim(\phi^{-1}(y) \cap U) = \dim(X) - \dim(Y)$.

EXERCISE III.104. Show that theorem III.88 is equivalent to theorem III.89.

EXERCISE III.105. Show that it suffices to prove (22) in order to prove that the projection $\Bbbk \times \Bbbk^m \to \Bbbk^m$ maps constructible sets to constructible sets. [Hint: use induction on dimension of the constructible subset of $\Bbbk \times \Bbbk^m$.]

13. Tangent space, singularities, local ring at a point

13.1. The case of affine varieties. Consider a straight line $L = \{a + tv : t \in \Bbbk\}$ through a point $a = (a_1, \dots, a_N) \in \Bbbk^N$, where $v = (v_1, \dots, v_N) \in \Bbbk^N$ determines the "direction" of L. Assume $f(a) = 0$, where f is a polynomial in $(x_1, \dots, x_N)$. We say that L is *tangent* to $V(f)$ at a if $\operatorname{ord}_t(f(a + tv)) > 1$, or equivalently, if

$$\sum_{i=1}^{N} v_i \frac{\partial f}{\partial x_i}(a) = 0. \tag{23}$$

More generally, L is *tangent at* a to an *affine variety* X containing a if (23) holds for *all* f vanishing on X. The *tangent space* $T_a(X)$ to X at a is the union of all tangent lines to X at a. It is straightforward to check (exercise III.106) that

$$T_a(X) = V\left(\sum_{i=1}^{N} (x_i - a_i) \frac{\partial f}{\partial x_i}(a) : f \in I(X)\right), \tag{24}$$

where $I(X)$ is the ideal in $\Bbbk[x_1, \dots, x_N]$ of polynomials vanishing on X. It is clear from (24) that $T_a(X)$ is of the form $V + a$ where V is a linear subspace (through the origin) of $\Bbbk^N$; the *dimension* of $T_a(X)$ is simply the dimension of V (as a vector space over $\Bbbk$). Let $f_1, \dots, f_s$ be a set of generators of $I(X)$, then exercise III.106 implies that

$$\dim T_a(X) = N - \operatorname{Rank}\left(\frac{\partial f_i}{\partial x_j}(a)\right)_{\substack{1 \le i \le s \\ 1 \le j \le N}}, \tag{25}$$

where $\operatorname{Rank}(\cdot)$ is the rank over $\Bbbk$ of the corresponding matrix.

Proposition III.91. *For each integer k, the set $X_{\geq k} := \{a \in X : \dim T_a(X) \geq k\}$ is Zariski closed in X; in other words, the map $X \mapsto \mathbb{Z}$ given by $a \mapsto \dim T_a(X)$ is* upper semicontinuous.

PROOF. [Mum95, Section 1A]. Let $\mathfrak{q}_k$ be the ideal of $\Bbbk[x_1, \dots, x_N]$ generated by determinants of $(N - k + 1) \times (N - k + 1)$-minors of the matrix $(\frac{\partial f_i}{\partial x_j}(a))$. Identity (25) implies that $X_{\geq k} = V(I(X) + \mathfrak{q}_k)$. □

Let $d := \min\{\dim T_a(X) : a \in X\}$. Assume X is irreducible. Then we say that $a \in X$ is a *singular* (respectively *nonsingular*) point if $\dim T_a(X) > d$ (respectively, $\dim T_a(X) = d$). Proposition III.91 implies that the set of nonsingular points of X is a nonempty Zariski open subset of X.

Example III.92. Every line or a quadric curve on $\Bbbk^2$ is everywhere nonsingular (exercise III.108). Both the curves $C_1 = \{x^2 = y^3\}$ and $C_2 = \{x^2 = y^2 - y^3\}$ are singular at the origin (exercise III.109); see fig. 6. The singularity of C_1 is called a *cusp*, and, when $\mathrm{char}(\Bbbk) \neq 2$, that of C_2 is called a *node*.

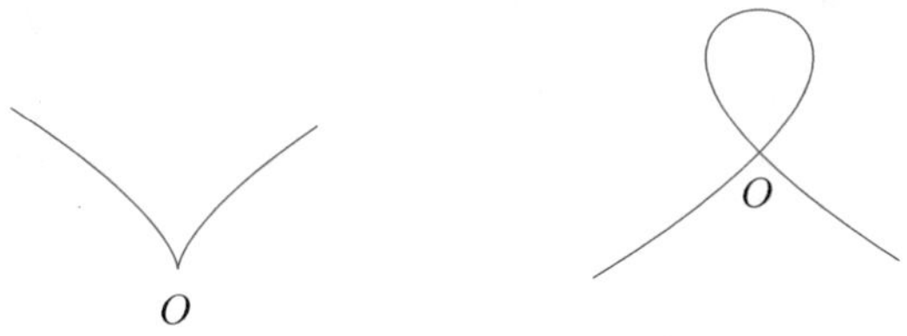

(A) $x^2 = y^3$ (B) $x^2 = y^2 - y^3$

FIGURE 6. Some curve singularities.

13.2. Intrinsicness of the tangent space; tangent spaces and singularities on arbitrary varieties. The definitions of tangent spaces and (non-)singular points given above apply only to irreducible affine varieties; moreover, they depend on the defining equations of the affine variety, and a priori it is not clear if they are preserved by isomorphisms. In this section we extend these notions to arbitrary varieties. We need two kinds of objects for this; the first one is a "derivation": given a point a of a subvariety X of $\Bbbk^N$, a map $D : \Bbbk[X] \to \Bbbk$ is called a *derivation centered at* a if D satisfied the following properties:

(a) D is $\Bbbk$-linear,
(b) $D(fg) = f(a)D(g) + g(a)D(f)$ for all $f, g \in \Bbbk[X]$,
(c) $D(\alpha) = 0$ for all $\alpha \in \Bbbk$.

The following is straightforward to see; we leave the proof as an exercise.

Proposition III.93. *Given* $a = (a_1, \ldots, a_N) \in \Bbbk^N$*, the derivations* $D : \Bbbk[x_1, \ldots, x_N] \to \Bbbk$ *centered at* a *are in one-to-one correspondence with* $\Bbbk^N$ *given by:*

(26) $$\lambda = (\lambda_1, \ldots, \lambda_N) \mapsto D_\lambda, \text{ where } D_\lambda(f) = \sum_{i=1}^{N} \lambda_i \frac{\partial f}{\partial x_i}(a), \text{ for all } f \in \Bbbk[x_1, \ldots, x_N]$$

□

The "local ring" of a variety X at $a \in X$ is the set of regular functions on arbitrarily small neighborhoods of a in X; more precisely, consider a binary relation $\sim$ on the collection of pairs (f, U), where U is an open neighborhood of a in X and f is a regular function on U, as follows:

$$(f, U) \sim (f', U') \text{ if and only if } f \text{ and } f' \text{ agree on } U \cap U'$$

(Note the similarity to the definition of rational functions.) It is clear that $\sim$ is an *equivalence relation*; the *local ring* $\mathcal{O}_{a,X}$ *of* X *at* a is the set of equivalence classes of $\sim$ with the natural $\Bbbk$-algebra structure. The following result compiles basic properties of $\mathcal{O}_{a,X}$—its proof is straightforward and left as exercise III.111.

Proposition III.94. *Let X be a quasiprojective variety and $a \in X$.*

(1) If U is an open neighborhood of a in X, then $\mathcal{O}_{a,X} \cong \mathcal{O}_{a,U}$.

(2) A morphism $\phi : X \to Y$ of varieties induces by pullback a $\Bbbk$-algebra morphism $\phi^* : \mathcal{O}_{\phi(a),Y} \to \mathcal{O}_{a,X}$. If ϕ is an isomorphism, then ϕ^* is an isomorphism of $\Bbbk$-algebras.

(3) Let $\mathfrak{m}_a$ be the ideal of $\mathcal{O}_{a,X}$ generated by all regular functions on neighborhoods of a on X which vanish at a. Then $\mathfrak{m}_a$ is the unique *maximal ideal of $\mathcal{O}_{a,X}$; in other words, $\mathcal{O}_{a,X}$ is a* local ring.

(4) Assume $a \in Z \subseteq X$, where Z is a subvariety of X. Let $\mathfrak{q}_Z$ be the ideal of $\mathcal{O}_{a,X}$ generated by all elements of the form (f, U) such that $f|_{Z \cap U} \equiv 0$. Then $\mathcal{O}_{a,Z} \cong \mathcal{O}_{a,X}/\mathfrak{q}_Z$. □

Example III.95. Given $a = (a_1, \dots, a_n) \in \Bbbk^n$, $\mathcal{O}_{a,\Bbbk^n} = \{f/g : f, g \in \Bbbk[x_1, \dots, x_n],\ g(a) \neq 0\}$; in other words, $\mathcal{O}_{a,\Bbbk^n}$ is the *localization* of $\Bbbk[x_1, \dots, x_n]$ at the ideal generated by polynomials vanishing at a. Assertion (1) of proposition III.94 implies that $\mathcal{O}_{a,\mathbb{P}^n} \cong \mathcal{O}_{a,\Bbbk^n}$. Given any other point $b = (b_1, \dots, b_n) \in \Bbbk^n$, the automorphism of $\Bbbk^n$ given by the translation $x \mapsto x + (b - a)$ induces a $\Bbbk$-algebra isomorphism between $\mathcal{O}_{a,\Bbbk^n}$ and $\mathcal{O}_{b,\Bbbk^n}$.

Example III.96. Consider the map $\phi : \Bbbk^2 \to \Bbbk^2$ from example III.16 given by $(x, y) \mapsto (x, xy)$. Then ϕ maps the origin to itself and $\phi^* : \mathcal{O}_{0,\Bbbk^2} \to \mathcal{O}_{0,\Bbbk^2}$ is not surjective, in particular it is *not* an isomorphism. However, ϕ^* does induce an isomorphism between $\mathcal{O}_{\phi(a_1,a_2),\Bbbk^2}$ and $\mathcal{O}_{(a_1,a_2),\Bbbk^2}$ whenever $a_1 \neq 0$ (exercise III.112). Note that ϕ restricts to an automorphism on $\Bbbk^2 \setminus V(x)$. In general the following is true: for $a \in X$, $b \in Y$, $\mathcal{O}_{a,X} \cong \mathcal{O}_{b,Y}$ as $\Bbbk$-algebras if and only if there are open neighborhoods U of a in X and V of b in Y such that $U \cong V$ (exercise III.113).

Example III.95, together with assertion (4) of proposition III.94, immediately implies that local rings of affine varieties are *localizations* of the coordinate ring.

Proposition III.97. *Assume X is a subvariety of $\Bbbk^N$ with coordinates $(x_1, \dots, x_N)$. Then $\mathcal{O}_{a,X}$ is the ring of rational functions $\{f/g : f, g \in \Bbbk[x_1, \dots, x_N],\ g(a) \neq 0\}$ modulo the ideal generated by polynomials vanishing on X; in other words $\mathcal{O}_{a,X}$ is the localization of $\Bbbk[X]$ with respect to the multiplicative set of regular functions* not *vanishing at a.* □

Corollary III.98. *The local ring of an irreducible variety at a point is an integral domain.*

PROOF. Since the coordinate ring of an irreducible affine variety is an integral domain (exercise III.19), the claim follows immediately from proposition III.97. □

The following result describes the relation among derivations, local rings and tangent spaces:

THEOREM III.99. *Given $a \in X$, where X is an affine variety, the following spaces are isomorphic as vector spaces over $\mathbb{k}$:*

(1) $T_a(X)$ with the vector space structure on $T_a(X)$ induced from that of $T_a(X) - a$ (in particular, a is the origin of $T_a(X)$),
(2) the space of derivations $\mathbb{k}[X] \to \mathbb{k}$ centered at a,
(3) the space $(\mathfrak{m}_a/\mathfrak{m}_a^2)^*$ of linear functions from $\mathfrak{m}_a/\mathfrak{m}_a^2$ to $\mathbb{k}$, where $\mathfrak{m}_a$ is the maximal ideal of $\mathcal{O}_{a,X}$ generated by functions vanishing at a.

PROOF. [Mum95, Section 1A]. A derivation $\mathbb{k}[X] \to \mathbb{k}$ centered at a is the same as a derivation $\mathbb{k}[x_1, \ldots, x_N] \to \mathbb{k}$ centered at a which vanishes on $I(X)$. Derivations $\mathbb{k}[x_1, \ldots, x_N] \to \mathbb{k}$ centered at a are of the form D_λ from the map defined in (26). Given $\lambda \in \mathbb{k}^n$, identity (24) implies that D_λ vanishes on all $f \in I(X)$ if and only if $\lambda + a \in T_a(X)$, which proves the isomorphisms between (1) and (2). Exercise III.114 below shows that every derivation $\mathbb{k}[X] \to \mathbb{k}$ centered at a defines an element of $(\mathfrak{m}_a/\mathfrak{m}_a^2)^*$. Conversely, given a linear map $\phi : \mathfrak{m}_a \to \mathbb{k}$ such that $\phi|_{\mathfrak{m}_a^2} \equiv 0$, let $\lambda_\phi := (\phi((x_1 - a_1)|_X), \ldots, \phi((x_N - a_N)|_X)) \in \mathbb{k}^N$, and define $D_\phi := D_{\lambda_\phi}$ as in (26). Let $f \in I(X)$. The Taylor series expansion of f shows that $f = f(a) + \sum_{i=1}^n (x_i - a_i)\frac{\partial f}{\partial x_i}(a) + f'$, where f' is of order two or higher in the $(x_i - a_i)$. Since $f(a) = 0$ and $\phi(f|_X) = 0 = \phi(f'|_X)$ [why?], it follows that $\sum_{i=1}^N \phi((x_i - a_i)|_X)\frac{\partial f}{\partial x_i}(a) = 0$, i.e. $D_\phi(f) = 0$. It follows that D_ϕ is a derivation $\mathbb{k}[X] \to \mathbb{k}$. It is straightforward to check that the maps we defined between (2) and (3) are inverse to each other, and induce an isomorphism of vector spaces. □

The *tangent space* to a variety X at a point $a \in X$ is by definition $T_a(X')$ for any affine open neighborhood X' of a in X. Theorem III.99 shows that $T_a(X) = (\mathfrak{m}_a/\mathfrak{m}_a^2)^*$, where $\mathfrak{m}_a$ is the maximal ideal of $\mathcal{O}_{a,X}$ generated by functions vanishing at a; in particular, $T_a(X)$ is well defined for an arbitrary quasiprojective variety.

Proposition III.100. *Assume X is an irreducible variety. Then* $\min\{\dim T_a(X) : a \in X\} = \dim(X)$.

PROOF. Let $d := \min\{\dim T_a(X) : a \in X\}$ and $U := \{x \in X : \dim T_a(X) = d\}$. Proposition III.91 implies that U is open and dense in X. Since X is birational to a hypersurface (exercise III.66), and since dimension is invariant under birational maps (exercise III.65), without loss of generality we may assume $X = V(f) \subset \mathbb{k}^N$ for some polynomial f in $(x_1, \ldots, x_N)$. If $f = 0$, then an easy computation shows that $U = \mathbb{k}^N$ (exercise III.107) and $d = N$, as required. Otherwise f is a non-constant irreducible polynomial, and since $\mathbb{k}$ is algebraically closed, not all the partial derivatives $\partial f/\partial x_j$ vanish identically on X (exercise III.115). It then follows the definition of the tangent space (or identity (25)) that $d = N - 1 = \dim(X)$ (the last equality uses proposition III.76). □

Now we can extend the notion of (non-)singular points to arbitrary (possibly reducible) varieties. Let X be a variety and $a \in X$. Define $\dim_a(X)$ to be the maximum of the dimensions of the irreducible components of X containing a. We say that a is a *singular* (respectively *nonsingular*) point of X if $\dim T_a(X) > \dim_a(X)$

(respectively, $\dim T_a(X) = \dim_a(X)$). A variety is called *singular* if it has a singular point; otherwise it is called *nonsingular*. The following is an immediate consequence of Propositions III.91 and III.100:

Proposition III.101. *The set of nonsingular points of a quasiprojective variety X is Zariski open and has a nonempty intersection with every irreducible component of X.* □

Example III.102. Proposition III.101 implies that a curve has at most finitely many singular points. Consider the affine curves $C_1 = V(x^2 - y^3)$ and $C_2 = V(x^2 - y^2 - y^3)$ from example III.92. Embedding $\mathbb{k}^2$ into $\mathbb{P}^2$ with homogeneous coordinates $[x : y : z]$ via the map $(x, y) \mapsto [x : y : 1]$ and homogenizing with respect to the z-coordinate shows that the closures of C_j in $\mathbb{P}^2$ are $\bar{C}_1 := V(x^2z - y^3)$ and $\bar{C}_2 := V(x^2z - y^2z - y^3)$ (example III.40). It follows that for both j, $\bar{C}_j \setminus C_j$ has only one point, namely, $P := [1 : 0 : 0]$. Identifying the basic open subset $U := \mathbb{P}^2 \setminus V(x)$ with $\mathbb{k}^2$ with coordinates $(u, v) := (y/x, z/x)$, we see that $C_1 \cap U = V(v - u^3)$ and $C_2 \cap U = V(v - u^2 - u^3)$ (exercise III.54). Since $P = (0, 0)$ with respect to (u, v)-coordinates, it follows that both $\bar{C}_j$ are nonsingular at P.

13.3. Equations near a nonsingular point. Recall that for a subvariety X of codimension k in $\mathbb{k}^N$, one needs at least k polynomials to generate the ideal $I(X)$ of polynomials vanishing on it (theorem III.80). We say that a subvariety X is a *complete intersection* if $I(X)$ can be generated by k polynomials. Not all varieties are complete intersections. Indeed, we have seen in example III.82 that the image X of the morphism $\mathbb{k} \to \mathbb{k}^3$ given by $t \mapsto (t^3, t^4, t^5)$ is a codimension two subvariety of $\mathbb{k}^3$ and it takes at least three polynomials to generate $I(X)$. On the other hand, exercise III.116 below shows that if $U = \mathbb{k}^3 \setminus V(x)$ (where (x, y, z) are coordinates on $\mathbb{k}^3$), then the ideal of $X \cap U$ in $\mathbb{k}[U] = \mathbb{k}[x, y, z, 1/x]$ can be generated by two regular functions on U, namely, $y - (y/x)^4$, $z - (y/x)^5$, and in addition, $X \cap U$ is *nonsingular*. In this section we show that in general every nonsingular point on a variety has an affine neighborhood which is a complete intersection. We follow the approach of [Mum95, Proof of Theorem 1.16]. Let $R := \mathcal{O}_{0,\mathbb{k}^N}$ and $\hat{R} := \mathbb{k}[[x_1, \ldots, x_N]]$ be the ring of formal power series in $(x_1, \ldots, x_N)$ over $\mathbb{k}$. Recall that $\hat{R}$ has only one maximal ideal, namely, the ideal $\hat{\mathfrak{m}}$ generated by all polynomials with zero constant term, and we can view R as a subring of $\hat{R}$ via the expansion $(1 - \sum_{i=1}^N x_i g_i)^{-1} = 1 + \sum_{j \geq 1} (x_i g_i)^j$ for any $g_1, \ldots, g_N \in \mathbb{k}[x_1, \ldots, x_N]$ (see appendix B.13).

Lemma III.103. *If $\mathfrak{q}$ is an ideal of R, then $\mathfrak{q}\hat{R} \cap R = \mathfrak{q}$.*

PROOF. Given $f \in \mathfrak{q}\hat{R} \cap R$, it suffices to show that $f \in \mathfrak{q}$. Indeed, write $f = \sum_j \phi_j f_j$, where f_j are polynomials which generate $\mathfrak{q}$ and ϕ_j are power series in $(x_1, \ldots, x_n)$. For each $k > \deg(f)$, if $\phi_{j,k}$ are the *polynomials* consisting of all monomial terms of ϕ_j of order at most k, then $g_k := \sum_j f_j \phi_{j,k} \in \mathfrak{q}$, so that $f = g_k + \sum_j f_j(\phi - \phi_{j,k}) \in \mathfrak{q} + \hat{\mathfrak{m}}^k$, where $\hat{\mathfrak{m}}$ is the (unique) maximal ideal of

$\hat{R}$. Let $\mathfrak{m} := \hat{\mathfrak{m}} \cap R$ be the (unique) maximal ideal of R. Since $\hat{\mathfrak{m}}^k \cap R = \mathfrak{m}^k$ (proposition B.44), it follows that $f \in \bigcap_{k \geq 0}(\mathfrak{q} + \mathfrak{m}^k) = \mathfrak{q}$ (theorem A.2). □

Corollary III.104. *Let $f_1, \ldots, f_r$ be polynomials in $(x_1, \ldots, x_N)$ with no constant term and linearly independent (over $\mathbb{k}$) linear terms. Then the ideal $\mathfrak{p}$ generated by $f_1, \ldots, f_r$ in $\mathcal{O}_{0,\mathbb{k}^N}$ is prime.*

PROOF. $f_1, \ldots, f_r$ generate a prime ideal $\mathfrak{q}$ in $\hat{R}$ (corollary B.42). Now apply lemma III.103. □

THEOREM III.105. *Every nonsingular point on a variety has an affine open neighborhood which is irreducible and a complete intersection.*

PROOF. Let a be a nonsingular point of a variety X, with $\dim_a(X) = n$. We may assume without loss of generality that X is a subvariety of $\mathbb{k}^m$ for some $m \geq n$, and a is the origin in $\mathbb{k}^m$. Identity (25) implies that there are $f_1, \ldots, f_{m-n} \in I(X)$ with no constant term and linearly independent linear terms. Let $\mathfrak{q}$ be the ideal of $\mathbb{k}[x_1, \ldots, x_m]$ generated by $f_1, \ldots, f_{m-n}$. Corollary III.104 implies that the ideal $\mathfrak{q}\mathcal{O}_{a,\mathbb{k}^m}$ generated by $\mathfrak{q}$ in $\mathcal{O}_{a,\mathbb{k}^m}$ is prime, so that $\mathfrak{q}' := (\mathfrak{q}\mathcal{O}_{a,\mathbb{k}^m}) \cap \mathbb{k}[x_1, \ldots, x_m]$ is also prime. Let $Z := V(\mathfrak{q})$ and $Z' := V(\mathfrak{q}') \subseteq \mathbb{k}^m$. Note that Z' is irreducible and $a \in Z' \subseteq Z$.

Claim III.105.1. $\dim(Z') = n$. *There is a polynomial g such that $g(a) \neq 0$ and $X \setminus V(g) = Z' \setminus V(g) = Z \setminus V(g)$.*

PROOF. Let $h'_1, \ldots, h'_s$ be a set of generators of $\mathfrak{q}'$. Then each h'_j can be expressed as as h_j/g_j for some polynomials g_j, h_j such that $h_j \in \mathfrak{q}$ and $g_j(a) \neq 0$. If $g := \prod_j g_j$, then it follows that $Z' \setminus V(g) = Z \setminus V(g) \supseteq X \setminus V(g)$. Since Z is defined by $m - n$ equations in $\mathbb{k}^m$, it follows that $\dim(Z' \setminus V(g)) = \dim(Z \setminus V(g)) \geq n$ (theorem III.80). On the other hand the assumptions on linear parts of f_j and proposition III.100 imply that $\dim(Z' \setminus V(g)) \leq n$. It follows that $\dim(Z' \setminus V(g)) = n \leq \dim(X \setminus V(g))$. On the other hand, since Z' is irreducible, every proper subvariety of $Z' \setminus V(g)$ has dimension smaller than n (theorem III.78). It follows that $Z' \setminus V(g) = X \setminus V(g)$, which completes the proof. □

Since $Z \setminus V(g)$ is isomorphic to the subvariety of $\mathbb{k}^{m+1}$ defined by $f_1, \ldots, f_{m-n}, gx_{m+1} - 1$ (example III.37), it is a complete intersection. The proof is now complete due to claim III.105.1. □

Corollary III.106. *The local ring of a variety at a nonsingular point is an integral domain.*

PROOF. Since the local ring of an irreducible variety at a point is an integral domain (corollary III.98), this follows directly from theorem III.105. □

13.4. Parametrizations of a curve at a nonsingular point. The local ring at the origin of $\mathbb{k}$ is $\mathcal{O}_{0,\mathbb{k}} = \{f/g : f, g \in \mathbb{k}[t],\ g(0) \neq 0\}$. Recall that the *order* of a polynomial $f \in \mathbb{k}[t]$, denoted $\operatorname{ord}(f)$, is the smallest integer d such that the coefficient of t^d in f is nonzero. One can extend ord uniquely to $\mathbb{k}(t)$ by defining

$$\operatorname{ord}(f/g) := \operatorname{ord}(f) - \operatorname{ord}(g)$$

It is straightforward to check that ord satisfies the following properties: $\operatorname{ord}(fg) = \operatorname{ord}(f) + \operatorname{ord}(g)$ and $\operatorname{ord}(f+g) \geq \min\{\operatorname{ord}(f), \operatorname{ord}(g)\}$, and $\mathcal{O}_{0,\mathbb{k}} = \{h \in \mathbb{k}(t) : \operatorname{ord}(h) \geq 0\}$. In other words, ord is a *discrete valuation* on $\mathbb{k}(t)$ and $\mathcal{O}_{0,\mathbb{k}}$ is a *discrete valuation ring* (see appendix B.8 for a discussion on discrete valuations). In this section we will see that the local ring of any curve at a nonsingular point is a discrete valuation ring. Let a be a point on a curve C and $\mathfrak{m}_a$ be the (unique) maximal ideal of $\mathcal{O}_{a,C}$.

Proposition III.107. *If $f \in \mathcal{O}_{a,C}$ is not identically zero on any of the irreducible components of C containing a, then the radical of the ideal generated by f in $\mathcal{O}_{a,C}$ is either $\mathcal{O}_{a,C}$ itself or $\mathfrak{m}_a$.*

PROOF. Without loss of generality we may assume C is affine. Then $f = f_1/f_2$ for some $f_1, f_2 \in \mathbb{k}[C]$, $f_2(a) \neq 0$ (proposition III.97). If $f_1(a) \neq 0$, then f is invertible in $\mathcal{O}_{a,C}$. Otherwise theorem III.78 implies that $V(f_1) \subset C$ consists of finitely many points excluding a. Choose a polynomial g which does not vanish at a but vanishes at every other point of $V(f_1)$. For any $h \in \mathbb{k}[C]$ such that $h(a) = 0$, the Nullstellensatz (theorem B.8) implies that $gh \in \sqrt{f_1} \subseteq \mathbb{k}[C]$. Since g is invertible in $\mathcal{O}_{a,C}$, it follows that $h \in \sqrt{f} \subseteq \mathcal{O}_{a,C}$. This implies that $\sqrt{f} = \mathfrak{m}_a$, as required.
□

Proposition III.108. *Assume C is nonsingular at a. Fix $t \in \mathfrak{m}_a \setminus \mathfrak{m}_a^2$ (that $\mathfrak{m}_a \setminus \mathfrak{m}_a^2$ is nonempty is a consequence of theorem III.99).*

(1) $\mathfrak{m}_a$ is the principal ideal generated by t.
(2) Let $\nu : \mathcal{O}_{a,C} \to \mathbb{Z}_{\geq 0} \cup \{\infty\}$ be the map given by $g \mapsto \inf\{m \geq 0 : t^m \in \langle g \rangle\}$. Then for all $g \in \mathcal{O}_{a,C} \setminus \{0\}$,
  (a) $g = ut^{\nu(g)}$ for some unit $u \in \mathcal{O}_{a,C}$,
  (b) $g = ct^{\nu(g)} + g'$ for some $c \in \mathbb{k} \setminus \{0\}$ and $g' \in \mathfrak{m}_a$ such that $\nu(g') > \nu(g)$.
(3) ν extends to a discrete valuation on the field of fractions of $\mathcal{O}_{a,C}$ (recall that $\mathcal{O}_{a,C}$ is an integral domain due to corollary III.106) and its valuation ring is $\mathcal{O}_{a,C}$.

PROOF. Since $\dim(T_a(C)) = 1$, the image of t generates $\mathfrak{m}_a/\mathfrak{m}_a^2$ over $\mathbb{k}$ (assertion (3) of theorem III.99). Assertion (1) then follows directly from corollary B.13, which is a corollary of Nakayama's lemma (lemma B.12). For assertion (2) pick $g \in \mathcal{O}_{a,C}$. Proposition III.107 implies that $\nu(g)$ is well defined. Let $m := \nu(g)$. By definition of ν, $t^m = u'g$ for some $u \in \mathcal{O}_{a,C}$. If $u' \in \mathfrak{m}_a$, then assertion (1) would imply that $u' = th$ for some $h \in \mathcal{O}_{a,C}$, which would in turn imply that $t^{m-1} = hg$ (since $\mathcal{O}_{a,C}$ is an integral domain), contradicting the minimality of $\nu(g)$. Therefore u' is a unit in $\mathcal{O}_{a,C}$, proving (2a) with $u := u'^{-1}$. Let $c := u(a) \neq 0$. Then $g = ut^m = ct^m + (u-c)t^m$. Since $u - c \in \mathfrak{m}_a$, assertion

(1) implies that $u - c \in \langle t \rangle$, which in turn implies that $\nu((u-c)t^m) > m$, proving (2b). Assertion (3) then follows in a straightforward way by extending ν to the field of fractions of $\mathcal{O}_{a,C}$ by defining $\nu(f/g) := \nu(f) - \nu(g)$ for $f, g \in \mathcal{O}_{a,C}$; we leave the details as an exercise. □

If C is nonsingular at a, proposition III.108 implies that any $t \in \mathfrak{m}_a \setminus \mathfrak{m}_a^2$ is a *parameter* of the discrete valuation ring $\mathcal{O}_{a,X}$; we say that t is a *parameter* of C at a.

Corollary III.109. *Let C be an irreducible curve and $f : C \dashrightarrow \mathbb{P}^N$ be a rational map. Assume there is $C' \subseteq C$ such that $f|_{C'}$ is a morphism and C is nonsingular at every point of $C \setminus C'$. Then f extends to a morphism $C \to \mathbb{P}^N$.*

PROOF. Note that $C \setminus C'$ is finite. Fix $a \in C \setminus C'$. It suffices to show that f can be extended to a morphism on a neighborhood of a. Without loss of generality we may assume that

(i) C is an (affine) subvariety of $\mathbb{k}^n$ with coordinates $(x_1, \ldots, x_n)$, and
(ii) there is an open neighborhood U of a in C such that f is a morphism from $U \setminus \{a\} \to \mathbb{P}^N$ given by $x \mapsto [h_0(x) : \cdots : h_N(x)]$.

Pick a parameter t of C at a. Then each $h_j = u_j t^{m_j}$ for some units $u_j \in \mathcal{O}_{a,C}$ and $m_j := \nu(u_j)$. Choose an open neighborhood U' of a in U such that each u_j is a regular function on U'. If $m := \min\{m_j\}_j$, then f uniquely extends to $U' :\longrightarrow \mathbb{P}^N$ given by $[u_0 t^{m_0 - m} : \cdots : u_N t^{m_N - m}]$. □

Example III.110. Let X be the image of $\phi : \mathbb{k} \to \mathbb{k}^3$ given by $t \mapsto (t^3, t^4, t^5)$. Exercise III.116 shows that $X \setminus \{(0,0,0)\}$ is nonsingular. We now use corollary III.109 to show that X is singular at $O := (0,0,0)$[21]. Indeed, ϕ is a birational map, and by the usual identification of $\mathbb{k}$ with $\mathbb{P}^1 \setminus V(x_0)$, we see that $\phi^{-1} : X \setminus \{O\} \dashrightarrow \mathbb{P}^1$ is a well-defined morphism given by $(x, y, z) \mapsto [1 : y/x]$. Now assume X is nonsingular at O. Then ϕ^{-1} extends to a morphism $\psi : X \to \mathbb{P}^1$. Since $\psi \circ \phi : \mathbb{k} \to \mathbb{P}^1$ is a morphism which is identity on $\mathbb{k} \setminus \{0\}$, the Zariski-continuity of morphisms (exercise III.50) implies that it is identity everywhere on $\mathbb{k}$. This implies that ϕ induces an isomorphism $\mathbb{k} \cong X$, and consequently, an isomorphism $\mathcal{O}_{O,X} \cong \mathcal{O}_{0,\mathbb{k}}$. However, it is clear that for any polynomial f in (x, y, z), the order of $f \circ \phi$ in $\mathbb{k}(t)$ is ≥ 3. This shows that ϕ^* can *not* be an isomorphism between $\mathcal{O}_{O,X}$ and $\mathcal{O}_{0,\mathbb{k}}$, which gives the required contradiction.

Corollary III.109 in general fails if $C \setminus C'$ has singular points—see exercise III.117 for an example.

13.5. Exercises.

EXERCISE III.106. Prove identity (24). If $f_1, \ldots, f_s$ generate $I(X)$, then show that $\sum_{i=1}^N (x_i - a_i) \frac{\partial f_j}{\partial x_i}(a)$, $j = 1, \ldots, s$, generate the ideal of polynomials vanishing on $T_a(X)$.

[21] To prove this directly using the definitions would require computation of $I(X)$ on an open neighborhood of O, which is a relatively complicated task.

EXERCISE III.107. Show that $\mathbb{k}^N$ is nonsingular everywhere. Given $a \in \mathbb{k}^N$, compute $T_a(\mathbb{k}^N)$.

EXERCISE III.108. Let $C = V(f) \subset \mathbb{k}^2$, where f is an irreducible polynomial of degree 1 or 2. Show that C is everywhere nonsingular. [Hint: since $\mathbb{k}$ is algebraically closed, a homogeneous polynomial of degree 2 can be written as ax^2+by^2, $a, b \in \mathbb{k}$, after an appropriate change of coordinates on $\mathbb{k}^2$. If $\deg(f) = 2$, then use this fact to reduce to the following cases: (1) $f = ax^2 + by^2 + c$, $a, b, c \neq 0$, and (2) $f = ax^2 + by$, $a, b \neq 0$.]

EXERCISE III.109. Let $C = V(f) \subset \mathbb{k}^2$. If either $f = x^2 - y^3$ or $f = x^2 - y^2 + y^3$, show that the origin is the only singular point of C.

EXERCISE III.110. Prove proposition III.93.

EXERCISE III.111. Prove proposition III.94. [Hint: for assertion (4) it suffices to show that the map $\mathcal{O}_{a,X} \to \mathcal{O}_{a,Z}$ given by restriction to Z is surjective. Any open neighborhood of a in Z is of the form $U \cap Z$ for some open neighborhood of a in X. Choose an open *affine* neighborhood U' of a in X such that $U' \subseteq U$. Then $U' \cap Z$ is also affine and therefore all regular functions on $U' \cap Z$ are restrictions of regular functions on $U' \cap X$.]

EXERCISE III.112. Let $\phi : \mathbb{k}^2 \to \mathbb{k}^2$ from example III.16 given by $(x, y) \mapsto (x, xy)$. Given $a = (a_1, a_2) \in \mathbb{k}^2$, consider the induced map $\phi^* : \mathcal{O}_{\phi(a),\mathbb{k}^2} \to \mathcal{O}_{a,\mathbb{k}^2}$. Show that

(1) ϕ^* is not surjective when a is the origin.
(2) ϕ^* is an isomorphism if $a_1 \neq 0$. [Hint: x is invertible in $\mathcal{O}_{(b_1,b_2),\mathbb{k}^n}$ if $b_1 \neq 0$.]

EXERCISE III.113. Given varieties X, Y and points $a \in X$, $b \in Y$, show that the following are equivalent:

(1) $\mathcal{O}_{a,X} \cong \mathcal{O}_{b,Y}$ as $\mathbb{k}$-algebras,
(2) there are open neighborhoods U of a in X and V of b in Y such that $U \cong V$.

[Hint: The $(\Leftarrow)$ implication follows directly from proposition III.94. For the $(\Rightarrow)$ implication, suffices to consider the case that X, Y are subvarieties respectively of $\mathbb{k}^m$ with coordinates $(x_1, \ldots, x_m)$ and $\mathbb{k}^n$ with coordinates $(y_1, \ldots, y_n)$. Given a $\mathbb{k}$-algebra isomorphism $\Phi : \mathcal{O}_{b,Y} \to \mathcal{O}_{a,X}$ there are $g, f_1, \ldots, f_n \in \mathbb{k}[x_1, \ldots, x_m]$ such that $g(a) \neq 0$ and Φ maps $y_j \mapsto f_j/g \in \mathcal{O}_{a,X}$. Then $\phi : x \mapsto (f_1(x)/g(x), \ldots, f_n(x)/g(x))$ is a morphism from $X \setminus V(g)$ to Y. Similarly, Φ^{-1} induces a morphism $\psi : Y \setminus V(q) \to X$ for some $q \in \mathbb{k}[y_1, \ldots, y_n]$ such that $q(b) \neq 0$. Then $\psi \circ \phi$ and $\phi \circ \psi$ must be identity near respectively a and b.]

EXERCISE III.114. Let X be an affine variety, $a \in X$, and $D : \mathbb{k}[X] \to \mathbb{k}$ be a derivation centered at a. If $\mathfrak{m}_a$ is the maximal ideal of $\mathcal{O}_{a,X}$ generated by polynomials vanishing at a, show that

(a) D extends to a linear map $\mathfrak{m}_a \to \Bbbk$ given by

$$D(f/g) := D(f)/g(a) \text{ for all } f \in \mathfrak{m}_a,\ g \in \Bbbk[X]$$

(b) $D(h) = 0$ for all $h \in \mathfrak{m}_a^2$.

EXERCISE III.115. Let f be an irreducible (nonzero) polynomial in $(x_1, \ldots, x_N)$.

(1) Show that there is i such that $\partial f/\partial x_i$ does *not* identically vanish on $V(f)$. [Hint: $I(V(f)) = \langle f \rangle$.]
(2) Give examples to show that the preceding assertion may not hold if f is not irreducible or if $\Bbbk$ is not algebraically closed.

EXERCISE III.116. Let $\phi : \Bbbk \to \Bbbk^3$ be the morphism given by $t \mapsto (t^3, t^4, t^5)$. Exercise III.91 shows that $X := \phi(\Bbbk)$ is a one-dimensional subvariety of $\Bbbk^3$. Let $U = \Bbbk^3 \setminus V(x)$ (where (x, y, z) are coordinates on $\Bbbk^3$), so that$\Bbbk[U] = \Bbbk[x, y, z, 1/x]$.

(1) Show that the ideal of $X \setminus V(x)$ in $\Bbbk[U]$ is generated by $g_1 := x - (y/x)^3$, $g_2 := y - (y/x)^4$, and $g_3 := z - (y/x)^5$. [Hint: modulo the ideal generated by g_1, g_2, g_3, every polynomial f in (x, y, z) restricts to a polynomial $\bar{f}$ in y/x. Show that $f|_{X \cap U} \equiv 0$ if and only if $\bar{f} \equiv 0$.]
(2) Show that g_1 is in the ideal generated by g_2 and g_3 in $\Bbbk[U]$. Deduce that the ideal of $X \setminus V(x)$ in $\Bbbk[U]$ is generated by g_2 and g_3.
(3) Show that $X \setminus V(x) \cong \Bbbk \setminus \{0\}$; in particular, $X \setminus V(x)$ is nonsingular.

EXERCISE III.117. This exercise shows that the conclusion of corollary III.109 might fail if $\bar{C} \setminus C$ has singular points with "more than one branch." Assume $\mathrm{char}(\Bbbk) \neq 2$. Let C be the curve on $\Bbbk^2$ defined by the equation $y^3 = x(y^2 - 1)$. Let $[x_0 : x_1 : x_2]$ be homogeneous coordinates on $\mathbb{P}^2$. The map $(x, y) \mapsto [1 : x : y]$ identifies $\Bbbk^2$ with the basic open subset $U_0 := \mathbb{P}^2 \setminus V(x_0)$ of $\mathbb{P}^2$. Let $\bar{C}$ be the closure of C in $\mathbb{P}^2$.

(1) Show that $\bar{C} = V(x_2^3 - x_1(x_2^2 - x_0^2)) \subset \mathbb{P}^2$ [Hint: use example III.40] and $\bar{C} \setminus C = \{[0 : 1 : 1], [0 : 1 : 0]\}$.
(2) Show that the projection from C to y-axis is one-to-one, and the inverse of this map extends to a morphism $\phi : \mathbb{P}^1 \to \bar{C}$ which is generically one-to-one, and $\phi^{-1}([0 : 1 : 0])$ consists of two points.
(3) Conclude that the conclusion of corollary III.109 fails with $C, \bar{C}$ and $f := y$.
(4) Show that $O := [0 : 1 : 0]$ has an affine neighborhood in $\bar{C}$ isomorphic to the plane curve $v^3 = (v - w)(v + w)$. A drawing of this curve makes apparent the two "branches" (see section VI.8.1) at O with "tangents" $v - w = 0$ and $v + w = 0$ (see fig. 6b).

14. Completion of the local ring at a point

To study local properties of a variety X near a point, sometimes one needs to pass to the ring of formal power series in affine coordinates at the point. We have seen an example of this in the proof of corollary III.104. In section III.15 and chapter IV we study different notions of *multiplicites* at a point, and power series expansions in coordinates at the point play a fundamental role in our study. The usefulness of these computations depends on the fact that they do *not* depend on the chosen coordinates, i.e. the "rings of formal power series associated to a point on a variety" are isomorphic under local isomorphisms. One way to see this is through the theory of "completions of local rings," which we describe now. Given an ideal I of a ring R and $f \in R$, consider the following property of subsets S of R:

$$S \supseteq f + I^m \text{ for some } m \geq 0. \tag{27}$$

It is straightforward to check that there is a unique topology on R in which a subset S of R is an open neighborhood of $f \in R$ if and only if satisfies condition (27) (exercise III.118); this is called the *I-adic topology* on R. A *Cauchy sequence* in R is a sequence of elements $(f_j)_{j\geq 0}$ of R such that for any open neighborhood U of 0, there is $m \geq 0$ with the property that $f_i - f_j \in U$ for all $i, j \geq m$. Two Cauchy sequences $(f_i)_i$ and $(g_j)_j$ are *equivalent* if the sequence $(f_i - g_i)_i$ converges[22] to 0 in R. The *I-adic completion* $\hat{R}$ of R is the set of equivalence classes of Cauchy sequences.

Example III.111. If I is the zero ideal, then all Cauchy sequences $(f_j)_{j\geq 0}$ are eventually constant, i.e. equivalent to a constant sequence of the form $(f, f, \ldots)$ for some $f \in R$, and moreover, $(f, f, \ldots)$ is equivalent to $(g, g, \ldots)$ if and only if $f = g$. It follows that $\hat{R} \cong R$.

Example III.112. On the other extreme, if $I = R$, then all Cauchy sequences are equivalent to $(0, 0, \ldots)$, so that $\hat{R}$ is the zero ring.

Example III.113. Let $R := \Bbbk[x_1, \ldots, x_n]$ and I be the ideal of R generated by all polynomials vanishing at some point $a = (a_1, \ldots, a_n) \in \Bbbk^n$. Let $(f_j)_j$ in R be a Cauchy sequence with respect to the I-adic topology. We treat each f_j as a polynomial in $y_i := x_i - a_i$, $i = 1, \ldots, n$. For each $k \geq 0$, there is M_k such that the degree (with respect to $(y_1, \ldots, y_n)$-coordinates) of $f_i - f_j$ is greater than k for each $i, j \geq M_k$. In particular, with respect to $(y_1, \ldots, y_n)$-coordinates the homogeneous components $f_{j,k}$ of degree k of all f_j agree with each other whenever $j \geq M_k$; write $F_k := f_{M_k,k}$. It is straightforward to check that the map

$$(f_j)_j \mapsto \sum_k F_k$$

induces a $\Bbbk$-algebra isomorphism between $\hat{R}$ and $\Bbbk[[x_1 - a_1, \ldots, x_n - a_n]]$.

Example III.114. We have seen (in the discussion preceding lemma III.103) that the local ring $\mathcal{O}_{a,\Bbbk^n}$ of $\Bbbk^n$ at a point $a = (a_1, \ldots, a_n)$ can be treated as a subring of $\Bbbk[[x_1 - a_1, \ldots, x_n - a_n]]$. We leave it as an exercise (exercise III.119) to check that $\hat{\mathcal{O}}_{a,\Bbbk^n} \cong \Bbbk[[x_1 - a_1, \ldots, x_n - a_n]]$; in particular, it follows from example III.113

[22]Recall that on a topological space X, a sequence $(x_i)_i$ *converges* to a point x if for every open neighborhood U of x in X there is an integer N such that $x_i \in U$ for all $i \geq N$.

that the completion of the local ring at a point of $\mathbb{k}^n$ with respect to its maximal ideal is isomorphic to the completion of the coordinate ring of $\mathbb{k}^n$ with respect to the maximal ideal at a point.

Proposition III.115. *Let $\hat{R}$ be the completion of a ring R with respect to an ideal I. Given $f \in R$, write $\hat{f}$ for the equivalence class of the constant sequence $(f, f, \ldots)$. Let $\phi : R \to \hat{R}$ be the map which sends $f \mapsto \hat{f}$.*

(1) $\ker \phi = \bigcap_{m \geq 1} I^m$.
(2) If R is a Noetherian local ring, and I is a proper ideal of R, then ϕ is injective.

PROOF. Since the zero element of $\hat{R}$ is the equivalence class of $(0, 0, \ldots)$, the first assertion is immediate from the definition of completion. The second assertion then follows from theorem A.2. □

Remark III.116. Example III.112 shows that assertion (2) of proposition III.115 does not hold if $I = R$.

If X is a variety and $a \in X$, then $\mathcal{O}_{a,X}$ has a unique maximal ideal $\mathfrak{m}_a$, namely, the ideal generated by polynomials vanishing at a (proposition III.94). We write $\hat{\mathcal{O}}_{a,X}$ for the $\mathfrak{m}_a$-adic completion of $\mathcal{O}_{a,X}$. The ring $\hat{\mathcal{O}}_{a,X}$ captures "very local" information about X at a. Example III.114 shows that if X is the affine space, then $\hat{\mathcal{O}}_{a,X}$ is the ring of formal power series expansions centered at a; in general it is a quotient of a power series ring:

Proposition III.117. *Let $a = (a_1, \ldots, a_N)$ be a point of a subvariety X of $\mathbb{k}^N$, and let $\hat{R} := \mathbb{k}[[x_1 - a_1, \ldots, x_N - a_N]]$. Then $\hat{\mathcal{O}}_{a,X} \cong \hat{R}/(I(X)\hat{R})$.*

PROOF. Since $\mathcal{O}_{a,X} \cong \mathcal{O}_{a,\mathbb{k}^N}/(I(X)\mathcal{O}_{a,\mathbb{k}^N})$ (assertion (4) of proposition III.94), the result follows from example III.114 and the exactness of completion (theorem A.3). □

THEOREM III.118. *Let a be a nonsingular point of an irreducible variety X of dimension n. Then*

(1) $\hat{\mathcal{O}}_{a,X} \cong \mathbb{k}[[x_1, \ldots, x_n]]$.

Assume X is a subvariety of $\mathbb{k}^N$ with coordinates $(x_1, \ldots, x_N)$. Pick $g_1, \ldots, g_n \in \mathbb{k}[x_1, \ldots, x_N]$ such that $g_i(a) = 0$ for each i, and $(\frac{\partial g_i}{\partial x_1}(a), \ldots, \frac{\partial g_i}{\partial x_N}(a))$, $i = 1, \ldots, n$, generate $T_a(X)$ as a vector space over $\mathbb{k}$ (see assertion (1) of theorem III.99). Then

(2) it is possible to choose the isomorphism from assertion (1) such that $\hat{\bar{g}}_i \to x_i$, $i = 1, \ldots, n$ (where $\bar{g}_i := g_i|_X \in \mathbb{k}[X]$ and $\hat{\bar{g}}_i$ are defined as in proposition III.115).

PROOF. Taking an appropriate open neighborhood of a we may assume that X is an irreducible subvariety of $\mathbb{k}^N$, and $I(X)$ is generated by $f_1, \ldots, f_{N-n}$ (the second property can be ensured due to theorem III.105). By a change of coordinates if necessary, we may assume a is the origin in $\mathbb{k}^N$. Then $\hat{\mathcal{O}}_{a,X} \cong \hat{R}/(I(X)\hat{R})$, where $\hat{R} := \Bbbk[[x_1, \ldots, x_N]]$ (proposition III.117). The nonsingularity of X at a implies that the linear parts of the f_j are linearly independent over $\Bbbk$, and we may choose $g_1, \ldots, g_n \in \Bbbk[X]$ which satisfy the hypothesis of assertion (2) (namely, take polynomials $g_1, \ldots, g_n$ which vanish at a and are such that the linear parts of $f_1, \ldots, f_{N-n}, g_1, \ldots, g_n$ are linearly independent over $\Bbbk$). It then follows from corollary B.42 that $\hat{R}/(I(X)\hat{R})$ is isomorphic to the ring of power series in n-variables over $\Bbbk$ via an isomorphism that maps $\hat{\bar{g}}_i \to x_i$, $i = 1, \ldots, n$. □

Corollary III.119. *Let a be a nonsingular point of an irreducible curve C, and t be a parameter of $\mathcal{O}_{a,C}$. Then $\hat{\mathcal{O}}_{a,C} \cong \Bbbk[[t]]$.*

PROOF. Pick g_1 as in theorem III.118. By definition of a parameter, $g_1 = ut^k$ for some $k \geq 0$ and a unit u of $\mathcal{O}_{a,C}$. The properties of g_1 implies that $k = 1$. Proposition III.108 then implies that $g_1 = ct + g'_1$ for some $g'_1 \in \mathfrak{m}_a^2$ (where $\mathfrak{m}_a$ is the maximal ideal of $\mathcal{O}_{a,C}$), it follows that $\partial g'_1/\partial x_i = 0$ and $\partial g_1/\partial x_i = c\partial t/\partial x_i$ for each $i = 1, \ldots, N$ (where $(x_1, \ldots, x_N)$ are coordinates on an affine open neighborhood of a), i.e. the linear parts of g_1 and t are proportional. The corollary then follows from the arguments of the proof of theorem III.118. □

14.1. Exercises.

EXERCISE III.118. Let R be a ring, $f \in R$, and I be an ideal of R. Let $\{S_j\}_{j \in \mathcal{J}}$ be a collection of subsets of R such that each S_j satisfies (27). Show that

(1) $\bigcup_{j \in \mathcal{J}} S_j$ satisfies (27).
(2) If $\mathcal{J}$ is finite, then $\bigcap_{j \in \mathcal{J}} S_j$ satisfies (27).

EXERCISE III.119. Let $a = (a_1, \ldots, a_n) \in \Bbbk^n$, and $\mathfrak{m}_a$ be the unique maximal ideal of $\mathcal{O}_{a,\Bbbk^n}$ generated by polynomials vanishing at a. Show that the $\mathfrak{m}_a$-adic completion $\hat{\mathcal{O}}_{a,\Bbbk^n}$ of $\mathcal{O}_{a,\Bbbk^n}$ is isomorphic as a $\Bbbk$-algebra to $\Bbbk[[x_1 - a_1, \ldots, x_n - a_n]]$.

EXERCISE III.120. Let $a \in X$ and $\mathfrak{q}$ be an ideal of $\mathcal{O}_{a,X}$ such that $\mathcal{O}_{a,X}/\mathfrak{q}\mathcal{O}_{a,X}$ is a finite-dimensional vector space over k. Show that $\mathcal{O}_{a,X}/\mathfrak{q}\mathcal{O}_{a,X} \cong \hat{\mathcal{O}}_{a,X}/\mathfrak{q}\hat{\mathcal{O}}_{a,X}$. [Hint: if $\mathfrak{m}$ is the maximal ideal of $\mathcal{O}_{a,X}$, then $\mathfrak{q} \supset \mathfrak{m}^q\mathcal{O}_{a,X}$ for some $q > 0$. Then it follows from the definition of completion that $(\mathcal{O}_{a,X}/\mathfrak{m}^q)/(\mathfrak{q}\mathcal{O}_{a,X}/\mathfrak{m}^q) \cong (\hat{\mathcal{O}}_{a,X}/\mathfrak{m}^q\hat{\mathcal{O}}_{a,X})/(\mathfrak{q}\hat{\mathcal{O}}_{a,X}/\mathfrak{m}^q\hat{\mathcal{O}}_{a,X})$]

15. Degree of a dominant morphism

Let $\phi : X \to Y$ be a dominant morphism of irreducible varieties. Theorem III.85 implies that

(i) if $\dim(X) > \dim(Y)$, then $|\phi^{-1}(y)|$ is infinite for each y in a dense open subset of Y, and
(ii) if $\dim(X) = \dim(Y)$, then $|\phi^{-1}(y)|$ is finite for each y in a dense open subset of Y.

Whenever case (ii) arises for a continuous (or differentiable) map in topology, it turns out that for "almost all" $y \in Y$, some "measure"[23] of the number of elements in $\phi^{-1}(y)$ is constant, and that number is called the "degree" of ϕ. In this section we will see that this remains true for morphisms of algebraic varieties as well. Indeed, since ϕ is a dominant, it induces an inclusion $\mathbb{k}(Y) \hookrightarrow \mathbb{k}(X)$ (proposition III.65). In case (ii) $\mathbb{k}(Y)$ is a *finite extension*[24] of $\mathbb{k}(X)$; the *degree* $\deg(\phi)$ of ϕ is the degree $[\mathbb{k}(X) : \mathbb{k}(Y)]$ of the induced extension of fields.

Example III.120. Let $\phi : \mathbb{k} \to \mathbb{k}$ be the morphism $x \mapsto x^d$, $d > 0$. If t is the coordinate on the target, then the induced extension $\mathbb{k}(t) \hookrightarrow \mathbb{k}(x)$ is given by $t \mapsto x^d$, so that $\deg(\phi) = [\mathbb{k}(x) : \mathbb{k}(x^d)] = d$. If $p := \operatorname{char}(\mathbb{k}) = 0$, then indeed $|\phi^{-1}(a)| = d = \deg(\phi)$ for all $a \in \mathbb{k} \setminus \{0\}$. On the other hand, if $p > 0$ and $d = qp^k$, where $k \geq 1$ and q is relatively prime to p, then $|\phi^{-1}(a)| = q < d$ for each $a \in \mathbb{k} \setminus \{0\}$. However, in this case the extension $\mathbb{k}(x)/\mathbb{k}(t)$ is *not* separable, and the *separable degree* of $\mathbb{k}(x)/\mathbb{k}(t)$ is precisely q (example B.39).

Motivated by example III.120 we define the *separable degree* $\deg_{\text{sep}}(\phi)$ (respectively, *inseparable degree* $\deg_{\text{insep}}(\phi)$) for the separable (respectively, inseparable) degree of the field extension $\mathbb{k}(X)/\mathbb{k}(Y)$ induced by ϕ. Note that

- $\deg(\phi) = \deg_{\text{sep}}(\phi) \deg_{\text{insep}}(\phi)$, and
- if $\operatorname{char}(\mathbb{k}) = 0$, then $\deg_{\text{insep}}(\phi) = 1$ and $\deg(\phi) = \deg_{\text{sep}}(\phi)$.

Theorem III.123 below states that if $\phi : X \to Y$ is a dominant morphism between irreducible varieties of the same dimension, then for all y in a dense open subset of Y,

- $|\phi^{-1}(y)| = \deg_{\text{sep}}(\phi)$, and
- for each $x \in \phi^{-1}(y)$, the "multiplicity" of ϕ at x is $\deg_{\text{insep}}(\phi)$, so that
- the sum over all $x \in \phi^{-1}(y)$ of the multiplicity of ϕ is precisely $\deg(\phi)$.

Before we state and prove theorem III.123 we give some applications.

Example III.121. In exercise III.18 we used the following fact: "Given an algebraically closed field $\mathbb{k}$ and a positive integer n, there is a nonempty Zariski open subset U of $\mathbb{k}^{n+1}$ such that for each $(c_0, \ldots, c_n) \in U$, the polynomial $f := c_0\lambda^n + c_1\lambda^{n-1} + \cdots + c_n$ has n distinct roots in $\mathbb{k}$." We now prove this fact. Indeed, let X be the hypersurface $V(f)$ on the affine space $\mathbb{k}^{n+2}$ with coordinates $(c_0, \ldots, c_n, \lambda)$. It is straightforward to check that f is irreducible as a polynomial in $(c_0, \ldots, c_n, \lambda)$, so that X is irreducible. The arguments of proposition III.76 implies that $\dim(X) = n + 1$ and $1, \lambda, \ldots, \lambda^{n-1}$ is a basis of $\mathbb{k}(X)$ over $\mathbb{k}(c_0, \ldots, c_n)$, so that $[\mathbb{k}(X) : \mathbb{k}(c_0, \ldots, c_n)] = n$. Since $\mathbb{k}$ is algebraically closed, the projection $\pi : X \to \mathbb{k}^{n+1}$ in $(c_0, \ldots, c_n)$-coordinates is dominant. On the other hand, the derivative of f with respect to λ is $c_{n-1} + 2c_{n-2}\lambda + \cdots$, which is *not* identically zero in $\mathbb{k}(c_0, \ldots, c_n)[\lambda]$. Therefore $\mathbb{k}(X)$ is separable over $\mathbb{k}(c_0, \ldots, c_n)$ (proposition B.33). It follows that $\deg_{\text{sep}}(\pi) = \deg(\pi) = n$ and theorem III.123 implies that $|\pi^{-1}(c_0, \ldots, c_n)| = n$ for all $(c_0, \ldots, c_n)$ on a dense open subset of $\mathbb{k}^{n+1}$, which proves the "fact."

---

[23] E.g. $|\phi^{-1}(y)|$, or $|\phi^{-1}(y)|$ modulo 2, etc.

[24] See appendices B.4 and B.12 for a discussion of field extensions and related notions.

Example III.122. (Degree of a projective variety). Let X be a subvariety of $\mathbb{P}^n$. If $d := \dim(X)$, we will show that for "almost all" $(n-d)$-dimensional linear subspaces L of $\mathbb{P}^n$, the number of points in the intersection $L \cap X$ (counted with appropriate multiplicity) is constant (this number is called the *degree* of X). Indeed, denote the homogeneous coordinates of $\mathbb{P}^n$ by $x := [x_0, \ldots, x_n]$, and consider another d copies of $\mathbb{P}^n$ with homogeneous coordinates $\xi^i := [\xi^i_0 : \cdots : \xi^i_n]$, $i = 1, \ldots, d$. Consider the subset Z of $X \times (\mathbb{P}^n)^d$ (where $(\mathbb{P}^n)^d$ is the d-fold Segre product $\mathbb{P}^n \times \cdots \times \mathbb{P}^n$) consisting of all $(x, \xi^1, \ldots, \xi^n)$ such that $\sum_j \xi^i_j x_j = 0$, $i = 1, \ldots, d$. It is straightforward to check that Z is Zariski closed in $X \times (\mathbb{P}^n)^d$. Consider irreducible components Z_k of Z such that the projection $\pi : Z \to (\mathbb{P}^n)^d$ in $(\xi^1, \ldots, \xi^d)$-coordinates maps Z_k dominantly to $(\mathbb{P}^n)^d$. Since Z is complete, any such Z_k, if exists, must get mapped *surjectively* by π. Exercise III.92 implies that such Z_k exists, and in addition, there are $(\xi^1, \ldots, \xi^d)$ such that $|(\pi|_{Z_k})^{-1}(\xi^1, \ldots, \xi^d)| < \infty$. It then follows due to theorem III.85 that $\dim(Z_k) = \dim((\mathbb{P}^n)^d) = nd$, so that theorem III.123 applies, and shows that for all $(\xi^1, \ldots, \xi^d)$ in a dense open subset of $(\mathbb{P}^n)^d$, the number of elements in $\pi^{-1}(\xi^1, \ldots, \xi^d)$ counted with appropriate multiplicity is precisely $\sum_k \deg(\pi|_{Z_k})$.

THEOREM III.123. *Let $\phi : X \to Y$ be a dominant morphism between irreducible varieties of same dimension. Then there is a nonempty Zariski open subset U of Y such that for each $y \in U$,*

(1) Y is nonsingular at y;

(2) $|\phi^{-1}(y)| = \deg_{\text{sep}}(\phi)$;

(3) for each $x \in \phi^{-1}(y)$

(a) X is nonsingular at x,

(b) $\dim_{\mathbb{k}}(\hat{\mathcal{O}}_{x,X}/\mathfrak{m}_y\hat{\mathcal{O}}_{x,X}) = \deg_{\text{insep}}(\phi)$, where $\mathfrak{m}_y$ is the maximal ideal of $\mathcal{O}_{y,Y}$ and $\hat{\mathcal{O}}_{x,X}$ is the completion of $\mathcal{O}_{x,X}$ with respect to its maximal ideal;

(4) in particular

$$\sum_{x \in \phi^{-1}(y)} \dim_{\mathbb{k}}(\hat{\mathcal{O}}_{x,X}/\mathfrak{m}_y\hat{\mathcal{O}}_{x,X}) = \deg(\phi)$$

PROOF. We may assume without loss of generality that X and Y are affine and nonsingular. Let L be the separable closure of $\mathbb{k}(Y)$ in $\mathbb{k}(X)$. Pick regular functions $z_1, \ldots, z_k$ on X such that L is the field of fractions of $T := \mathbb{k}[Y][z_1, \ldots, z_k]$. There is (up to isomorphism) a unique affine variety Z with coordinate ring T (exercise III.26). The chain of inclusions $\mathbb{k}[Y] \hookrightarrow T \hookrightarrow \mathbb{k}[X]$ induces a factorization of ϕ of the form:

$$X \xrightarrow{\phi_i} Z \xrightarrow{\phi_s} Y \subseteq \mathbb{k}^N.$$

By theorem B.35 there is $g \in T$ which generates L over $\mathbb{k}(Y)$. We can factor ϕ_s as

$$Z \xrightarrow{\psi} Y \times \mathbb{k} \xrightarrow{\pi} Y,$$

where ψ maps $z \mapsto (\phi_s(z), g(z))$ and π is the projection onto Y. Let $G(y,t) = \sum_{i=0}^{d_s} a_i(y)t^{d_s-i} \in \Bbbk[Y][t]$, where t is an indeterminate and $d_s := \deg_{\text{sep}}(\phi)$, be the minimal polynomial of g over $\Bbbk(Y)$. The separability of g over $\Bbbk(Y)$ implies that $(\partial G/\partial t)|_{t=g}$ is a nonzero element of $\Bbbk[Z]$. Let U_0 be a nonempty Zariski open subset of Y contained in $Y \setminus (V(a_0) \cup \phi_s(V((\partial G/\partial t)|_{t=g})))$ [why does such U_0 exist?] and $U_0' := \{(y,t) : \sum_{i=0}^{d_s} a_i(y)t^{d_s-i} = 0\} \subset U_0 \times \Bbbk$. Then U_0' is irreducible (since G is irreducible in $\Bbbk(Y)[t]$), and ψ induces a birational map from Z to U_0' [why?]. Let Y_0 be a nonempty Zariski open subset of U_0 such that ψ induces an isomorphism $\phi_s^{-1}(Y_0) \cong Y_0' := (\pi|_{U_0'})^{-1}(Y_0)$. Let $y_0 \in Y_0$. Then Y_0' contains (y_0, t_0) for all the roots t_0 of $G(y_0, t)$. Let $z_0 := \psi^{-1}(y_0, t_0)$. Since $(\partial G/\partial t)(y_0, t_0)$ equals $(\partial G/\partial t)|_{t=g}$ evaluated at z_0, it follows that $(\partial G/\partial t)(y_0, t_0) \neq 0$ for every root t_0 of $G(y_0, t)$. Consequently, $|\pi^{-1}(y_0)| = \deg(G(y_0, t)) = d_s = \deg_{\text{sep}}(\phi)$.

Claim III.123.1. *For each* $(y_0, t_0) \in \pi^{-1}(y_0)$, $\dim_{\Bbbk}(\hat{\mathcal{O}}_{(y_0,t_0),Y_0'}/\mathfrak{m}_{y_0}\hat{\mathcal{O}}_{(y_0,t_0),Y_0'}) = 1$.

PROOF. Pick $(y_0, t_0) \in \pi^{-1}(y_0)$. The image of $G(y,t)$ in $\hat{\mathcal{O}}_{(y_0,t_0),\Bbbk^{N+1}} = \Bbbk[[y_1 - y_{0,1}, \dots, y_N - y_{0,N}, t - t_0]]$ (where $(y_{0,1}, \dots, y_{0,N})$ are coordinates of y_0 in $\Bbbk^N$) is

$$G(y_0, t_0) + \sum_{j=1}^{N} (y_j - y_{0,j}) \frac{\partial G}{\partial y_j}(y_0, t_0) + (t - t_0)\frac{\partial G}{\partial t}(y_0, t_0) + \text{h.o.t.},$$

where h.o.t. denotes terms with order (in $(y - y_0, t - t_0)$) greater than one. Since $G(y_0, t_0) = 0$ and $(\partial G/\partial t)(y_0, t_0) \neq 0$, theorem B.41 implies that $t - t_0$ is in the ideal of $\Bbbk[[y_1 - y_{0,1}, \dots, y_N - y_{0,N}, t - t_0]]$ generated by $G(y,t)$ and $y_j - y_{0,j}$, $j = 1, \dots, N$. Since $\hat{\mathcal{O}}_{(y_0,t_0),Y_0'}$ is the quotient of $\Bbbk[[y_1 - y_{0,1}, \dots, y_N - y_{0,N}, t - t_0]]$ modulo the ideal generated by $G(y,t)$ (proposition III.117), it follows that $t - t_0$ is in the ideal of $\hat{\mathcal{O}}_{(y_0,t_0),Y_0'}$ generated by the $y_j - y_{0,j}$, which implies the claim. □

Note that the above claim and the sentence preceding it proves theorem III.123 in the case that $\Bbbk(X)$ is separable over $\Bbbk(Y)$, in particular when $p := \text{char}(\Bbbk) = 0$. It remains to consider the case that $p > 0$ and $\Bbbk(X)$ is not separable over $\Bbbk(Y)$. Pick $x_1, \dots, x_q \in \Bbbk[X]$ such that $\Bbbk[X] = T[x_1, \dots, x_q]$. Set $T_0 := T$ and $T_j := T_{j-1}[x_j]$ for $j = 1, \dots, q$. For each j, let X_j be the unique affine variety with coordinate ring T_j. Note that each X_j is irreducible (since T_j is an integral domain). The inclusions $T_{j-1} \hookrightarrow T_j$ induces a factorization of $\phi_i : X \to Z$ as follows:

$$X = X_q \xrightarrow{\phi_{i,q}} X_{q-1} \xrightarrow{\phi_{i,q-1}} \cdots \xrightarrow{\phi_{i,2}} X_1 \xrightarrow{\phi_{i,1}} X_0 = Z.$$

The minimal equation of x_j over the field L_{j-1} of fractions of T_{j-1} is of the form $a_{j,0}x_j^{p^{e_j}} - a_{j,1} = 0$ for some $a_{j,0}, a_{j,1} \in T_{j-1}$ (proposition B.38). It follows that L_j is generated by $1, x_j, \dots, x_j^{p^{e_j}-1}$ as a vector space over L_{j-1}; in particular,

$[L_j : L_{j-1}] = p^{e_j}$. Consequently,

$$\deg_{\text{insep}}(\phi) = [L_q : L_0] = p^{\sum_{j=1}^q e_j}. \tag{28}$$

Choose a nonempty open affine subset W_0 of X_0 such that

(i) W_0 is nonsingular,
(ii) $a_{1,0}(x) \neq 0$ for each $x \in W_0$,
(iii) $W_j := (\phi_{i,1} \circ \cdots \circ \phi_{i,j})^{-1}(W_0)$ is nonsingular for each j,
(iv) $a_{j+1,0}(x) \neq 0$ for each $x \in W_j$.

Then each W_j is isomorphic to the hypersurface $V(x_j^{p^{e_j}} - a_{j,1}/a_{j,0})$ of $W_{j-1} \times \Bbbk$, and $\phi_j|_{W_j}$ is simply the restriction of the projection $W_{j-1} \times \Bbbk \to W_{j-1}$. It follows that $\phi_j|_{W_j}$ is one-to-one for each j, and consequently so is $\phi_i|_{W_q} : W_q \to W_0$. Fix $z_0 \in W_0$ and $z_q := (\phi_i|_{W_q})^{-1}(z_0) \in W_q$. Due to (28) in order to complete the proof of theorem III.123 it suffices to show that

$$\dim_\Bbbk(\hat{\mathcal{O}}_{z_q,W_q}/\mathfrak{m}_{z_0}\hat{\mathcal{O}}_{z_q,W_q}) = p^{\sum_{j=1}^q e_j}. \tag{29}$$

Choose coordinates $(w_1, \ldots, w_m)$ on W_0 such that z_0 becomes the origin on $\Bbbk^m$, $m \geq 1$. Replacing each x_j by $x_j - c_j$ for some appropriate $c_j \in \Bbbk$ if necessary, we may in addition assume that x_j vanishes at $z_j := (\phi_{i,1} \circ \cdots \circ \phi_{i,j})^{-1}(z_0)$ for each j. This implies that

$$\begin{aligned} \hat{\mathcal{O}}_{z_0,W_0} &\cong R_0/\mathfrak{p}_0 R_0 \\ \hat{\mathcal{O}}_{z_j,W_j} &\cong R_0\,[[x_1, \ldots, x_j]]\,/\langle \mathfrak{p}_0, x_1^{p^{e_1}} - a_{1,1}/a_{1,0}, \ldots, x_j^{p^{e_j}} - a_{j,1}/a_{j,0}\rangle \end{aligned} \tag{30}$$

where $R_0 := \Bbbk[[w_1, \ldots, w_m]]$ and $\mathfrak{p}_0 \subseteq \Bbbk[w_1, \ldots, w_m]$ is the ideal of polynomials vanishing on W (proposition III.117).

Claim III.123.2. *Each element of $\hat{\mathcal{O}}_{z_q,W_q}$ can be represented by a linear combination of $\mathcal{G} := \{\prod_j x_j^{i_j} : 0 \leq i_j < p^{e_j}\}$ with coefficients in R_0.*

PROOF. Pick $\rho \in \hat{\mathcal{O}}_{z_q,W_q}$ and a power series f in $(w_1, \ldots, w_m, x_1, \ldots, x_q)$ which represents ρ. Replacing $(x_q)^{ip^{e_q}}$ by $(a_{q,1}/a_{q,0})^i$ for each i yields a power series f_1 such that all powers of x_q in f_1 are smaller than p^{e_q}, and f_1 also represents $\rho \in \hat{\mathcal{O}}_{z_q,W_q}$. Continuing this process with x_{q-1} and so on yields a power series as claimed. □

Claim III.123.2 implies that $\mathcal{G}$ spans $\hat{\mathcal{O}}_{z_q,W_q}/\mathfrak{m}_{z_0}\hat{\mathcal{O}}_{z_q,W_q}$ over $\Bbbk$. On the other hand, using (30) it is straightforward to check that the elements of $\mathcal{G}$ are linearly independent over $\Bbbk$ in $\hat{\mathcal{O}}_{z_q,W_q}/\mathfrak{m}_{z_0}\hat{\mathcal{O}}_{z_q,W_q}$. Therefore $\mathcal{G}$ is a basis of $\hat{\mathcal{O}}_{z_q,W_q}/\mathfrak{m}_{z_0}\hat{\mathcal{O}}_{z_q,W_q}$ over $\Bbbk$. Since $|\mathcal{G}| = p^{\sum_j e_j}$, this completes the proof of (29) and consequently the theorem. □

CHAPTER IV

*Intersection multiplicity

1. Introduction

In[1] this chapter we define the intersection multiplicity of n hypersurfaces at a point on a nonsingular variety X of dimension n, and prove some of its basic properties. As fig. 1 suggests, nontrivial considerations prop up even in the intersection of a parabola and a line. However, we do have a natural candidate for the intersection multiplicity, namely, if f_j are regular functions on X, then for each $a \in \bigcap_j \{f_j = 0\}$, we can consider the "multiplicity" at a of the morphism $X \to \mathbb{k}^n$ given by $x \mapsto (f_1(x), \ldots, f_n(x))$ suggested by theorem III.123, i.e. the quantity

$$\dim_{\mathbb{k}}(\hat{\mathcal{O}}_{a,X}/\langle f_1, \ldots, f_n\rangle) \tag{31}$$

This is indeed the definition we are going to use (see [Ful89, Section 3.3] for a wonderful axiomatic motivation for this definition). However, in chapters VII to XI we would in addition need to use a method of computing the intersection multiplicity via "parametrization." Consider, e.g. the case of two plane curves $f = 0$ and $g = 0$ on $\mathbb{k}^2$. Their intersection multiplicity at the origin, as given by (31), is the dimension (as a vector space) over $\mathbb{k}$ of the quotient of the power series ring $\mathbb{k}[[x, y]]$ by the ideal generated by f, g. The "parametric" procedure on the other hand is as follows: find a "parametrization" $\phi(t)$ of the curve $g = 0$ such that $\phi(0) = 0$, and compute the order of $f(\phi(t))$, which measures how fast f is vanishing along the curve $g = 0$.

[1]The asterisk in the chapter name is to indicate that most of the material of this section might be skipped in the first reading and/or in a first course of algebraic geometry. Only a small part of chapter VI uses the results of this chapter. For the proof of Bernstein's theorem and its applications one would mainly need lemma IV.29, theorems IV.24, IV.31 and IV.32, corollary IV.25 and propositions IV.27 and IV.28—which might be explained without proof in a first course of algebraic geometry.

P. Mondal, *How Many Zeroes?*, CMS/CAIMS Books in Mathematics 2,
https://doi.org/10.1007/978-3-030-75174-6_IV

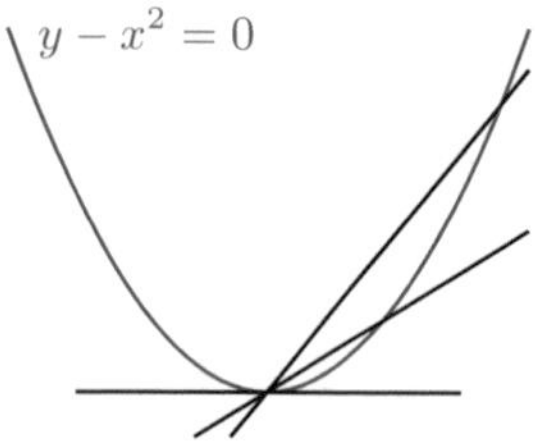
$y - x^2 = 0$
As secants approach the tangent at O more and more closely, both of the two points of intersection move arbitrarily close to O.

FIGURE 1. A tangent line intersects a parabola at a point with multiplicity two.

For example, consider the situation of fig. 1, i.e. $f = y - x^2$ and $g = y - mx$. Then $\phi(t) := (t, mt)$ parametrizes the line $g = 0$, and $f(\phi(t)) = mt - t^2$. Consequently $\operatorname{ord}_t(f(\phi(t))) = 1$ if $m \neq 0$. If $m = 0$, i.e. $g = 0$ is a horizontal line, then $\operatorname{ord}_t(f(\phi(t))) = 2$, as expected. However, to use this approach in practice one needs to define order, parametrization etc. even in the case that the equations are "not reduced"; consider e.g. the case that $f = y - x^2$ and $g = (y - mx)^2$. The algebraic quantity (31) gives the expected answer (which is 2 if $m \neq 0$, and 4 if $m = 0$), but what is geometrically the object $(y - mx)^2 = 0$? The underlying space is still the same line $y = mx$, but the defining equation is different, and the "coordinate ring" $\mathbb{k}[x, y]/\langle (y - mx)^2 \rangle$ is "non-reduced," (i.e. it has a nonzero nilpotent, namely, the image of $y - mx$). In particular, in order to make the geometric approach more generally applicable one needs to

(1) build a theory of "non-reduced varieties," and
(2) define the notion of order at a point of a "non-reduced curve."

In section IV.2 we introduce the notion of "closed subschemes" which are the correct candidates for "non-reduced subvarieties," and in section IV.3 we extend the notion of order to non-reduced curves. In sections IV.4 and IV.5 we apply these notions to the study of intersection multiplicity.

2. Closed subschemes of a variety

2.1. Closed subschemes of an affine variety. Let X be an affine variety and $\mathfrak{q}$ be a (not necessarily radical) ideal of $\mathbb{k}[X]$. The *closed subscheme* of X determined by $\mathfrak{q}$, which by an abuse of notation[2] we denote by $V(\mathfrak{q})$ is the pair (Z', R), where Z' is the *subvariety* of X determined by $\mathfrak{q}$, and $R = \mathbb{k}[X]/\mathfrak{q}$. We say that Z' is the *support* of $V(\mathfrak{q})$. One can picture $V(\mathfrak{q})$ as a "thickened" version of Z'. For example, if $X = \mathbb{k}^2$ and $\mathfrak{q} = \langle x, y^2 \rangle \subseteq \mathbb{k}[x, y]$, then $V(\mathfrak{q})$ is supported at the origin. The image of $g = a + bx + cy + dx^2 + exy + fy^2 + \cdots$ in $\mathbb{k}[x, y]/\mathfrak{q}$ is determined by $a = g(0, 0)$ and $c = \frac{\partial g}{\partial y}(0, 0)$ so that $V(\mathfrak{q})$ is the origin coupled with the vertical tangent line at the origin (fig. 2). In general one can picture $V(\mathfrak{q})$ as a union of thickened varieties corresponding to *primary decompositions* (appendix B.10) of $\mathfrak{q}$,

[2]This is an abuse of notation since we also use $V(\cdot)$ to denote *subvarieties* of a given variety. We will try to ensure that the intended meaning is clear from the context.

and different primary decompositions may lead to different pictures of the same closed subscheme—see [Eis95, Section 3.8] for an illuminating exposition.

FIGURE 2. The subscheme of $\Bbbk^2$ corresponding to $\mathfrak{q} = \langle x, y^2 \rangle$ is the origin coupled with a vertical tangent.

Example IV.1. If Z' is a subvariety of X, then in general there are infinitely many closed subschemes of X supported at Z'. However, there is a canonical one among these, namely, the subscheme $V(I(Z'))$, where $I(Z')$ is the ideal in $\Bbbk[X]$ consisting of all regular functions that vanish on Z'; this is called the *reduced* subscheme structure on Z' (since the "coordinate ring" $\Bbbk[X]/I(Z')$ is reduced).

Example IV.2. Following an answer on *MathOverflow* [dh] we now present an example where different closed subschemes with the same support appear "naturally." Recall that an $n \times n$ matrix A over $\Bbbk$ is *nilpotent* if $A^k = 0$ for some $k \geq 1$. The space $\mathcal{M}_n$ of $n \times n$ matrices can be naturally identified with $\Bbbk^{n^2}$ with coordinates x_{ij}, $1 \leq i, j \leq n$. For $A \in \mathcal{M}_n$ it is a standard result from linear algebra that each of the following properties are equivalent to A being nilpotent:

(i) $A^n = 0$,
(ii) $\det(A - \lambda \mathbb{1}_n) = \lambda^n$, where $\mathbb{1}_n$ is the $n \times n$ identity matrix, and λ is an indeterminate over the x_{ij}.

Each of these conditions identifies the set $\mathcal{N}_n$ of nilpotent matrices as the set of zeroes of a system of polynomials in $(x_{ij})_{i,j}$. Let $\mathfrak{q}_1, \mathfrak{q}_2$ be the ideals generated by respectively these two systems of polynomials. Then both $V(\mathfrak{q}_j)$ are supported at $\mathcal{N}_n$. We now show that $\mathfrak{q}_1 \neq \mathfrak{q}_2$ when $n > 1$. Indeed, since the trace of a nilpotent matrix is zero, it follows that $f := x_{11} + \cdots + x_{nn} \in I(\mathcal{N}_n) = \sqrt{\mathfrak{q}_1} = \sqrt{\mathfrak{q}_2}$. Since all polynomials that arise from condition (i) are *homogeneous* of degree n, it follows that $f \notin \mathfrak{q}_1$ if $n > 1$. On the other hand, the coefficient of λ^{n-1} in $\det(A - \lambda \mathbb{1}_n)$ is $\pm f$, so that $f \in \mathfrak{q}_2$.

2.2. Closed subschemes of a quasiprojective variety. Defining subschemes on an arbitrary quasiprojective variety is a bit more complicated than the case of affine varieties, since the same set of equations can look different in different coordinate charts. We want the subscheme to be a notion which would keep track of equations. Giving a set of equations on a neighborhood of a point x is essentially same as giving an ideal of the local ring $\mathcal{O}_{x,X}$ of X at x. However, the equations at different points need to be compatible, i.e. the equations at all points on a sufficiently small neighborhood must "come from the same set of equations." This leads to the following definition: a *sheaf $\mathcal{I}$ of ideals* on a quasiprojective variety X is a product $\prod_{x \in X} \mathcal{I}_x$, where each $\mathcal{I}_x$ is an ideal of $\mathcal{O}_{x,X}$, such that

(32) each $x \in X$ has a nonempty open affine neighborhood U in X and an ideal I of $\Bbbk[U]$ such that $\mathcal{I}_{x'} = I\mathcal{O}_{x',X}$ for each $x' \in U$.

For each $x \in X$, we say that $\mathcal{I}_x$ is the *stalk of* $\mathcal{I}$ *at* x. For us the *closed subscheme* $V(\mathcal{I})$ of X determined by $\mathcal{I}$ would be the product $\prod_{x \in X} \mathcal{O}_{x,X}/\mathcal{I}_x$ of quotient rings[3]. An *embedded affine chart* of $V(\mathcal{I})$ is an affine open subset U of X which satisfies condition (32). Let $\{U_j\}$ be an open covering of X by embedded affine charts of $V(\mathcal{I})$, i.e. for each j, condition (32) is satisfied with $U = U_j$ and $I = I_j$ for some ideal I_j of $\Bbbk[U_j]$. Then the union of the subvarieties Z'_j of X determined by I_j is in fact a subvariety Z' of X. We say that Z' is the *support* of $V(\mathcal{I})$, and write $Z' = \mathrm{Supp}(V(\mathcal{I}))$.

Example IV.3. Let X be an affine variety. Every ideal $\mathfrak{q}$ of $\Bbbk[X]$ canonically corresponds to the sheaf $\mathcal{I}_\mathfrak{q} := \prod_{x \in X}(\mathfrak{q}\mathcal{O}_{X,x})$ of ideals on X. We identify the closed subscheme $V(\mathfrak{q})$ defined in section IV.2.1 with the closed subscheme $V(\mathcal{I}_\mathfrak{q})$ of X.

Example IV.4. Every variety can be regarded as a closed subscheme of itself corresponding to the sheaf of "zero ideals."

Example IV.5. Let X be a subvariety of $\mathbb{P}^n$ with homogeneous coordinates $[x_0 : \cdots : x_n]$. If f is any homogeneous polynomial in $(x_0, \ldots, x_n)$ of degree d, then f/x_j^d is a regular function on the basic open subset $U_j := \mathbb{P}^n \setminus V(x_j)$ for each $j = 0, \ldots, n$. Moreover, on $U_i \cap U_j$, $f/x_j^d = u_{ji} f/x_i^d$, where $u_{ji} = x_i^d/x_j^d$ is a unit in $\Bbbk[U_i \cap U_j]$. It follows that f defines a sheaf of ideals on $\mathbb{P}^n$ whose stalk at $x \in X \cap U_j$ is the ideal of $\mathcal{O}_{x,X}$ generated by f/x_j^d; we write $V(f)$ for the corresponding closed subscheme of X. It is straightforward to check that $\mathrm{Supp}(V(f))$ is precisely the subvariety of X determined by f, and $X \cap U_j$ is an embedded affine chart of $V(f)$ for each basic open subset U_j of $\mathbb{P}^n$.

Example IV.6. The arguments from example IV.5 can be generalized in a straightforward way to show that any finite collection $f_1, \ldots, f_N$ of homogeneous polynomials in $(x_0, \ldots, x_n)$ determines a sheaf of ideals $\mathcal{I}(f_1, \ldots, f_N)$ on a quasiprojective subset X of $\mathbb{P}^n$ such that for each $x \in X \cap U_j$, the stalk of $\mathcal{I}(f_1, \ldots, f_N)$ at x is the ideal of $\mathcal{O}_{x,X}$ generated by $f_1/x_j^{\deg(f_1)}, \ldots, f_N/x_j^{\deg(f_N)}$; the support of the corresponding closed subscheme $V(f_1, \ldots, f_N)$ of X is precisely the subvariety of X determined by $f_1, \ldots, f_N$. If I is a homogeneous ideal of $\Bbbk[x_0, \ldots, x_n]$, and $f_1, \ldots, f_N$ are homogeneous generators of I, then it is straightforward to check that the sheaf of ideals $\mathcal{I}(f_1, \ldots, f_N)$ does *not* depend on $f_1, \ldots, f_N$ (i.e. if $g_1, \ldots, g_M$ are homogeneous generators of I, then $\mathcal{I}(f_1, \ldots, f_N) = \mathcal{I}(g_1, \ldots, g_M)$); we denote the corresponding closed subschme on X by $V(I)$.

Given a sheaf $\mathcal{I}$ of ideals on a variety X, let $Z := V(\mathcal{I})$, and $Z' := \mathrm{Supp}(Z)$. Given a point $x \in Z'$, we often abuse the notation and say $x \in Z$. The *local*

[3]If you are already familiar with schemes you will note that we are identifying a closed subscheme with its structure sheaf.

ring $\mathcal{O}_{x,Z}$ of Z at x is the quotient $\mathcal{O}_{x,X}/\mathcal{I}_x$. We say that Z has (pure) dimension k if and only if Z' has (pure) dimension k. If U is an open subset of X, then $\mathcal{I}|_U := \prod_{x \in U} \mathcal{I}_x$ is a sheaf of ideals on U; we denote the corresponding closed subscheme of U by the "scheme-theoretic intersection" $Z \cap U$, and say that it is an *open subscheme* of Z. There is also a scheme-theoretic intersection of two closed subschemes: if $Y = V(\mathcal{J})$ is a closed subscheme of X corresponding to a sheaf $\mathcal{J}$ of ideals on X, then the scheme-theoretic intersection $Y \cap Z$ is the closed subscheme of X corresponding to the sheaf of ideals $\mathcal{I} + \mathcal{J} := \prod_{x \in X}(\mathcal{I}_x + \mathcal{J}_x)$. We identify the variety X with its closed subscheme defined by the sheaf of zero ideals. This in particular implies that the scheme-theoretic intersection $Z \cap X$ is simply Z, as expected.

Example IV.7. Given a quasiprojective subset X of $\mathbb{P}^n$ and homogeneous polynomials $f_1, \ldots, f_N$ in $(x_0, \ldots, x_n)$, if $V(f_j)$ are closed subschemes of X constructed in example IV.5, then the scheme-theoretic intersection $\bigcap V(f_j)$ is precisely $V(f_1, \ldots, f_N)$ constructed in example IV.6.

2.3. Rational functions. Usually the notion of rational functions is considered only for irreducible varieties. Exercise III.62 gives an hint that defining a rational function on a reducible variety can get tricky due to the presence of nonzero regular functions which are *zero-divisors*, i.e. which vanish identically on some irreducible component. Consider e.g. $X = V(xy) \subset \mathbb{k}^2$, i.e. X is the union of x and y-axes on $\mathbb{k}^2$. In this case both x and y are zero-divisors (in fact $xy = 0$ on X), and it would be difficult to give geometric interpretation of a ring containing $1/x$ and $1/y$ (what would be the "value" of $1/x + 1/y$ at a point on X?). A standard solution is therefore not to allow zero-divisors in the denominator: let Z be a subscheme of X. For each $x \in Z$, let S_x be the localization of $\mathcal{O}_{x,Z}$ at the set of its non zero-divisors, i.e. $S_x := \{f_1/f_2 : f_1, f_2 \in \mathcal{O}_{x,Z},\ f_2 \text{ is } \textit{not} \text{ a zero-divisor in } \mathcal{O}_{x,Z}\}$. Let $Z' := \operatorname{Supp}(Z)$. A *rational function* on Z is an element $f = (f_x : x \in Z') \in \prod_{x \in Z'} S_x$ such that

(33) each $x \in Z'$ has a nonempty open affine neighborhood U in Z' and $f_1, f_2 \in \mathbb{k}[U]$ such that $f_{x'} = f_1/f_2 \in S_{x'}$ for each $x' \in U$.

A rational function $f = (f_x : x \in Z')$ on Z is a *regular function* if each $f_x \in \mathcal{O}_{x,Z}$, and it is an *invertible rational* function if $1/f$ is also a rational function, i.e. if (33) holds with the additional condition that no f_i is a zero-divisor in $\mathcal{O}_{x',Z}$ for any $x' \in U \cap Z'$.

Example IV.8. Let X be a variety. Regardless of whether we treat X as a variety or the closed subscheme of itself determined by the zero ideal, the set of regular functions on X remain the same, and in addition, if X is irreducible, then the set of rational functions on X remains the same (recall that in section III.7 we did *not* define the rational functions on a reducible variety).

Example IV.9. Let X be the *subvariety* $V(xy)$ of $\mathbb{k}^2$, so that $\mathbb{k}[X] = \mathbb{k}[x, y]/\langle xy \rangle$. Since both x and y are zero-divisors in $\mathbb{k}[X]$, it follows that the set of rational functions on X (when we treat X as its closed subscheme defined by the zero ideal) can be identified with $\{f/g : f, g \in \mathbb{k}[X],\ g(0) \neq 0\} \cong \mathcal{O}_{0,\mathbb{k}^2}/\langle xy \rangle$.

2.4. Completeness and compactification of schemes. Let $\phi : Y \to X$ be a morphism of varieties. If $\mathcal{I} = \prod_{x \in X} \mathcal{I}_x$ is a sheaf of ideals on X, then $\phi^*\mathcal{I} := \prod_{y \in Y} \phi^*(\mathcal{I}_{\phi(y)})\mathcal{O}_{y,Y}$ is a sheaf of ideals on Y. If ϕ is an *isomorphism of varieties*, then for each $y \in Y$, $\mathcal{I}_{\phi(y)}$ is naturally isomorphic as an $\mathcal{O}_{x,X}$-module to $(\phi^*\mathcal{I})_y = \phi^*(\mathcal{I}_{\phi(y)})\mathcal{O}_{y,Y}$; we say that $\phi^* : V(\mathcal{I}) \to V(\phi^*(\mathcal{I}))$ is an *embedded isomorphism of closed subschemes*. A basic example of embedded isomorphism arises in the following context.

Example IV.10. Given a polynomial $g \in \mathbb{k}[x_1, \ldots, x_n]$, recall that $X := \mathbb{k}^n \setminus V(g)$ is isomorphic to the subvariety $Y := V(x_{n+1}g - 1)$ of $\mathbb{k}^{n+1}$ (example III.37). Now assume Z is the closed subscheme of $\mathbb{k}^n$ determined by an ideal I of $\mathbb{k}[x_1, \ldots, x_n]$ such that g does not identically vanish on $\operatorname{Supp}(Z)$, i.e. $g \notin \sqrt{I}$. Then the open subscheme $V := Z \cap X$ of Z is a closed subscheme of X with nonempty support. The isomorphism $Y \to X$ induces an embedded isomorphism between V and the closed subscheme of Y defined by the ideal generated by I.

A *compactification* of $Z := V(\mathcal{I})$ is a closed subscheme $\bar{Z}$ of a compactification $\bar{X}$ of X such that Z is embedded isomorphic to an open subscheme of $\bar{Z}$. A closed subscheme of a variety is *complete* if its support is complete; note that $\operatorname{Supp}(\bar{Z})$, being a closed subvariety of a complete variety, is complete (proposition III.60). A fundamental result of Nagata states that every closed subscheme Z of a given variety X can be compactified to a closed subscheme of a given compactification $\bar{X}$ of X. In this chapter we will use Nagata's result for the case that $\dim(Z) = 1$, which we now prove.

Remark IV.11. Our proof of theorem IV.12 below would have been much shorter if we had used the fact that if C is an irreducible curve and S is a finite nonempty subset of C, then $C \setminus S$ is an affine curve. But with the tools developed in chapter III we could prove only an approximate version of it, namely, $C \setminus S'$ is affine for some $S' \supseteq S$, where S' is a finite set, and given any finite set $S'' \subseteq C \setminus S$, we can ensure that $S' \cap S'' = \emptyset$.

THEOREM IV.12. *Let Z be a one-dimensional closed subscheme of a quasiprojective variety X and $\bar{X}$ be a projective compactification of X. Then there is a closed subscheme $\bar{Z}$ of $\bar{X}$ such that*

(1) Z is embedded isomorphic to $\bar{Z} \cap X$.
(2) $\operatorname{Supp}(\bar{Z})$ is the closure in $\bar{X}$ of $\operatorname{Supp}(Z)$.
(3) every rational function on Z extends to a rational function on $\bar{Z}$.
(4) every invertible rational function on Z extends to an invertible rational function on $\bar{Z}$.

PROOF. Since $Z' := \operatorname{Supp}(Z)$ has dimension one, $S := \bar{Z}' \setminus Z'$, where $\bar{Z}'$ is the closure of Z' in $\bar{X}$, is *finite*.

Claim IV.12.1. *There is an affine open subset $\bar{W}$ of $\bar{X}$ containing S and there is a non zero-divisor $g \in \Bbbk[\bar{W}]$ such that $V(g) \cap \bar{Z}' = S$, $W := \bar{W} \setminus V(g) \subset X$. In addition one can ensure that*

(a) *W is an* embedded affine chart *of Z, and*
(b) *the ideal I of $\Bbbk[W]$ defining $Z \cap W$ is* unmixed, *i.e. has no zero-dimensional*[4] *associated prime ideals.*

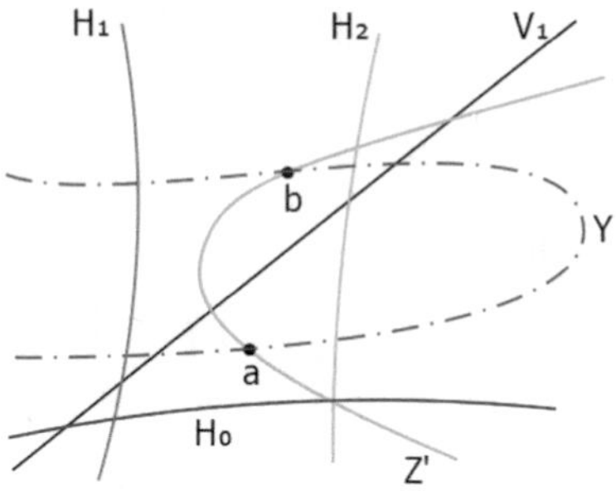

FIGURE 3. $Y := \bar{X} \setminus X$, $S = \{a, b\}$, $\bar{W} := \bar{X} \setminus (V_1 \cup H_1 \cup H_2)$, $W := \bar{W} \setminus (Y \cup H_0)$.

PROOF. Choose a closed embedding $\bar{X} \hookrightarrow \mathbb{P}^N$ with homogeneous coordinates $[x_0 : \cdots : x_N]$. After composing ϕ with an appropriate Veronese embedding (see section III.8.2) followed by a linear change of coordinates if necessary we may ensure that

(i) $\bar{X} \setminus X \subseteq V(x_0)$,
(ii) $V(x_0)$ does *not* contain any irreducible component of $\bar{X}$ or any irreducible component of $\bar{Z}'$,
(iii) $S \cap V(x_1) = \emptyset$.

(In fig. 3 $V(x_1)$ is denoted by V_1, and $V(x_0)$ is the union of $Y := \bar{X} \setminus X$ and possibly some other variety, say H_0, not containing any irreducible component of $\bar{Z}'$.) Let $U_1 := \mathbb{P}^N \setminus V(x_1)$. Property (iii) implies that $Z' \cap U_1 \neq \emptyset$, so that we can choose an affine open subset W_1 of $X \cap U_1$ such that

(iv) W_1 is also an embedded affine chart of Z, and
(v) the ideal I_1 of $\Bbbk[W_1]$ defining $Z \cap W_1$ is unmixed.

(In fig. 3 W_1 is the complement of $V_1 \cup H_1$.) Choose a regular function h on U_1 such that

(vi) $V(h) \supseteq ((Z' \cap W_1 \cap V(x_0)) \cup ((X \cap U_1) \setminus W_1))$.
(vii) h does *not* vanish at any point of $S \subset U_1$.

(In fig. 3 $V(h) \cap U_1$ is the union of H_1 and possibly some other variety, say H_2, not containing any point of S, but containing all points of $H_0 \cap Z'$.) Let $\bar{W} := (\bar{X} \cap U_1) \setminus V(h)$ and $W := \bar{W} \setminus V(x_0)$. Then $\bar{W}$ is an affine open subset of $\bar{X}$

[4]The dimension of an ideal J of $\Bbbk[W]$ is the dimension of the subvariety $V(J)$ of W.

(exercise III.54). Property (vi) implies that $W = W_1 \setminus V(hx_0/x_1)$ (note that x_0/x_1 is a regular function on $U_1 \supset W_1$). Since W_1 is an embedded affine chart of Z, it is straightforward to check that so is W. Properties (ii), (v), (vi) and (vii) then imply that the claim holds with $g = x_0/x_1$. □

Choose $\bar{W}, W, g$ as in claim IV.12.1. Let I be the ideal of $\Bbbk[W]$ defining $Z \cap W$. Since g is a non zero-divisor in $\Bbbk[\bar{W}]$,

(34) the natural map $\Bbbk[\bar{W}] \to \Bbbk[\bar{W}]_g = \Bbbk[W]$ is injective.

Define $\bar{I} := I \cap \Bbbk[\bar{W}]$. Then I is generated by $\bar{I}$ in $\Bbbk[W]$ (proposition B.14) and g is a non zero-divisor in $\Bbbk[\bar{W}]/\bar{I}$ (corollary B.27). If $\mathcal{I}$ is the sheaf of ideals on X defining Z, then it follows that

$$\prod_{x \in \bar{W}} \bar{I}\mathcal{O}_{x,\bar{X}} \times \prod_{x \in X \setminus W} \mathcal{I}_x \times \prod_{x \in Y \setminus \bar{W}} \mathcal{O}_{x,\bar{X}} = \prod_{x \in X} \mathcal{I}_x \times \prod_{x \in S} \bar{I}\mathcal{O}_{x,\bar{X}} \times \prod_{x \in Y \setminus X} \mathcal{O}_{x,\bar{X}}$$

(where $Y := \bar{X} \setminus X$) is a sheaf of ideals on $\bar{X}$ and the corresponding subscheme $\bar{Z}$ of $\bar{X}$ satisfies assertions (1) and (2). Now let $f = (f_x : x \in Z')$ be a rational function on Z.

Claim IV.12.2. *There is an open subset U' of Z' which intersects every irreducible component of Z', and in addition* (33) *is satisfied with $U = U'$.*

PROOF. For each irreducible component Z'_j of Z', choose an open subset U'_j of Z' such that (33) is satisfied with $U = U'_j$, and U'_j does *not* intersect any irreducible component of Z' other than Z'_j. Then $U' := \bigcup_j U'_j$ satisfies the claim. □

Let U' be as in claim IV.12.2. Due to assertion (b) of claim IV.12.1 one can find a regular function q on W such that $Z' \cap W \setminus V(q) \subseteq U$, and q is a non zero-divisor in both $\Bbbk[W]$ and $\Bbbk[W]/I$. Note that Z is defined in $W^* := W \setminus V(q)$ by the ideal generated by I. Moreover, there is $f_1, f_2 \in \Bbbk[W^*]$ such that f_2 is a non zero-divisor in $\Bbbk[W^*]/I\Bbbk[W^*]$, and $f_x = f_1/f_2 \in \mathcal{O}_{x,X}/I\mathcal{O}_{x,X}$ for each $x \in W^* \cap Z$. Since $\Bbbk[W^*] = \Bbbk[W]_q = \Bbbk[\bar{W}]_{gq}$, it follows that f_1/f_2 can be represented in the total quotient ring of $\Bbbk[\bar{W}]/\bar{I}$ as $g^q\bar{q}^b\bar{f}_1/\bar{f}_2$ for some integers a, b and $\bar{q}, \bar{f}_1, \bar{f}_2 \in \Bbbk[\bar{W}]$ such that $\bar{q}$ and $\bar{f}_2$ are non zero-divisors in $\Bbbk[\bar{W}]/\bar{I}$. This proves assertion (3). For assertion (4) note that if f_1 is a non zero-divisor in $\Bbbk[W^*]/I\Bbbk[W^*]$, then $\bar{f}_1$ is also a non zero-divisor in $\Bbbk[\bar{W}]/\bar{I}$. □

2.5. Irreducible components, local rings. Let Z be a closed subscheme of a variety X. The *irreducible components* of Z are simply the irreducible components of $\mathrm{Supp}(Z)$. Note that every irreducible component of a closed subscheme Z of X is a *subvariety* of X. Let Y be a closed irreducible *subvariety* of $\mathrm{Supp}(Z)$. The *local ring $\mathcal{O}_{Y,Z}$ of Z at Y* is the set of the equivalence classes of pairs (h, U), where U is an open subset of X such that $U \cap Y \neq \emptyset$ and h is a *regular function* on the open subscheme $Z \cap U$ of Z, and the equivalence relation $\sim$ is defined as follows: $(h_1, U_1) \sim (h_2, U_2)$ if and only if $h_1 = h_2$ in $\mathcal{O}_{x,Z}$ for each $x \in U_1 \cap U_2$.

It is straightforward to check that $\mathcal{O}_{Y,Z}$ is a $\Bbbk$-algebra. A more explicit realization of $\mathcal{O}_{Y,Z}$ is as follows: pick an *embedded affine chart* U of Z such that $U \cap Y \neq \emptyset$. Then $U \cap Z$ is the closed subscheme of U defined by an ideal $\mathfrak{q}$ of $\Bbbk[U]$. On the other hand, since $U \cap Y$ is an irreducible subvariety of $\text{Supp}(Z) \cap U$, it corresponds to a prime ideal $\mathfrak{p}$ of $\Bbbk[U]$ containing $\mathfrak{q}$. Then $\mathcal{O}_{Y,Z}$ can be identified with the localization $(\Bbbk[U]/\mathfrak{q})_{\mathfrak{p}}$ of $\Bbbk[U]/\mathfrak{q}$ at the ideal generated by $\mathfrak{p}$. This in particular implies that $\mathcal{O}_{Y,Z}$ is a *local ring*. If $Y = a$ is a point of Z, then it is straightforward to check that this definition of $\mathcal{O}_{Y,Z}$ agrees with the earlier definition of $\mathcal{O}_{a,Z}$ from section IV.2.2.

Lemma IV.13. *Let Z be a closed subscheme of an affine variety X, and Y be an irreducible subvariety of* $\text{Supp}(Z)$. *Given $f \in \Bbbk[X]$, if the image of f is invertible in $\mathcal{O}_{Y,\text{Supp}(Z)}$, then it is also invertible in $\mathcal{O}_{Y,Z}$.*

PROOF. This immediately follows from the following observation: if $\phi : S \to T$ is a ring homomorphism such that $\ker(\phi)$ is contained in the nilradical of S, and if $u \in S$ is such that $\phi(u)$ is invertible in T, then u is invertible in S. □

2.6. Cartier divisors. A *Cartier divisor* is a closed subscheme generated locally by single *non zero-divisors*; it is the natural scheme-theoretic analogue of a "hypersurface." More precisely, a closed subscheme $Z = V(\mathcal{I})$ of a variety X is called a Cartier divisor if each $x \in X$ has a nonempty open affine neighborhood U in X and an element $g \in \Bbbk[U]$ which is not a zero-divisor in $\Bbbk[U]$ such that $\mathcal{I}_{x'} = g\mathcal{O}_{x',X}$ for each $x' \in U$. It is straightforward to check that defining a Cartier divisor on X is equivalent to prescribing a collection $\{(U_i, g_i)\}_i$ of pairs such that

(a) $\{U_i\}$ is an open affine covering of X, and
(b) $g_i \in \Bbbk[U_i]$ are such that
  (1) g_i is a non zero-divisor in $\Bbbk[U_i]$ for each i, and
  (2) g_i/g_j is invertible in $\Bbbk[U_i \cap U_j]$ for each i, j.

If X is of pure dimension n, it follows from theorem III.80 that the support of a Cartier divisor on X is either empty or has pure dimension $n - 1$.

Example IV.14. Let $X = V(I)$ be the *irreducible subvariety* of $\mathbb{P}^n$ determined by a *prime* homogeneous ideal I of $\Bbbk[x_0, \ldots, x_n]$. If f is a homogeneous polynomial which is *not* in I, then the closed subscheme $V(f)$ of X constructed in example IV.5 is a Cartier divisor on X.

3. Possibly non-reduced curves

This section is devoted to task (2) outlined in section IV.1. A *reduced curve* is a variety of pure dimension one. By a *possibly non-reduced curve* we mean a pure dimension one closed subscheme Z of a variety. Unless explicitly stated otherwise, a *curve* will mean a reduced curve. Our convention of identifying a variety with its subscheme defined by the zero ideal sheaf implies that a "curve" is indeed a special case of a "possibly non-reduced curve." In section IV.3.1 we describe some properties of curves we are going to use without proof. In section IV.3.2 we define the notion of order at a point of a possibly non-reduced curve and describe some of its properties whose proofs are deferred to appendix C.2.

3.1. (Reduced) Curves. Curves are in a sense the simplest nontrivial algebraic varieties. Theorem IV.15 below states one of their basic properties, namely, that they can be *desingularized*. In particular, the map $\pi : \tilde{C} \to C$ from theorem IV.15 is called the *desingularization* of C. It is the one-dimensional case of *resolution of singularities*, which is still an open problem for dimension greater than 3 in nonzero characteristics. Proofs of theorem IV.15 can be found in many introductory algebraic geometry texts; in particular [Ful89] gives an elementary (but long) proof, and [Kol07, Chapter 1] contains an illuminating exposition of many different proofs.

THEOREM IV.15. ([Sha94, Theorems II.5.6 and II.5.7]). *Let C be an irreducible curve. Then there is a nonsingular irreducible curve $\tilde{C}$ and a surjective morphism $\pi : \tilde{C} \to C$ such that*

(1) *for each nonsingular point $a \in C$, π restricts to an isomorphism near $\pi^{-1}(a)$;*
(2) *if $\phi : D \to C$ is any surjective morphism of curves with D nonsingular, then there is a morphism $\tilde{\phi} : D \to \tilde{C}$ such that the following diagram commutes.*

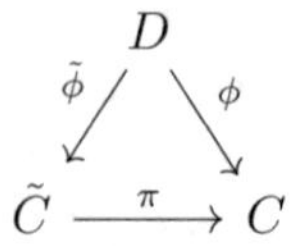

The curve $\tilde{C}$ is unique up to isomorphism. If C is projective, then so is $\tilde{C}$. Moreover, condition (2) *is automatically satisfied if $\tilde{C}$ is projective and π satisfies condition* (1).

Let a be a nonsingular point on a curve C. Identify an affine neighborhood of a in C with a curve in some $\mathbb{k}^n$ with coordinates $(x_1, \ldots, x_n)$ defined by an ideal I of $\mathbb{k}[x_1, \ldots, x_n]$. Identity (25) implies that there is $f \in \mathbb{k}[x_1, \ldots, x_n]$ such that $f(a) = 0$ and $(\nabla f)(a) := ((\partial f/\partial x_1)(a), \ldots, (\partial f/\partial x_n)(a))$ is *not* in the span of $(\nabla g_1)(a), \ldots, (\nabla g_k)(a)$, where $g_1, \ldots, g_k$ are any set of generators of I. We say that f is a *parameter* of C at a. The local ring $\mathcal{O}_{a,C}$ of C at a is a *discrete valuation ring* with parameter f (section III.13.4); we denote the discrete valuation of $\mathcal{O}_{a,C}$ by $\mathrm{ord}_a(\cdot)$, and given a rational function g on C, we say that $\mathrm{ord}_a(g)$ is the *order* of *g at a.*

Example IV.16. Let a be the origin and C be the parabola $V(y - x^2) \subset \mathbb{k}^2$. Since $\nabla(y - x^2)|_a = (0, 1)$, it follows that x is a parameter of $\mathcal{O}_{a,C}$. Since y/x^2 is invertible on C, it follows that $\mathrm{ord}_a(y|_C) = 2$.

THEOREM IV.17. ([Sha94, Corollary to Theorem III.2.1]). *Let g be a nonzero rational function on a nonsingular curve C. Then there are only finitely many points a on C such that* $\mathrm{ord}_a(g) \neq 0$. *If C is in addition projective, then*

$$\sum_{a \in C} \mathrm{ord}_a(g) = 0. \tag{35}$$

Note that identity (35) may not hold if C is not projective, consider, e.g. any non-constant polynomial on the affine line. The following result lists two basic properties of parameters.

Proposition IV.18. *Let C be a curve on $\mathbb{k}^n$ with coordinates $(x_1, \dots, x_n)$. Then*

(1) Given a nonsingular point $a = (a_1, \dots, a_n)$ of C, there is j, $1 \leq j \leq n$, such that $x_j - a_j$ is a parameter of C at a.
(2) Given a polynomial f in $(x_1, \dots, x_n)$, the property of f being a parameter at a nonsingular point of C is "Zariski open," i.e. if f is a parameter of C at some nonsingular point of C, then it is a parameter of C at each nonsingular point on a nonempty Zariski open subset of C.

PROOF. The first assertion follows immediately from the definition of parameters and identity (25). For the second assertion note that given a nonsingular point a of C, due to theorem III.105 one can assume, after replacing C by an appropriate open neighborhood of a on C, that the ideal of C in $\mathbb{k}[x_1, \dots, x_{n-1}]$ defined by $n-1$ polynomials $g_1, \dots, g_{n-1}$. If f is a parameter of C at a, identity (25) then implies that the determinant of the $n \times n$-matrix of partial derivatives of $g_1, \dots, g_{n-1}, f$ is nonzero at a; therefore it is nonzero on a nonempty Zariski open subset U of C. Then f is a parameter of C at each point of $U \cap C$, as required. □

The following is a standard result covered in most introductory books in algebraic geometry. We use it in the proof of the main result (theorem IV.24) of next section.

THEOREM IV.19. ([Sha94, Theorem II.5.8]). *Every non-constant morphism $\phi : C \to D$ between two* irreducible *projective curves is* finite, *i.e. if U is any affine open subset of D, then $\phi^{-1}(U)$ is affine and $\mathbb{k}[\phi^{-1}(U)]$ is a finite module over $\mathbb{k}[U]$.*

Consider the embedding of $\mathbb{k} \setminus \{0\} \hookrightarrow \mathbb{k}$. The coordinate ring of $\mathbb{k} \setminus \{0\}$ is $\mathbb{k}[x, x^{-1}]$, which is *not* a finite module over $\mathbb{k}[x]$. This shows that the condition that C and D are projective is crucial in theorem IV.19. Also, if C is the union of the closures of x and y axes in $\mathbb{P}^2$, then the projection map from C to the x-axis is not finite; i.e. theorem IV.19 may fail to hold if C is reducible.

3.2. Order at a point on a possibly non-reduced curve. Let a be a point on a possibly non-reduced curve C and $f \in \mathcal{O}_{a,C}$. The *order* $\mathrm{ord}_a(f)$ of f at a is the dimension of $\mathcal{O}_{a,C}/f\mathcal{O}_{a,C}$ as a vector space over $\mathbb{k}$. Note that $\mathrm{ord}_a(f) = \infty$ if f vanishes on any irreducible component of $\mathrm{Supp}(C)$ containing a.

Example IV.20. Let C be the parabola $V(y - x^2) \subset \mathbb{k}^2$ and $f = y$. Since $y \equiv x^2$ on C, it follows that $\mathcal{O}_{0,C}/y\mathcal{O}_{0,C}$ is a 2-dimensional vector space over $\mathbb{k}$ generated by 1 and x, so that $\mathrm{ord}_0(y|_C) = 2$. Note that this agrees with the computation from example IV.16. More generally, part (4) of proposition B.17 shows that when C is a nonsingular (reduced) curve, the two definitions of order agree.

Proposition IV.21. *Let a be a point on a possibly non-reduced curve C and $f \in \mathcal{O}_{a,C}$.*

(1) If f is a non zero-divisor in $\mathcal{O}_{a,C}$, then $\operatorname{ord}_a(f) < \infty$.
(2) $\operatorname{ord}_a(f) = 0$ *if and only if f is invertible in $\mathcal{O}_{a,C}$.*
(3) If f is a non zero-divisor in $\mathcal{O}_{a,C}$ and $g \in \mathcal{O}_{a,C}$, then $\operatorname{ord}_a(fg) = \operatorname{ord}_a(f) + \operatorname{ord}_a(g)$.

The proof of proposition IV.21 is given in appendix C.2. If h is a *rational function* on C, then $h = f/g$ for some $f, g \in \mathcal{O}_{a,C}$ such that g is a non zero-divisor in $\mathcal{O}_{a,C}$. We define $\operatorname{ord}_a(h)$ to be $\operatorname{ord}_a(f) - \operatorname{ord}_a(g)$. Proposition IV.21 shows that $\operatorname{ord}_a(h)$ does not depend on the choice of f or g. As example IV.20 suggests, it is straightforward to check using basic properties of discrete valuation rings that this definition of order agrees with the definition from section IV.3.1 when both are applicable, i.e. C is a nonsingular (reduced) curve.

Example IV.22. Let $C' = V(x^2 - y^2 + y^3) \subset \mathbb{k}^2$. Assume $\operatorname{char}(\mathbb{k}) \neq 2$. We saw in example III.92 that C' is singular at the origin. It is straightforward to check that $\mathcal{O}_{0,C'}/y\mathcal{O}_{0,C'}$ is a 2-dimensional vector space over $\mathbb{k}$ generated by 1 and x, so that $\operatorname{ord}_0(y|_{C'}) = 2$. Now we compute the order of $\pi^*(y)$ for a desingularization π of C'. Consider the map $\pi : \mathbb{k} \to C'$ defined as follows—given $t \in \mathbb{k} \setminus \{0\}$, the straight line $\{(tu, u) : u \in \mathbb{k}\}$ through the origin with slope $1/t$ intersects C' at a single point other than the origin—define $\pi(t)$ to be that point (see fig. 4). It is straightforward to compute that $\pi(t) = (t - t^3, 1 - t^2)$. It can be checked[5] that π is a desingularization of C'. Note that $\pi^{-1}(0)$ consists of two points $t = \pm 1$. It is straightforward to check that $\pi^*(y|_{C'}) = 1 - t^2$ is a *parameter* at each $\tilde{a} \in \pi^{-1}(0)$, so that $\operatorname{ord}_{\tilde{a}}(\pi^*(y|_{C'})) = 1$. It follows that

$$\operatorname{ord}_0(y|_{C'}) = \sum_{\tilde{a} \in \pi^{-1}(0)} \operatorname{ord}_{\tilde{a}}(\pi^*(y|_{C'}))$$

Example IV.23. Assume $\operatorname{char}(\mathbb{k}) \neq 2$. Let C be the closed subscheme of $\mathbb{k}^2$ corresponding to the ideal of $\mathbb{k}[x, y]$ generated by $(x^2 - y^2 + y^3)^2$. Note that C is *non-reduced* and $\operatorname{Supp}(C)$ is the singular curve C' from example IV.22. It is not hard to see that $\mathcal{O}_{0,C}/y\mathcal{O}_{0,C}$ is a 4-dimensional vector space over $\mathbb{k}$ generated by $1, x, x^2, x^3$, so that $\operatorname{ord}_0(y|_C) = 4$. Combining this with the observation from example IV.22 yields that

$$\operatorname{ord}_0(y|_C) = 2 \sum_{\tilde{a} \in \pi^{-1}(0)} \operatorname{ord}_{\tilde{a}}(\pi^*(y|_{C'}))$$

here the factor 2 on the right-hand side is the *multiplicity* of C' in C, which we now define.

[5]Checking condition (1) of theorem IV.15 is straightforward. To verify condition (2), note that π extends to a map from $\mathbb{P}^1$ to the closure $\bar{C}'$ of C' in $\mathbb{P}^2$, and use the last assertion of theorem IV.15.

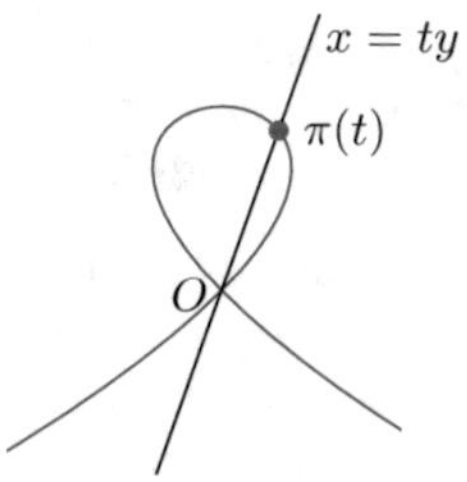

FIGURE 4. Desingularization of the nodal cubic $x^2 = y^2 - y^3$.

Let C be a closed subscheme of a variety and D be an irreducible component of $\mathrm{Supp}(C)$. Pick an embedded affine chart U of C such that $U \cap D \neq \emptyset$. Recall that if $\mathfrak{q}$ (respectively, $\mathfrak{p}$) is the ideal of $\Bbbk[U]$ defining C (respectively, D), then the local ring $\mathcal{O}_{D,C}$ of C at D can be identified with the localization $(\Bbbk[U]/\mathfrak{q})_{\mathfrak{p}}$ of $\Bbbk[U]/\mathfrak{q}$ at the ideal generated by $\mathfrak{p}$. Since D is an irreducible component of C, it follows that $\mathfrak{p}$ is a *minimal* prime ideal containing $\mathfrak{q}$. Therefore the *length* of $\mathcal{O}_{D,C}$ as a module over itself is finite (proposition B.32); we call it the *multiplicity* of D in C and denote it by $\mu_D(C)$. In example IV.23, $U = \Bbbk^2$, $\mathfrak{q}$ (respectively, $\mathfrak{p}$) is the ideal of $\Bbbk[x, y]$ generated by $(x^2 - y^2 + y^3)^2$ (respectively, $x^2 - y^2 + y^3$). It is straightforward to check directly that $\mathcal{O}_{D,C} \supsetneq \mathfrak{p}\mathcal{O}_{D,C} \supsetneq 0$ is a maximal chain of ideals of $\mathcal{O}_{D,C}$, which implies that $\mu_D(C) = 2$. The following result, whose proof is given in appendix C.2, is the key relation between orders at a point on a possibly non-reduced curve and the desingularizations of its irreducible components; it shows that our observation from example IV.23 holds in general.

THEOREM IV.24. *Let a be a point on a possibly non-reduced curve C. Let $C_1, \ldots, C_s$ be the irreducible components of* $\mathrm{Supp}(C)$ *containing a and $\pi_i : \tilde{C}_i \to C_i$ be the desingularizations of C_i. If f is a non zero-divisor in $\mathcal{O}_{a,C}$, then*

$$\operatorname{ord}_a(f) = \sum_i \mu_{C_i}(C) \operatorname{ord}_a(f|_{C_i}) = \sum_i \mu_{C_i}(C) \sum_{\tilde{a} \in \pi_i^{-1}(a)} \operatorname{ord}_{\tilde{a}}(\pi_i^*(f|_{C_i})) \tag{36}$$

Corollary IV.25. *Let C be a possibly non-reduced curve and h be an* invertible rational function *(see section IV.2.3) on C. If* $\mathrm{Supp}(C)$ *is a* projective *curve, then* $\sum_{a \in C} \operatorname{ord}_a(h) = 0$.

PROOF. Let $C_1, \ldots, C_s$ be the irreducible components of C containing a and $\pi_i : \tilde{C}_i \to C_i$ be the desingularizations of C_i. Theorem IV.24 implies that

$$\sum_{a \in C} \operatorname{ord}_a(h) = \sum_i \mu_{C_i}(C) \sum_{\tilde{a} \in C_i} \operatorname{ord}_{\tilde{a}}(\pi_i^*(h|_{C_i})).$$

Corollary B.28 implies that each i, $h|_{C_i}$ is a well-defined rational function on C_i. It then follows from theorem IV.17 that $\sum_{\tilde{a} \in C_i} \operatorname{ord}_{\tilde{a}}(\pi_i^*(h|_{C_i})) = 0$ for each i, as required. □

Corollary IV.26. *Let X be a projective variety and C be a projective (reduced) curve on $X \times \mathbb{P}^1$. Assume no component of C is contained in $X \times \{a\}$ for any $a \in \mathbb{P}^1$. Then each component of C intersects $X \times \{a\}$ for every $a \in \mathbb{P}^1$.*

PROOF. We may assume without loss of generality that C is irreducible. Fix a point (x_0, a_0), where $x_0 \in X$ and $a_0 \in \mathbb{P}^1$, on C. Choose an arbitrary point $a \neq a_0 \in \mathbb{P}^1$. We will show that C intersects $X \times \{a\}$. Pick a point $\infty \in \mathbb{P}^1 \setminus \{a, a_0\}$. Identify $\mathbb{P}^1 \setminus \{\infty\}$ with $\Bbbk$, so that we can treat a, a_0 as elements of $\Bbbk$. Let t be a coordinate on $\Bbbk$. Corollary IV.25 implies that

$$\sum_{(x,b)\in C} \operatorname{ord}_{(x,b)}((t-a)|_C) = \sum_{(x,b)\in C} \operatorname{ord}_{(x,b)}((t-a_0)|_C) = 0.$$

Since $f := (t-a)/(t-a_0)$ is regular and nonzero at all points of $X \times \{\infty\}$ [check that $f(x,\infty) = 1$ for each $x \in X$], proposition IV.21 implies that

$$\sum_{(x,b)\in C\cap(X\times\{\infty\})} \operatorname{ord}_{(x,b)}((t-a)|_C) = \sum_{(x,b)\in C\cap(X\times\{\infty\})} \operatorname{ord}_{(x,b)}((t-a_0)|_C)$$

Note that $C \setminus (X \times \{\infty\}) = C \cap (X \times \Bbbk)$, and for each $(x,b) \in C \cap (X \times \Bbbk)$,

$$\operatorname{ord}_{(x,b)}((t-a)|_C) = \begin{cases} \text{positive} & \text{if } b = a, \\ 0 & \text{otherwise.} \end{cases}$$

It follows that

$$\begin{aligned} &\sum_{(x,a)\in C\cap(X\times\{a\})} \operatorname{ord}_{(x,a)}((t-a)|_C) \\ &= \sum_{(x,a_0)\in C\cap(X\times\{a_0\})} \operatorname{ord}_{(x,a_0)}((t-a_0)|_C) \geq \operatorname{ord}_{(x_0,a_0)}((t-a_0)|_C) > 0. \end{aligned}$$

This implies that $C \cap (X \times \{a\}) \neq \emptyset$, as required. □

4. Intersection multiplicity at a nonsingular point of a variety

4.1. Intersection multiplicity of power series. The *intersection multiplicity at the origin* of $f_1, \ldots, f_n \in \Bbbk[[x_1, \ldots, x_n]]$ is

$$[f_1, \ldots, f_n]_0 := \dim_{\Bbbk}(\Bbbk[[x_1, \ldots, x_n]]/\langle f_1, \ldots, f_n \rangle). \tag{37}$$

Every power series can be approximated up to arbitrarily high order by polynomials. The following result shows that the intersection multiplicity of power series can also be approximated up to arbitrarily high order by (sufficiently close) polynomial approximations. Recall that the *order* $\operatorname{ord}(f)$ of a power series f is the smallest m for which there is there is a monomial $x_1^{\alpha_1} \cdots x_n^{\alpha_n}$ with nonzero coefficient in f such that $\sum_j \alpha_j = m$.

Proposition IV.27. *Let* $f_1, \dots, f_n \in \Bbbk[[x_1, \dots, x_n]]$.

(1) *If* $[f_1, \dots, f_n]_0 < \infty$, *then there exists* $m \geq 0$ *such that* $[g_1, \dots, g_n]_0 = [f_1, \dots, f_n]_0$ *for all* $g_1, \dots, g_n \in \Bbbk[x_1, \dots, x_n]$ *such that* $\operatorname{ord}(f_j - g_j) \geq m$.

(2) *If* $[f_1, \dots, f_n]_0 = \infty$, *then for each* $N \geq 0$, *there exists* $m \geq 0$ *such that* $[g_1, \dots, g_n]_0 \geq N$ *for all* $g_1, \dots, g_n \in \Bbbk[x_1, \dots, x_n]$ *such that* $\operatorname{ord}(f_j - g_j) \geq m$.

PROOF. It follows immediately from theorem B.56, by taking $\preceq$ e.g. to be the graded lexicographic order (see example B.50). □

4.2. Intersection multiplicity of regular functions. Let a be a nonsingular point of a variety X of dimension n. Recall that the completion $\hat{\mathcal{O}}_{a,X}$ of the local ring $\mathcal{O}_{a,X}$ of X at a is isomorphic to $\Bbbk[[x_1, \dots, x_n]]$ (theorem III.118). If $f_1, \dots, f_n$ are regular functions on a neighborhood of a in X, the *intersection multiplicity at* a of $f_1, \dots, f_n$ is

$$[f_1, \dots, f_n]_a := \dim_{\Bbbk}(\hat{\mathcal{O}}_{a,X}/\langle f_1, \dots, f_n \rangle), \tag{38}$$

where we identify each f_j with its natural image in $\hat{\mathcal{O}}_{a,X}$ (see proposition III.115). Exercise III.120 implies that

$$[f_1, \dots, f_n]_a = \dim_{\Bbbk}(\mathcal{O}_{a,X}/\langle f_1, \dots, f_n \rangle). \tag{39}$$

Now we deduce some basic properties of intersection multiplicity using results of the preceding sections. In addition, the "unmixedness theorem" of F. S. Macaulay (theorem C.3) is used in a fundamental way in all the upcoming results of this and the following section. Given a subset S of X and $b \in S$, we say that b is an *isolated* point of S if there is an open neighborhood U of b in X such that b is the only point of $S \cap U$.

Proposition IV.28. *Let* $V := V(f_1, \dots, f_n) \subseteq X$.

(1) $[f_1, \dots, f_n]_a = 0$ *if and only if* $a \notin V$.

(2) $0 < [f_1, \dots, f_n]_a < \infty$ *if and only if* a *is an isolated point of* V.

(3) *If* $0 < [f_1, \dots, f_n]_a < \infty$, *then there is a Zariski open neighborhood* U *of* a *in* X *such that* $V(f_2, \dots, f_n) \cap U$ *has pure dimension one.*

(4) *If there is a Zariski open neighborhood* U *of* a *in* X *such that* $C := V(f_2, \dots, f_n) \cap U$ *is a pure dimension one closed subscheme of* U, *then* $[f_1, \dots, f_n]_a = \operatorname{ord}_a(f_1|_C)$.

(5) $[f_1 f_1', f_2, \dots, f_n]_a = [f_1, \dots, f_n]_a + [f_1', f_2, \dots, f_n]_a$.

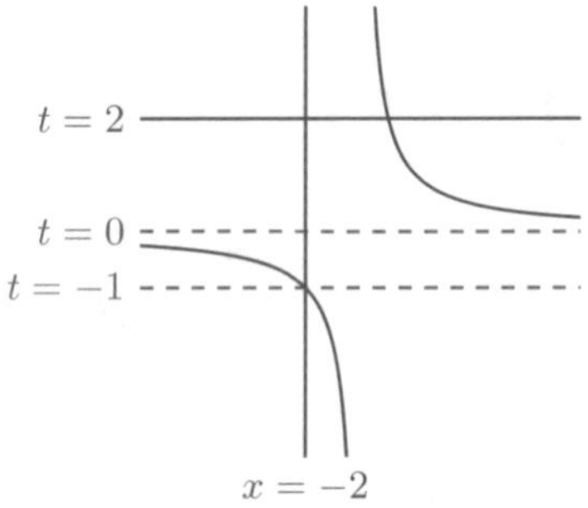

FIGURE 5. The set of zeroes of $h(x,t) = (x+2)(t-2)(xt-2)$.

PROOF. Assertion (1) is clear. Assertion (2) follows from identity (39) and the fact that a is an isolated point of V if and only if the radical of the ideal generated by $f_1, \ldots, f_n$ in $\mathcal{O}_{a,X}$ is the maximal ideal of $\mathcal{O}_{a,X}$. The third assertion follows from assertion (2) and theorem III.80. Since $\mathcal{O}_{a,X}/\langle f_1, \ldots, f_n \rangle \cong \mathcal{O}_{a,C}/f_1\mathcal{O}_{a,C}$, the fourth assertion follows from identity (39) and the definition of order. The fifth assertion is obvious in the case that either $[f_1, \ldots, f_n]_a$ or $[f_1', f_2, \ldots, f_n]_a$ is zero or infinite. Otherwise assertion (3) implies that $C := V(f_2, \ldots, f_n)$ is a possibly non-reduced curve near a, and then assertion (2) and lemma IV.29 below imply that $f_1 f_1'$ is a non zero-divisor in $\mathcal{O}_{a,C}$. Therefore assertion (4) and proposition IV.21 imply that $[f_1 f_1', f_2, \ldots, f_n]_a = \operatorname{ord}_a((f_1 f_1')|_C) = \operatorname{ord}_a(f_1|_C) + \operatorname{ord}_a(f_1'|_C) = [f_1, \ldots, f_n]_a + [f_1', f_2, \ldots, f_n]_a$, as required. □

Lemma IV.29. *Let a be a nonsingular point of a variety X of dimension n. Let $f_1, \ldots, f_m$, $m \leq n$, be regular functions on a neighborhood U of a on X such that $a \in V(f_1, \ldots, f_m)$ and $V(f_1, \ldots, f_m) \cap U$ has pure dimension $n - m$. Then for every $j = 1, \ldots, m$, f_j is a non zero-divisor in $\mathcal{O}_{a,X}/\langle f_1, \ldots, f_{j-1} \rangle$.*

PROOF. If some f_j is a zero-divisor in $\mathcal{O}_{a,X}/\langle f_1, \ldots, f_{j-1} \rangle$, then there is an open neighborhood U' of a in X such that f_j is a zero-divisor in $\mathbb{k}[U']/\langle f_1, \ldots, f_{j-1} \rangle$. Due to proposition III.38 and theorem III.105 we can choose a neighborhood U'' of a in U' such that U'' is an affine variety in $\mathbb{k}^{n+r}$ and the ideal $I(U'')$ of U'' in $\mathbb{k}[x_1, \ldots, x_{n+r}]$ is generated by m polynomials $g_1, \ldots, g_r$. Applying Macaulay's unmixedness theorem (theorem C.3) to $g_1, \ldots, g_r, f_1, \ldots, f_m$, we see that f_j is a non zero-divisor in $\mathbb{k}[U'']/\langle f_1, \ldots, f_{j-1} \rangle$, which is a contradiction. □

4.3. Intersection multiplicity in a family. Let X be a nonsingular affine variety of dimension n. Let $h_i := \sum_j \phi_{i,j}(t) g_{i,j}$, $i = 1, \ldots, n$, where $\phi_{i,j}$ are rational functions in an indeterminate t and $g_{i,j}$ are regular functions on X. Let T be the set of all $\epsilon \in \mathbb{k}$ such that each $\phi_{i,j}$ is defined at ϵ (in other words, T is the complement in $\mathbb{k}$ of the poles of $\prod_{i,j} \phi_{i,j}$). For each $\epsilon \in T$, $h_{\epsilon,i} := h_i|_{t=\epsilon}$ are regular functions on X which can be thought of "deformations" of $h_{\epsilon_0,i}$ for some fixed $\epsilon_0 \in T$. In this section we describe how, as ϵ varies in T, the multiplicities of $h_{\epsilon,i}$, $i = 1, \ldots, n$, change *locally* at a point of X, and *globally* on all of X.

Example IV.30. Let $h(x,t) := (x+2)(t-2)(xt-2)$ (see fig. 5) over a field $\Bbbk$ of characteristic different from 2 or 3, and let $b_0 := -2 \in \Bbbk$. We compute the multiplicity at $x = b_0$ of $h(x,t)$ for different values of t. If $\epsilon \notin \{-1, 2\}$, the multiplicity at $x = -2$ of $h|_{t=\epsilon}$ is the same as the multiplicity of $x + 2$, which is 1. For $\epsilon = 2$, $h|_{t=\epsilon}$ is identically zero, and therefore has infinite multiplicity everywhere. On the other hand, for $\epsilon = -1$, the multiplicity at $x = -2$ is

$$[h|_{t=-1}]_{x=-2} = [3(x+2)^2]_{x=-2} = 2[x+2]_{x=-2} = 2.$$

Note that the point $(x, \epsilon) = (-2, -1)$ is on two distinct irreducible components of the curve $h(x,t) = 0$, namely, $x = -2$ and $xt = 2$.

In example IV.30 the multiplicity of $h(x,t)$ at $x = b_0$ is the same for almost all values of t, and it jumps to higher values for $t = \epsilon$ only when either the point (b_0, ϵ) is non-isolated in $V(h(x,t) = 0)$, or (b_0, ϵ) is also on an irreducible component of $V(h(x,t) = 0)$ different from the "vertical line" $\{b_0\} \times \Bbbk$. In theorem IV.31 below we show that this is true in general, and then in theorem IV.32 we describe the "global" analogue of theorem IV.31. In the statements and proof of these results, t will be a coordinate on $\Bbbk$, and we will treat the $h_j(x,t)$ as regular functions on $X \times T$. Let $V := V(h_1, \ldots, h_n)$ be the closed subscheme of $X \times T$ determined by $h_1, \ldots, h_n$, and for each $\epsilon \in T$, let Z_ϵ be the subvariety $\operatorname{Supp}(V) \cap (X \times \{\epsilon\})$; in other words, Z_ϵ is the set of zeroes of $h_1, \ldots, h_n, t - \epsilon$.

THEOREM IV.31. *Let $b_0 \in X$, and $m : T \to \mathbb{Z} \cup \{\infty\}$ be the function given by $\epsilon \mapsto [h_{\epsilon,1}, \ldots, h_{\epsilon,n}]_{b_0}$. Let $m^* := \min\{m(\epsilon) : m \in T\}$ and $\tilde{T} := \{\epsilon \in T :$ either $(b_0, \epsilon) \notin Z_\epsilon$ or (b_0, ϵ) is isolated in $Z_\epsilon\}$. Then*

(1) $\tilde{T}$ is a Zariski open subset of T.
(2) $m^* < \infty$ if and only if $\tilde{T}$ is nonempty.
(3) $m^* = 0$ if and only if there is $\epsilon \in T$ such that $(b_0, \epsilon) \notin Z_\epsilon$.
(4) Assume $m^* < \infty$. Then $\{\epsilon \in T : m(\epsilon) = m^*\}$ is a nonempty Zariski open subset of $\tilde{T}$.
(5) Assume $m^* < \infty$. Then for each $\epsilon \in T$, $m(\epsilon) > m^*$ if and only if one of the following is true:
   (a) $\epsilon \notin \tilde{T}$, i.e. (b_0, ϵ) is a non-isolated zero of $h_{\epsilon,1}, \ldots, h_{\epsilon,n}$, or
   (b) there is an irreducible component of $V(h_1, \ldots, h_n) \subset X \times \Bbbk$ containing (b_0, ϵ) other than the "vertical line" $\{b_0\} \times \Bbbk$.

PROOF. The first assertion follows from the observation (due to theorem III.80) that (b_0, ϵ) is isolated in Z_ϵ if and only if V has pure dimension one near (b_0, ϵ). Assertions (2) and (3) are immediate consequences of the definition of intersection multiplicity. Now we prove the last two assertions. If there is $\epsilon \in T$ such that b_0 is not a common zero of $h_{\epsilon,1}, \ldots, h_{\epsilon,n}$, then $m^* = 0$, and both assertions (4) and (5) are true. So assume $0 < m^* < \infty$. Then proposition IV.28 and theorem III.80 imply that $C_0 := \{b_0\} \times T$ is an irreducible component of the subscheme V of $X \times T$. Let

$$T_0 := \{\epsilon \in T : (b_0, \epsilon) \text{ is isolated in } Z_\epsilon \text{ and } C_0 \text{ is the only component of } V \text{ containing } (b_0, \epsilon)\}.$$

Then $T_0 \subset \tilde{T}$ is a nonempty Zariski open subset of T, and there is a Zariski open neighborhood U_0 of $C_0 \cap (X \times T_0) = \{b_0\} \times T_0$ in $X \times T_0$ such that the scheme-theoretic intersection $D_0 := V \cap U_0$ has pure dimension one, and $\mathrm{Supp}(D_0) = \{b_0\} \times T_0$. Pick distinct points $\epsilon_1, \epsilon_2 \in T_0$. For each j, lemma IV.29 implies that $t - \epsilon_j$ is not a zero-divisor in $\mathcal{O}_{(b_0,\epsilon_j),D_0} = \mathcal{O}_{(b_0,\epsilon_j),V}$, which implies that $f := (t-\epsilon_1)/(t-\epsilon_2)$ is an *invertible rational function* on D_0. Embed $\mathbb{k} \hookrightarrow \mathbb{P}^1$, and let $\bar{C}_0 := \{b_0\} \times \mathbb{P}^1$ be the closure of C_0 in $X \times \mathbb{P}^1$. Theorem IV.12 implies that each of $f, t-\epsilon_1, t-\epsilon_2$ can be extended to an invertible rational function on a closed subscheme $\bar{D}_0$ of $X \times \mathbb{P}^1$ such that $\mathrm{Supp}(\bar{D}_0) = \bar{C}_0$ and D_0 is (isomorphic to) an open subscheme of $\bar{D}_0$. The following is an immediate consequence of lemma IV.13.

Claim IV.31.1. *f is invertible in $\mathcal{O}_{(b_0,\infty),\bar{D}_0}$, where ∞ is the only point of $\mathbb{P}^1 \setminus \mathbb{k}$. For each j, $t - \epsilon_j$ is invertible in $\mathcal{O}_{(b_0,\epsilon),\bar{D}_0}$ for each $\epsilon \in \mathbb{k} \setminus \{\epsilon_j\}$.* □.

Proposition IV.21 and claim IV.31.1 imply that

$$\mathrm{ord}_{(b_0,\infty)}((t-\epsilon_1)|_{\bar{D}_0}) = \mathrm{ord}_{(b_0,\infty)}((t-\epsilon_2)|_{\bar{D}_0}).$$

Theorem IV.24, corollary IV.25, proposition IV.28, and claim IV.31.1 then imply that for each $\epsilon \in T_0$,

$$\begin{aligned}[h_{\epsilon,1}, \ldots, h_{\epsilon,n}]_{b_0} &= [t-\epsilon, h_1, \ldots, h_n]_{(b_0,\epsilon)} = \mathrm{ord}_{(b_0,\epsilon)}((t-\epsilon)|_{\bar{D}_0}) \\ &= \mu_{\bar{C}_0}(\bar{D}_0)\,\mathrm{ord}_{(b_0,\epsilon)}((t-\epsilon)|_{\bar{C}_0}) = \mu_{\bar{C}_0}(\bar{D}_0).\end{aligned}$$

To complete the proof it suffices to show that $[h_{\epsilon,1}, \ldots, h_{\epsilon,n}]_{b_0} > \mu_{\bar{C}_0}(\bar{D}_0)$ whenever $\epsilon \in T \setminus T_0$. This inequality is clear if (b_0, ϵ) is non-isolated in Z_ϵ, so assume (b_0, ϵ) is an isolated point of Z_ϵ and there are irreducible components of V containing (b_0, ϵ) other than C_0; denote them by $C_1, \ldots, C_k$. Choose a Zariski open neighborhood W of (b_0, ϵ) in $X \times T$ such that $C_0 \cap W, \ldots, C_k \cap W$ are the only irreducible components of the open subscheme $V \cap W$ of V. Then

$$\begin{aligned}[h_{\epsilon,1}, \ldots, h_{\epsilon,n}]_{b_0} &= [t-\epsilon, h_1, \ldots, h_n]_{(b_0,\epsilon)} = \mathrm{ord}_{(b_0,\epsilon)}((t-\epsilon)|_{V\cap W}) \\ &= \sum_{i=0}^{k} \mu_{C_i \cap W}(V \cap W)\,\mathrm{ord}_{(b_0,\epsilon)}((t-\epsilon)|_{C_i\cap W}) > \mu_{C_0\cap W}(V \cap W).\end{aligned}$$

Since $\mathcal{O}_{C_0 \cap W, V\cap W} = \mathcal{O}_{C_0,\bar{D}_0}$ (see section IV.2.5), it follows that $\mu_{C_0\cap W}(V \cap W) = \mu_{\bar{C}_0}(\bar{D}_0)$, which completes the proof. □

Now we prove a global counterpart of theorem IV.31. To motivate the statement of this result, we compute for different values of ϵ the number (counted with multiplicities) of isolated solutions of $h|_{t=\epsilon}$, for $h(x,t) := (x+2)(t-2)(xt-2)$ from example IV.30. It is straightforward to check (see fig. 5) that if $\epsilon \not\in \{0, 2\}$, then this number is 2; indeed, if $\epsilon \not\in \{-1, 0, 2\}$, then $h|_{t=\epsilon}$ has two solutions of multiplicity one: $(-2, \epsilon), (2/\epsilon, \epsilon)$, and if $\epsilon = -1$, then $h|_{t=\epsilon}$ has one solution of multiplicity two: $(-2, -1)$. On the other hand, if $\epsilon = 2$, the polynomial $h|_{t=\epsilon}$ is identically zero on $\mathbb{k}$, and therefore has *zero* isolated solutions. Finally, for $\epsilon = 0$ there is only one point of multiplicity one on $h|_{t=\epsilon} = 0$, namely, the point $(-2, \epsilon)$; the other solution

$(2/\epsilon, \epsilon)$ "goes to infinity" at $t = 0$. In particular, the total number of isolated solutions of $h|_{t=\epsilon}$ is equal for almost all values of ϵ, and can only drop in exceptional cases when some of the solutions become non-isolated or run to infinity. Theorem IV.32 below states that this is also the case in general; in particular, "minimum" in the local case (i.e. theorem IV.31) becomes "maximum" in the global case (i.e. theorem IV.32). The notation (in particular the meaning of V and Z_ϵ) below remains unchanged from theorem IV.31.

THEOREM IV.32. *Let $C \subset X \times T$ be the union of all irreducible components of V containing at least one isolated point of Z_ϵ for some $\epsilon \in T$.*

(1) Either C is empty, or it has pure dimension one.
(2) $C_\epsilon := C \cap (X \times \{\epsilon\})$ is finite for every $\epsilon \in T$.

Now assume C is not empty.

(3) Let T^* be the set of all $\epsilon \in T$ such that all points on C_ϵ are isolated in Z_ϵ. Then T^* is a nonempty Zariski open subset of $\mathbb{k}$.
(4) For each $\epsilon \in T$, let $\tilde{C}_\epsilon := \{(b, \epsilon) \in X \times T : (b, \epsilon)$ is isolated in $Z_\epsilon\}$. The function $M : T \to \mathbb{Z}$ given by $\epsilon \mapsto \sum_{(b,\epsilon)\in\tilde{C}_\epsilon} [h_{\epsilon,1}, \dots, h_{\epsilon,n}]_b$ achieves the maximum on a nonempty Zariski open subset of T^*.
(5) If $\epsilon \in T$, then $M(\epsilon)$ fails to attain the maximum if and only if at least one of the following is true:
   (a) $\epsilon \notin T^*$, i.e. there is a point on C_ϵ which is a non-isolated zero of $h_{\epsilon,1}, \dots, h_{\epsilon,n}$, or
   (b) C "has a point at infinity at $t = \epsilon$", i.e. if $\bar{X}$ is a projective compactification of X and $\bar{C}$ is the closure of C in $\bar{X} \times \mathbb{P}^1$, then $\bar{C} \cap ((\bar{X} \setminus X) \times \{\epsilon\}) \neq \emptyset$.

PROOF. If (b, ϵ) is an isolated point of Z_ϵ, which is defined by $n + 1$ regular functions $h_1, \dots, h_n, t - \epsilon$ on a variety of dimension $n + 1$, theorem III.80 implies that $V = V(h_1, \dots, h_n)$ has pure dimension one near (b, ϵ), which proves assertion (1). If assertion (2) does not hold, then assertion (1) implies that there is $\epsilon \in T$ such that $X \times \{\epsilon\}$ contains an irreducible component of C. But then no point on this component is isolated in Z_ϵ, contradicting the definition of C. This proves assertion (2). For assertion (3), let Y be the union of the irreducible components of V not contained in C. Since $\dim(C) = 1$, it follows that $C \cap Y$ is a finite set. If $\{(b'_j, \epsilon'_j)\}_j$ are the points in this intersection, then note that $T^* = T \setminus \{\epsilon'_j\}_j$. It remains to prove the last two assertions. Let

$$\tilde{C} := \bigcup_\epsilon \tilde{C}_\epsilon = \{(b, \epsilon) \in C : (b, \epsilon) \text{ is isolated in } Z_\epsilon\}.$$

Then $\tilde{C}$ is a Zariski open subset of C, i.e. there is a Zariski open subset U of $X \times T$ such that $\tilde{C} = U \cap C$. Let $\tilde{D} := V \cap U$ be the corresponding open subscheme of V. Now choose a projective compactification $\bar{X}$ of X. Let $\bar{C}$ be the closure of C in $\bar{X} \times \mathbb{P}^1$. For every $\epsilon \in \mathbb{k}$ and every $(b, \epsilon) \in \tilde{C}$, lemma IV.29 implies that $t - \epsilon$ is a non zero-divisor in $\mathcal{O}_{(b,\epsilon),\tilde{D}} = \mathcal{O}_{(b,\epsilon),V}$. If ϵ_1, ϵ_2 are distinct elements of $\mathbb{k}$, then theorem IV.12 implies that there is a closed subscheme $\bar{D}$ of $\bar{X} \times \mathbb{P}^1$ containing $\tilde{D}$ as an open subscheme such that $\operatorname{Supp}(\bar{D}) = \bar{C}$, and each of $t - \epsilon_1, t - \epsilon_2, f := (t - \epsilon_1)/(t - \epsilon_2)$ extends to an *invertible* rational function on $\bar{D}$. The following is an immediate consequence of lemma IV.13.

Claim IV.32.1. *Denote the only point of $\mathbb{P}^1 \setminus \mathbb{k}$ by ∞. For each $b \in \bar{X}$ such that $(b, \infty) \in \bar{C}$, f is invertible in $\mathcal{O}_{(b,\infty),\bar{D}}$. For each j, $t - \epsilon_j$ is invertible in $\mathcal{O}_{(b,\epsilon),\bar{D}}$ for each $(b, \epsilon) \in \bar{C}$ such that $\epsilon \in \mathbb{k} \setminus \{\epsilon_j\}$.* □

Proposition IV.21 and claim IV.32.1 imply that for each $\epsilon_1, \epsilon_2 \in \mathbb{k}$,

$$\sum_{b:(b,\infty)\in\bar{C}} \operatorname{ord}_{(b,\infty)}((t - \epsilon_1)|_{\bar{D}}) = \sum_{b:(b,\infty)\in\bar{C}} \operatorname{ord}_{(b,\infty)}((t - \epsilon_2)|_{\bar{D}}).$$

Theorem IV.24, corollary IV.25, proposition IV.28, and claim IV.32.1 then imply that for each $\epsilon \in \mathbb{k}$,

$$M^* := \sum_{b:(b,\epsilon)\in\bar{C}} \operatorname{ord}_{(b,\epsilon)}((t - \epsilon)|_{\bar{D}}) = \sum_{b:(b,\epsilon)\in\bar{C}} \sum_i \mu_{C_i}(\bar{D}) \operatorname{ord}_{(b,\epsilon)}((t - \epsilon)|_{C_i})$$

is constant, where the C_i are irreducible components of C. On the other hand, theorem IV.24 and proposition IV.28 imply that for all $\epsilon \in T$,

$$\begin{aligned} M(\epsilon) &= \sum_{b:(b,\epsilon)\in\tilde{C}_\epsilon} [h_{\epsilon,1}, \ldots, h_{\epsilon,n}]_b = \sum_{b:(b,\epsilon)\in\tilde{C}_\epsilon} [t - \epsilon, h_1, \ldots, h_n]_{(b,\epsilon)} \\ &= \sum_{b:(b,\epsilon)\in\tilde{C}_\epsilon} \operatorname{ord}_{(b,\epsilon)}((t - \epsilon)|_D) = \sum_{b:(b,\epsilon)\in\tilde{C}_\epsilon} \sum_i \mu_{C_i}(\bar{D}) \operatorname{ord}_{(b,\epsilon)}((t - \epsilon)|_{C_i}). \end{aligned}$$

Since $t - \epsilon$ is regular and has a zero at each point of $\bar{C} \cap (X \times \{\epsilon\})$, it follows that $M^* \geq M(\epsilon)$; moreover, $M^* > M(\epsilon)$ if and only if $\tilde{C}_\epsilon \subsetneqq \bar{C} \cap (\bar{X} \times \{\epsilon\})$. By construction of $\bar{C}$, the latter condition is true if and only if at least one of the conditions of assertion (5) holds. Since these conditions hold at most finitely many points of C, assertion (4) also holds. □

5. Intersection multiplicity of complete intersections

Let $f_1, \ldots, f_k$ be regular functions on a nonsingular variety X of dimension $n \geq k$ such that $V(f_1, \ldots, f_k)$ has an irreducible component Z of dimension $n - k$. Then there is a Zariski open subset U of X such that $Z \cap U = V(f_1, \ldots, f_k) \cap U$, and $Z \cap U$ has pure dimension $n - k$ (in particular, $Z \cap U$ is nonempty). Let V be the *closed subscheme* of U defined by (the ideal generated by) $f_1, \ldots, f_k$. Then $Z \cap U = \operatorname{Supp}(V)$, so that we can define the *multiplicity* $\mu_{Z\cap U}(V)$ of $Z \cap U$ in V as in section IV.3.2 (in the paragraph preceding theorem IV.24). It is straightforward to check that $\mu_{Z\cap U}(V)$ does not depend on U; we say that $\mu_{Z\cap U}(V)$ is the *intersection multiplicity* $[f_1, \ldots, f_k]_Z$ *of* $f_1, \ldots, f_k$ *along* Z. If $k = n$, then Z is a singleton $\{a\}$, and $\mathcal{O}_{Z,V} \cong \mathcal{O}_{a,X}/\langle f_1, \ldots, f_n \rangle$, and corollary B.31 implies that $\mu_Z(V) =$

$\dim_{\mathbb{k}}(\mathcal{O}_{a,X}/\langle f_1,\dots,f_n\rangle)$, so that the definition of $[f_1,\dots,f_n]_Z$ from this section agrees with the definition from the preceding section.

Proposition IV.33. *Let $f_1,\dots,f_{n-1} \in \mathbb{k}[x_1,\dots,x_n]$. Let Y be the coordinate subspace $x_1 = \cdots = x_k = 0$ of $\mathbb{k}^n$. Assume*

(1) Y is an irreducible component of $V(f_1,\dots,f_k) \subset \mathbb{k}^n$.
(2) $V(f_{k+1}|_Y,\dots,f_{n-1}|_Y)$ has a one-dimensional irreducible component Z.
(3) Z is not contained in any irreducible component of $V(f_1,\dots,f_k)$ other than Y.

For each $\epsilon = (0,\dots,0,\epsilon_{k+1},\dots,\epsilon_n) \in Y$, and each $j = 1,\dots,k$, we write $f_{j,\epsilon}$ for the polynomial in $(x_1,\dots,x_k)$ obtained by substituting ϵ_i for x_i for $i = k+1,\dots,n$. Then

$$[f_1,\dots,f_{n-1}]_Z = [f_{1,\epsilon},\dots,f_{k,\epsilon}]_0[f_{k+1}|_Y,\dots,f_{n-1}|_Y]_Z$$

for generic $\epsilon \in Y$.

PROOF. We prove this by induction on $n-k$. At first consider the case that $n-k=1$. Then Z is the x_n-axis, and for generic $\epsilon \in \mathbb{k}$, Z is the only irreducible component of $V(f_1,\dots,f_{n-1})$ containing $a_\epsilon := (0,\dots,0,\epsilon)$. Since $\operatorname{ord}_{a_\epsilon}((x_n - \epsilon)|_Z) = 1$, theorem IV.24 implies that

$$\begin{aligned}
[f_1,\dots,f_{n-1}]_Z &= [x_n-\epsilon, f_1,\dots,f_{n-1}]_{a_\epsilon} \\
&= \dim_{\mathbb{k}}(\mathbb{k}[[x_1,\dots,x_{n-1},x_n-\epsilon]]/\langle x_n-\epsilon, f_1,\dots,f_{n-1}\rangle) \\
&= \dim_{\mathbb{k}}(\mathbb{k}[[x_1,\dots,x_{n-1}]]/\langle f_1|_{x_n=\epsilon},\dots,f_{n-1}|_{x_n=\epsilon}\rangle) \\
&= [f_{1,\epsilon},\dots,f_{n-1,\epsilon}]_0
\end{aligned}$$

as required. In the general case, pick a nonsingular point $z = (0,\dots,0,z_{k+1},\dots,z_n)$ of Z. Then there is j, $k+1 \le j \le n$, such that $(x_j - z_j)|_Z$ has order one at z (proposition IV.18, assertion (1)). Pick a generic $\epsilon_j \in \mathbb{k}$. Assertion (2) of proposition IV.18 implies that the set $V(x_j - \epsilon_j) \cap Z$ is nonempty and contains a nonsingular point a of Z such that a is *not* in any other irreducible component of $V(f_1,\dots,f_{n-1})$, and $\operatorname{ord}_a((x_j-\epsilon_j)|_Z) = 1$. Since a is an isolated zero of $x_j - \epsilon_j, f_1,\dots,f_{n-1}$, theorem III.80 implies that $x_j-\epsilon_j, f_1,\dots,f_{n-2}$ defines a possibly non-reduced curve W near a. Let $W_1,\dots,W_s$ be the irreducible components of W and $\pi_i : \tilde{W}_i \to W_i$ be the desingularization. Theorem IV.24 and proposition IV.28 imply that

$$\begin{aligned}
[f_1,\dots,f_{n-1}]_Z &= [x_j-\epsilon_j, f_1,\dots,f_{n-1}]_a = \operatorname{ord}_a(f_{n-1}|_W) \\
&= \sum_i \mu_{W_i}(W) \sum_{\tilde{a}\in\pi_i^{-1}(a)} \operatorname{ord}_{\tilde{a}}(\pi_i^*(f_{n-1}|_{W_i})) \\
&= \sum_i [f_1|_{x_j=\epsilon_j},\dots,f_{n-2}|_{x_j=\epsilon_j}]_{W_i} \sum_{\tilde{a}\in\pi_i^{-1}(a)} \operatorname{ord}_{\tilde{a}}(\pi_i^*(f_{n-1}|_{W_i})).
\end{aligned}$$

Let $Y_{\epsilon_j} := Y \cap V(x_j - \epsilon_j)$. Then the inductive hypothesis implies that

$$[f_1, \ldots, f_{n-1}]_Z = [f_{1,\epsilon}, \ldots, f_{k,\epsilon}]_0 \sum_i [f_{k+1}|_{Y_{\epsilon_j}}, \ldots, f_{n-2}|_{Y_{\epsilon_j}}]_{W_i} \sum_{\tilde{a} \in \pi_i^{-1}(a)} \operatorname{ord}_{\tilde{a}}(\pi_i^*(f_{n-1}|_{W_i}))$$
$$= [f_{1,\epsilon}, \ldots, f_{k,\epsilon}]_0 \operatorname{ord}_a(f_{n-1}|_V),$$

where V is the closed subscheme $W \cap Y$ of Y. It follows that

$$\begin{aligned}[f_1, \ldots, f_{n-1}]_Z &= [f_{1,\epsilon}, \ldots, f_{k,\epsilon}]_0 [x_j - \epsilon_j, f_{k+1}|_Y, \ldots, f_{n-1}|_Y]_a \\ &= [f_{1,\epsilon}, \ldots, f_{k,\epsilon}]_0 [f_{k+1}|_Y, \ldots, f_{n-1}|_Y]_Z\end{aligned}$$

as required. □

Corollary IV.34. *Let $a \in Y$ be such that $V := V(f_1, \ldots, f_{n-1})$ is purely one dimensional near a and no irreducible component of V containing a is contained in any irreducible component of $V(f_1, \ldots, f_k)$ other than Y. Then for all $f_n \in \mathbb{k}[x_1, \ldots, x_n]$,*

$$[f_1, \ldots, f_n]_a = [f_{1,\epsilon}, \ldots, f_{k,\epsilon}]_0 [f_{k+1}|_Y, \ldots, f_n|_Y]_a$$

for generic $\epsilon \in Y$.

PROOF. Follows from propositions IV.28 and IV.33. □

CHAPTER V

Convex polyhedra

A "polytope" has two equivalent definitions: a convex hull of finitely many points or a bounded intersection of finitely many "half-spaces." In sections V.1 and V.2 we prove the equivalence of these definitions after introducing the basic terminology. The rest of the chapter is devoted to different properties of polytopes which are implicitly or explicitly used in the forthcoming chapters.

1. Basic notions

In this chapter we treat the spaces $\mathbb{R}^n$, $n \geq 0$, as vector spaces over $\mathbb{R}$ equipped with the Euclidean topology, and deal only with "affine maps" between these spaces. Recall that a map $\phi : \mathbb{R}^n \to \mathbb{R}^m$ is *affine* if there is $\beta \in \mathbb{R}^m$ such that $\phi(\cdot) = \beta + \phi_0(\cdot)$ for some *linear* map $\phi_0 : \mathbb{R}^n \to \mathbb{R}^m$. Given $S \subset \mathbb{R}^n$ and $T \subset \mathbb{R}^m$, and a map $\phi : S \to T$, we say that ϕ is *affine* if it is the restriction of an affine map from $\mathbb{R}^n \to \mathbb{R}^m$; we say that ϕ is an *affine isomorphism* if ϕ is affine and bijective. An *affine subspace* A of $\mathbb{R}^n$ is a subset of $\mathbb{R}^n$ which is the image of an affine map; in other words, it is simply a translation of a linear subspace L of $\mathbb{R}^n$. The *dimension* $\dim(A)$ of A is the dimension of L as a vector space over $\mathbb{R}$. A *hyperplane* in $\mathbb{R}^n$ is an affine subspace of dimension $n - 1$. The *affine hull* $\operatorname{aff}(S)$ of a set S of $\mathbb{R}^n$ is the smallest affine subspace of $\mathbb{R}^n$ containing S; alternatively, if L is the linear subspace of $\mathbb{R}^n$ spanned by all elements of the form $\alpha - \beta$ such that $\alpha, \beta \in S$, then $\operatorname{aff}(S) = L + \alpha$ for any $\alpha \in S$. In fig. 1, it is straightforward to check that $\operatorname{aff}(S) = \mathbb{R}^2$, $\operatorname{aff}(\{C\}) = \{C\}$, and since A, B, E are collinear, $\operatorname{aff}(\{A, B, E\})$ is the (unique) line L through these points.

P. Mondal, *How Many Zeroes?*, CMS/CAIMS Books in Mathematics 2,
https://doi.org/10.1007/978-3-030-75174-6_V

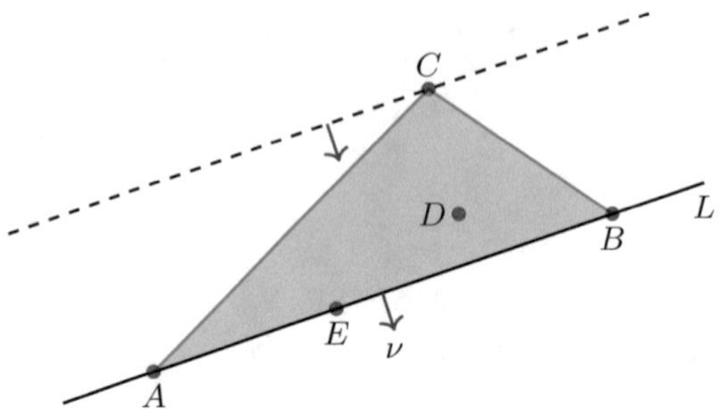

FIGURE 1. $S = \{A, B, C, D, E\}$, $\mathrm{In}_\nu(S) = \{C\}$, $\mathrm{ld}_\nu(S) = \{A, E, B\}$.

It is straightforward to check that an affine map preserves affine subspaces, affine hulls, and if the map is injective, then also the dimension of affine hulls (exercise V.1). Given $\alpha \in \mathbb{R}^n$ and $\nu \in (\mathbb{R}^n)^*$, we write $\langle \nu, \alpha \rangle$ for the "value of ν at α," and write $\nu^\perp := \{\alpha \in \mathbb{R}^n : \langle \nu, \alpha \rangle = 0\}$. Define

$$\min_S(\nu) := \min\{\langle \nu, \alpha \rangle : \alpha \in S\}, \text{ provided the minimum exists.}$$
$$\max_S(\nu) := \max\{\langle \nu, \alpha \rangle : \alpha \in S\}, \text{ provided the maximum exists.}$$
$$\mathrm{In}_\nu(S) := \{\alpha \in S : \langle \nu, \alpha \rangle = \min_S(\nu)\}, \text{ provided } \min_S(\nu) \text{ exists.}$$
$$\mathrm{ld}_\nu(S) := \{\alpha \in S : \langle \nu, \alpha \rangle = \max_S(\nu)\}, \text{ provided } \max_S(\nu) \text{ exists.}$$

See fig. 1 for an illustration of these notions for a planar set. Note that in fig. 1 we depicted $\nu \in (\mathbb{R}^2)^*$ on $\mathbb{R}^2$ by identifying it with an element (modulo "parallel translations") in $\mathbb{R}^2$ via the "dot product." A set is *convex* if it contains the line segment joining any two points in it. The *convex hull* $\mathrm{conv}(S)$ of S is the smallest convex set containing S. In fig. 1 the convex hull of the 5 points is the green triangle. Given $\alpha_1, \ldots, \alpha_k \in \mathbb{R}^n$, an expression of the form $\sum_{j=1}^k \epsilon_j \alpha_j$, where the ϵ_j are nonnegative numbers whose sum is 1, is called a *convex combination* of the α_j. The set of convex combinations of two elements in $\mathbb{R}^n$ is precisely the line segment joining them, and from this observation it can be shown that the convex hull of a set consists of the convex combinations of its points (exercise V.2), i.e.

$$\mathrm{conv}(S) = \{\sum_{j=1}^k \epsilon_j \alpha_j : k \geq 0,\ \alpha_j \in S,\ \epsilon_j \geq 0 \text{ for each } j,\ 1 \leq j \leq k, \text{ and } \sum_{j=1}^k \epsilon_j = 1\}. \tag{40}$$

Given a nonnegative real number r and $S \subset \mathbb{R}^n$, we define $rS := \{r\alpha : \alpha \in S\}$, i.e. rS is the "dilation of S by a factor of r." We say that S is a *cone* if it is "dilation invariant," i.e. if $rS \subseteq S$ for each $r \geq 0$. The *convex cone generated by* S is defined to be

$$\mathrm{cone}(S) := \{\sum_{j=1}^k \epsilon_j \alpha_j : k \geq 0,\ \alpha_j \in S \text{ and } \epsilon_j \geq 0 \text{ for each } j,\ 1 \leq j \leq k\} \tag{41}$$

We say that $\operatorname{cone}(S)$ is *finitely generated* if S is finite. A *convex polyhedron* $\mathcal{P}$ is a subset of $\mathbb{R}^n$ defined by finitely many linear inequalities, i.e. inequalities of the form $a_0 + a_1x_1 + \cdots + a_nx_n \geq 0$. Geometrically, a polyhedron is a finite intersection of "half-spaces," where a "half-space" is the set of all points on one side of a hyperplane, see fig. 2. If $\mathcal{P}$ is bounded, we call it a convex *polytope*, and if it is a cone, we call it a convex *polyhedral cone*. A convex polyhedral cone is equivalently a set defined by finitely many linear inequalities with *zero* constant term (exercise V.6). In this book we only consider convex polyhedra, and therefore we will simply write "polyhedra," "polytopes," "cones," "polyhedral cones" to mean "convex polyhedra," "convex polytopes," "convex cones," "convex polyhedral cones," respectively. The *dimension* $\dim(\mathcal{P})$ of a polyhedron $\mathcal{P}$ is the dimension of its affine hull. Figure 2 depicts a few two-dimensional convex polyhedra.

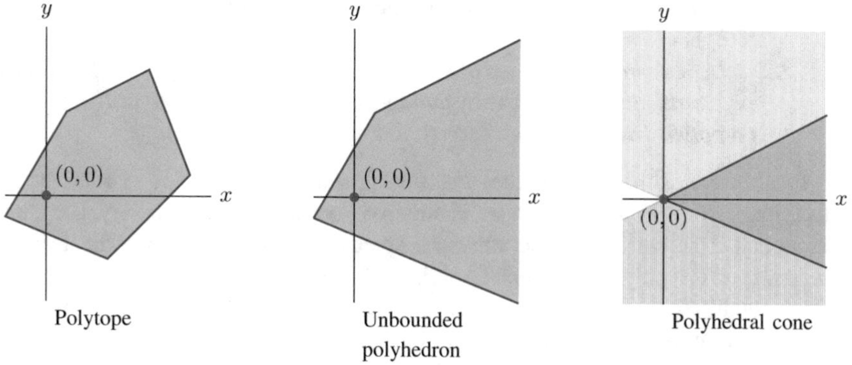

FIGURE 2. Some planar convex polyhedra; "half-planes" defining the cone are also depicted.

A *strongly convex cone* $\mathcal{C}$ is a convex cone which does not contain any line through the origin, equivalently, for all $\alpha \in \mathcal{C} \setminus \{0\}$, $-\alpha \notin \mathcal{C}$. The *Minkowski sum* of two subsets $\mathcal{P}, \mathcal{Q}$ of $\mathbb{R}^n$ is $\mathcal{P} + \mathcal{Q} := \{\alpha + \beta : \alpha \in \mathcal{P},\ \beta \in \mathcal{Q}\}$. It is straightforward to check that the Minkowski sum of convex sets is also convex (exercise V.8); see fig. 3 for some examples in $\mathbb{R}^2$. We now show that every convex polyhedral cone or finitely generated cone[1] can be represented as the Minkowski sum of a linear subspace and a strongly convex cone.

Proposition V.1. *Let $\mathcal{C}$ be a convex cone in $\mathbb{R}^n$.*

1. *If $\mathcal{C}$ is a polyhedral cone, then there is a strongly convex polyhedral cone $\mathcal{C}'$ and a linear subspace L of $\mathbb{R}^n$ such that $\mathcal{C} = \mathcal{C}' + L$ and* $\operatorname{aff}(\mathcal{C}') \cap L = \{0\}$.

[1]We will see in section V.2 that convex polyhedral cones and finitely generated cones are the same. However, we will use proposition V.1 in the proof of this equivalence.

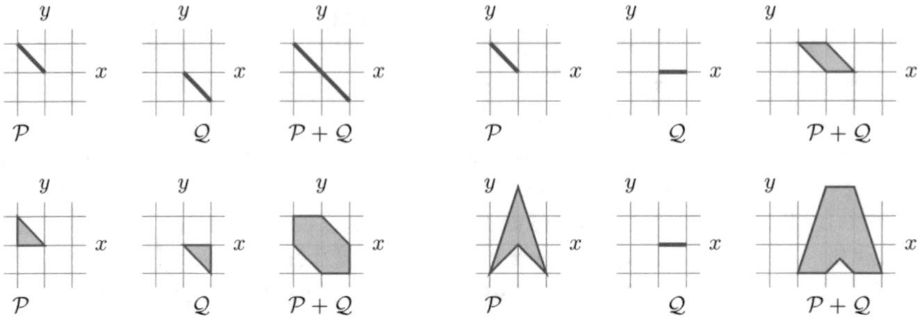

FIGURE 3. Minkowski sums of planar sets.

2. *If $\mathcal{C}$ is a finitely generated cone, then there is a strongly convex finitely generated cone $\mathcal{C}'$ and a linear subspace L of $\mathbb{R}^n$ such that $\mathcal{C} = \mathcal{C}' + L$ and* $\operatorname{aff}(\mathcal{C}') \cap L = \{0\}$.
3. *Let L' be a linear subspace of $\mathbb{R}^n$ such that $L' \cap \operatorname{aff}(C) = \{0\}$. Then $\mathcal{C}$ is a polyhedral cone (respectively, finitely generated cone) if and only if $\mathcal{C} + L'$ is a polyhedral cone (respectively, finitely generated cone).*

PROOF. Let L be the (unique) maximal linear subspace of $\mathbb{R}^n$ contained in $\mathcal{C}$ (see exercise V.9). After a linear change of coordinates if necessary, we may assume that L is the coordinate subspace spanned by the first k coordinates. At first assume $\mathcal{C}$ is a polyhedral cone. Exercise V.6 implies that it is defined by finitely many inequalities of the form $a_{i,1}x_1 + \cdots + a_{i,n}x_n \geq 0$, $1 \leq i \leq m$. For each $j = 1, \ldots, n$, write e_j for the j-th standard unit vector in $\mathbb{R}^n$. Since $re_j \in L \subseteq \mathcal{C}$ for each $j = 1, \ldots, k$, and each $r \in \mathbb{R}$, it follows that $a_{i,1} = \cdots = a_{i,k} = 0$ for each i. Let $\mathcal{C}'$ be the polyhedral cone on the $(n-k)$-dimensional coordinate subspace of $\mathbb{R}^n$ spanned by $e_{k+1}, \ldots, e_n$ defined by the inequalities $a_{i,k+1}x_{k+1} + \cdots + a_{i,n}x_n = 0$. Then assertion (1) holds with $\mathcal{C}'$ and L. Now assume $\mathcal{C}$ is the cone generated by finitely many elements $\alpha_1, \ldots, \alpha_m \in \mathbb{R}^n$. Let $\alpha_i := (\alpha_{i,1}, \ldots, \alpha_{i,n})$. Since $\tilde{\alpha}_i := (-\alpha_{i,1}, \ldots, -\alpha_{i,k}, 0, \ldots, 0) \in L \subseteq \mathcal{C}$, it follows that $\alpha'_i := \alpha_i + \tilde{\alpha}_i = (0, \ldots, 0, \alpha_{i,k+1}, \ldots, \alpha_{i,n}) \in \mathcal{C}$ for each i. Then assertion (2) holds with L and $\mathcal{C}' := \operatorname{cone}(\alpha'_1, \ldots, \alpha'_m)$. The proof of assertion (3) follows via the same arguments as in the proof of the preceding assertions by choosing a system of coordinates on $\mathbb{R}^n$ such that L' is the coordinate subspace spanned by the first k' coordinates, where $k' := \dim(L')$. □

Remark V.2. If $\mathcal{A} \subset \mathbb{R}^n$ and d is a positive integer, then $d\mathcal{A}$ has two natural interpretations: one is the dilation (which is how we defined it), and the other is the Minkowski sum of d copies of $\mathcal{A}$. If $\mathcal{A}$ is convex, then these are equivalent (exercise V.10).

1.1. Exercises.

EXERCISE V.1. Let $\phi : \mathbb{R}^n \to \mathbb{R}^m$ be an affine map and let $S \subseteq \mathbb{R}^n$. Show that

(1) If A is an affine subspace of $\mathbb{R}^n$, then $\phi(A)$ is an affine subspace of $\mathbb{R}^m$, and $\dim(\phi(A)) \leq \dim(A)$.
(2) $\phi(\text{aff}(S)) = \text{aff}(\phi(S))$ and $\phi(\text{conv}(S)) = \text{conv}(\phi(S))$.
(3) Assume ϕ is in addition a *linear map*, i.e. $\phi(0) = 0$. Then $\phi(\text{cone}(S)) = \text{cone}(\phi(S))$.

Now assume ϕ is in addition injective. Then show that

(4) The inequality in assertion (1) holds with equality.
(5) The converses of assertions (1), (2) and (3) are also true, i.e.
   (a) if $A \subseteq \mathbb{R}^n$ is such that $\phi(A)$ is an affine subspace of $\mathbb{R}^m$, then A is also an affine subspace of $\mathbb{R}^n$;
   (b) $\phi^{-1}(\text{aff}(S)) = \text{aff}(S)$ and $\phi^{-1}(\text{conv}(\phi(S)) = \text{conv}(S)$;
   (c) if ϕ is a linear map, then $\phi^{-1}(\text{cone}(\phi(S)) = \text{cone}(S)$.

EXERCISE V.2. Show that the line segment joining $\alpha, \beta \in \mathbb{R}^n$ is precisely the set of their convex combinations. Use it to prove identity (40). If $\nu \in (\mathbb{R}^n)^*$ and $S \subset \mathbb{R}^n$ are such that $\min_S(\nu)$ exists, then use identity (40) to show that $\min_{\text{conv}(S)}(\nu) = \min_S(\nu)$.

EXERCISE V.3. Check that a convex polyhedron as defined in section V.1 is actually convex.

EXERCISE V.4. Let $S \subseteq \mathbb{R}^n$.

(1) Show that $\text{aff}(S)$ is zero dimensional if and only if S is a "singleton" (i.e. it consists of a single point). Deduce that zero-dimensional polyhedra are precisely singletons.
(2) Assume S is convex and closed in $\mathbb{R}^n$, and $\text{aff}(S)$ is one dimensional. Show that there is an affine isomorphism $\phi : \mathbb{R} \to \text{aff}(S)$ such that $\phi^{-1}(S)$ is a (possibly unbounded) closed interval of $\mathbb{R}$.

EXERCISE V.5. Show that every finitely generated two-dimensional cone S in $\mathbb{R}^2$ can be generated as a cone by two nonzero elements of $\mathbb{R}^2$ (these correspond to "edges" of S). If S is in addition strongly convex, show that the generators of S are linearly independent.

EXERCISE V.6. Let $\mathcal{P} \subseteq \mathbb{R}^n$.

(1) Assume $\mathcal{P}$ is "dilation invariant," i.e. for every $\alpha \in \mathcal{P}$ and every $r \geq 0$, $r\alpha \in \mathcal{P}$. Show that for every $\nu \in (\mathbb{R}^n)^*$, $\inf\{\langle \nu, \alpha \rangle : \alpha \in \mathcal{P}\}$ is either 0 or $-\infty$.
(2) Deduce that $\mathcal{P}$ is a convex polyhedral cone if and only if it can be defined by finitely many inequalities with zero constant term, i.e. inequalities of the form $a_1x_1 + \cdots + a_nx_n \geq 0$.

EXERCISE V.7. If $S, T \subseteq \mathbb{R}^m$ and $\phi : \mathbb{R}^m \to \mathbb{R}^n$ is a linear map, show that $\phi(S + T) = \phi(S) + \phi(T)$.

EXERCISE V.8. Let $\mathcal{P}$, $\mathcal{Q}$ be subsets of $\mathbb{R}^n$.

(1) If $\mathcal{P} = \text{conv}(S)$ and $\mathcal{Q} = \text{conv}(T)$, then show that $\mathcal{P} + \mathcal{Q} = \text{conv}(S + T)$. [Hint: if $\sum_i \delta_i = \sum_j \epsilon_j = 1$, then $\sum_i \delta_i a_i + \sum_j \epsilon_j b_j = \sum_{i,j} \delta_i \epsilon_j (a_i + b_j)$.]

(2) Deduce that if $\mathcal{P}$ and $\mathcal{Q}$ are convex (respectively, convex hulls of finitely many points, finitely generated cones) then so is $\mathcal{P} + \mathcal{Q}$.

EXERCISE V.9. Let $\mathcal{C}$ be a convex subset of $\mathbb{R}^n$ containing the origin. Show that $\mathcal{C}$ contains a unique maximal linear subspace of $\mathbb{R}^n$, i.e. there is a linear subspace L of $\mathbb{R}^n$ such that $L \subseteq \mathcal{C}$, and if L' is any linear subspace of $\mathbb{R}^n$ contained in $\mathcal{C}$, then $L' \subseteq L$. [Hint: if L_1, L_2 are two linear subspaces of $\mathbb{R}^n$ contained in $\mathcal{C}$, then $L_1 + L_2 \subseteq \mathcal{C}$.]

EXERCISE V.10. Let $\mathcal{A} \subset \mathbb{R}^n$ and d be a positive integer.

(1) If $\mathcal{A}$ is convex, then show that $\{d\alpha : \alpha \in \mathcal{A}\} = \{\alpha_1 + \cdots + \alpha_d : \alpha_j \in \mathcal{A}$ for each $j\}$.
(2) Give an example to show that the above equality may not hold if $\mathcal{A}$ is not convex.

EXERCISE V.11. Let $S \subset \mathbb{R}^m$ and $n := \dim(\text{aff}(S))$.

(1) Show that there is $T \subseteq \mathbb{R}^n$ and an injective affine map $\phi : \mathbb{R}^n \to \mathbb{R}^m$ such that $\text{aff}(T)$ is $\mathbb{R}^n$, and $\phi(T) = S$. [Hint: $\text{aff}(S)$ is a translation of an n-dimensional linear subspace of $\mathbb{R}^m$.]
(2) If S contains the origin (this is the case when, e.g. S is a cone) then show that it is possible to ensure in assertion (1) that ϕ is in addition a *linear* map.

EXERCISE V.12. Show that each of the following properties is invariant under injective affine maps, i.e. if $\phi : \mathbb{R}^{n_1} \to \mathbb{R}^{n_2}$ is an injective affine map then each of the following properties holds with $S = S_1$ and $n = n_1$ if and only if it holds with $S = \phi(S_1)$ and $n = n_2$.

(1) S is an affine subspace of $\mathbb{R}^n$,
(2) S is a convex subset of $\mathbb{R}^n$,
(3) S is the convex hull of finitely many points in $\mathbb{R}^n$,
(4) S is convex polyhedron in $\mathbb{R}^n$,
(5) S is convex polytope in $\mathbb{R}^n$.

EXERCISE V.13. Show that each of the following properties is invariant under injective linear maps, i.e. if $\phi : \mathbb{R}^{n_1} \to \mathbb{R}^{n_2}$ is an injective linear map then each of the following properties holds with $S = S_1$ and $n = n_1$ if and only if it holds with $S = \phi(S_1)$ and $n = n_2$.

(1) S is a linear subspace of $\mathbb{R}^n$,
(2) S is a convex cone in $\mathbb{R}^n$,
(3) S is a strongly convex cone in $\mathbb{R}^n$,
(4) S is a finitely generated convex cone in $\mathbb{R}^n$,
(5) S is a convex polyhedral cone in $\mathbb{R}^n$.

EXERCISE V.14. Let H be a linear subspace of $\mathbb{R}^n$.

(1) Show that every linear map $\nu : H \to \mathbb{R}$ can be extended to a map $\mathbb{R}^n \to \mathbb{R}$.
(2) Assume $H \subseteq \eta^\perp$ for some $\eta \in (\mathbb{R}^n)^*$. Let $S \subset \mathbb{R}^n$ be a finite set such that $\langle \eta, \alpha \rangle > 0$ for each $\alpha \in S$, and let c be an arbitrary real number. Show

that in assertion (1) it can be ensured that $\langle \nu', \alpha \rangle > c$ for each $\alpha \in S$. [Hint: after a linear change of coordinates on $\mathbb{R}^n$ we may assume that H is spanned by the first k-coordinates and η is the projection onto the $(k+1)$-th coordinate.]

EXERCISE V.15. Given $S \subset \mathbb{R}^n$, consider the following property:

$$\text{there is } \nu \in (\mathbb{R}^n)^* \text{ such that } \langle \nu, \alpha \rangle > 0 \text{ for each } \alpha \in S \setminus \{0\}. \tag{42}$$

Let $\phi : \mathbb{R}^{n_1} \to \mathbb{R}^{n_2}$ be an injective linear map and $S_1 \subseteq \mathbb{R}^{n_1}$. Show that property (42) holds with $S = S_1$ and $n = n_1$ if and only if it holds with $S = \phi(S_1)$ and $n = n_2$. [Hint: use assertion (1) of exercise V.14.]

EXERCISE V.16. Let $S \subseteq \mathbb{R}^n \setminus \{0\}$. Show that the origin is in the convex hull of S if and only if $\operatorname{cone}(S)$ contains a line through the origin. [Hint: 0 can be written as a convex combination of elements from S if and only if there is $\alpha \in \operatorname{cone}(S)$, $\alpha \neq 0$, such that $-\alpha$ is also in $\operatorname{cone}(S)$.]

EXERCISE V.17. Let $\mathcal{C}$ be a convex cone in $\mathbb{R}^n$ and $\pi : \mathbb{R}^n \to \mathbb{R}^m$ be a linear map such that $\ker(\pi) \cap \mathcal{C} = \{0\}$. Show that $\mathcal{C}$ is a strongly convex cone in $\mathbb{R}^n$ if and only if $\pi(\mathcal{C})$ is a strongly convex cone in $\mathbb{R}^m$. [Hint: Use exercise V.16.]

2. Characterization of convex polyhedra

In this section we describe a characterization of convex polytopes and polyhedral cones, and use it to show that every polyhedron is the Minkowski sum of a polytope and a polyhedral cone. The proofs we give are elementary and geometric, but not the most "efficient"; see, e.g. [Sch98, Section 7.2] for a quicker proof (which is somewhat less intuitive in the beginning), and [Zie95, Section 1.3] for a more algorithmic proof.

THEOREM V.3. (Farkas (1898, 1902), Minkowski (1896), Weyl (1935)). *Let $\mathcal{P}$ be a convex subset of $\mathbb{R}^n$.*

(1) $\mathcal{P}$ is a polytope if and only if it is the convex hull of finitely many points.
(2) $\mathcal{P}$ is a polyhedral cone if and only if it is a finitely generated cone.
(3) Assume $\mathcal{P}$ is a polyhedral cone. Then it is strongly convex if and only if there is $\nu \in (\mathbb{R}^n)^*$ such that $\langle \nu, \alpha \rangle > 0$ for each $\alpha \in \mathcal{P} \setminus \{0\}$.

PROOF. We are going to prove assertions (2), (3) and ($\Leftarrow$) implication of assertion (1). The ($\Rightarrow$) direction of assertion (1) follows from these assertions; it is left as an exercise (exercise V.19). We start with the proof of assertion (3). For the ($\Leftarrow$) direction note that if $\mathcal{P}$ contains both α and $-\alpha$ for some nonzero $\alpha \in \mathbb{R}^n$, then for all $\nu \in (\mathbb{R}^n)^*$, either $\langle \nu, \alpha \rangle \leq 0$ or $\langle \nu, -\alpha \rangle < 0$. For the ($\Rightarrow$) direction we proceed by induction on $\dim(\mathcal{P})$. Due to exercises V.11, V.13 and V.15 we may assume without loss of generality that $\dim(\mathcal{P}) = n$. Since the only strongly positive cones in $\mathbb{R}$ are $\mathbb{R}_{\geq 0}$ and $\mathbb{R}_{\leq 0}$, it holds for $n = 1$. In the general case $\mathcal{P}$ is defined by finitely many inequalities of the form $\langle \nu, x \rangle \geq 0$ for nonzero $\nu \in (\mathbb{R}^n)^*$ (exercise V.6). Take one such ν. Since $\nu^\perp \cap \mathcal{P}$ is a smaller dimensional strongly convex polyhedral cone, it follows by induction that there is $\eta \in (\mathbb{R}^n)^*$ which is positive on $(\nu^\perp \cap \mathcal{P}) \setminus \{0\}$.

Claim V.3.1. *For sufficiently small $\epsilon > 0$, $\nu + \epsilon\eta$ is positive on $\mathcal{P} \setminus \{0\}$.*

PROOF. If the claim is false, then we can find a sequence of positive numbers $\epsilon_k \to 0$ and $\alpha_k \in S^{n-1} \cap \mathcal{P}$, where S^{n-1} is the unit sphere centered at the origin in $\mathbb{R}^n$, such that $\langle \nu + \epsilon_k\eta, \alpha_k \rangle < 0$. Since $\mathcal{P}$ is closed and S^{n-1} is compact, we may assume that the α_k converge to $\alpha \in \mathcal{P} \cap S^{n-1}$. Then $\langle \nu, \alpha \rangle = \lim_{k\to\infty} \langle \nu + \epsilon_k\eta, \alpha_k \rangle \leq 0$, so that $\langle \nu, \alpha \rangle = 0$. It follows that $\alpha \in S^{n-1} \cap \nu^\perp \cap \mathcal{P}$, so that $\langle \eta, \alpha \rangle > 0$. By continuity η is positive on an open neighborhood U of α, which means that $\langle \nu + \epsilon_k\eta, \alpha_k \rangle > 0$ for sufficiently large k. This contradiction proves the claim. □

Claim V.3.1 finishes the proof of assertion (3). We now prove the ($\Rightarrow$) direction of assertion (2) by induction on $\dim(\mathcal{P})$. Due to proposition V.1 and exercises V.11 and V.13 we may assume without loss of generality that $\mathcal{P}$ is an n-dimensional strongly convex polyhedral cone in $\mathbb{R}^n$. As in the proof of assertion (3), the $n = 1$ case follows directly from the observation that the only strongly convex polyhedral cones in $\mathbb{R}$ are $\mathbb{R}_{\geq 0}$ and $\mathbb{R}_{\leq 0}$. Now consider the case $n \geq 2$. Assertion (3) implies that there is $\nu \in (\mathbb{R}^n)^*$ such that $\langle \nu, \alpha \rangle > 0$ for each $\alpha \in \mathcal{P} \setminus \{0\}$.

Claim V.3.2. *$\mathcal{P}' := \{\alpha \in \mathcal{P} : \langle \nu, \alpha \rangle = 1\}$ is bounded.*

PROOF. Indeed, otherwise take a sequence $\alpha_k \in \mathcal{P}'$ such that $||\alpha_k||$ is unbounded, where $||\cdot||$ is the Euclidean norm on $\mathbb{R}^n$. Note that each $\alpha_k/||\alpha_k||$ is in the intersection of $\mathcal{P}$ and the $(n-1)$-dimensional unit sphere S^{n-1}. Since S^{n-1} is compact and $\mathcal{P}$ is closed, we may assume without loss of generality that $\alpha_k/||\alpha_k||$ converge to $\alpha \in \mathcal{P} \cap S^{n-1}$. But then $\langle \nu, \alpha \rangle = \lim_{k\to\infty} \langle \nu, \alpha_k/||\alpha_k|| \rangle = 0$, which contradicts the choice of ν. □

Exercise V.6 implies that $\mathcal{P}$ is defined by inequalities of the form $\langle \nu_j, x \rangle \geq 0$, $j = 1, \ldots, N$, for some $\nu_1, \ldots, \nu_N \in (\mathbb{R}^n)^*$. For each j, let $\mathcal{P}_j := \mathcal{P} \cap \nu_j^\perp$. By the inductive hypothesis, each $\mathcal{P}_j$ is generated by finite sets $S_j \subset \mathbb{R}^n$. We claim that $\mathcal{P}$ is generated by $\bigcup_j S_j$. Indeed, take $\alpha \in \mathcal{P} \setminus \{0\}$. There is $r > 0$ such that $r\alpha \in \mathcal{P}'$. Take any straight line through $r\alpha$ on the hyperplane $\{\alpha : \langle \nu, \alpha \rangle = 1\}$. Claim V.3.2 implies that either $r\alpha \in \mathcal{P}_j$ for some j, or each end of the line intersects one of the $\mathcal{P}_j$. In any event, the inductive hypothesis implies that α is a nonnegative linear combination of elements from $\bigcup_j S_j$, as required to complete the proof of ($\Rightarrow$) implication of assertion (2). Now we prove ($\Leftarrow$) implications of assertions (1) and (2) by induction on $\dim(\mathrm{aff}(\mathcal{P}))$. We will first show that it suffices to prove only the implication from assertion (2).

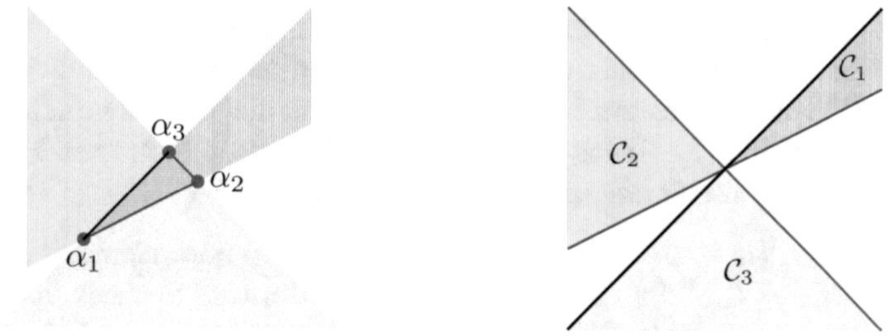

FIGURE 4. Convex hull of finitely many points is a finite intersection of translations of cones.

Claim V.3.3. *Let $k \geq 1$. Assume the $(\Leftarrow)$ implication of assertion (2) holds whenever $\dim(\text{aff}(\mathcal{P})) \leq k$. Then the $(\Leftarrow)$ implication of assertion (1) holds whenever $\dim(\text{aff}(\mathcal{P})) \leq k$.*

PROOF. Let $\mathcal{P}$ be the convex hull of a finite set S, and $m := \dim(\text{aff}(\mathcal{P})) \leq k$. Without loss of generality we may assume that $\mathcal{P} \subset \mathbb{R}^m$ (exercises V.11 and V.12). Let $\alpha_1, \ldots, \alpha_s$ be the elements of S. For each j, let $\mathcal{C}_j$ be the cone generated by $S - \alpha_j = \{\alpha_i - \alpha_j : 1 \leq i \leq s\}$. The hypothesis of the claim implies that each $\mathcal{C}_j$ is a polyhedral cone, which implies in turn that $\mathcal{C}_j + \alpha_j$ is a polyhedron (exercise V.12). It follows that $\mathcal{P}' := \bigcap_j (\mathcal{C}_j + \alpha_j)$ is also a polyhedron. Therefore it suffices to show that $\mathcal{P}' = \mathcal{P}$ (fig. 4). By (40) every element $\alpha \in \mathcal{P}$ can be expressed as $\sum_i \epsilon_i \alpha_i$ where ϵ_i are nonnegative real numbers with $\sum_i \epsilon_i = 1$. But then $\alpha = \alpha_j + \sum_i \epsilon_i(\alpha_i - \alpha_j) \in \mathcal{C}_j + \alpha_j$ for each j. Therefore $\mathcal{P} \subseteq \mathcal{P}'$. Now take $\alpha \notin \mathcal{P}$. To complete the proof of the claim it suffices to show that $\alpha \notin \mathcal{P}'$. Since $\alpha \notin \mathcal{P}$, it follows that $\text{cone}(\mathcal{P} - \alpha)$ is strongly convex (exercise V.16). By the hypothesis of the claim $\text{cone}(\mathcal{P} - \alpha)$ is polyhedral, so that due to assertion (3) there is $\nu \in (\mathbb{R}^n)^*$ which is positive on $\mathcal{P} - \alpha$. Pick j such that $\langle \nu, \alpha_j \rangle = \min\{\langle \nu, \alpha_i \rangle : 1 \leq i \leq s\}$. Then ν is nonnegative on $\mathcal{C}_j$, whereas $\langle \nu, \alpha - \alpha_j \rangle$ is negative. Therefore $\alpha \notin \mathcal{C}_j + \alpha_j$, and consequently $\alpha \notin \mathcal{P}'$, as required. □

Now we start the proof of $(\Leftarrow)$ direction of assertion (2) by induction on $\dim(\text{aff}(\mathcal{P}))$. Due to proposition V.1 and exercises V.11 and V.13 we may assume without loss of generality that $\mathcal{P}$ is a strongly convex cone generated by a finite set $S \subset \mathbb{R}^n$ and $n = \dim(\text{aff}(\mathcal{P}))$. For $n = 1$ the possibilities for $\mathcal{P}$ are $\mathbb{R}_{\geq 0}$ and $\mathbb{R}_{\leq 0}$, both of which are polyhedral. For general n we proceed by induction on $|S|$. The case $|S| = 1$ is also covered by the case $n = 1$. So assume $|S| \geq 2$.

Claim V.3.4. *There is $\nu \in (\mathbb{R}^n)^*$ which is positive on $S \setminus \{0\}$.*

PROOF. Pick $\alpha \in S \setminus \{0\}$. By the inductive hypothesis $\mathcal{P}_1 := \text{cone}(S \setminus \{\alpha\})$ is polyhedral, so that $\mathcal{P}_1$ is defined by finitely many inequalities of the form $\langle \nu_j, x \rangle \geq 0$, $j = 1, \ldots, N$ (exercise V.6). We claim that $\langle \nu_j, \alpha \rangle > 0$ for some j. Indeed, otherwise $\langle \nu_j, -\alpha \rangle \geq 0$ for each j, so that $-\alpha \in \mathcal{P}_1$. But then $\mathcal{P}$ contains the line through the origin and α, contradicting the strong convexity of $\mathcal{P}$. So we can pick

j such that $\langle \nu_j, \alpha \rangle > 0$. Now let $\mathcal{P}_2 := \mathcal{P} \cap \nu_j^\perp$. Then $\dim(\operatorname{aff}(\mathcal{P}_2)) < n$, so that $\mathcal{P}_2$ is polyhedral by the inductive hypothesis. Assertion (3) then implies that there is $\nu \in (\mathbb{R}^n)^*$ which is positive on $\mathcal{P}_2 \setminus \{0\}$. We claim that if ϵ is a sufficiently small positive number, then $\nu_j + \epsilon\nu$ is positive on $\mathcal{P} \setminus \{0\}$. Indeed, simply take ϵ such that $\epsilon|\langle \nu, \beta \rangle| < \langle \nu_j, \beta \rangle$ for all (the finitely many) elements of $S \setminus \mathcal{P}_2$. □

Let $\mathcal{P}' := \{\alpha \in \mathcal{P} : \langle \nu, \alpha \rangle = 1\}$. If $\alpha_1, \ldots, \alpha_s$ are elements of S, then $\alpha_j' := \alpha_j / \langle \nu, \alpha_j \rangle \in \mathcal{P}'$ for each j, and it is straightforward to check that $\mathcal{P}'$ is the convex hull of $\alpha_1', \ldots, \alpha_s'$. Since $\dim(\operatorname{aff}(\mathcal{P}')) < n$, the inductive hypothesis and claim V.3.3 imply that $\mathcal{P}'$ is a polytope. It is then straightforward to check that $\mathcal{P}$ is a polyhedral cone (exercise V.18), as required. □

Corollary V.4. *The Minkowski sum of two convex polytopes (respectively, polyhedral cones) is a convex polytope (respectively, polyhedral cone).*

PROOF. Follows immediately from theorem V.3 and exercise V.8. □

The next corollary of theorem V.3 shows that each convex polyhedron $\mathcal{Q}$ has a representation of the form $\mathcal{Q} = \mathcal{P} + \mathcal{C}$, where $\mathcal{C}$ is a polyhedral cone and $\mathcal{P}$ is a polytope. In general the decomposition is *not* unique, see fig. 5. However, $\mathcal{C}$ is uniquely determined from $\mathcal{Q}$ (exercise V.21). It is also possible to find $\mathcal{P}$ which is "minimal" (modulo translations in the case that $\mathcal{C}$ is not strongly convex), but we will not get into that.

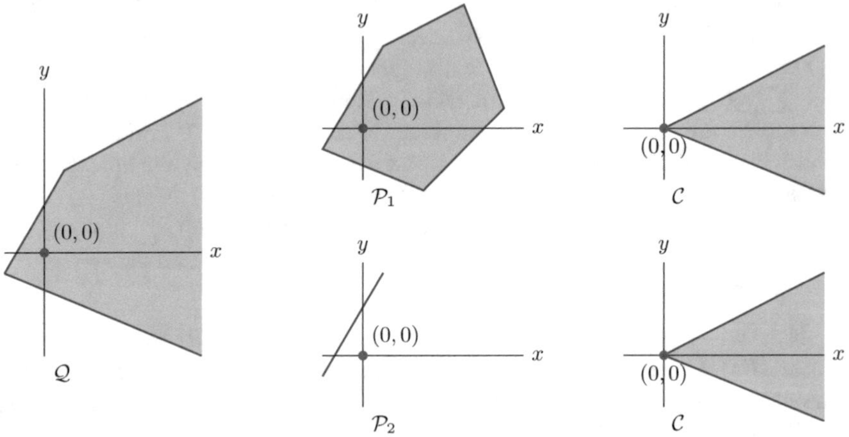

FIGURE 5. $\mathcal{Q} = \mathcal{P} + \mathcal{C}$ for $\mathcal{P} = \mathcal{P}_1$ and $\mathcal{P} = \mathcal{P}_2$. The "minimal" possible choice for $\mathcal{P}$ is the line segment $\mathcal{P}_2$.

Corollary V.5. (Motzkin (1936)). *A subset of $\mathbb{R}^n$ is a convex polyhedron if and only if it is the Minkowski sum of a convex polyhedral cone and a convex polytope.*

PROOF. At first we prove the $(\Rightarrow)$ implication. Let $\mathcal{Q}$ be a polyhedron in $\mathbb{R}^n$ defined by inequalities $a_{q,0} + a_{q,1}x_1 + \cdots + a_{q,n}x_n \geq 0$, $q = 1, \ldots, N$. We will show that it is the sum of a polyhedral cone and a polytope. Consider the polyhedral cone $\mathcal{Q}'$ in $\mathbb{R}^{n+1}$ defined by the "homogenizations" $a_{q,0}x_0 + a_{q,1}x_1 + \cdots + a_{q,n}x_n \geq 0$, $q = 1, \ldots, N$, and $x_0 \geq 0$. Note that $\mathcal{Q} = \mathcal{Q}' \cap (\{1\} \times \mathbb{R}^n)$. Theorem V.3 implies that $\mathcal{Q}'$ is a finitely generated cone. We may choose a set of generators of $\mathcal{Q}'$ of the form $(1, \alpha_1), \ldots, (1, \alpha_k), (0, \beta_1), \ldots, (0, \beta_l)$, where $\alpha_1, \ldots, \alpha_k, \beta_1, \ldots, \beta_l \in \mathbb{R}^n$. Let $\mathcal{P}$ be the convex hull of $\alpha_1, \ldots, \alpha_k$, and $\mathcal{C}$ be the cone in $\mathbb{R}^n$ generated by $\beta_1, \ldots, \beta_l$. Theorem V.3 implies that $\mathcal{P}$ is a polytope and $\mathcal{C}$ is a polyhedral cone.

Claim V.5.1. $\mathcal{Q} = \mathcal{P} + \mathcal{C}$.

PROOF. Let $\gamma \in \mathcal{Q}$. Then $(1, \gamma) \in \mathcal{Q}'$, and therefore $(1, \gamma) = \sum_i a_i(1, \alpha_i) + \sum_j b_j(0, \beta_j)$ where the a_i, b_j are nonnegative numbers. Then it follows that $\sum_i a_i = 1$ and $\gamma = \alpha + \beta$, where $\alpha = \sum_i a_i\alpha_i$ and $\beta = \sum_j b_j\beta_j$. It is clear that $\beta \in \mathcal{C}$. Since $\sum_i a_i = 1$, it follows that α is a convex combination of the α_i, so that $\alpha \in \mathcal{P}$. Therefore $\mathcal{Q} \subset \mathcal{P} + \mathcal{C}$. Now we check the opposite inclusion. Let $\alpha \in \mathcal{P}$ and $\beta \in \mathcal{C}$. Then $\alpha = \sum_i a_i\alpha_i$ where the a_i are nonnegative numbers such that $\sum_i a_i = 1$. Since $(1, \alpha_i) \in \mathcal{Q}'$ for each i, it follows that $(1, \alpha) = \sum_i a_i(1, \alpha_i) \in \mathcal{Q}'$. On the other hand, it follows from the construction of $\mathcal{C}$ that $\{0\} \times \mathcal{C} = \mathcal{Q}' \cap (\{0\} \times \mathbb{R}^n)$, so that $(0, \beta) \in \mathcal{Q}'$. Since the sum of two elements of a convex cone is also in that cone, it follows that $(1, \alpha + \beta) = (1, \alpha) + (0, \beta) \in \mathcal{Q}'$. Therefore, $\alpha + \beta \in \mathcal{Q}$, as required. □

Claim V.5.1 finishes the proof of $(\Rightarrow)$ implication of the corollary. Now we prove the $(\Leftarrow)$ implication. Let $\mathcal{P}$ be a convex polytope and $\mathcal{C}$ be a convex polyhedral cone in $\mathbb{R}^n$. We will show that $\mathcal{P} + \mathcal{C}$ is a convex polyhedron. By exercise V.12 we may assume without loss of generality that $\mathcal{P}$ contains the origin. Identify $\mathbb{R}^{n+1}$ with $\mathbb{R} \times \mathbb{R}^n$. Let $\mathcal{P}'$ be the cone in $\mathbb{R}^{n+1}$ generated by $\{1\} \times \mathcal{P}$, and $\mathcal{C}'$ be the closure of the cone in $\mathbb{R}^{n+1}$ generated by $\{1\} \times \mathcal{C}$. Either theorem V.3 or exercise V.18 implies that $\mathcal{P}'$ is a polyhedral cone, and exercise V.20 implies that $\mathcal{C}'$ is a polyhedral cone. Corollary V.4 then implies that $\mathcal{P}' + \mathcal{C}'$ is a polyhedral cone, so that $\mathcal{Q} := (\mathcal{P}' + \mathcal{C}') \cap (\{1\} \times \mathbb{R}^n)$ is a polyhedron.

Claim V.5.2. $\mathcal{Q} = \{1\} \times (\mathcal{P} + \mathcal{C})$.

PROOF. Let the inequalities defining $\mathcal{P}$ be $a_{i,0} + a_{i,1}x_1 + \cdots + a_{i,n}x_n \geq 0$, $i = 1, \ldots, k$, and the inequalities defining $\mathcal{C}$ be $b_{j,1}x_1 + \cdots + b_{j,n}x_n \geq 0$, $j = 1, \ldots, l$. Exercise V.18 shows that $\mathcal{P}'$ is defined by the inequalities $a_{i,0}x_0 + a_{i,1}x_1 + \cdots + a_{i,n}x_n \geq 0$, $i = 1, \ldots, k$, and $x_0 \geq 0$. On the other hand exercise V.20 implies that $\mathcal{C}'$ is defined by the same inequalities as $\mathcal{C}$ together with the inequality $x_0 \geq 0$. If $\alpha \in \mathcal{P}$ and $\beta \in \mathcal{C}$, then it follows that $(1, \alpha) \in \mathcal{P}'$ and $(0, \beta) \in \mathcal{C}'$, so that $(1, \alpha + \beta) \in \mathcal{P}' + \mathcal{C}'$, which proves that $\mathcal{Q} \supset \{1\} \times (\mathcal{P} + \mathcal{C})$. For the opposite inclusion, pick $(1, \gamma) \in \mathcal{Q}$. We will show that $\gamma \in \mathcal{P} + \mathcal{C}$. Write $(1, \gamma) = (a, \alpha) + (b, \beta)$, where $(a, \alpha) \in \mathcal{P}'$ and $(b, \beta) \in \mathcal{C}'$, and a, b are nonnegative numbers such that $a + b = 1$. An examination of the inequalities defining $\mathcal{C}'$ shows that $\beta \in \mathcal{C}$. If $a = 0$, then assertion (1) of exercise V.18 implies that $\alpha = 0$. Since the origin is in $\mathcal{P}$, it follows that $\gamma = 0 + \beta \in \mathcal{P} + \mathcal{C}$. On the other hand, if $a \neq 0$, then $(1/a)\alpha \in \mathcal{P}$. Since $0 < a \leq 1$, and since $0 \in \mathcal{P}$, it follows by convexity of $\mathcal{P}$ that $\alpha \in \mathcal{P}$. Therefore $\gamma = \alpha + \beta \in \mathcal{P} + \mathcal{Q}$, as required. □

Claim V.5.2 finishes the proof of ($\Leftarrow$) implication, and therefore the proof of corollary V.5. □

Corollary V.6. *Let $\mathcal{P}$ be a polyhedron in $\mathbb{R}^n$.*

(1) $\mathcal{P}$ has a decomposition of the form $\mathcal{P} = \mathcal{P}_0 + \mathcal{C} + L$, where $\mathcal{P}_0$ is a polytope, $\mathcal{C}$ is strongly convex polyhedral cone and L is a linear subspace of $\mathbb{R}^n$ such that $\mathcal{C} \cap L = \{0\}$. Moreover, L and $\mathcal{C} + L$ are uniquely determined by $\mathcal{P}$.
(2) If $\mathcal{P}$ is defined by inequalities $a_{i,0} + a_{i,1}x_1 + \cdots + a_{i,n}x_n \geq 0$, $i = 1, \ldots, N$, then
  (a) $L = \{x : a_{i,1}x_1 + \cdots + a_{i,n}x_n = 0,\ i = 1, \ldots, N\}$.
  (b) $\mathcal{C} + L = \{x : a_{i,1}x_1 + \cdots + a_{i,n}x_n \geq 0,\ i = 1, \ldots, N\}$.
  (c) For each $r \gg 1$, the set $\mathcal{P}_r := \{x : a_{i,1}x_1 + \cdots + a_{i,n}x_n = 0,\ i = 1, \ldots, N$, and $-r \leq x_j \leq r,\ j = 1, \ldots, n\}$ is a polytope, and we can take $\mathcal{P}_0 = \mathcal{P}_r$ for any such r.

PROOF. Combine proposition V.1 and corollary V.5. □

Corollary V.7. *The Minkowski sum of finitely many convex polyhedra is a convex polyhedron.*

PROOF. Follows directly from corollary V.4 and the arguments of the proof of corollary V.5. □

Corollary V.8. *The image of a polyhedron under an affine map is also a polyhedron.*

PROOF. By exercise V.12 it suffices to consider the case of linear maps. Let $\mathcal{P}$ be a polyhedron in $\mathbb{R}^n$, and $\phi : \mathbb{R}^n \to \mathbb{R}^m$ be a linear map. Write $\mathcal{P} = \mathcal{P}_0 + \mathcal{C} + L$ as in corollary V.6. Then $\phi(\mathcal{P}) = \phi(\mathcal{P}_0) + \phi(\mathcal{C}) + \phi(L)$ (exercise V.7). Theorem V.3 and exercise V.1 imply that $\phi(\mathcal{P})$ is a polytope, and $\phi(\mathcal{C})$ is a polyhedral cone. Since $\phi(L)$ is a linear subspace of $\mathbb{R}^m$ (due to linearity of ϕ), corollary V.7 implies that $\phi(\mathcal{P})$ is a polyhedron. □

2.1. Exercises.

EXERCISE V.18. Let $\mathcal{P}$ be a nonempty polytope in $\mathbb{R}^n$ defined by inequalities $a_{i,0} + a_{i,1}x_1 + \cdots + a_{i,n}x_n \geq 0$, $i = 1, \ldots, N$. Let $\mathcal{C}_0 := \{x \in \mathbb{R}^n : a_{i,1}x_1 + \cdots + a_{i,n}x_n \geq 0,\ 1 \leq i \leq N\}$, and $\mathcal{C}$ be the cone generated by $\{1\} \times \mathcal{P}$ in $\mathbb{R}^{n+1} = \mathbb{R} \times \mathbb{R}^n$.

(1) Show that $\mathcal{C}_0 = \{0\}$. [Hint: $\mathcal{C}_0$ is a cone, $\mathcal{C}_0 + \mathcal{P} \subset \mathcal{P}$, and $\mathcal{P}$ is bounded.]
(2) Deduce that $\mathcal{C}$ is a polyhedral cone in $\mathbb{R}^{n+1}$ defined by the inequalities $x_0 \geq 0$ and $a_{i,0}x_0 + a_{i,1}x_1 + \cdots + a_{i,n}x_n \geq 0$, $i = 1, \ldots, N$.
(3) Show that the assumption "$\mathcal{P}$ is nonempty" is necessary in the preceding statements. In particular, show by an example that $\mathcal{C}_0$ may be a nontrivial cone if $\mathcal{P} = \emptyset$.

EXERCISE V.19. Show that if the ($\Rightarrow$) implication of assertion (2) of theorem V.3 holds for all n, then the ($\Rightarrow$) direction of assertion (1) of theorem V.3 also holds for all n. [Hint: Use exercise V.18.]

EXERCISE V.20. Let $\mathcal{C}$ be a polyhedral cone in $\mathbb{R}^n$. Exercise V.6 implies that $\mathcal{C}$ is defined by finitely many inequalities of the form $a_1x_1 + \cdots + a_nx_n \geq 0$. Show that the closure in $\mathbb{R}^{n+1}$ of the cone generated by $\{1\} \times \mathcal{C}$ is a polyhedral cone defined by the same inequalities as the ones defining $\mathcal{C}$ coupled with $x_0 \geq 0$.

EXERCISE V.21. Let $\mathcal{P} \subset \mathbb{R}^n$.

(1) Assume $\mathcal{P} = \mathcal{P}_1 + \mathcal{C}_1 = \mathcal{P}_2 + \mathcal{C}_2$, where $\mathcal{P}_j$ are bounded and $\mathcal{C}_j$ are closed cones. Prove that $\mathcal{C}_1 = \mathcal{C}_2$. Show by examples that it is possible to have $\mathcal{P}_1 \neq \mathcal{P}_2$.
(2) Assume $\mathcal{P} = \mathcal{C}_1 + L_1 = \mathcal{C}_2 + L_2$ where $\mathcal{C}_j$ are strongly convex polyhedral cones and L_j are linear subspaces of $\mathbb{R}^n$ such that $\mathcal{C}_j \cap L_j = \{0\}$. Prove that $L_1 = L_2$. Show by examples that it is possible to have $\mathcal{C}_1 \neq \mathcal{C}_2$.
(3) Assume $\mathcal{P} = \mathcal{P}_1 + \mathcal{C}_1 + L_1 = \mathcal{P}_2 + \mathcal{C}_2 + L_2$ where $\mathcal{P}_j$ are polytopes, $\mathcal{C}_j$ are strongly convex polyhedral cones and L_j are linear subspaces of $\mathbb{R}^n$ such that $\mathcal{C}_j \cap L_j = \{0\}$. Prove that $L_1 = L_2$ and $\mathcal{C}_1 + L_1 = \mathcal{C}_2 + L_2$.

[Hint: We may assume without loss of generality that $\mathcal{P}$ contains origin. Then for the first assertion pick $\alpha \in \mathcal{P}$ such that $r\alpha \in \mathcal{P}$ for each $r \geq 0$. Express α as a sum of elements of $\frac{1}{r}\mathcal{P}_j$ and $\mathcal{C}_j$, and then take the limit as $r \to \infty$ to show that each $\mathcal{C}_j$ must be the largest cone contained in $\mathcal{P}$. For the second assertion prove that each L_j is the largest linear subspace of $\mathbb{R}^n$ contained in $\mathcal{P}$ as follows: pick $\alpha \in \mathbb{R}^n$ such that both α and $-\alpha$ are in $\mathcal{P}$. Express both α and $-\alpha$ as elements of $\mathcal{C}_j + L_j$, add them up and use the condition that $\mathcal{C}_j \cap L_j = \{0\}$ to show that $\alpha \in L_j$.]

3. Basic properties of convex polyhedra

In this section we establish a few basic properties of convex polyhedra. Throughout this section $\mathcal{P}$ will denote a *nonempty* convex polyhedron in $\mathbb{R}^n$. The first property we state follows directly from the definition of a polyhedron:

Proposition V.9. *For each $\alpha \in \mathbb{R}^n \setminus \mathcal{P}$, there is $\nu \in (\mathbb{R}^n)^*$ such that $\min_{\mathcal{P}}(\nu)$ exists and $\langle \nu, \alpha \rangle < \min_{\mathcal{P}}(\nu)$.* □

Geometrically, proposition V.9 states that every point in the complement of $\mathcal{P}$ is separated from $\mathcal{P}$ by a hyperplane; see fig. 6. A *face* of $\mathcal{P}$ is a subset of the form $\mathrm{In}_\nu(\mathcal{P})$ for some $\nu \in (\mathbb{R}^n)^*$.

Proposition V.10. *Every face of $\mathcal{P}$ is a convex polyhedron. If $\mathcal{P}$ is a polytope (respectively, polyhedral cone), then every face of $\mathcal{P}$ is also a polytope (respectively, polyhedral cone).*

PROOF. The case of a convex polyhedron and that of a polytope are direct consequences of the definitions of polyhedra, polytopes and faces. The case of a polyhedral cone follows from combining exercises V.6 and V.23. □

We also note the following property whose proof is left as an exercise:

Proposition V.11. *Let $\phi : \mathcal{P} \to \mathcal{Q}$ be an affine isomorphism of convex polyhedra. Then ϕ induces a bijection of faces, i.e. $\mathcal{R} \subseteq \mathcal{P}$ is a face of $\mathcal{P}$ if and only if $\phi(\mathcal{R})$ is a face of $\mathcal{Q}$.*

PROOF. This is exercise V.24. □

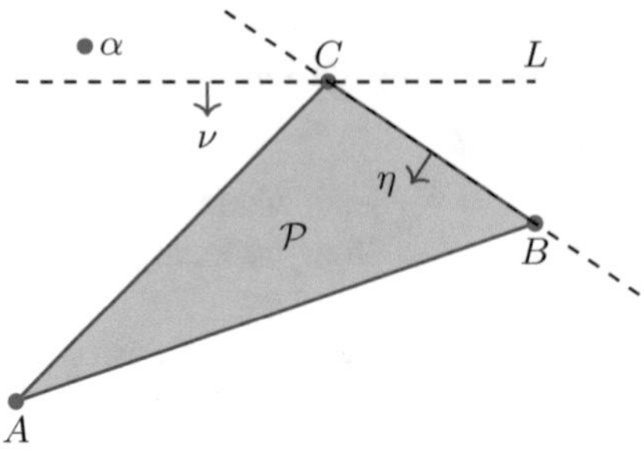

FIGURE 6. $\mathrm{In}_\nu(\mathcal{P}) = \{C\}$, $\mathrm{In}_\eta(\mathcal{P}) = BC$. The line L separates $\mathcal{P}$ from α.

Proposition V.10 implies that every face of $\mathcal{P}$ is a polyhedron, and therefore has a well-defined dimension. A *vertex* (respectively, *edge, facet*) is a face of dimension zero (respectively one, $\dim(\mathcal{P}) - 1$). In fig. 6 the vertices of $\mathcal{P}$ are A, B, C, and the edges are the three sides. Proposition V.12 gives a more precise description of vertices and edges. The part of the proposition about finiteness of numbers of vertices and edges is a special case of the more general result (corollary V.17) that a polyhedron can have only finitely many faces. We will also see later that every polytope is the convex hull of its vertices (as it is evident for $\mathcal{P}$ from fig. 6).

Proposition V.12. *Every vertex of a polyhedron is a singleton, and up to an affine isomorphism, every edge is a (possibly unbounded) closed interval of* $\mathbb{R}$. *Every polytope has finitely many vertices. Every strongly convex polyhedral cone has only one vertex (namely, the origin) and finitely many edges.*

PROOF. The first statement follows directly from exercise V.4. By theorem V.3 every polytope $\mathcal{P}$ is the convex hull of a finite set. The first statement together with exercise V.2 then implies that $\mathcal{P}$ has only finitely many vertices. Now let $\mathcal{C}$ be a strongly convex polyhedral cone. Theorem V.3 implies that the origin is a vertex of $\mathcal{C}$. Since every face of $\mathcal{C}$ is also a cone (proposition V.10), and since the only zero-dimensional cone is the origin, it follows that the origin is the only vertex of $\mathcal{C}$. The finiteness of the numbers of edges of $\mathcal{C}$ follows from exercise V.26 and finiteness of the numbers of vertices of polytopes. □

Note in fig. 6 every pair of edges of $\mathcal{P}$ intersects in a vertex. The following result shows that this corresponds to a general phenomenon:

Proposition V.13. *The intersection of finitely many faces of* $\mathcal{P}$ *is also a face of* $\mathcal{P}$*, provided the intersection is nonempty.*

PROOF. Let $\mathcal{Q} := \mathrm{In}_\nu(\mathcal{P})$ and $\mathcal{R} := \mathrm{In}_\eta(\mathcal{P})$ be faces of $\mathcal{P}$ with $\mathcal{Q} \cap \mathcal{R} \neq \emptyset$. It is then straightforward to check that $\mathcal{Q} \cap \mathcal{R} = \mathrm{In}_{\nu+\eta}(\mathcal{P})$. [Where is it used that $\mathcal{Q} \cap \mathcal{R} \neq \emptyset$? See exercise V.27.] □

Every polyhedron $\mathcal{P}$ is a face of itself, since $\mathcal{P} = \mathrm{In}_0(\mathcal{P})$. A *proper face* of $\mathcal{P}$ is a face which is properly contained in $\mathcal{P}$, and the *relative interior* $\mathrm{relint}(\mathcal{P})$ of $\mathcal{P}$ is the complement in $\mathcal{P}$ of the union of its proper faces. In fig. 6 the relative interior is the complement in $\mathcal{P}$ of the union of the three edges.

Proposition V.14. *Let $\mathcal{Q}, \mathcal{R}$ be faces of $\mathcal{P}$.*

(1) If $\mathcal{Q}$ is a proper face of $\mathcal{P}$, then $\dim(\mathcal{Q}) < \dim(\mathcal{P})$.
(2) If $\mathcal{Q} \cap \mathrm{relint}(\mathcal{R}) \neq \emptyset$, then $\mathcal{Q} \supset \mathcal{R}$.

PROOF. The first assertion is an immediate consequence of the observation that if $\nu \in (\mathbb{R}^n)^*$ is non-constant on an affine subspace H of $\mathbb{R}^n$, then for every $r \in \mathbb{R}$, $\{\alpha \in H : \langle \nu, \alpha \rangle = r\}$ is an affine subspace of $\mathbb{R}^n$ whose dimension is one less than that of H. For the second assertion, note that if $\mathcal{Q} \cap \mathcal{R} \neq \emptyset$, then by proposition V.13 $\mathcal{Q} \cap \mathcal{R}$ is a face of $\mathcal{R}$, and it cannot be a proper face of $\mathcal{R}$ if it contains a point from $\mathrm{relint}(\mathcal{R})$. □

Note that in fig. 6 the relative interior of $\mathcal{P}$ is also its *topological interior*[2] in $\mathbb{R}^2$. We will now see that this is a manifestation of a general property of polyhedra.

Proposition V.15. *$\mathcal{P}$ is the closure of its relative interior. In particular,* $\mathrm{relint}(\mathcal{P})$ *is nonempty, and it is precisely the topological interior of $\mathcal{P}$ in* $\mathrm{aff}(\mathcal{P})$ *(where* $\mathrm{aff}(\mathcal{P})$ *is equipped with the relative topology from $\mathbb{R}^n$). In particular,* $\mathrm{relint}(\mathcal{P})$ *is a nonempty relatively open subset of* $\mathrm{aff}(\mathcal{P})$*. In the case that $\mathcal{P} \subset \mathbb{R}^n$ is "full dimensional" (i.e. $\dim(\mathcal{P}) = n$), and $a_{i,0} + a_{i,1}x_1 + \cdots + a_{i,n}x_n \geq 0$, $i = 1, \ldots, N$, are a set of "nontrivial inequalities"*[3] *defining $\mathcal{P}$, then* $\mathrm{relint}(\mathcal{P})$ *is the nonempty open set of $\mathbb{R}^n$ defined by the strict inequalities $a_{i,0} + a_{i,1}x_1 + \cdots + a_{i,n}x_n > 0$, $i = 1, \ldots, N$.*

PROOF. Due to exercises V.11, V.12 and V.24 it suffices to consider the case that $\mathcal{P}$ is full dimensional. For each i, let H_i be the hyperplane $a_{i,0} + a_{i,1}x_1 + \cdots + a_{i,n}x_n = 0$. Let $\mathcal{P}^0 := \{x \in \mathbb{R}^n : a_{i,0} + a_{i,1}x_1 + \cdots + a_{i,n}x_n > 0,\ i = 1, \ldots, N\} = \mathcal{P} \setminus \bigcup_i H_i$. Exercise V.25 implies that $\mathcal{P}^0$ is the topological interior of $\mathcal{P}$. We claim that $\mathcal{P}^0 \neq \emptyset$. Indeed, for each $i = 1, \ldots, N$, there is $\alpha_i \in \mathcal{P}$ such that $a_{i,0} + a_{i,1}x_1 + \cdots + a_{i,n}x_n > 0$, since otherwise $\mathcal{P}$ would be contained in H_i, and $\dim(\mathcal{P})$ would be less than n. Since $\mathcal{P}$ is convex, $\alpha := (1/N)\sum_{i=1}^{N} \alpha_i \in \mathcal{P}$. It is straightforward then to check that $\alpha \in \mathcal{P}^0$, as required. We now show that $\mathcal{P}^0 = \mathrm{relint}(\mathcal{P})$. The full dimensionality of $\mathcal{P}$ also implies that $H_i \cap \mathcal{P}$ is a proper face of $\mathcal{P}$ for each i, so that $\mathcal{P}^0 \supset \mathrm{relint}(\mathcal{P})$. On the other hand, if $\alpha \in \mathcal{P}^0$, then by openness of $\mathcal{P}^0$, for every $\nu \in (\mathbb{R}^n)^* \setminus \{0\}$, there is $\alpha' \in \mathcal{P}^0$ such that $\langle \nu, \alpha' \rangle < \langle \nu, \alpha \rangle$, so that $\alpha \notin \mathrm{In}_\nu(\mathcal{P})$. This shows that $\mathcal{P}^0 \subset \mathrm{relint}(\mathcal{P})$, and therefore $\mathcal{P}^0 = \mathrm{relint}(\mathcal{P})$. Finally, to see that $\mathcal{P}$ is the closure of its relative interior, let $\beta := (\beta_1, \ldots, \beta_n) \in \mathrm{relint}(\mathcal{P})$. Given $\gamma = (\gamma_1, \ldots, \gamma_n) \in \mathcal{P}$, let $L := \{(1-\epsilon)\beta + \epsilon\gamma : 0 \leq \epsilon \leq \beta\}$ be the line segment from β to γ. We claim that $L \setminus \{\gamma\} \subseteq \mathrm{relint}(\mathcal{P})$. Indeed, for each $i = 1, \ldots, N$, let

[2]A point $x \in S \subseteq \mathbb{R}^n$ is in the topological interior of S if and only if S contains an open neighborhood of x in $\mathbb{R}^n$.

[3]"Nontrivial inequalities" means that we do not allow inequalities with $a_{i,0} = \cdots = a_{i,n} = 0$.

$$\epsilon_i := \sup\{\epsilon : 0 \leq \epsilon \leq 1,\ a_{i,0} + a_{i,1}((1-\epsilon)\beta_1 + \epsilon\gamma_1) + \cdots + a_{i,n}((1-\epsilon)\beta_n + \epsilon\gamma_n) > 0\}.$$

Note that each of the functions $a_{i,0}+a_{i,1}((1-\epsilon)\beta_1+\epsilon\gamma_1)+\cdots+a_{i,n}((1-\epsilon)\beta_n+\epsilon\gamma_n)$ is linear in ϵ, and it is strictly positive at $\epsilon = 0$. Therefore if $\epsilon_i < 1$, then it must be zero at ϵ_i and negative on the interval $(\epsilon_i, 1]$, contradicting the fact that $L \subset \mathcal{P}$. Therefore $\epsilon_i = 1$ for each i, and consequently, $L \setminus \{\gamma\} \subset \mathcal{P}^0 = \text{relint}(\mathcal{P})$. This implies that $\mathcal{P}$ is the closure of $\text{relint}(\mathcal{P})$, as required. □

Proposition V.16. *Let $\mathcal{P}_1, \ldots, \mathcal{P}_s$ be polyhedra in $\mathbb{R}^n$, and $\nu \in (\mathbb{R}^n)^*$. Then*

(1) $\min_{\sum_j \mathcal{P}_j}(\nu)$ exists if and only if $\min_{\mathcal{P}_j}(\nu)$ exists for each j.

Now assume $\min_{\sum_j \mathcal{P}_j}(\nu)$ exists. Then

(2) $\min_{\sum_j \mathcal{P}_j}(\nu) = \sum_j \min_{\mathcal{P}_j}(\nu)$.

(3) $\text{In}_\nu(\sum_j \mathcal{P}_j) = \sum_j \text{In}_\nu(\mathcal{P}_j)$.

(4) $\text{In}_\nu(\mathcal{P}_j)$ are the "unique maximal" subsets of $\mathcal{P}_j$ whose sum is $\text{In}_\nu(\sum_j \mathcal{P}_j)$, i.e. if $\alpha_j \in \mathcal{P}_j$ are such that $\sum_j \alpha_j \in \text{In}_\nu(\sum_j \mathcal{P}_j)$, then $\alpha_j \in \text{In}_\nu(\mathcal{P}_j)$ for each j.

(5) If each $\mathcal{P}_j$ is a polytope, then $\text{In}_\nu(\mathcal{P}_j)$ are, in fact, unique faces of $\mathcal{P}_j$ whose sum is $\text{In}_\nu(\sum_j \mathcal{P}_j)$, i.e. if $\mathcal{Q}_j$ are faces of $\mathcal{P}_j$ such that $\sum_j \mathcal{Q}_j = \text{In}_\nu(\sum_j \mathcal{P}_j)$, then $\mathcal{Q}_j = \text{In}_\nu(\mathcal{P}_j)$ for each j.

(6) Assume there is j such that

(a) either $\mathcal{P}_j$ is a cone, or

(b) $\mathcal{P}_j$ is a linear subspace of $\mathbb{R}^n$.

Then $\text{In}_\nu(\mathcal{P}_j)$ contains the origin and $\min_{\mathcal{P}_j}(\nu) = 0$. In addition in case (6b), $\mathcal{P}_j \subset \nu^\perp$, and $\text{In}_\nu(\mathcal{P}_j) = \mathcal{P}_j$.

PROOF. It suffices to treat the case $s = 2$. Assume at first $\min_{\mathcal{P}_1}(\nu)$ does not exist. Then there are $\alpha_k \in \mathcal{P}_1$ such that $\langle \nu, \alpha_k \rangle \to -\infty$. If β is an arbitrary element in $\mathcal{P}_2$, then $\langle \nu, \alpha_k + \beta \rangle \to -\infty$ as well, so that $\min_{\sum_j \mathcal{P}_j}(\nu)$ does not exist. On the other hand, if $\min_{\mathcal{P}_j}(\nu)$ exists for each j, then pick $\alpha_j \in \text{In}_\nu(\mathcal{P}_j)$ for each j, and note that for all $\beta_j \in \mathcal{P}_j$, $\langle \nu, \sum_j \beta_j \rangle \geq \sum_j \langle \nu, \alpha_j \rangle = \langle \nu, \sum_j \alpha_j \rangle$. This simultaneously proves assertions (1) to (3). For assertion (4) note that if $\alpha_1 \notin \text{In}_\nu(\mathcal{P}_1)$, then $\langle \nu, \alpha_1 \rangle > \min_{\mathcal{P}_1}(\nu)$. It follows that for each $\alpha_2 \in \mathcal{P}_2$, $\langle \nu, \alpha_1 + \alpha_2 \rangle > \min_{\mathcal{P}_1}(\nu) + \min_{\mathcal{P}_2}(\nu)$, so that $\alpha_1 + \alpha_2 \notin \text{In}_\nu(\mathcal{P}_1 + \mathcal{P}_2)$. For assertion (5) assume $\mathcal{Q}_1 \subsetneq \text{In}_\nu(\mathcal{P}_1)$. Since $\mathcal{Q}_1$ is a polytope, there is $\eta \in (\mathbb{R}^n)^*$ such that $\min_{\mathcal{Q}_1}(\eta) > \min_{\text{In}_\nu(\mathcal{P}_1)}(\eta)$. On the other hand, $\text{In}_\nu(\mathcal{P}_j)$ is a polytope for each j, and therefore by assertion (1),

$$\min_{\mathcal{Q}_1+\text{In}_\nu(\mathcal{P}_2)}(\eta) = \min_{\mathcal{Q}_1}(\eta) + \min_{\text{In}_\nu(\mathcal{P}_2)}(\eta) > \min_{\text{In}_\nu(\mathcal{P}_1)}(\eta) + \min_{\text{In}_\nu(\mathcal{P}_2)}(\eta) = \min_{\text{In}_\nu(\mathcal{P}_1)+\text{In}_\nu(\mathcal{P}_2)}(\eta).$$

It follows that $\mathcal{Q}_1+\text{In}_\nu(\mathcal{P}_2) \neq \text{In}_\nu(\mathcal{P}_1)(\eta)+\text{In}_\nu(\mathcal{P}_2)$, as required. Finally, assertion (6) follows directly from exercise V.23. □

The example from fig. 5 shows that assertion (5) of proposition V.16 may not be true if some of the $\mathcal{P}_j$ are not bounded [what goes wrong with the proof?]. On the other hand, if $\mathcal{P}_j$ are polytopes, then it is not too hard to show that it remains true even if $\mathcal{Q}_j$ are allowed to be arbitrary convex subsets of $\mathcal{P}_j$, i.e. $\mathrm{In}_\nu(\mathcal{P}_j)$ are unique *convex* subsets of $\mathcal{P}_j$ whose sum is $\mathrm{In}_\nu(\sum_j \mathcal{P}_j)$. Exercise V.28 asks you to show that convexity is necessary for uniqueness.

Corollary V.17. *Every polyhedron has finitely many distinct faces.*

PROOF. Let $\nu \in (\mathbb{R}^n)^*$. If $\mathcal{P}_0 = \mathrm{conv}(S)$ for some finite set $S \subset \mathbb{R}^n$, exercise V.29 implies that $\mathrm{In}_\nu(\mathcal{P}_0) = \mathrm{conv}(\mathrm{In}_\nu(S))$, in particular, every face of the polytope $\mathcal{P}_0$ is the convex hull of a subset of S, and therefore $\mathcal{P}_0$ has finitely many distinct faces. Now consider the case that $\mathcal{C} = \mathrm{cone}(T)$ for a finite subset T of $\mathbb{R}^n$. If $\min_{\mathcal{C}}(\nu)$ exists, then proposition V.16 implies that $\min_{\mathcal{C}}(\nu) = 0$ and $\mathrm{In}_\nu(\mathcal{C}) = \nu^\perp \cap \mathcal{C} = \mathrm{cone}(\nu^\perp \cap T)$. Therefore, the number of distinct faces of $\mathcal{C}$ is bounded by the number of subsets of T. By corollary V.6 every polyhedron $\mathcal{P}$ be can be expressed as $\mathcal{P}_0 + \mathcal{C} + L$, where $\mathcal{P}_0$ is a polytope, $\mathcal{C}$ is a strongly convex polyhedral cone and L is a linear subspace of $\mathbb{R}^n$. Proposition V.16 then implies that every face of $\mathcal{P}$ is of the form $\mathcal{P}_0' + \mathcal{C}' + L$, where $\mathcal{P}_0'$ (respectively, $\mathcal{C}'$) is a face of $\mathcal{P}_0$ (respectively, $\mathcal{C}$). The result then follows from the cases of polytopes and polyhedral cones. □

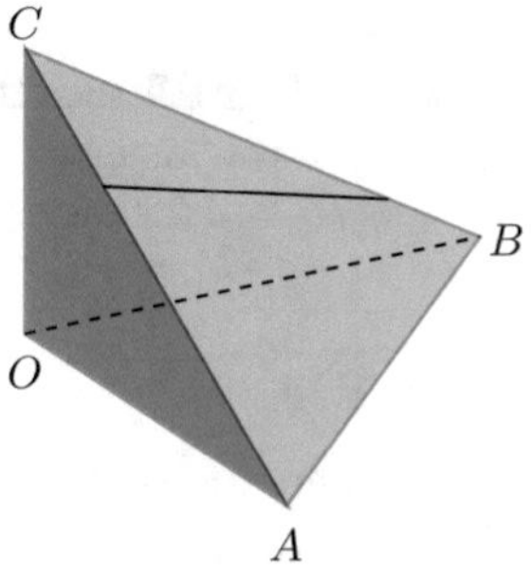

FIGURE 7. ABC is the minimal face containing edges AC and BC. The relative interior of the line segment from a point in $\mathrm{relint}(AC)$ to a point in $\mathrm{relint}(BC)$ is contained in $\mathrm{relint}(ABC)$.

As a corollary of the finiteness of the number of faces, we now prove a more technical result that we use in section V.4. It shows that given two faces $\mathcal{Q}_1, \mathcal{Q}_2$ of a polyhedron $\mathcal{P}$, there is a "smallest" face $\mathcal{Q}$ of $\mathcal{P}$ that contains both $\mathcal{Q}_j$, and given two points in the relative interiors of the $\mathcal{Q}_j$, the relative interior of the line segment joining these points is contained in the relative interior of $\mathcal{Q}$, see fig. 7.

Corollary V.18. *Let $\mathcal{Q}_1, \mathcal{Q}_2$ be faces of a convex polyhedron $\mathcal{P}$.*

(1) There is a (unique) face $\mathcal{Q}$ of $\mathcal{P}$ such that $\mathcal{Q}_1 \cup \mathcal{Q}_2 \subset \mathcal{Q}$, and $\mathcal{Q} \subset \mathcal{Q}'$ for every face $\mathcal{Q}'$ of $\mathcal{P}$ such that $\mathcal{Q}_1 \cup \mathcal{Q}_2 \subset \mathcal{Q}'$.

(2) Pick $\alpha_i \in \mathrm{relint}(\mathcal{Q}_i)$, $i = 1, 2$. Then for every $\epsilon \in (0, 1)$, the positive convex combination $\epsilon\alpha_1 + (1-\epsilon)\alpha_2$ of α_1 and α_2 is in the relative interior of $\mathcal{Q}$.

PROOF. Let $\mathcal{Q}$ be the intersection of all faces of $\mathcal{P}$ containing both $\mathcal{Q}_j$. Proposition V.13 and corollary V.17 imply that $\mathcal{Q}$ is a face of $\mathcal{P}$, and it is clearly as in the first assertion. The second assertion follows from exercise V.29 and the definition of relative interior. □

The following proposition shows that the property of being a face is transitive. Note that this is clear in figs. 6 and 7: every vertex of every edge of $\mathcal{P}$ is also a vertex of $\mathcal{P}$.

Proposition V.19. *Every face of a face of $\mathcal{P}$ is also a face of $\mathcal{P}$.*

PROOF. Let $\mathcal{Q} := \mathrm{In}_\nu(\mathcal{P})$ and $\mathcal{R} := \mathrm{In}_\eta(\mathcal{Q})$, $\nu, \eta \in (\mathbb{R}^n)^*$. Let $\mathcal{P} = \mathcal{P}_0 + \mathcal{C} + L$ be a decomposition as in corollary V.6. Proposition V.16 then implies that

(i) $\mathcal{Q} = \mathrm{In}_\nu(\mathcal{P}_0) + \mathrm{In}_\nu(\mathcal{C}) + L$, where $\nu|_L \equiv 0$ and $\min_{\mathcal{C}}(\nu) = 0$;
(ii) $\mathcal{R} = \mathrm{In}_\eta(\mathrm{In}_\nu(\mathcal{P}_0)) + \mathrm{In}_\eta(\mathrm{In}_\nu(\mathcal{C})) + L$, where $\eta|_L \equiv 0$, and $\min_{\mathrm{In}_\nu(\mathcal{C})}(\eta) = 0$.

Due to proposition V.11 after a translation of $\mathcal{P}$ if necessary we may further assume that $\mathrm{In}_\nu(\mathcal{P}_0)$ contains the origin, which implies that

(iii) $\min_{\mathcal{P}_0}(\nu) = 0$ and
(iv) $\mathcal{Q} \supset \mathrm{In}_\nu(\mathcal{P}_0) \cup \mathrm{In}_\nu(\mathcal{C}) \cup L$.

By theorem V.3 there are finite sets $S, T \subset \mathbb{R}^n$ such that $\mathcal{P}_0 = \mathrm{conv}(S)$ and $\mathcal{C} = \mathrm{cone}(T)$. By observations (i) and (iii), ν is nonnegative on $S \cup T$, and therefore we may choose $r > 0$ such that $\langle r\nu + \eta, \alpha \rangle > 0$ for each $\alpha \in (S \cup T) \setminus \nu^\perp$.

Claim V.19.1. $\mathcal{R} = \mathrm{In}_{r\nu+\eta}(\mathcal{P})$.

PROOF. This is left as exercise V.33. □

The proposition follows immediately from claim V.19.1. □

Proposition V.20. *Let $\mathcal{P}$ be a polytope and $\mathcal{V}$ be the set of vertices of $\mathcal{P}$.*

(1) $\mathcal{P} = \mathrm{conv}(\mathcal{V})$.
(2) $\mathcal{V}$ is the unique minimal set whose convex hull is $\mathcal{P}$.
(3) $\mathrm{relint}(\mathcal{P})$ *is the set of* positive *convex combinations of its vertices, i.e.* $\alpha \in \mathrm{relint}(\mathcal{P})$ *if and only if* $\alpha = \sum_{\beta \in \mathcal{V}} \epsilon_\beta \beta$, *where ϵ_β are positive real numbers such that* $\sum_{\beta \in \mathcal{V}} \epsilon_\beta = 1$.

PROOF. For the first assertion proceed by induction on $\dim(\mathcal{P})$. It is evident when $\dim(\mathcal{P}) = 0$. Now consider the case that $\dim(\mathcal{P}) \geq 1$. Take $\alpha \in \mathcal{P}$. If $\alpha \notin \mathrm{relint}(\mathcal{P})$, then it is on a proper face $\mathcal{Q}$ of $\mathcal{P}$. Since $\dim(\mathcal{Q}) < \dim(\mathcal{P})$, by induction α is in the convex hull of vertices of $\mathcal{Q}$. Proposition V.19 implies that every vertex of $\mathcal{Q}$ is also a vertex of $\mathcal{P}$, so that $\alpha \in \mathrm{conv}(\mathcal{V})$, as required. If $\alpha \in \mathrm{relint}(\mathcal{P})$, then take a line L through α on $\mathrm{aff}(\mathcal{P})$. Since $\mathcal{P}$ is bounded, proposition V.15 implies that each end of L intersects a proper face of $\mathcal{P}$. Since we already showed that every proper face of $\mathcal{P}$ is in $\mathrm{conv}(\mathcal{P})$, it follows that $\alpha \in \mathrm{conv}(\mathcal{V})$ and completes the proof of the first assertion. The second assertion and the $(\Leftarrow)$ implication of the third assertion follow from the first assertion and exercise V.29. Now pick $\alpha \in \mathrm{relint}(\mathcal{P})$. It remains to show that α is a positive convex combination of the vertices of $\mathcal{P}$. By proposition V.12 we may assume $\dim(\mathcal{P}) \geq 1$. Since $\operatorname{relint}(\mathcal{P})$ is relatively open in $\operatorname{aff}(\mathcal{P})$ (proposition V.15), $\alpha - \epsilon \sum_{\beta \in \mathcal{V}} \beta \in \operatorname{relint}(\mathcal{P})$ for sufficiently small positive ϵ. By the first assertion then $\alpha - \epsilon \sum_{\beta \in \mathcal{V}} \beta \in \operatorname{conv}(\mathcal{V})$. It then follows that α is a positive convex combination of elements from $\mathcal{V}$, as required. □

Corollary V.21. *If $\mathcal{C}$ is a strongly convex polyhedral cone, then*

(3) $\mathcal{C}$ is the cone generated by its edges.

(4) $\operatorname{relint}(\mathcal{C})$ *is the set of* positive *linear combinations of nonzero elements of its edges, i.e. if $\mathcal{E}$ is a set consisting of one nonzero element from each of the edges of $\mathcal{C}$, then $\alpha \in \operatorname{relint}(\mathcal{C})$ if and only if $\alpha = \sum_{\beta \in \mathcal{E}} r_\beta \beta$, where r_β are positive real numbers.*

PROOF. Combine proposition V.20 and exercise V.26. □

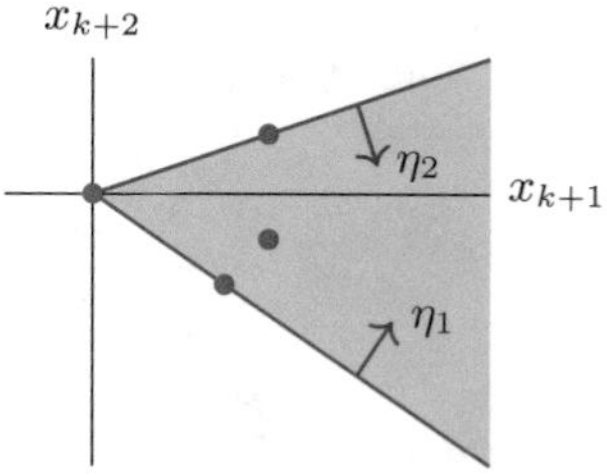

FIGURE 8. $\operatorname{cone}(\pi(S \cup T))$ is a two-dimensional cone contained in the half-plane $x_{k+1} \geq 0$, and intersects x_{k+2}-axis only at the origin.

Proposition V.22. *Every proper face $\mathcal{Q}$ of a polyhedron $\mathcal{P}$ is the intersection of facets of $\mathcal{P}$ containing $\mathcal{Q}$. In particular, every proper face is contained in a facet.*

PROOF. Let $\mathcal{P} = \mathcal{P}_0 + \mathcal{C} + L$ be a decomposition of $\mathcal{P}$ as in corollary V.6. Let $\mathcal{Q} = \operatorname{In}_\nu(\mathcal{P})$, $\nu \in (\mathbb{R}^n)^*$, be a proper face of $\mathcal{P}$. Proposition V.16 implies that $\mathcal{Q} = \operatorname{In}_\nu(\mathcal{P}_0) + \operatorname{In}_\nu(\mathcal{C}) + L$, where $\nu|_L \equiv 0$ and $\min_{\mathcal{C}}(\nu) = 0$. Due to proposition V.11 after an affine isomorphism of $\mathcal{P}$ if necessary we may further assume that $\dim(\mathcal{P}) = n$, and $\operatorname{aff}(\mathcal{Q})$ is the coordinate subspace of $\mathbb{R}^n$ spanned by the first k-coordinates (in particular, $\min_{\mathcal{P}}(\nu) = \min_{\mathcal{P}_0}(\nu) = 0$), where $k := \dim(\mathcal{Q}) < n$, and ν is the projection on to the $(k+1)$-th coordinate. If $n = k+1$, then $\mathcal{Q}$ is a facet, and we are done. So assume $n-k \geq 2$. Then after a linear isomorphism of $\mathcal{P}$ we can also ensure that all points of $\mathcal{Q}$ has zero $(k+2)$-th coordinate. Choose finite sets S, T such that $\mathcal{P}_0 = \operatorname{conv}(S)$ and $\mathcal{C} = \operatorname{cone}(T)$. Let π be the projection map from $\mathbb{R}^n$ onto the two-dimensional coordinate subspace spanned by the $(k+1)$-th and $(k+2)$-th coordinates. By construction every nonzero element of $\pi(S \cup T)$ has *positive* x_{k+1}-coordinate. We claim that the cone generated by $\pi(S \cup T)$ is two dimensional, i.e. it is as shown in fig. 8. Indeed, by construction $\pi(L) = 0$, so that exercises V.1 and V.7 imply that $\pi(\mathcal{P}) = \pi(\mathcal{P}_0) + \pi(\mathcal{C}) = \operatorname{conv}(\pi(S)) + \operatorname{cone}(\pi(T)) \subseteq$

$\mathrm{cone}(\pi(S\cup T))$. Therefore, if $\mathrm{cone}(\pi(S\cup T))$ is contained in a line L, and $\mathcal{P}$ would be contained in $\pi^{-1}(L)$, contradicting the full dimensionality of $\mathcal{P}$. It follows that $\dim(\mathrm{cone}(\pi(S\cup T))) = 2$, and $\mathrm{cone}(\pi(S\cup T))$ has two edges (exercise V.5). As shown in fig. 8, let η_1, η_2 be the elements in $(\mathbb{R}^2)^*$ which attains their minima on the two edges of $\mathrm{cone}(\pi(S\cup T))$. It is straightforward to check that

- for each j, $\mathcal{R}_j := \mathrm{In}_{\pi^*(\eta_j)}(\mathcal{P})$ is a face of $\mathcal{P}$ such that $\mathcal{R}_j \supset \mathcal{Q}$ and $\dim(\mathcal{R}_j) > \dim(\mathcal{Q})$,
- If $n = k+2$, then each $\mathcal{R}_j$ is a facet of $\mathcal{P}$, and $\mathcal{Q} = \mathcal{R}_1 \cap \mathcal{R}_2$.

The proposition follows from these observations by a straightforward induction on $\dim(\mathcal{P}) - \dim(\mathcal{Q})$. □

Corollary V.23. *Let $\mathcal{P}$ be an n-dimensional polyhedron in $\mathbb{R}^n$. Assume $\mathcal{P} \neq \mathbb{R}^n$. Then $\mathcal{P}$ is determined by the affine hyperplanes corresponding to facets of $\mathcal{P}$. More precisely,*

(1) up to multiplications by nonzero real numbers, there is a unique minimal set of inequalities determining $\mathcal{P}$.

Let $a_{i,0}+a_{i,1}x_1+\cdots+a_{i,n}x_n \geq 0$, $i = 1,\ldots,N$, be the minimal set of inequalities defining $\mathcal{P}$. Then

(2) $\mathcal{P} \cap \{x : a_{i,0} + a_{i,1}x_1 + \cdots + a_{i,n}x_n = 0\}$ is a facet of $\mathcal{P}$ for each i;
(3) for every face $\mathcal{Q}$ of $\mathcal{P}$, there is $I \subseteq \{1,\ldots,N\}$ such that $\mathcal{Q} = \mathcal{P} \cap \{x : a_{i,0} + a_{i,1}x_1 + \cdots + a_{i,n}x_n = 0$ for each $i \in I\}$.

PROOF. Let $a_{i,0} + a_{i,1}x_1 + \cdots + a_{i,n}x_n \geq 0$, $i = 1,\ldots,N$, be a minimal set of inequalities defining $\mathcal{P}$. It suffices to show that $\mathcal{Q}_i := \mathcal{P} \cap \{x : a_{i,0} + a_{i,1}x_1 + \cdots + a_{i,n}x_n = 0\}$, $i = 1,\ldots,N$, are the facets of $\mathcal{P}$. Indeed, if $\mathcal{Q}$ is a facet of $\mathcal{P}$, then proposition V.15 implies that $\mathcal{Q} \subset \bigcup_i \mathcal{Q}_i$. Exercise V.31 and assertion (1) of proposition V.14 then imply that $\mathcal{Q} = \mathcal{Q}_i$ for some i. It remains to show that every $\mathcal{Q}_i$ is a facet. Indeed, reorder the inequalities so that $\mathcal{Q}_i$ are facets for $i = 1,\ldots,M$. Let $\mathcal{P}'$ be the polytope determined by $a_{i,0}+a_{i,1}x_1+\cdots+a_{i,n}x_n \geq 0$, $i = 1,\ldots,M$. If $M < N$, then $\mathcal{P} \subsetneq \mathcal{P}'$ and proposition V.15 implies that there is $\alpha \in \mathrm{relint}(\mathcal{P}') \setminus \mathcal{P}$. Pick $\beta \in \mathrm{relint}(\mathcal{P})$ and let L be the line segment from α to β. Then propositions V.14 and V.15 imply that

- for each $i = 1,\ldots,M$, and each $x \in L$, $a_{i,0}+a_{i,1}x_1+\cdots+a_{i,n}x_n > 0$.
- L contains a point γ on the *topological boundary*[4] of $\mathcal{P}$.

But then proposition V.14 implies that γ is a point of a face of $\mathcal{P}$ which is not on any facet of $\mathcal{P}$, contradicting proposition V.22. □

The theory of toric varieties (to be discussed in chapter VI) exploits many similarities between polytopes and algebraic varieties. Here we note some of the more obvious analogues between a polyhedron $\mathcal{P}$ and an irreducible variety X.

[4]Topological boundary of $S \subset \mathbb{R}^n$ is the complement in S of the topological interior of S.

The dimension of every proper irreducible subvariety of X is less than $\dim(X)$ (theorem III.78).	$\leftrightarrow$	The dimension of every proper face of $\mathcal{P}$ is less than $\dim(\mathcal{P})$ (proposition V.14).
If an irreducible subvariety of X is contained in the union of finitely many subvarieties of X, then it is contained in one of them (proposition III.6).	$\leftrightarrow$	If a face of $\mathcal{P}$ is contained in the union of finitely many faces of $\mathcal{P}$, then it is contained in one of them (exercise V.31).

3.1. Exercises.

EXERCISE V.22. Let $\mathcal{Q}$ be a face of $\mathcal{P}$. Show that $\mathcal{Q} = \mathcal{P} \cap \mathrm{aff}(\mathcal{Q})$. [Hint: If $\mathcal{Q} = \mathrm{In}_\nu(\mathcal{P})$, then $\langle \nu, \cdot \rangle$ is constant on $\mathrm{aff}(\mathcal{Q})$.]

EXERCISE V.23. Let $\mathcal{P}$ be a convex polyhedron and $\nu \in (\mathbb{R}^n)^*$. Assume $\inf\{\langle \nu, \alpha \rangle : \alpha \in \mathcal{P}\} > -\infty$, and

(1) either $\mathcal{P}$ is a cone, or
(2) $\mathcal{P}$ is a linear subspace of $\mathbb{R}^n$.

Then $\mathrm{In}_\nu(\mathcal{P})$ contains the origin and $\min_{\mathcal{P}}(\nu) = 0$. In addition, in case (2) $\mathcal{P} \subset \nu^\perp$, and $\mathrm{In}_\nu(\mathcal{P}) = \mathcal{P}$. [Hint: In either case $\mathcal{P}$ contains the origin, so that $\min_{\mathcal{P}}(\nu) \leq 0$. If there is $\alpha \in \mathcal{P}$ such that $\langle \nu, \alpha \rangle < 0$, then $\langle \nu, r\alpha \rangle$ approaches $-\infty$ as $r \to \infty$.]

EXERCISE V.24. Prove proposition V.11. [Hint: Use assertion (1) of exercise V.14.]

EXERCISE V.25. Let $S \subset \mathbb{R}^n$ and ν be a nonzero element in $(\mathbb{R}^n)^*$ such that $m := \inf\{\langle \nu, \alpha \rangle : \alpha \in S\} > -\infty$. Show that no point of the hyperplane $\{x \in \mathbb{R}^n : \langle \nu, \alpha \rangle = m\}$ is in the topological interior of S in $\mathbb{R}^n$.

EXERCISE V.26. Let $\mathcal{C}$ be a positive dimensional strongly convex polyhedral cone. By theorem V.3 there is $\nu \in (\mathbb{R}^n)^*$ such that ν is positive on $\mathcal{C} \setminus \{0\}$. Let $\mathcal{P} := \{\alpha \in \mathcal{P} : \langle \nu, \alpha \rangle = 1\}$. Claim V.3.2 implies that $\mathcal{P}$ is a polytope.

(1) If $\mathcal{Q}$ is a face of $\mathcal{P}$, show that $\mathrm{cone}(\mathcal{Q})$ is a face of $\mathcal{C}$ of dimension $\dim(\mathcal{Q}) + 1$.
(2) Show that the correspondence $\mathcal{Q} \mapsto \mathrm{cone}(\mathcal{Q})$ induces a bijection between faces of $\mathcal{P}$ and positive dimensional faces of $\mathcal{C}$.
(3) Show that the above correspondence also preserves the relative interiors of the faces, i.e. If $\mathcal{Q}$ is a face of $\mathcal{P}$, then $\mathrm{relint}(\mathrm{cone}(\mathcal{Q})) = \mathrm{cone}(\mathrm{relint}(\mathcal{Q}))$.

[Hint: Change coordinates on such that ν is the projection onto the last coordinate.]

EXERCISE V.27. Let $\mathcal{P}$ be a polyhedron, and $\mathcal{Q} = \mathrm{In}_\nu(\mathcal{P})$ and $\mathcal{R} = \mathrm{In}_\eta(\mathcal{P})$ be faces of $\mathcal{P}$. If $\mathcal{Q} \cap \mathcal{R} = \emptyset$, show by an example that it may not be true that $\mathrm{In}_{\nu+\eta}(\mathcal{P}) = \mathcal{Q} \cap \mathcal{R}$.

EXERCISE V.28. Find examples of polytopes $\mathcal{P}_1, \mathcal{P}_2$ in $\mathbb{R}^n$ and faces $\mathcal{Q}_j$ of $\mathcal{P}_j$ such that $\mathcal{Q}_1 + \mathcal{Q}_2$ is a face of $\mathcal{P}_1 + \mathcal{P}_2$, and there are $\mathcal{Q}'_j \subsetneq \mathcal{Q}_j$ such that

$\mathcal{Q}'_1 + \mathcal{Q}'_j = \mathcal{Q}_1 + \mathcal{Q}_2$. [Hint: There are examples with $n = 1$ with $\mathcal{Q}_2 = \mathcal{P}_j$ for each j.]

EXERCISE V.29. Let $\alpha = \sum_{j=1}^{k} \epsilon_j \alpha_j$ be a convex combination of $\alpha_1, \ldots, \alpha_k \in \mathbb{R}^n$. Assume each ϵ_j is positive.

(1) If $\nu \in (\mathbb{R}^n)^*$, then show that $\langle \nu, \alpha \rangle \geq \min_j \langle \nu, \alpha_j \rangle$, with equality if and only if $\langle \nu, \alpha_1 \rangle = \cdots = \langle \nu, \alpha_k \rangle$.
(2) Conclude that if the α_j are points of a polyhedron $\mathcal{P}$, and $\mathcal{Q}$ is a face of $\mathcal{P}$, then $\mathcal{Q}$ contains α if and only if $\mathcal{Q}$ contains each α_j.

EXERCISE V.30. Let S be a finite subset of $\mathbb{R}^n$. Let $\mathcal{P} := \operatorname{conv}(S)$ and $\mathcal{C}$ be the cone generated by S. Show that

(1) $\alpha \in \operatorname{relint}(\mathcal{P})$ if and only if $\alpha = \sum_{\beta \in S} \epsilon_\beta \beta$, where ϵ_β are positive real numbers such that $\sum_{\beta \in S} \epsilon_\beta = 1$.
(2) $\alpha \in \operatorname{relint}(\mathcal{C})$ if and only if $\alpha = \sum_{\beta \in S} r_\beta \beta$, where r_β are positive real numbers.

[Hint: Follow the arguments from the proof of assertion (3) of proposition V.20.]

EXERCISE V.31. Let $\mathcal{Q}_1, \ldots, \mathcal{Q}_k$ be faces of a polyhedron $\mathcal{P}$. If $\mathcal{Q}$ is a convex subset (e.g. a face) of $\mathcal{P}$ such that $\mathcal{Q} \subseteq \bigcup_j \mathcal{Q}_j$, then show that $\mathcal{Q} \subseteq \mathcal{Q}_j$ for some j. [Hint: Otherwise for each j, there is $\alpha_j \in \mathcal{Q} \setminus \mathcal{Q}_j$. Apply exercise V.29 to a positive convex combination of the α_j.]

EXERCISE V.32. Given distinct vertices α, α' of a polytope $\mathcal{P}$, show that there are vertices $\beta_0 = \alpha, \beta_2, \ldots, \beta_k = \alpha'$ of $\mathcal{P}$ such that there is an edge of $\mathcal{P}$ connecting β_{j-1} to β_j for each $j = 1, \ldots, k$; in other words, the "edge-graph" of $\mathcal{P}$ is connected. [Hint: Reduce to the case that $\mathcal{P}$ is full dimensional. Pick $\nu, \nu' \in (\mathbb{R}^n)^*$ such that $\operatorname{In}_\nu(\mathcal{P}) = \{\alpha\}$ and $\operatorname{In}_{\nu'}(\mathcal{P}) = \{\alpha'\}$. Consider $\mathcal{P}_\epsilon := \operatorname{In}_{\nu+\epsilon\nu'}(\mathcal{P})$ for $\epsilon \geq 0$. As ϵ goes from 0 to ∞, $\mathcal{P}_\epsilon$ transitions from $\{\alpha\}$ to $\{\alpha'\}$ in finitely many steps. Apply induction on dimension to each $\mathcal{P}_\epsilon$.]

EXERCISE V.33. Prove claim V.19.1. [Hint: Every $\alpha \in \mathcal{P}$ can be expressed as $\sum_{\alpha \in S} \epsilon_\alpha \alpha + \sum_{\beta \in T} r_\beta \beta + \gamma$, where $\gamma \in L$, r_β are nonnegative real numbers, and ϵ_α are nonnegative real numbers such that $\sum_{\alpha \in S} \epsilon_\alpha = 1$. Define $\alpha' := \sum_{\alpha \in S \cap \nu^\perp} \epsilon_\alpha \alpha$, $\alpha'' := \sum_{\alpha \in S \setminus \nu^\perp} \epsilon_\alpha \alpha$, $\beta' := \sum_{\beta \in T \cap \nu^\perp} r_\beta \beta$ and $\beta'' := \sum_{\beta \in T \setminus \nu^\perp} r_\beta \beta$. Examine the value of $r\nu + \eta$ on each of $\alpha', \alpha'', \beta', \beta''$. Show that $\langle r\nu + \eta, \alpha \rangle \geq \min_{\operatorname{In}_\nu(\mathcal{P}_0)}(\eta)$, with equality if and only if $\alpha'' = \beta'' = 0$, $\alpha' \in \operatorname{In}_\eta(\operatorname{In}_\nu(\mathcal{P}_0))$ and $\beta' \in \operatorname{In}_\eta(\operatorname{In}_\nu(\mathcal{C}))$.]

4. Normal fan of a convex polytope

A *fan*[5] in $\mathbb{R}^n$ is a collection Σ of convex polyhedral cones in $\mathbb{R}^n$ such that

(1) Each face of a cone in Σ is also a cone in Σ.
(2) The intersection of any two cones in Σ is a face of each of them.

[5]Our definition of a fan differs from the definition in standard texts on toric varieties (e.g.[Ful93, CLS11]) in that we do not require the cones in a fan to be strongly convex.

Any (finite dimensional) vector space V over real numbers can be identified with $\mathbb{R}^n$ after choosing a basis $\mathcal{B}$, and thus the notions of convex polyhedra, cones, polytopes, fans, etc., can be extended to subsets of V. In this section we take $V = (\mathbb{R}^n)^*$ and $\mathcal{B}$ to be the basis dual to the standard basis of $\mathbb{R}^n$. Exercise V.1 implies that convex polyhedra (and therefore cones, polytopes, fans, etc.) in $(\mathbb{R}^n)^*$ which are defined in this way remain so after linear changes of coordinates on $\mathbb{R}^n$. Let $\mathcal{P}$ be a convex polytope in $\mathbb{R}^n$. For each face $\mathcal{Q}$ of $\mathcal{P}$, define

$$\sigma_{\mathcal{Q}} := \{\nu \in (\mathbb{R}^n)^* : \operatorname{In}_\nu(\mathcal{P}) \supseteq \mathcal{Q}\} \subset (\mathbb{R}^n)^*. \tag{43}$$

It is straightforward to see that $\sigma_{\mathcal{Q}}$ is a convex polyhedral cone in $(\mathbb{R}^n)^*$ (exercise V.35); it is called the *normal cone* of $\mathcal{Q}$. Let $\Sigma_{\mathcal{P}} := \{\sigma_{\mathcal{Q}} : \mathcal{Q}$ is a face of $\mathcal{P}\}$. We will show in corollary V.27 that $\Sigma_{\mathcal{P}}$ is a fan in $(\mathbb{R}^n)^*$; this is called the *normal fan* of $\mathcal{P}$. Given a face $\mathcal{Q}$ of $\mathcal{P}$, let $\sigma^0_{\mathcal{Q}} := \{\nu \in (\mathbb{R}^n)^* : \operatorname{In}_\nu(\mathcal{P}) = \mathcal{Q}\} \subset \sigma_{\mathcal{Q}}$. We show in corollary V.26 that $\sigma^0_{\mathcal{Q}}$ is the relative interior of $\sigma_{\mathcal{Q}}$ (fig. 9).

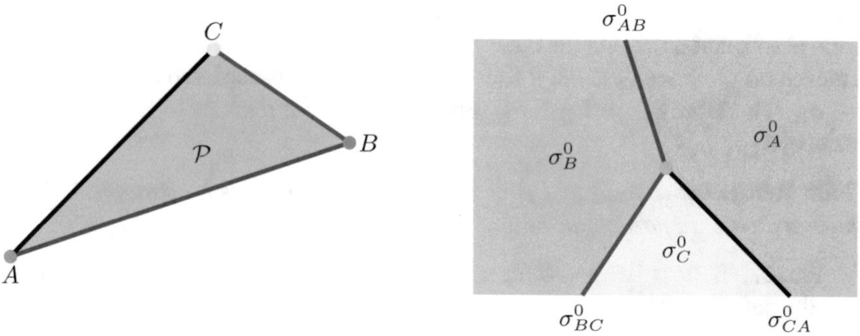

FIGURE 9. Normal fan of a triangle $\mathcal{P}$ in $\mathbb{R}^2$. $\sigma_{\mathcal{P}} = \sigma^0_{\mathcal{P}}$ is the origin.

Proposition V.24. *Let $\mathcal{R}_1, \ldots, \mathcal{R}_k$ be the faces of $\mathcal{P}$ containing $\mathcal{Q}$. Then* $\sigma^0_{\mathcal{Q}} = \sigma_{\mathcal{Q}} \setminus \bigcup_j \sigma_{\mathcal{R}_j}$.

PROOF. This follows directly from the definition of a face and the definition of $\sigma_{\mathcal{Q}}$. □

Let $L_{\mathcal{Q}}$ be the linear subspace of $\mathbb{R}^n$ spanned by all vectors of the form $\alpha - \beta$ with $\alpha, \beta \in \mathcal{Q}$ and $L^\perp_{\mathcal{Q}}$ be the linear subspace of $(\mathbb{R}^n)^*$ consisting of all $\nu \in (\mathbb{R}^n)^*$ such that $\nu|_{L_{\mathcal{Q}}} \equiv 0$.

Proposition V.25. $L^\perp_{\mathcal{Q}} = \operatorname{aff}(\sigma_{\mathcal{Q}}) = \operatorname{aff}(\sigma^0_{\mathcal{Q}})$. *In particular,* $\dim(\sigma_{\mathcal{Q}}) = n - \dim(\mathcal{Q})$.

PROOF. Since $\mathcal{Q}$ is a face of $\mathcal{P}$, there is $\nu' \in (\mathbb{R}^n)^*$ such that $\mathcal{Q} = \min_{\mathcal{P}}(\nu')$. It follows that $\nu' \in \sigma^0_{\mathcal{Q}}$, in particular, $\sigma^0_{\mathcal{Q}}$ is nonempty.

Claim V.25.1. *Let $\nu \in \sigma^0_{\mathcal{Q}}$. Then for each $\mu \in L^\perp_{\mathcal{Q}}$, $\nu + \epsilon\mu \in \sigma^0_{\mathcal{Q}}$ for all sufficiently small positive ϵ.*

PROOF. Indeed, let $m := \min_{\mathcal{P}}(\nu)$ and pick ϵ such that for all vertices α of $\mathcal{P}$ not on $\mathcal{Q}$, $\epsilon\langle -\mu, \alpha\rangle < \langle \nu, \alpha\rangle - m$. □

Claim V.25.1 implies that $\operatorname{aff}(\sigma_{\mathcal{Q}}^0) \supset L_{\mathcal{Q}}^{\perp}$. On the other hand, if $\nu \in \sigma_{\mathcal{Q}}^0$, then for each $\alpha, \beta \in \mathcal{Q}$, $\langle \nu, \alpha - \beta\rangle = 0$, so that $\nu \in L_{\mathcal{Q}}^{\perp}$. It follows that $\operatorname{aff}(\sigma_{\mathcal{Q}}^0) = L_{\mathcal{Q}}^{\perp}$. Since $\sigma_{\mathcal{Q}}$ is the union of the $\sigma_{\mathcal{R}}^0$ over all faces $\mathcal{R}$ of $\mathcal{P}$ containing $\mathcal{Q}$, and since $L_{\mathcal{R}}^{\perp} \subset L_{\mathcal{Q}}^{\perp}$ whenever $\mathcal{R}$ contains $\mathcal{Q}$, it follows that $\operatorname{aff}(\sigma_{\mathcal{Q}}) = L_{\mathcal{Q}}^{\perp}$ as well, as required. □

Corollary V.26. *Let $\mathcal{Q}$ be a face of $\mathcal{P}$.*

(1) If $\mathcal{R}$ is a face of $\mathcal{P}$ containing $\mathcal{Q}$, then $\sigma_{\mathcal{R}} = L_{\mathcal{R}}^{\perp} \cap \sigma_{\mathcal{Q}}$. In particular, $\sigma_{\mathcal{R}}$ is a face of $\sigma_{\mathcal{Q}}$.
(2) $\sigma_{\mathcal{Q}}^0$ is the relative interior of $\sigma_{\mathcal{Q}}$.
(3) Every face of $\sigma_{\mathcal{Q}}$ is of the form $\sigma_{\mathcal{R}}$ for some face $\mathcal{R}$ of $\mathcal{P}$ containing $\mathcal{Q}$.

PROOF. In this proof we will identify $\mathbb{R}^n$ with $((\mathbb{R}^n)^*)^*$ in the usual way by treating $\alpha \in \mathbb{R}^n$ as the linear function on $(\mathbb{R}^n)^*$ given by $\nu \mapsto \langle \nu, \alpha\rangle$. Proposition V.25 implies that $\sigma_{\mathcal{R}} \subset L_{\mathcal{R}}^{\perp} \cap \sigma_{\mathcal{Q}}$. Now pick $\nu \in L_{\mathcal{R}}^{\perp} \cap \sigma_{\mathcal{Q}}$. Since $\nu \in L_{\mathcal{R}}^{\perp}$, it follows that ν is constant on $\mathcal{R}$. On the other hand, since $\nu \in \sigma_{\mathcal{Q}}$, it follows that $\min_{\mathcal{P}}(\nu)$ is achieved on $\mathcal{Q}$. Since $\mathcal{Q} \subset \mathcal{R}$, it follows that $\min_{\mathcal{P}}(\nu)$ is achieved on all of $\mathcal{R}$, and $\nu \in \sigma_{\mathcal{R}}$. Therefore $\sigma_{\mathcal{R}} = L_{\mathcal{R}}^{\perp} \cap \sigma_{\mathcal{Q}}$. For assertion (1) it remains to show that $\sigma_{\mathcal{R}}$ is a face of $\sigma_{\mathcal{Q}}$.

Claim V.26.1. *Let $\alpha \in \mathcal{Q}$ and $\beta \in \mathcal{R}$. Then $\beta - \alpha$ induces a nonnegative linear function on $\sigma_{\mathcal{Q}}$. In particular, $\min_{\sigma_{\mathcal{Q}}}(\beta - \alpha) = 0$.*

PROOF. If $\langle \nu, \beta - \alpha\rangle < 0$ for some $\nu \in \sigma_{\mathcal{Q}}$, then $\alpha \notin \operatorname{In}_\nu(\mathcal{P})$, which contradicts the definition of $\sigma_{\mathcal{Q}}$.

Choose a basis of $L_{\mathcal{R}}$ of the form $\beta_j - \alpha$, $j = 1, \ldots, \dim(L_{\mathcal{R}})$, where $\alpha \in \mathcal{Q}$, and each $\beta_j \in \mathcal{R}$. Then claim V.26.1 implies that $\sigma_{\mathcal{R}} = L_{\mathcal{R}}^{\perp} \cap \sigma_{\mathcal{Q}} = \bigcap_j(\sigma_{\mathcal{Q}} \cap (\beta_j - \alpha)^{\perp}) = \bigcap_j(\operatorname{In}_{\beta_j - \alpha}(\sigma_{\mathcal{Q}}))$ is an intersection of finitely many faces of $\sigma_{\mathcal{Q}}$, and therefore is also a face of $\sigma_{\mathcal{Q}}$. This completes the proof of assertion (1), and due to proposition V.24 also implies that $\sigma_{\mathcal{Q}}^0$, being the complement of a union of faces of $\sigma_{\mathcal{Q}}$, contains $\operatorname{relint}(\sigma_{\mathcal{Q}})$. On the other hand, claim V.25.1 implies that for each $\nu \in \sigma_{\mathcal{Q}}$, one can fit inside $\sigma_{\mathcal{Q}}$ a $\dim(\sigma_{\mathcal{Q}})$-dimensional ball B_ν centered at ν. Proposition V.15 then implies that $\sigma_{\mathcal{Q}}^0 \subset \operatorname{relint}(\sigma_{\mathcal{Q}})$. Consequently, $\sigma_{\mathcal{Q}}^0 = \operatorname{relint}(\sigma_{\mathcal{Q}})$, which proves assertion (2). By assertions (1) and (2), and corollary V.17 and exercise V.31, every proper face of $\sigma_{\mathcal{Q}}$ is a face of $\sigma_{\mathcal{R}}$ for some face $\mathcal{R}$ of $\mathcal{P}$ properly containing $\mathcal{Q}$. Assertion (3) therefore follows from a straightforward induction on $\dim(\mathcal{P}) - \dim(\mathcal{Q})$. □

Corollary V.27. *$\Sigma_{\mathcal{P}}$ is a fan in $(\mathbb{R}^n)^*$.*

PROOF. Corollary V.26 shows that $\Sigma_{\mathcal{P}}$ satisfies property (1) of a fan. Let $\mathcal{Q}_1$ and $\mathcal{Q}_2$ be faces of $\mathcal{P}$, and let $\mathcal{Q}$ from corollary V.18 be the smallest face of $\mathcal{P}$ containing both $\mathcal{Q}_j$. It is clear that $\sigma_{\mathcal{Q}} \subset \sigma_{\mathcal{Q}_1} \cap \sigma_{\mathcal{Q}_2}$. On the other hand, since $\mathcal{Q}$ is contained in every face of $\mathcal{P}$ containing both $\mathcal{Q}_j$, it follows that for every $\nu \in \sigma_{\mathcal{Q}_1} \cap \sigma_{\mathcal{Q}_2}$, $\operatorname{In}_\nu(\mathcal{P}) \supset \mathcal{Q}$, so that $\nu \in \sigma_{\mathcal{Q}}$. Therefore $\sigma_{\mathcal{Q}_1} \cap \sigma_{\mathcal{Q}_2} = \sigma_{\mathcal{Q}}$, which is a face of each $\sigma_{\mathcal{Q}_j}$ (corollary V.26). This shows $\Sigma_{\mathcal{P}}$ satisfies property (2) of a fan as well. □

Recall that in our definition the cones in a fan do not have to be strongly convex. They do however turn out to be so in an important case.

Proposition V.28. *If $\mathcal{P}$ is an n-dimensional convex polytope in $\mathbb{R}^n$, then each cone of $\Sigma_{\mathcal{P}}$ is strongly convex.*

PROOF. Indeed, if $\dim(\mathcal{P}) = n$, then for every $\nu \in (\mathbb{R}^n)^* \setminus \{0\}$, $\mathcal{P} \setminus \mathrm{In}_\nu(\mathcal{P})$ is nonempty. If $\nu \in \sigma_{\mathcal{Q}}$ for some face $\mathcal{Q}$ of $\mathcal{P}$, it then follows from (43) that $\mathrm{In}_{-\nu}(\mathcal{P}) \cap \mathcal{Q} = \emptyset$, i.e. $-\nu \notin \sigma_{\mathcal{Q}}$, as required. □

4.1. Exercises.

EXERCISE V.34. Let $\mathcal{P}$ be an n-dimensional polyhedron in $\mathbb{R}^n$, $\mathcal{Q}$ be a facet of $\mathcal{P}$ and $\alpha \in \mathrm{relint}(\mathcal{Q})$. Let $\mathcal{Q} = \mathrm{In}_\nu(\mathcal{P})$, $\nu \in (\mathbb{R}^n)^*$. If β is an arbitrary element of $\mathbb{R}^n$ such that $\langle \nu, \beta \rangle \geq 0$, then show that $r\alpha + \beta \in r\mathcal{P}$ for all $r \gg 1$. [Hint: Use corollary V.23 to show that $\alpha + \epsilon\beta \in \mathcal{P}$ if ϵ is a sufficiently small positive number.]

EXERCISE V.35. Let $\mathcal{Q}$ be a face of a polytope $\mathcal{P}$. Show that $\sigma_{\mathcal{Q}}$ defined in section V.4 is a convex polyhedral cone in $(\mathbb{R}^n)^*$. [Hint: Express $\mathcal{P}$ and $\mathcal{Q}$ as convex hulls of finite sets. Show that the condition defining $\sigma_{\mathcal{Q}}$ can be expressed in terms of finitely many linear inequalities as given in exercise V.6.]

5. Rational polyhedra

We say that a convex polyhedron $\mathcal{P}$ is *rational* if it can be defined in $\mathbb{R}^n$ by finitely many inequalities of the form $a_0 + a_1x_1 + \cdots + a_nx_n \geq 0$ where each a_i is a rational number (or equivalently, an integer). In particular, a *rational affine hyperplane* is the set of zeroes in $\mathbb{R}^n$ of an equation $a_0 + a_1x_1 + \cdots + a_nx_n = 0$, where each $a_i \in \mathbb{Q}$, and $(a_1, \ldots, a_n) \neq (0, \ldots, 0)$. Proposition V.29 gives a characterization of rational polyhedra. In particular, it implies that a polytope is rational if and only if its vertices have rational coordinates, and a polyhedral cone is rational if and only if each of its edges have a nonzero element with rational coordinates. We say that a polytope in $\mathbb{R}^n$ is *integral* if all its vertices are in $\mathbb{Z}^n$.

Proposition V.29. *Let $\mathcal{P} \subset \mathbb{R}^n$.*

(1) $\mathcal{P}$ is a rational polytope if and only if it is the convex hull of finitely many points with rational coordinates.
(2) $\mathcal{P}$ is a rational polyhedral cone if and only if it is the cone generated by finitely many points with rational coordinates.
(3) $\mathcal{P}$ is a rational polyhedron if and only if it is the Minkowski sum of a rational convex polyhedral cone and a rational convex polytope.

PROOF. At first assume $\mathcal{P} = \mathrm{conv}(S) + \mathrm{cone}(T)$ for $S, T \subset \mathbb{Q}^n$. We claim that $\mathcal{P}$ is a rational polyhedron. Due to exercise V.36 the claim reduces to the case that $\dim(\mathrm{aff}(\mathcal{P})) = n$. By the results from section V.3, $\mathcal{P}$ is an n-dimensional polyhedron in $\mathbb{R}^n$, and if $\mathcal{Q}$ is a facet of $\mathcal{P}$, then $\mathcal{Q} = \mathrm{conv}(S') + \mathrm{cone}(T')$ for $S' \subset S$

and $T' \subset T$. It is then straightforward to see that $\operatorname{aff}(\mathcal{Q})$ is a rational affine hyperplane. Since an n-dimensional polyhedron in $\mathbb{R}^n$ is determined by the hyperplanes containing its facets (corollary V.23), it follows that $\mathcal{P}$ is a rational polyhedron, as required, and proves the ($\Leftarrow$) implications of all three assertions of proposition V.29. We now prove the ($\Rightarrow$) implications of all three assertions simultaneously by induction on $\dim(\mathcal{P})$. Due to exercise V.36 we may assume without loss of generality that $\dim(\mathcal{P}) = n$. The cases of $\dim(\mathcal{P}) = 0$ and $\dim(\mathcal{P}) = 1$ are then obvious. Now assume $\dim(\mathcal{P}) \geq 2$. Since the vertices and edges of $\mathcal{P}$ are also rational polyhedra (corollary V.23), the inductive hypothesis and proposition V.20 and corollary V.21 imply that if $\mathcal{P}$ is a polytope, then $\mathcal{P} = \operatorname{conv}(S)$ for a finite $S \subset \mathbb{Q}^n$, and if $\mathcal{P}$ is a strongly convex polyhedral cone, then $\mathcal{P} = \operatorname{cone}(T)$ for a finite $T \subset \mathbb{Q}^n$. In general, corollary V.6 implies that $\mathcal{P} = \mathcal{P}_0 + \mathcal{C}$, where $\mathcal{P}_0$ is a rational convex polytope and $\mathcal{C}$ is a rational convex polyhedron. Due to what we already proved, it suffices to show that $\mathcal{C} = \operatorname{cone}(T)$ with finite $T \subset \mathbb{Q}^n$, even if $\mathcal{C}$ is *not* strongly convex. Indeed, if $\mathcal{C}$ is not strongly convex, then proposition V.1 and corollary V.6 imply that $\mathcal{C} = \mathcal{C}' + L$ where $\mathcal{C}'$ is a cone and L is a *positive* dimensional rational linear subspace of $\mathbb{R}^n$. Let $k := \dim(L)$. Then we can choose a surjective linear map $\pi : \mathbb{R}^n \to \mathbb{R}^{n-k}$ such that $\ker(\pi) = L$, and π maps $\mathbb{Q}^n$ onto $\mathbb{Q}^{n-k}$. Exercise V.37 implies that $\pi(\mathcal{C})$ is a rational polyhedral cone, and therefore, by the inductive hypothesis $\pi(\mathcal{C}) = \operatorname{cone}(T')$ for some finite $T' \subseteq \mathbb{Q}^{n-k}$. Another application of exercise V.37 then implies that $\mathcal{C} = \operatorname{cone}(T)$ for a finite $T \subseteq \mathbb{Q}^n$, as required. □

Corollary V.30. *If $\mathcal{P}$ is a convex rational polytope, then each cone in the normal fan of $\mathcal{P}$ is also rational (with respect to the basis on $(\mathbb{R}^n)^*$ which is dual to the standard basis of $\mathbb{R}^n$).*

PROOF. Let $\mathcal{Q}$ be a face of $\mathcal{P}$. We will show that the cone $\sigma_{\mathcal{Q}}$ from section V.4 is rational. Corollary V.23 implies that $\mathcal{Q}$ is also a convex rational polytope. Therefore by proposition V.29 there are finite subsets S, T of $\mathbb{Q}^n$ such that each $\mathcal{P}$ is $\mathcal{P} = \operatorname{conv}(S)$ and $\mathcal{Q} = \operatorname{conv}(T)$. Since $\sigma_{\mathcal{Q}} = \{\nu \in (\mathbb{R}^n)^* : \langle \nu, \beta - \beta' \rangle = 0$ for all $\beta, \beta' \in T$, and $\langle \nu, \alpha - \beta \rangle \geq 0$ for all $\alpha \in S$, $\beta \in T\}$, it follows that $\sigma_{\mathcal{Q}}$ is a rational polyhedral cone, as required. □

Recall that the *Hausdorff distance* of $\mathcal{P}, \mathcal{Q} \subset \mathbb{R}^n$ is $\sup_{\alpha \in \mathcal{P}}\{\inf_{\beta \in \mathcal{Q}} ||\alpha - \beta||\}$, where $|| \cdot ||$ is the Euclidean norm on $\mathbb{R}^n$.

Corollary V.31. *Every polytope can be approximated arbitrarily closely (with respect to the Hausdorff distance) by rational polytopes.*

PROOF. This follows from theorem V.3 and proposition V.29, since to approximate the convex hull of $\alpha_1, \ldots, \alpha_N \in \mathbb{R}^n$, it suffices to take the convex hull of $\beta_1, \ldots, \beta_N$, where each β_j has rational coordinates, and is sufficiently close to α_j. □

Many results from the theory of rational polyhedra (including proposition V.29) are ultimately based on the following basic fact from linear algebra.

Lemma V.32. *Consider a linear system of equations of the form* $a_{i,0} + a_{i,1}x_1 + \cdots + a_{i,n}x_n = 0$, $i = 1, \ldots, k$. *Assume each* $a_{i,j} \in \mathbb{Q}$. *Let* $H_{\mathbb{Q}}$ *and* $H_{\mathbb{R}}$ *be the set of common solutions of this system, respectively, in* $\mathbb{Q}^n$ *and* $\mathbb{R}^n$. *Then* $\dim_{\mathbb{Q}}(H_{\mathbb{Q}}) = \dim_{\mathbb{R}}(H_{\mathbb{R}})$. *In particular, the system has a common solution over* $\mathbb{R}$ *if and only if it has a common solution over* $\mathbb{Q}$.

PROOF. If the system has a solution, then it can be found by Gaussian elimination, which produces a solution in $\mathbb{Q}$ (since each $a_{i,j} \in \mathbb{Q}$). Moreover, in that case the dimension of the space of solutions is precisely n minus the rank of the $k \times n$ matrix $(a_{i,j})$, $1 \leq i \leq k$, $1 \leq j \leq n$. Since the rank over $\mathbb{Q}$ is the same as the rank over $\mathbb{R}$, this completes the proof. □

Corollary V.33. (Cf. proposition V.20). *Let* $\mathcal{P}$ *be a convex rational polyhedron and* α *be a point with rational coordinates in the relative interior of* $\mathcal{P}$.

(1) *If* $\mathcal{P}$ *is a polytope with vertices* $\alpha_1, \ldots, \alpha_k$, *then there are positive rational numbers* $q_1, \ldots, q_k$ *such that* $\sum_j q_j = 1$ *and* $\alpha = \sum_j q_j \alpha_j$.
(2) *If* $\mathcal{P}$ *is a cone generated by* $\alpha_1, \ldots, \alpha_k \in \mathbb{Z}^n$, *then there are positive rational numbers* $q_1, \ldots, q_k$ *such that* $\alpha = \sum_j q_j \alpha_j$.

PROOF. Let $\phi : \mathbb{R}^k \to \mathbb{R}^n$ be the map which sends $(r_1, \ldots, r_k) \mapsto \sum_{j=1}^k r_j \alpha_j - \alpha$. If $\mathcal{P}$ is a cone, then corollary V.21 implies that $\phi^{-1}(0) \cap \mathbb{R}^k_{>0} \neq \emptyset$. Lemma V.32 then implies that $\phi^{-1}(0) \cap \mathbb{Q}^k_{>0} \neq \emptyset$, which together with corollary V.21 proves assertion (2). Assertion (1) follows by the same arguments from proposition V.20 and lemma V.32 by considering the map $\phi' : \mathbb{R}^k \to \mathbb{R}^{n+1}$ given by $(r_1, \ldots, r_k) \mapsto (\phi(r_1, \ldots, r_k), r_1 + \cdots + r_k - 1)$. □

The following result is one of the foundations of the theory of toric varieties we will encounter in chapter VI.

Lemma V.34. (Gordan's lemma). *If* σ *is a rational convex polyhedral cone in* $\mathbb{R}^n$, *then* $\sigma \cap \mathbb{Z}^n$ *is a finitely generated semigroup.*

PROOF. It is straightforward to check that $\sigma \cap \mathbb{Z}^n$ is a semigroup. Pick $\alpha_1, \ldots, \alpha_s \in S_\sigma$ which generate σ as a cone. Let $K := \{\sum_{i=1}^s t_i \alpha_i : t_i \in \mathbb{R}, 0 \leq t_i \leq 1\} \subset \mathbb{R}^n$. Since K is compact, $K \cap \mathbb{Z}^n$ is a finite set. Now pick $\alpha \in \sigma \cap \mathbb{Z}^n$. Then $\alpha = \sum_{i=1}^s r_i \alpha_i$, where each $r_i \geq 0$. Write $\alpha = \sum_{i=1}^s \lfloor r_i \rfloor \alpha_i + \beta$, where $\lfloor r_i \rfloor$ is the greatest integer less than or equal to r_i for each i, and $\beta := \sum_i (r_i - \lfloor r_i \rfloor)\alpha_i \in K \cap \mathbb{Z}^n$. It follows that $\sigma \cap \mathbb{Z}^n$ is generated as a semigroup by all the α_j together with $K \cap \mathbb{Z}^n$. □

5.1. Exercises.

EXERCISE V.36. Assume either $\mathcal{P}$ is a rational polyhedron or $\mathcal{P} = \mathrm{conv}(S) + \mathrm{cone}(T)$, where $S, T \subset \mathbb{Q}^n$. If $\dim(\mathrm{aff}(\mathcal{P})) < n$, then show that there is a rational affine hyperplane in $\mathbb{R}^n$ containing $\mathcal{P}$. [Hint: If $\mathcal{P}$ is a polyhedron in $\mathbb{R}^n$ with $\dim(\mathcal{P}) < n$, then the set of points for which each of the inequalities defining

$\mathcal{P}$ is strict must be empty, for otherwise $\dim(\mathcal{P})$ would be n. In the other case, $\mathcal{P} = \text{conv}(S) + \text{cone}(T)$, with $S, T \subset \mathbb{Q}^n$. Let $L_{\mathcal{P}}$ be the linear subspace of $\mathbb{R}^n$ generated by all elements of the form $\alpha - \beta$, where $\alpha - \beta \in \mathcal{P}$. It is possible to choose a generator of $L_{\mathcal{P}}$ consisting of elements of the form $\alpha_1 - \alpha_2 + \beta_1 - \beta_2$, where $\alpha_1, \alpha_2 \in S$ and β_1, β_2 are scalar multiples of elements from T. Since each such element is in $\mathbb{Q}^n$, it follows that $L_{\mathcal{P}}$ is contained in a rational hyperplane. $\mathcal{P}$ is contained in a translation of $L_{\mathcal{P}}$ by an element of $\mathbb{Q}^n$.]

EXERCISE V.37. Let $\pi : \mathbb{R}^n \to \mathbb{R}^m$ be a linear map such that π maps $\mathbb{Q}^n$ into $\mathbb{Q}^m$. Given $\mathcal{C} \subseteq \mathbb{R}^m$, show that

(1) $\mathcal{C} = \text{cone}(T)$ for some (finite) $T \subset \mathbb{Q}^m$ if and only if $\pi^{-1}(\mathcal{C}) = \text{cone}(T')$ for some (finite) $T' \subset \mathbb{Q}^n$.

(2) $\mathcal{C}$ is a rational polyhedral cone in $\mathbb{R}^m$ if and only if $\pi^{-1}(\mathcal{C})$ is a rational polyhedral cone in $\mathbb{R}^n$. [Hint: Use exercise V.6.]

6. *Volume of convex polytopes

[6]Given linearly independent elements $\alpha_1, \ldots, \alpha_d \in \mathbb{R}^n$, the *parallelotope generated by the α_j* is the set

$$\mathcal{P} := \{\lambda_1\alpha_1 + \cdots + \lambda_d\alpha_d : 0 \leq \lambda_j \leq 1,\ 1 \leq j \leq d\}.$$

It is straightforward to check that $\mathcal{P}$ is a d-dimensional polytope in $\mathbb{R}^n$ (exercise V.38). We denote by Vol_n the usual n-dimensional volume (i.e. the Lebesgue measure) on $\mathbb{R}^n$. Given an affine subspace H of $\mathbb{R}^n$, we write Vol_H for the measure induced on H by Vol_n. Here is a precise definition of Vol_H: let H_0 be the (unique) linear subspace of $\mathbb{R}^n$ which is a translate of H (i.e. $H_0 = H - \alpha$ for any $\alpha \in H$), $d := \dim(H)$, and $\beta_1, \ldots, \beta_{n-d}$ be an orthonormal set (with respect to the dot product) of elements in $\mathbb{R}^n$ such that H_0 is precisely the set of elements in $\mathbb{R}^n$ whose dot product with each β_j is zero. Let $\mathcal{Q}$ be the parallelotope generated by $\beta_1, \ldots, \beta_d$. If $\mathcal{R}$ is any subset of H, then $\mathcal{R}$ is measurable with respect to Vol_H if and only if $\mathcal{Q} + \mathcal{R}$ is measurable with respect to Vol_n, and $\text{Vol}_H(\mathcal{R}) = \text{Vol}_n(\mathcal{Q} + \mathcal{R})$ (fig. 10).

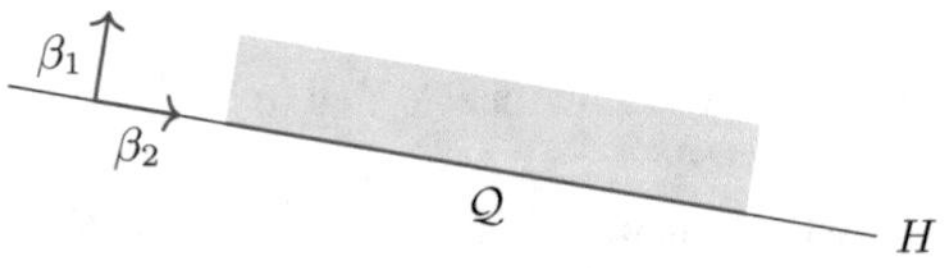

FIGURE 10. If $\mathcal{Q}$ is a segment of a line $H \subset \mathbb{R}^2$, then $\text{Vol}_H(\mathcal{Q})$ is the length of $\mathcal{Q}$, and equals the area of the rectangle with base $\mathcal{Q}$ and height one.

If $\mathcal{P} \subset \mathbb{R}^n$ is such that the $\dim(\text{aff}(\mathcal{P})) \leq d$, then we write $\text{Vol}_d(\mathcal{P})$ for $\text{Vol}_H(\mathcal{P})$, where H is any d-dimensional affine subspace of $\mathbb{R}^n$ containing $\mathcal{P}$; this

[6]The asterisk in the section name is to indicate that the material of this section is not going to be used in chapter VI. It is only used in the proof of Bernstein's theorem in chapter VII.

is well defined since $\mathrm{Vol}_H(\mathcal{P})$ does *not* depend on H (exercise V.39). The following properties of Vol follow from basic analysis; we use these without proof.

THEOREM V.35. *Let $\mathcal{P}$ be a polytope in $\mathbb{R}^n$.*

(1) If $\dim(\mathcal{P}) \leq d \leq n$, then the map $\lambda \in \mathbb{R}_{\geq 0} \mapsto \mathrm{Vol}_d(\lambda\mathcal{P})$ is homogeneous of order *d, i.e.* $\mathrm{Vol}_d(\lambda\mathcal{P}) = \lambda^d \mathrm{Vol}_d(\mathcal{P})$ *for all* $\lambda \geq 0$.

(2) As a real-valued function from the set of the polytopes in $\mathbb{R}^n$, Vol_n is continuous with respect to the Hausdorff distance.

Let $\mathcal{P}$ be an n-dimensional polytope, and O be an arbitrary point of $\mathbb{R}^n$. For each facet $\mathcal{Q}$ of $\mathcal{P}$, let $\mathcal{S}_{\mathcal{Q},O}$ be the convex hull of $\mathcal{Q} \cup \{O\}$, i.e. $\mathcal{S}_{\mathcal{Q},O}$ is the "cone with base $\mathcal{Q}$ and apex O." Then it is intuitively clear that the volume of $\mathcal{P}$ can be computed in terms of the volumes of the $\mathcal{S}_{\mathcal{Q},O}$, see fig. 11. This leads to our next result. To state it we introduce a notation: if $\mathcal{Q}$ is a facet of an n-dimensional polytope $\mathcal{P} \subset \mathbb{R}^n$, then (up to a positive multiple) there is a unique $\nu \in (\mathbb{R}^n)^* \setminus \{0\}$ such that $\mathcal{Q} = \mathrm{In}_\nu(\mathcal{P})$. Define

$$\mathrm{sign}_{\mathcal{Q}}(\mathcal{P}, O) = \begin{cases} 1 & \text{if } \langle \nu, O\rangle > \min_{\mathcal{P}}(\nu), \\ -1 & \text{otherwise.} \end{cases}$$

Geometrically, $\mathrm{sign}_{\mathcal{Q}}(\mathcal{P}, O)$ is 1 if and only if O and $\mathcal{P}$ are on the same side of $\mathrm{aff}(\mathcal{Q})$, see fig. 11.

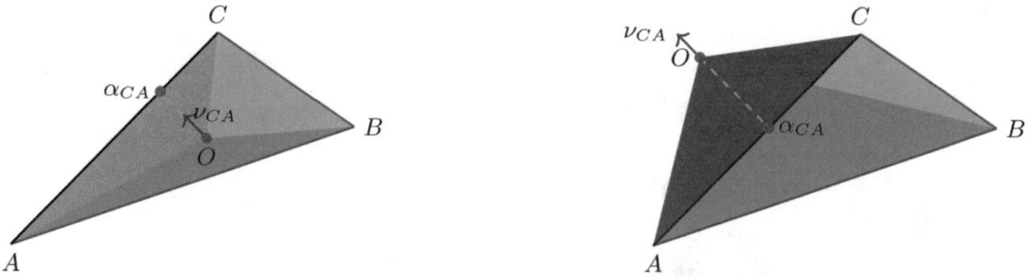

(A) $\mathrm{Vol}_2(ABC) = \mathrm{Vol}_2(OAB) + \mathrm{Vol}_2(OBC) + \mathrm{Vol}_2(OCA)$

(B) $\mathrm{Vol}_2(ABC) = \mathrm{Vol}_2(OAB) + \mathrm{Vol}_2(OBC) - \mathrm{Vol}_2(OCA)$

FIGURE 11. If $\mathcal{P}$ is a triangle in $\mathbb{R}^2$ and $O \in \mathbb{R}^2$, then for each side $\mathcal{Q}$ of $\mathcal{P}$, $\mathcal{S}_{\mathcal{Q},O}$ is the triangle formed by $\mathcal{Q}$ and O. In fig. 11b $\mathrm{sign}_{CA}(ABC, O) = -1$ since O and ABC are on different sides of the line containing CA.

THEOREM V.36. *Let $\mathcal{P}$ be an n-dimensional polytope in $\mathbb{R}^n$ and $O \in \mathbb{R}^n$. For each facet $\mathcal{Q}$ of $\mathcal{P}$, let $d(O, \mathrm{aff}(\mathcal{Q}))$ denote the distance between O and the affine hull of $\mathcal{Q}$. Then*

$$\mathrm{Vol}_n(\mathcal{P}) = \frac{1}{n} \sum_{\mathcal{Q}} \mathrm{sign}_{\mathcal{Q}}(\mathcal{P}, O) d(O, \mathrm{aff}(\mathcal{Q})) \mathrm{Vol}_{n-1}(\mathcal{Q}), \tag{44}$$

where the sum is over all facets $\mathcal{Q}$ of $\mathcal{P}$. In particular,

(45) $$\mathrm{Vol}_n(\mathcal{P}) = \frac{1}{n}\sum_{\substack{\nu\in(\mathbb{R}^n)^* \\ ||\nu||=1}} \max_{\mathcal{P}}(\nu)\,\mathrm{Vol}_{n-1}(\mathrm{ld}_\nu(\mathcal{P})) = -\frac{1}{n}\sum_{\substack{\nu\in(\mathbb{R}^n)^* \\ ||\nu||=1}} \min_{\mathcal{P}}(\nu)\,\mathrm{Vol}_{n-1}(\mathrm{In}_\nu(\mathcal{P})).$$

Remark V.37. The norm $||\cdot||$ on $(\mathbb{R}^n)^*$ in (45) is the Euclidean norm induced from $\mathbb{R}^n$ upon identification of $(\mathbb{R}^n)^*$ and $\mathbb{R}^n$ via the basis dual to the standard basis of $\mathbb{R}^n$.

PROOF. Let $\mathcal{I}_+$ (respectively, $\mathcal{I}_-$) be the collection of facets $\mathcal{Q}$ of $\mathcal{P}$ such that $\mathrm{sign}_{\mathcal{Q}}(\mathcal{P}, O) = 1$ (respectively, -1). The results of section V.3 imply that

(1) $\mathcal{P} \subset \bigcup_{\mathcal{Q}\in\mathcal{I}_+} \mathcal{S}_{\mathcal{Q}}(\mathcal{P}, O)$ (exercise V.40);
(2) if $\mathcal{Q} \in \mathcal{I}_-$ then $\mathcal{S}_{\mathcal{Q},O} \cap \mathrm{relint}(\mathcal{P}) = \emptyset$ and $\mathcal{S}_{\mathcal{Q},O} \subset \bigcup_{\mathcal{Q}'\in\mathcal{I}_+} \mathcal{S}_{\mathcal{Q}'}(\mathcal{P}, O)$ (exercise V.41);
(3) if $\mathcal{Q} \in \mathcal{I}_+$ then $\mathcal{S}_{\mathcal{Q},O} \setminus \mathcal{P} \subset \bigcup_{\mathcal{Q}'\in\mathcal{I}_-} \mathcal{S}_{\mathcal{Q}'}(\mathcal{P}, O)$ (exercise V.42);
(4) if $\mathcal{Q}$ and $\mathcal{Q}'$ are distinct facets of $\mathcal{P}$ such that $\mathrm{sign}_{\mathcal{Q}}(\mathcal{P}, O) = \mathrm{sign}_{\mathcal{Q}'}(\mathcal{P}, O)$, then $\dim(\mathcal{S}_{\mathcal{Q},O} \cap \mathcal{S}_{\mathcal{Q}',O}) \leq n - 1$ (exercise V.43).

These observations immediately imply that

$$\mathrm{Vol}_n(\mathcal{P}) = \sum_{\mathcal{Q}} \mathrm{sign}_{\mathcal{Q}}(\mathcal{P}, O)\,\mathrm{Vol}_n(\mathcal{S}_{\mathcal{Q},O}).$$

Since every cross section of $\mathcal{S}_{\mathcal{Q},O}$ parallel to the hyperplane $\mathrm{aff}(\mathcal{Q})$ is a dilation of $\mathcal{Q}$, it follows that

$$\begin{aligned}\mathrm{Vol}_n(\mathcal{S}_{\mathcal{Q},O}) &= \int_{r=0}^{d(O,\mathrm{aff}(\mathcal{Q}))} \mathrm{Vol}_{n-1}(r\mathcal{Q})dr = \mathrm{Vol}_{n-1}(\mathcal{Q}) \int_{r=0}^{d(O,\mathrm{aff}(\mathcal{Q}))} r^{n-1}dr \\ &= \frac{1}{n} d(O, \mathrm{aff}(\mathcal{Q}))\,\mathrm{Vol}_{n-1}(\mathcal{Q}),\end{aligned}$$

where the second equality follows from theorem V.35. This completes the proof of identity (44). Now for each facet $\mathcal{Q}$ of $\mathcal{P}$, let $\nu_{\mathcal{Q}} \in (\mathbb{R}^n)^*$ be the *outward facing* unit normal to $\mathcal{Q}$, i.e. $\mathcal{Q} = \mathrm{ld}_\nu(\mathcal{P})$ and $||\nu|| = 1$. We now apply identity (44) with O being the origin of $\mathbb{R}^n$. If we identify $(\mathbb{R}^n)^*$ with $\mathbb{R}^n$ via the basis dual to the standard basis of $\mathbb{R}^n$, then $\alpha_{\mathcal{Q}} := \mathrm{sign}_{\mathcal{Q}}(\mathcal{P}, O) d(O, \mathrm{aff}(\mathcal{Q}))\nu_{\mathcal{Q}}$ is a point on $\mathrm{aff}(\mathcal{Q})$ (see, e.g. fig. 11). It follows that $\max_{\mathcal{P}}(\nu_{\mathcal{Q}}) = \langle \nu_{\mathcal{Q}}, \alpha_{\mathcal{Q}}\rangle = \mathrm{sign}_{\mathcal{Q}}(\mathcal{P}, O) d(O, \mathrm{aff}(\mathcal{Q}))$. The first equality of identity (45) now follows from identity (44). The second equality follows from the first by replacing ν by $-\nu$. □

6.1. Exercises.

EXERCISE V.38. Let $\mathcal{P} := \{\lambda_1\alpha_1 + \cdots + \lambda_d\alpha_d : 0 \leq \lambda_j \leq 1 \text{ for each } j = 1, \ldots, d\}$, where $\alpha_1, \ldots, \alpha_d \in \mathbb{R}^n$. Show that

(1) $\mathcal{P}$ is the convex hull of the set consisting of the origin and all elements of the form $\alpha_{i_1} + \cdots + \alpha_{i_k}$, where $1 \leq i_1 < \cdots < i_k \leq d$.
(2) $\mathrm{aff}(\mathcal{P})$ is the linear subspace of $\mathbb{R}^n$ spanned by $\alpha_1, \ldots, \alpha_d$.

EXERCISE V.39. Let $\mathcal{P} \subset \mathbb{R}^n$ and $k := \dim(\mathcal{P})$. Fix d, $k \leq d \leq n$. If H_1, H_2 are d-dimensional affine subspaces of $\mathbb{R}^n$ containing $\mathcal{P}$, show that

(1) $\mathcal{P}$ is measurable with respect to Vol_{H_1} if and only if it is measurable with respect to Vol_{H_2}.
(2) If $\mathcal{P}$ is measurable with respect to either of them, then $\mathrm{Vol}_{H_1}(\mathcal{P}) = \mathrm{Vol}_{H_2}(\mathcal{P})$.

EXERCISE V.40. Prove observation (1) from the proof of theorem V.36. [Hint: It suffices to prove that $\mathrm{relint}(\mathcal{P}) \subset \bigcup_{\mathcal{Q} \in \mathcal{I}_+} \mathcal{S}_{\mathcal{Q}}(\mathcal{P}, O)$. If $\alpha \in \mathrm{relint}(\mathcal{P}) \setminus \{O\}$, the line through α and O intersects the facets at two points. One of these is contained in facets from $\mathcal{I}_+$.]

EXERCISE V.41. Prove observation (2) from the proof of theorem V.36. [Hint: If $\mathcal{I}_- \neq \emptyset$, then either $O \notin \mathcal{P}$ or O is on a facet of $\mathcal{P}$. Pick $\alpha \in \mathcal{Q}$, where $\mathcal{Q} \in \mathcal{I}_-$. If $\mathcal{Q} = \mathrm{In}_\nu(\mathcal{P})$, for the first part show that $\langle \nu, \beta \rangle \leq \min_{\mathcal{P}}(\nu)$ for each β on the line segment from O to α. For the second part it suffices to show $\mathrm{relint}(\mathcal{S}_{\mathcal{Q},O}) \subset \bigcup_{\mathcal{Q}' \in \mathcal{I}_+}$. Every point of $\mathrm{relint}(\mathcal{S}_{\mathcal{Q},O})$ is on the line segment between O and a point on $\mathrm{relint}(\mathcal{Q})$. Extending this line segment hits another point of the topological boundary of $\mathcal{P}$ which belongs to a facet from $\mathcal{I}_+$.]

EXERCISE V.42. Prove observation (3) from the proof of theorem V.36. [Hint: It suffices to consider the case that $O \notin \mathcal{P}$. If $\alpha \in \mathrm{relint}(\mathcal{Q})$ and $\mathcal{Q} \in \mathcal{I}_+$, then the line segment L from O to α intersects the boundary of $\mathcal{P}$ at a point β "in between" O and α. Any facet $\mathcal{R}$ of $\mathcal{P}$ containing β is in $\mathcal{I}_-$ and $L \setminus \mathcal{P} \subset \mathcal{S}_{\mathcal{R}}(\mathcal{P}, O)$.]

EXERCISE V.43. Prove observation (4) from the proof of theorem V.36. [Hint: Pick distinct facets $\mathcal{Q}_1, \mathcal{Q}_2$ of $\mathcal{P}$ such that $\mathcal{S}_{\mathcal{Q}_j,O}$ are full dimensional and relint $(\mathcal{S}_{\mathcal{Q}_1,O}) \cap \mathrm{relint}(\mathcal{S}_{\mathcal{Q}_2,O}) \neq \emptyset$. It suffices to show that $\mathrm{sign}_{\mathcal{Q}_j}(\mathcal{P}, O)$ have different signs for $j = 1$ and $j = 2$. Indeed, if α is a point in the intersection, then the line through O and α intersects $\mathcal{P}$ at $\beta_j \in \mathrm{relint}(\mathcal{Q}_j)$, $j = 1, 2$. Show that one of the β_j is "in between" O and the other β_j.]

EXERCISE V.44. Given positive $\omega_1, \ldots, \omega_n$ and nonnegative $d, m_1, \ldots, m_p$, where $0 \leq p \leq n$, let $\mathcal{Q}(\vec{\omega}, d, \vec{m})$ be the polytope in $\mathbb{R}^n$ determined by the following inequalities:

$$x_i \geq 0 \text{ for each } i = 1, \ldots, n,$$
$$\sum_{i=1}^n \omega_i x_i \leq d,$$
$$x_i \leq m_i,\ i = 1, \ldots, p.$$

Let $\mathcal{I}(\vec{\omega}, d, \vec{m})$ be the collection of subsets I of $\{1, \ldots, p\}$ such that $\sum_{i \in I} \omega_i m_i \leq d$.

(1) Show that the vertices of $\mathcal{Q}(\vec{\omega}, d, \vec{m})$ are precisely the elements $\alpha_I = (\alpha_{I,1}, \ldots, \alpha_{I,n})$ and $\beta_{I,j} = (\beta_{I,j,1}, \ldots, \beta_{I,j,k}) \in \mathbb{R}^n$, indexed by $I \in$

$\mathcal{I}(\vec{\omega}, d, \vec{m})$ and $j \in \{1, \dots, n\} \setminus I$, and defined as follows:

$$\alpha_{I,k} := \begin{cases} m_k & \text{if } k \in I, \\ 0 & \text{if } k \notin I, \end{cases}$$

$$\beta_{I,j,k} := \begin{cases} m_k & \text{if } k \in I, \\ \frac{d - \sum_{i \in I} \omega_i m_i}{\omega_j} & \text{if } k = j, \\ 0 & \text{if } k \notin I \cup \{j\}. \end{cases}$$

(2) Let $d_i, m_{i,1}, \dots, m_{i,p}$, $i = 1, 2$, be nonnegative real numbers such that $\mathcal{I}(\vec{\omega}, d_1, \vec{m}_1) = \mathcal{I}(\vec{\omega}, d_2, \vec{m}_2) = \mathcal{I}(\vec{\omega}, d_1 + d_2, \vec{m}_1 + \vec{m}_2)$. Then show that $\mathcal{Q}(\vec{\omega}, d_1 + d_2, \vec{m}_1 + \vec{m}_2) = \mathcal{Q}(\vec{\omega}, d_1, \vec{m}_1) + \mathcal{Q}(\vec{\omega}, d_2, \vec{m}_2)$.

(3) With $\vec{\omega} = (1, 1, 1)$, and $p = 2$, show that $\mathcal{Q}(\vec{\omega}, 3, (1, 1)) + \mathcal{Q}(\vec{\omega}, 3, (3, 3)) \subsetneq \mathcal{Q}(\vec{\omega}, 6, (4, 4))$, i.e. the conclusion of assertion (2) may not be true in the absence of its assumption.

EXERCISE V.45. (1) Given polytopes $\Delta_1, \dots, \Delta_p \subset \mathbb{R}^n$, show that

$$\operatorname{Vol}\left(\bigcup_{q=1}^{p} \Delta_q\right) = \sum_{q=1}^{p} (-1)^{q-1} \sum_{I \subseteq [p],\ |I| = q} \operatorname{Vol}\left(\bigcap_{i \in I} \Delta_i\right),$$

where $[p]$ denotes the set $\{1, \dots, p\}$ (this is the so-called "inclusion-exclusion principle," it holds for all "measurable" sets with finite volume, i.e. as long as the volume of each intersection on the right-hand side is well defined and finite).

(2) Given $\vec{\omega}, d, \vec{m}$ as in exercise V.44, define

$$\Delta_0 := \{(x_1, \dots, x_n) : x_i \geq 0,\ i = 1, \dots, n,\ \sum_{i=1}^{n} \omega_i x_i \leq d\}$$

$$\Delta_q := \{(x_1, \dots, x_n) : x_i \geq 0,\ i = 1, \dots, n,\ x_q \geq m_q, \sum_{i=1}^{n} \omega_i x_i \leq d\},\ q = 1, \dots, p.$$

Show that the polytope $\mathcal{Q}(\vec{\omega}, d, \vec{m})$ from exercise V.44 equals $\Delta_0 \setminus \bigcup_{q=1}^{p} \Delta_p$. Conclude that

$$\operatorname{Vol}(\mathcal{Q}(\vec{\omega}, d, \vec{m})) = \frac{1}{n! \omega_1 \cdots \omega_n} \sum_{q=0}^{p} (-1)^q \sum_{\substack{I \subseteq [p],\ |I| = q \\ \sum_{i \in I} \omega_i m_i < d}} (d - \sum_{i \in I} \omega_i m_i)^n.$$

7. *Volume of special classes of polytopes

[7] In section V.7.1 we study the dependence of the volume of Minkowski sums of polytopes on its summands, and in section V.7.2 we give a formula of the volume of rational polytopes in terms of "lattice volumes" of its facets.

7.1. Minkowski sums. Theorem V.35 implies that volume interacts well with Minkowski addition, in the sense that given compact convex subsets $\mathcal{P}$, $\mathcal{Q}$ of $\mathbb{R}^n$, the function from $\mathbb{R}_{\geq 0}$ to $\mathbb{R}_{\geq 0}$ given by $\lambda \mapsto \mathrm{Vol}_n(\mathcal{P} + \lambda\mathcal{Q})$ is continuous. However, it turns out that this function is much more than a continuous function, it is a *polynomial*. In this section we are going to prove this result for the case of polytopes. At first we need the following result.

Lemma V.38. *Let $\mathcal{P}_1, \ldots, \mathcal{P}_s$ be subsets of $\mathbb{R}^n$ and $\lambda := (\lambda_1, \ldots, \lambda_s) \in \mathbb{R}^s_{>0}$. Then for different λ, the affine hull A_λ of $\lambda_1\mathcal{P}_1 + \cdots + \lambda_s\mathcal{P}_s$ are translations of each other. In particular,* $\dim(A_\lambda)$ *is independent of* λ.

PROOF. Fix an arbitrary element α_i of $\mathcal{P}_i$, $i = 1, \ldots, n$. Without loss of generality we may replace $\mathcal{P}_i$ by $\mathcal{P}_i - \alpha_i$ and assume that each $\mathcal{P}_i$ contains the origin. For each $\lambda \in \mathbb{R}^s_{>0}$, it then follows that A_λ contains each $\mathcal{P}_i$, and therefore it is simply the linear subspace of $\mathbb{R}^n$ spanned by elements in $\bigcup_i \mathcal{P}_i$. □

THEOREM V.39. *Let $\mathcal{P}_1, \ldots, \mathcal{P}_s$ be convex polytopes in $\mathbb{R}^n$. Then there are nonnegative real numbers $v_\alpha(\mathcal{P}_1, \ldots, \mathcal{P}_s)$ for all $\alpha \in \mathcal{E}_s := \{(\alpha_1, \ldots, \alpha_s) \in \mathbb{Z}^s_{\geq 0} : \alpha_1 + \cdots + \alpha_s = n\}$ such that for all $\lambda = (\lambda_1, \ldots, \lambda_s) \in \mathbb{R}^s_{\geq 0}$,*

$$\mathrm{Vol}_n(\lambda_1\mathcal{P}_1 + \cdots + \lambda_s\mathcal{P}_s) = \sum_{\alpha \in \mathcal{E}_s} v_\alpha(\mathcal{P}_1, \ldots, \mathcal{P}_s)\lambda_1^{\alpha_1} \cdots \lambda_s^{\alpha_s},$$

where Vol_n *is the n-dimensional volume.*

PROOF. We proceed by induction on n. If $n = 1$ each $\mathcal{P}_i$ is of the form $[a_i, b_i]$, so that

$$\begin{aligned}\mathrm{Vol}_1(\lambda_1\mathcal{P}_1 + \cdots + \lambda_s\mathcal{P}_s) &= \mathrm{Vol}_1([\lambda_1 a_1 + \cdots + \lambda_s a_s, \lambda_1 b_1 + \cdots + \lambda_s b_s]) \\ &= \lambda_1(b_1 - a_1) + \cdots + \lambda_s(b_s - a_s) \\ &= \sum_i \lambda_i \mathrm{Vol}_1(\mathcal{P}_i).\end{aligned}$$

Now assume it is true for convex polytopes in $\mathbb{R}^{n-1}$. Pick convex polytopes $\mathcal{P}_1, \ldots, \mathcal{P}_n$ in $\mathbb{R}^n$. Since the volume is translation invariant, we may assume that

(∗) the origin is in the relative interior of each $\mathcal{P}_j$.

Let $P_\lambda := \lambda_1\mathcal{P}_1 + \cdots + \lambda_s\mathcal{P}_s$. Due to theorem V.35 it suffices to consider the case that each λ_i is *positive*. If $\dim(P_\lambda) \leq n - 1$, then due to lemma V.38 the result is true with all v_α being zero. So assume $\dim(P_\lambda) = n$. Then proposition V.16

[7] The asterisk in the section name is to indicate that the material of this section is not going to be used in chapter VI. It is only used in the proof of Bernstein's theorem in chapter VII.

and lemma V.38 imply that the number of facets of $\mathcal{P}_\lambda$ does not depend on λ, and, moreover, if $\mathcal{P}_{\lambda,1},\dots,\mathcal{P}_{\lambda,N}$ are the facets of $\mathcal{P}_\lambda$, then for each j, there are faces $\mathcal{P}_{i,j}$ of $\mathcal{P}_i$, $i=1,\dots,s$, such that

$$\mathcal{P}_{\lambda,j} = \lambda_1\mathcal{P}_{1,j} + \cdots + \lambda_s\mathcal{P}_{s,j}.$$

For each i,j, pick an arbitrary $\alpha_{i,j} \in \mathcal{P}_{i,j}$. Let ν_j be the outward pointing unit normal to $\mathcal{Q}_{\lambda,j}$. Identity (45) implies that

$$\begin{aligned}\mathrm{Vol}_n(\mathcal{P}_\lambda) &= \frac{1}{n}\sum_j \max_{\mathcal{P}_\lambda}(\nu_j)\,\mathrm{Vol}_{n-1}(\mathcal{P}_{\lambda,j})\\ &= \frac{1}{n}\sum_j \langle \nu_j, \lambda_1\alpha_{1,j} + \cdots + \lambda_s\alpha_{s,j}\rangle\,\mathrm{Vol}_{n-1}(\lambda_1\mathcal{P}_{1,j} + \cdots + \lambda_s\mathcal{P}_{s,j}).\end{aligned}$$

Condition (*) implies that $\langle \nu_j, \alpha_{i,j}\rangle$ is nonnegative for each i,j. Since for each j, all the $\mathcal{P}_{i,j}$ can be identified (via a volume-preserving affine map from $\mathrm{aff}(\mathcal{P}_{\lambda,j})$ to $\mathbb{R}^{n-1}$) with polytopes in $\mathbb{R}^{n-1}$, the result then follows from the inductive hypothesis. □

7.2. Rational polytopes. Let H be a d-dimensional rational affine subspace of $\mathbb{R}^n$. If $\beta \in H \cap \mathbb{Z}^n$, lemma V.32 implies that $G_H := (H-\beta)\cap\mathbb{Z}^n$ is isomorphic (as an abelian group) to $\mathbb{Z}^d$. A *fundamental lattice parallelotope* on H is a polytope of the form $\mathcal{P}+\beta$, where $\mathcal{P}$ is a (d-dimensional) parallelotope generated by d elements from G_H which generate G_H as an abelian group. We write $\mathrm{fund}(H) := \mathrm{Vol}_H(\mathcal{P})$. Proposition V.40 shows that $\mathrm{fund}(H)$ is well defined. In this section we identify $\mathbb{R}^n$ with $(\mathbb{R}^n)^*$ via the dot product, and given $\alpha,\beta \in \mathbb{R}^n$, write $\langle\alpha,\beta\rangle$ for the dot product of α and β. Similarly we write $\beta^\perp := \{\gamma \in \mathbb{R}^n : \langle \beta,\gamma\rangle = 0\}$.

Proposition V.40. *Let H be a rational affine subspace of $\mathbb{R}^n$. If $\mathcal{P}_1,\mathcal{P}_2$ are two fundamental lattice parallelotopes of H, then $\mathrm{Vol}_H(\mathcal{P}_1) = \mathrm{Vol}_H(\mathcal{P}_2)$.*

PROOF. By translating H and the $\mathcal{P}_j$ if necessary we may assume H is a linear subspace of $\mathbb{R}^n$, and each $\mathcal{P}_j$ is the parallelotope generated by $\alpha_{j,1},\dots,\alpha_{j,d} \in H\cap\mathbb{Z}^n$, where $d := \dim(H)$. Pick an orthonormal set $\beta_1,\dots,\beta_{n-d} \in \mathbb{R}^n$ such that $H = \bigcap_i \beta_i^\perp$. For each $j = 1,2$, let $\mathcal{B}_j$ be the basis of $\mathbb{R}^n$ consisting of $\beta_1,\dots,\beta_{n-d},\alpha_{j,1},\dots,\alpha_{j,d}$. By definition $\mathrm{Vol}_H(\mathcal{P}_j) = \mathrm{Vol}_n(\mathcal{Q}_j)$, where $\mathcal{Q}_j$ is the parallelotope generated by the elements of $\mathcal{B}_j$. Let $\phi : \mathbb{R}^n \to \mathbb{R}^n$ be the linear map which changes coordinates with respect to $\mathcal{B}_1$ to that of $\mathcal{B}_2$. Since $\alpha_{j,1},\dots,\alpha_{j,d}$ generate $H\cap\mathbb{Z}^n$ (as an abelian group), it follows that the matrices of both ϕ and ϕ^{-1} have only integer entries. This means that the determinant of the matrix of ϕ is ±1, and therefore ϕ preserves Vol_n. Since ϕ maps one of the $\mathcal{Q}_j$ to the other, this completes the proof. □

Let H' be a rational affine subspace of $\mathbb{R}^n$ such that $H' \supset H$ and $\dim(H') = d+1$. We now describe the relation between $\mathrm{fund}(H)$ and $\mathrm{fund}(H')$. Pick $\beta \in H$. Since $(H-\beta)\cap\mathbb{Q}^n \subset (H'-\beta)\cap\mathbb{Q}^n$ is an inclusion of vector spaces over $\mathbb{Q}$, it follows from the elementary theory of vector spaces that there is $\eta' \in (H'-\beta)\cap\mathbb{Q}^n$

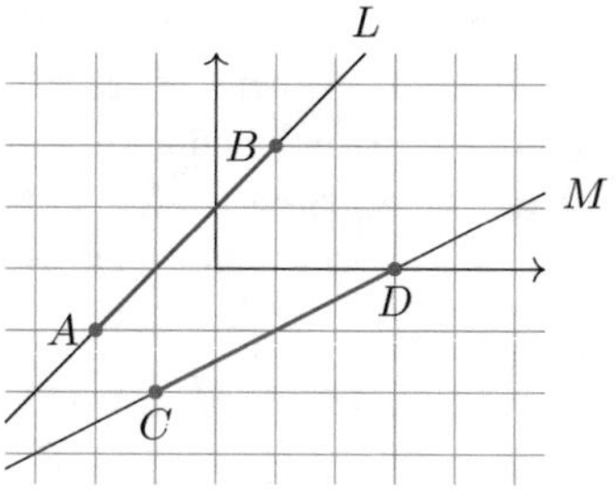

FIGURE 12. $\operatorname{fund}(L) = \sqrt{2}$, $\operatorname{fund}(M) = \sqrt{5}$. $\operatorname{Vol}'_L(AB) = 3$, $\operatorname{Vol}'_M(CD) = 2$.

such that $H - \beta = \eta'^{\perp} \cap (H' - \beta)$. Pick $r \in \mathbb{Q} \setminus \{0\}$ such that $\eta := r\eta'$ is "primitive," i.e. there is no integer $k > 1$ such that $\eta = k\eta''$ for some $\eta'' \in \mathbb{Z}^n$. Let $||\cdot||$ denote the Euclidean norm on $\mathbb{R}^n$.

Proposition V.41.

$$\operatorname{fund}(H) = \frac{||\eta||}{\min\{|\langle \eta, \alpha \rangle| : \alpha \in ((H' - \beta) \cap \mathbb{Z}^n) \setminus (H - \beta)\}} \operatorname{fund}(H').$$

In particular, if $d = n - 1$ and $H' = \mathbb{R}^n$, then $\operatorname{fund}(H) = ||\eta||$.

PROOF. Replacing H by $H - \beta$ if necessary we may assume that $\beta = 0$. Lemma B.57 implies that we may pick $\alpha_1, \ldots, \alpha_{d+1}, u_{d+1}, \ldots, u_n \in \mathbb{R}^n$ such that

- $\alpha_1, \ldots, \alpha_d$ generate $H \cap \mathbb{Z}^n$,
- $\alpha_1, \ldots, \alpha_{d+1}$ generate $H' \cap \mathbb{Z}^n$,
- $u_{d+1} \in H'$, $H = u_{d+1}^{\perp} \cap H'$, $||u_{d+1}|| = 1$,
- $u_{d+2}, \ldots, u_n$ are orthonormal (with respect to dot product), and $H' = \bigcap_{j=d+2}^{n} u_j^{\perp}$.

Let M be the matrix with column vectors $\alpha_1, \ldots, \alpha_d, u_{d+1}, \ldots, u_n$ and M' be the matrix with column vectors $\alpha_1, \ldots, \alpha_{d+1}, u_{d+2}, \ldots, u_n$. Write $u_{d+1} = c_{d+1}\alpha_{d+1} + u'$, where $u' \in H$. Then

$$\operatorname{fund}(H) = |\det(M)| = |c_{d+1}\det(M')| = |c_{d+1}| \operatorname{fund}(H').$$

Note that $u_{d+1} = \pm\eta/||\eta||$. It follows that $\langle \eta, u_{d+1} \rangle = \pm||\eta||$. On the other hand, $\langle \eta, u_{d+1} \rangle = \langle \eta, c_{d+1}\alpha_{d+1} + u' \rangle = c_{d+1}\langle \eta, \alpha_{d+1} \rangle$. Since for each $\alpha \in H' \cap \mathbb{Z}^n$, $\langle \eta, \alpha \rangle$ is an integer multiple of $\langle \eta, \alpha_{d+1} \rangle$, the result follows. $\square$

Definition V.42. Let H be a rational affine subspace of $\mathbb{R}^n$ of dimension d. The *H-normalized volume* is

$$\operatorname{Vol}'_H(\cdot) := \operatorname{Vol}_d(\cdot)/\operatorname{fund}(H),$$

see fig. 12 for some examples with $n = 2$. An *integral* element of $\mathbb{R}^n$ is an element with integral coordinates; an integral element of $(\mathbb{R}^n)^*$ is one which has integral coordinates with respect to the basis which is dual to the standard basis of $\mathbb{R}^n$. An integral element η of $\mathbb{R}^n$ or $(\mathbb{R}^n)^*$ is *primitive*, if it is not of the form $k\eta'$, where $k > 1$ and η' is also integral. If ν is an integral element of $(\mathbb{R}^n)^*$, then we write Vol'_ν for $\mathrm{Vol}'_{\nu^\perp}$.

Corollary V.43. *Let $\mathcal{P}$ be a convex rational polytope in $\mathbb{R}^n$. Then*

$$\text{(46)}\quad \mathrm{Vol}_n(\mathcal{P}) = \frac{1}{n}\sum_\nu \max_{\mathcal{P}}(\nu)\,\mathrm{Vol}'_\nu(\mathrm{ld}_\nu(\mathcal{P})) = -\frac{1}{n}\sum_\nu \min_{\mathcal{P}}(\nu)\,\mathrm{Vol}'_\nu(\mathrm{In}_\nu(\mathcal{P})),$$

where the sum is over all primitive integral $\nu \in (\mathbb{R}^n)^*$.

PROOF. Since $\mathcal{P}$ is rational, every facet of $\mathcal{P}$ is determined by integral elements of $(\mathbb{R}^n)^*$. Therefore the result follows from combining eq. (45) and proposition V.41. □

Remark V.44. If $\mathcal{Q}$ is a facet of an n-dimensional rational polytope $\mathcal{P}$ in $\mathbb{R}^n$, then the *primitive inner* (respectively, *outer*) *normal* to $\mathcal{F}$ is the unique primitive integral $\nu \in (\mathbb{R}^n)^*$ such that $\mathcal{Q} = \mathrm{In}_\nu(\mathcal{P})$ (respectively, $\mathcal{Q} = \mathrm{ld}_\nu(\mathcal{P})$). Note that the sum in (46) is practically finite: the only nonzero contributions come from those ν which are primitive outer normal to $(n-1)$-dimensional faces of $\mathcal{P}$.

Part 2

Number of zeroes on the torus

CHAPTER VI

Toric varieties over algebraically closed fields

This chapter introduces *toric varieties*, which are the setting of all the subsequent results of this book. Our treatment will be mostly based on the results from chapters III and V; only in section VI.7 we use the notion of *closed subschemes* discussed in section IV.2. Unless explicitly stated otherwise, from this chapter onward $\mathbb{k}$ denotes an algebraically closed field (of arbitrary characteristic), and $\mathbb{k}^*$ denotes $\mathbb{k} \setminus \{0\}$.

1. Algebraic torus

If $(x_1, \ldots, x_n)$ are coordinates on $\mathbb{k}^n$, $n \geq 1$, then $(\mathbb{k}^*)^n = \mathbb{k}^n \setminus V(x_1 \cdots x_n)$. This implies that $(\mathbb{k}^*)^n$ is an affine variety, and its coordinate ring is the ring $\mathbb{k}[x_1, x_1^{-1}, \ldots, x_n, x_n^{-1}]$ of *Laurent polynomials* in $x_1, \ldots, x_n$ over $\mathbb{k}$ (example III.37). An *(algebraic) torus* is a variety X isomorphic to $(\mathbb{k}^*)^n$ for some $n \geq 1$. A *system of coordinates* on X is an ordered collection $(x_1, \ldots, x_n)$ of regular functions on X such that the coordinate ring $\mathbb{k}[X]$ of X is $\mathbb{k}[x_1, x_1^{-1}, \ldots, x_n, x_n^{-1}]$. A basic property of a torus is that *every* morphism between two tori is a group homomorphism and also a *monomial* map with respect to *every* set of coordinates. Indeed, let $\phi : X \to Y \cong (\mathbb{k}^*)^N$ be a morphism. Choose coordinates $(y_1, \ldots, y_N)$ on Y. If $\phi(x) = (\phi_1(x), \ldots, \phi_N(x))$, then each ϕ_j must be a monomial in $(x_1, \ldots, x_n)$, for otherwise it will be zero at some point of X (exercise VI.1). This shows that ϕ is a monomial map. Write $\phi_j(x) = x^{\alpha_j}$, $\alpha_1, \ldots, \alpha_N \in \mathbb{Z}^n$. If $x = (x_1, \ldots, x_n), x' = (x'_1, \ldots, x'_n) \in X$, then it follows that $\phi(x \cdot x') = \phi(x_1 x'_1, \ldots, x_n x'_n) = ((x \cdot x')^{\alpha_1}, \ldots, (x \cdot x')^{\alpha_N}) = \phi(x) \cdot \phi(x')$, so that ϕ is indeed a homomorphism. This implies, in particular, that the multiplication with respect to every set of (algebraic) coordinates on a torus induces the same group structure on it, and the image of a morphism between two tori is a subgroup of the target. Proposition VI.1 shows that it is in addition a Zariski closed subset of the target[1]. We use the following notation in proposition VI.1: given a monomial map $\phi : x \mapsto (x^{\alpha_1}, \ldots, x^{\alpha_N})$ between tori

[1]Contrast this to the case of $\mathbb{k}^n$: the additive group structures on $\mathbb{k}^n$ with respect to different systems of algebraic coordinates are, in general, different, and the image of a morphism from $\mathbb{k}^n$ to $\mathbb{k}^n$ is, in general, neither a subgroup nor a closed subvariety of the target (see example III.16 and section III.12).

P. Mondal, *How Many Zeroes?*, CMS/CAIMS Books in Mathematics 2,
https://doi.org/10.1007/978-3-030-75174-6_VI

with fixed systems of coordinates, we write $[\phi]$ for the $N \times n$ matrix whose rows are the α_i. Some basic properties of $[\phi]$ are established in exercise VI.2.

Proposition VI.1. *Let G be the subgroup of $\mathbb{Z}^n$ generated by the α_i, and $\bar{G} := \{\alpha \in \mathbb{Z}^n : k\alpha \in G$ for some $k \geq 1\}$ be the "saturation" of G in $\mathbb{Z}^n$. Let q be the index of G in $\bar{G}$. Let r be the rank of $[\phi]$ as a matrix over $\mathbb{Q}$. Then*

(1) *$\phi(X)$ is a torus and a closed subvariety of Y of dimension r.*
(2) *$\bar{G}/G \cong \prod_{j=1}^r \mathbb{Z}/q_j\mathbb{Z}$ for positive integers $q_1, \ldots, q_r$ such that $q = \prod_j q_j$.*
(3) *$\ker(\phi)$ is an $(n-r)$-dimensional subgroup of X isomorphic to $(\bar{G}/G) \times (\Bbbk^*)^{n-r}$. In particular, if $r = n$, then the degree of ϕ (as a map from X to $\phi(X)$) is q.*
(4) *Pick a basis $(\beta_1, \ldots, \beta_{n-r})$ of $\ker[\phi] \subseteq \mathbb{Z}^n$, and let $\eta : (\Bbbk^*)^{n-r} \to (\Bbbk^*)^n$ be the morphism such that the column vectors of $[\eta]$ are the β_j. Then the irreducible component of $\ker(\phi)$ containing $(1, \ldots, 1)$ is the image of η.*

PROOF. Let $\phi' : \mathbb{Z}^n \to \mathbb{Z}^N$ be the map corresponding to multiplication by $[\phi]$. With respect to appropriate coordinates on $\mathbb{Z}^n$ and $\mathbb{Z}^N$, the matrix of ϕ' is of the form

$$\left[\begin{array}{c|c} D & 0 \\ \hline 0 & 0 \end{array}\right],$$

where D is a diagonal matrix with positive integers as diagonal entries (corollary B.58). This means that we can choose coordinates $(x_1, \ldots, x_n)$ on X and $(y_1, \ldots, y_N)$ on Y with respect to which ϕ takes the form $(x_1, \ldots, x_n) \mapsto (x_1^{q_1}, \ldots, x_r^{q_r}, 1, \ldots, 1)$ for positive integers $q_1, \ldots, q_r$. All the assertions are now straightforward; their proofs are left as exercise VI.3. □

1.1. Exercises.

EXERCISE VI.1. Show that every polynomial in $(x_1, \ldots, x_n)$ which is not a monomial vanishes at some points on $(\Bbbk^*)^n$. [Hint: Use exercise III.14.]

EXERCISE VI.2. Let $\phi : X \to Y$ be a morphism between two tori. Let $(x_1, \ldots, x_n)$ (respectively, $(y_1, \ldots, y_N)$) be coordinates on X (respectively, Y) and $[\phi]$ be the corresponding matrix of ϕ. Show that

(1) for each $\beta = (\beta_1, \ldots, \beta_N) \in \mathbb{Z}^N$, $(\phi(x))^\beta = x^{\beta[\phi]}$, where $\beta[\phi]$ is the product of β (regarded as a $1 \times N$ matrix) and $[\phi]$;
(2) ϕ is an isomorphism if and only if $N = n$ and $[\phi]$ is invertible over $\mathbb{Z}$;
(3) if Z is a torus and $\psi : Y \to Z$ is a morphism, then $[\psi \circ \phi] = [\psi][\phi]$.

EXERCISE VI.3. Complete the proof of proposition VI.1.

2. Toric varieties from finite subsets of $\mathbb{Z}^N$

A *toric variety* is a variety X which contains an algebraic torus as a dense open subset such that the (multiplicative) action of the torus on itself extends to an action on all of X. Given a finite subset $\mathcal{A} = \{\alpha_0, \ldots, \alpha_N\}$ of $\mathbb{Z}^n$, we write $\phi_{\mathcal{A}} : (\Bbbk^*)^n \to \mathbb{P}^N$ for the map given by

(47) $$x \mapsto [x^{\alpha_0} : \cdots : x^{\alpha_N}].$$

We write $X^0_{\mathcal{A}}$ for the image of $\phi_{\mathcal{A}}$ and $X_{\mathcal{A}}$ for the closure of $X^0_{\mathcal{A}}$ in $\mathbb{P}^N$. We will now show that $X_{\mathcal{A}}$ is a toric variety with torus $X^0_{\mathcal{A}}$. Denote the homogeneous coordinates of $\mathbb{P}^N$ by $[z_{\alpha_0} : \cdots : z_{\alpha_N}]$. Let $U_\alpha := \mathbb{P}^N \setminus V(z_\alpha)$, $\alpha \in \mathcal{A}$, be the basic open subsets of $\mathbb{P}^N$.

Proposition VI.2. *$X_{\mathcal{A}}$ is a toric variety. More precisely,*

(1) $X^0_{\mathcal{A}}$ is a torus and $X^0_{\mathcal{A}} = X_{\mathcal{A}} \cap \bigcap_{\alpha \in \mathcal{A}} U_\alpha$.
(2) For each $\alpha \in A$, $X_{\mathcal{A}} \cap U_\alpha$ is an affine variety with coordinate ring $\mathbb{k}[x^\beta : \beta \in S_\alpha]$, where S_α is the subsemigroup of $\mathbb{Z}^n$ generated by $\mathcal{A} - \alpha := \{\beta - \alpha : \beta \in \mathcal{A}\}$.
(3) The dimension of $X_{\mathcal{A}}$ (and equivalently, of $X^0_{\mathcal{A}}$) equals the dimension (as a polytope) of the convex hull of $\mathcal{A}$ in $\mathbb{R}^n$.
(4) $X^0_{\mathcal{A}}$ acts on $X_{\mathcal{A}}$ via the multiplicative action of $X^0_{\mathcal{A}}$ on $\mathbb{P}^N$ given by

(48) $$[y_{\alpha_0} : \cdots : y_{\alpha_N}] \cdot [z_{\alpha_0} : \cdots : z_{\alpha_N}] := [y_{\alpha_0} z_{\alpha_0} : \cdots : y_{\alpha_N} z_{\alpha_N}]$$

for all $[y_{\alpha_0} : \cdots : y_{\alpha_N}] \in X^0_{\mathcal{A}}$ and $[z_{\alpha_0} : \cdots : z_{\alpha_N}] \in \mathbb{P}^N$.

PROOF. Since $X^0_{\mathcal{A}} \subset U := \bigcap_{\alpha \in \mathcal{A}} U_\alpha$, and since $U \cong (\mathbb{k}^*)^N$ via the map $[z_{\alpha_0} : \cdots : z_{\alpha_N}] \mapsto (z_{\alpha_1}/z_{\alpha_0}, \ldots, z_{\alpha_N}/z_{\alpha_0})$, it follows that $X^0_{\mathcal{A}}$ is the image in $(\mathbb{k}^*)^N$ of the map $x \mapsto (x^{\alpha_1 - \alpha_0}, \ldots, x^{\alpha_n - \alpha_0})$. Proposition VI.1 then implies that $X^0_{\mathcal{A}}$ is a torus, and also implies assertion (3). It also says that $X^0_{\mathcal{A}}$ is a closed subset of U, which implies that $X^0_{\mathcal{A}} = X_{\mathcal{A}} \cap U$, and proves assertion (1). Assertion (2) follows directly from corollary III.73. Finally, since for a fixed $y \in X^0_{\mathcal{A}}$, the action of y on $\mathbb{P}^N$ given by (48) is an isomorphism (exercise VI.4) and since $X^0_{\mathcal{A}}$ is closed under this action, it follows that $X_{\mathcal{A}}$ is also closed under it (exercise VI.5), as required to prove assertion (4). □

Proposition VI.2 states, in particular, that $X_{\mathcal{A}}$ is "equivariantly embedded" in $\mathbb{P}^N$, i.e. the action of the torus on $X_{\mathcal{A}}$ extends to all of $\mathbb{P}^N$. Conversely every equivariantly embedded projective toric variety is essentially of the form $X_{\mathcal{A}}$ for some appropriate $\mathcal{A}$ (see, e.g. [GKZ94, Proposition 5.1.5]); we will not use this result. We now show that $X^0_{\mathcal{A}}$ and $X_{\mathcal{A}}$ depend only on the affine geometry of the set $\mathcal{A}$.

Proposition VI.3. ([GKZ94, Proposition 5.1.2]). *Let $\mathcal{A} \subset \mathbb{Z}^n$, $\mathcal{B} \subset \mathbb{Z}^m$, and $T : \mathbb{Z}^n \to \mathbb{Z}^m$ be an injective integer affine transformation such that $T(\mathcal{A}) = \mathcal{B}$. Then $X^0_{\mathcal{A}} = X^0_{\mathcal{B}}$ and $X_{\mathcal{A}} = X_{\mathcal{B}}$ as subsets of $\mathbb{P}^N$, where $N = |\mathcal{A}| - 1$.*

PROOF. Let $A = \{\alpha_0, \ldots, \alpha_N\}$, $\mathcal{B} = \{\beta_0, \ldots, \beta_N\}$, where $\beta_j = T(\alpha_j)$, $j = 0, \ldots, N$. By definition there is $\lambda = (\lambda_1, \ldots, \lambda_m) \in \mathbb{Z}^m$ and an $n \times m$ matrix M such that for each $\gamma = (\gamma_1, \ldots, \gamma_n) \in \mathbb{Z}^n$, $T(\gamma) = \lambda + \gamma M$. Let $\mu_j \in \mathbb{Z}^m$ be the j-th row vector of M, $j = 1, \ldots, n$, and $T^* : (\mathbb{k}^*)^m \to (\mathbb{k}^*)^n$ be the map defined by $T^*(y) = (y^{\mu_1}, \ldots, y^{\mu_n})$. Exercise VI.2 implies that $T^*(y)^{\alpha_j} = y^{\alpha_j M} = y^{\beta_j - \lambda}$ for each $j = 0, \ldots, N$. Since the rank of $M = [T^*]$ is n, proposition VI.1 implies that T^* is surjective, and therefore

$$X^0_{\mathcal{B}} = \{[y^{\beta_0} : \cdots : y^{\beta_N}] : y \in (\Bbbk^*)^m\} = \{[y^{-\lambda}y^{\beta_0} : \cdots : y^{-\lambda}y^{\beta_N}] : y \in (\Bbbk^*)^m\}$$
$$= \{[T^*(y)^{\alpha_0} : \cdots : T^*(y)^{\alpha_N}] : y \in (\Bbbk^*)^m\} = \{[x^{\alpha_0} : \cdots : x^{\alpha_N}] : x \in (\Bbbk^*)^n\} = X^0_{\mathcal{A}}$$

which completes the proof. □

2.1. Exercises.

EXERCISE VI.4. Show that for a fixed $y \in X^0_{\mathcal{A}}$, the action of y on $\mathbb{P}^N$ given by (48) is an isomorphism from $\mathbb{P}^N$ to itself.

EXERCISE VI.5. Let W be a subset of a topological space X and $\phi : X \to X$ be a continuous map. Show that $\phi(\overline{W}) \subseteq \overline{\phi(W)}$ (where the "bar" indicates closure in X). Deduce that if $\phi(W) \subseteq W$, then $\phi(\overline{W}) \subseteq \overline{W}$.

EXERCISE VI.6. In this exercise you will show that $X_{\mathcal{A}}$ is a *binomial variety*, i.e. $X_{\mathcal{A}}$ is defined in $\mathbb{P}^N$ by binomial equations. Given $q = (q_0, \ldots, q_N) \in \mathbb{Z}^N$, write $z^q := \prod_{j=0}^N z_{\alpha_j}^{q_j}$. Let J be the ideal of $R := \Bbbk[z_{\alpha_0}, \ldots, z_{\alpha_N}]$ generated by binomials of the form $z^{q_1} - z^{q_2}$, where $q_i := (q_{i,0}, \ldots, q_{i,N}) \in \mathbb{Z}^n$, $i = 1, 2$, are such that $\sum_{j=0}^N q_{1,j} = \sum_{j=0}^N q_{2,j}$ and $\sum_{j=0}^N q_{1,j}\alpha_j = \sum_{j=0}^N q_{2,j}\alpha_j$. Let I be the ideal of R consisting of all homogeneous polynomials that vanish on $X_{\mathcal{A}}$.

(1) Show that $J \subset I$.
(2) Let $f = \sum_{q \in \mathbb{Z}^N_{\geq 0}} c_q z^q \in I$. Use the fact that $f(x^{\alpha_0}, \ldots, x^{\alpha_N})$ is identically zero on $(\Bbbk^*)^n$ to show that for each $\alpha \in \mathbb{Z}^n$, $\sum_{q \in \mathcal{S}_\alpha} c_q = 0$, where $\mathcal{S}_\alpha := \{(q_0, \ldots, q_N) \in \mathbb{Z}^N_{\geq 0} : \sum_{j=0}^N q_j \alpha_j = \alpha\}$.
(3) Given $\alpha \in \mathbb{Z}^n$, write $f_\alpha := \sum_{q \in \mathcal{S}_\alpha} c_q z^q$. Use the preceding step to show that $f_\alpha \in J$. Deduce that $I = J$.

3. Examples of toric varieties

Given a subset Y of a topological space X, in this section we write $\mathrm{Cl}_X(Y)$ for the closure of Y in X. We also write $e_1, \ldots, e_n$ for the standard unit vectors of $\mathbb{R}^n$, i.e. for each i, j, the j-th coordinate of e_i is 0 if $j \neq i$, and 1 if $j = i$.

Example VI.4. If $\mathcal{A} = \{0, 1\} \subset \mathbb{Z}$, then $X_{\mathcal{A}} = \mathrm{Cl}_{\mathbb{P}^2}(\{[1 : x] : x \in \Bbbk^*\} = \mathbb{P}^1$. More generally, if $\mathcal{A} = \{0, e_1, \ldots, e_n\}$, then $X_{\mathcal{A}} = \mathrm{Cl}_{\mathbb{P}^n}(\{[1 : x_1 : \cdots : x_n] : (x_1, \ldots, x_n) \in (\Bbbk^*)^n\}) = \mathbb{P}^n$.

Example VI.5. If $\mathcal{A} = \{0, 1\}^2 = \{(0,0), (1,0), (0,1), (1,1)\}$, then $X_{\mathcal{A}} = \mathrm{Cl}_{\mathbb{P}^3}$ $(\{[1 : x : y : xy] : x, y \in \Bbbk^*\}) = V(z_1 z_2 - z_0 z_3) \subset \mathbb{P}^3$ (the last equality follows by a dimension count). Recall (from example III.54) that $X_{\mathcal{A}}$ is the ruled surface isomorphic to $\mathbb{P}^1 \times \mathbb{P}^1$. More generally, if $\mathcal{A} = \{0, 1\}^n$, assertion (2) of proposition VI.6 implies that $X_{\mathcal{A}} \cong (\mathbb{P}^1)^n$.

Given a set $\mathcal{A} \subset \mathbb{R}^n$ and a positive integer d, in this chapter we write $d\mathcal{A} := \{\alpha_1 + \cdots + \alpha_d : \alpha_j \in \mathcal{A} \text{ for each } j\} \subset \mathbb{R}^n$. If $\mathcal{A}, \mathcal{B}$ are finite subsets of $\mathbb{Z}^n$, we write $\delta_{\mathcal{A},\mathcal{B}}$ for the *diagonal* map from $\Bbbk^n$ to $\mathbb{P}^{|\mathcal{A}|-1} \times \mathbb{P}^{|\mathcal{B}|-1}$ given by

$$x \mapsto (\phi_{\mathcal{A}}(x), \phi_{\mathcal{B}}(x)), \tag{49}$$

where $\phi_{\mathcal{A}}$ and $\phi_{\mathcal{B}}$ are defined as in (47).

Proposition VI.6. *Let $\mathcal{A}_1, \mathcal{A}_2$ be finite subsets of $\mathbb{Z}^{n_i}$ and d_1, d_2 be positive integers.*

(1) $X_{\mathcal{A}_1} \cong X_{d_1\mathcal{A}_1}$.

(2) $X_{\mathcal{A}_1} \times X_{\mathcal{A}_2} \cong X_{d_1\mathcal{A}_1 \times d_2\mathcal{A}_2}$.

(3) Assume $n_1 = n_2 = n$. Let X_{d_1,d_2} be the closure in $X_{d_1\mathcal{A}_1} \times X_{d_2\mathcal{A}_2}$ of $\delta_{d_1\mathcal{A}_1, d_2\mathcal{A}_2}((\Bbbk^*)^n)$. Then $X_{\mathcal{A}_1+\mathcal{A}_2} \cong X_{d_1\mathcal{A}_1+d_2\mathcal{A}_2} \cong X_{d_1,d_2} \cong X_{1,1}$. □

PROOF. We may assume without loss of generality that each $\mathcal{A}_i$ contains the origin. If ν_{d_i} is the degree-d_i *Veronese map* (see section III.8.2), then it follows that $\phi_{d_i\mathcal{A}_i} = \pi \circ \nu_{d_i} \circ \phi_{\mathcal{A}_i}$, where π is a projection which omits "redundant" coordinates of $\nu_{d_i} \circ \phi_{\mathcal{A}_i}$ (i.e. for each $\alpha \in d_i\mathcal{A}_i$, π retains only one of the coordinates of $\nu_{d_i} \circ \phi_{\mathcal{A}_i}$ equaling x^α). Assertion (1) then follows from proposition III.55. For assertion (2), let $\phi_{\mathcal{A}_1} \times \phi_{\mathcal{A}_2} : (\Bbbk^*)^{n_1+n_2} \to X_{\mathcal{A}_1} \times X_{\mathcal{A}_2}$ be the morphism which maps $(x_1, x_2) \mapsto (\phi_{\mathcal{A}_1}(x_1), \phi_{\mathcal{A}_2}(x_2))$, where $x_i \in (\Bbbk^*)^{n_i}$, $i = 1, 2$. If $s : \mathbb{P}^{|\mathcal{A}_1|-1} \times \mathbb{P}^{|\mathcal{A}_2|-1} \to \mathbb{P}^{|\mathcal{A}_1||\mathcal{A}_2|-1}$ is the Segre map, then assertion (2) follows from assertion (1) and the observation that $s \circ (\phi_{\mathcal{A}_1} \times \phi_{\mathcal{A}_2}) = \phi_{\mathcal{A}_1 \times \mathcal{A}_2}$. The proof of assertion (3) is left as exercise VI.7. □

Example VI.7. Let d be a positive integer. If $\mathcal{A} = \{0, 1, 2, \ldots, d\} \subset \mathbb{Z}$, then assertion (1) of proposition VI.6 and example VI.4 imply that $X_\mathcal{A} \cong \mathbb{P}^1$. Note that $X_\mathcal{A}$ is the *rational normal curve of degree d* from example III.56.

Example VI.8. Let $\mathcal{A}_1 = \{0, e_1, \ldots, e_n\}$ and $\mathcal{A}_2 = \{e_1, \ldots, e_n\}$, so that $\mathcal{A}_1 + \mathcal{A}_2 = \{e_1, \ldots, e_n\} \cup \{e_i + e_j\}_{i,j}$. Since $X_{\mathcal{A}_1} \cong \mathbb{P}^n$ and $X_{\mathcal{A}_2} \cong \mathbb{P}^{n-1}$ (example VI.4), it follows from proposition VI.6 that $X_{\mathcal{A}_1+\mathcal{A}_2}$ is the closure in $\mathbb{P}^n \times \mathbb{P}^{n-1}$ of the image of the map from $(\Bbbk^*)^n \to \mathbb{P}^n \times \mathbb{P}^{n-1}$ given by

$$(x_1, \ldots, x_n) \mapsto ([1 : x_1 : \cdots : x_n], [x_1 : \cdots : x_n]).$$

It follows that $X_{\mathcal{A}_1+\mathcal{A}_2}$ is precisely the *blow up* of $\mathbb{P}^n$ at the point $[1 : 0 : \cdots : 0]$ (see example III.71). Figure 1 shows the convex hull of $\mathcal{B} := \mathcal{A}_1 + \mathcal{A}_2$ for $n = 2, 3$.

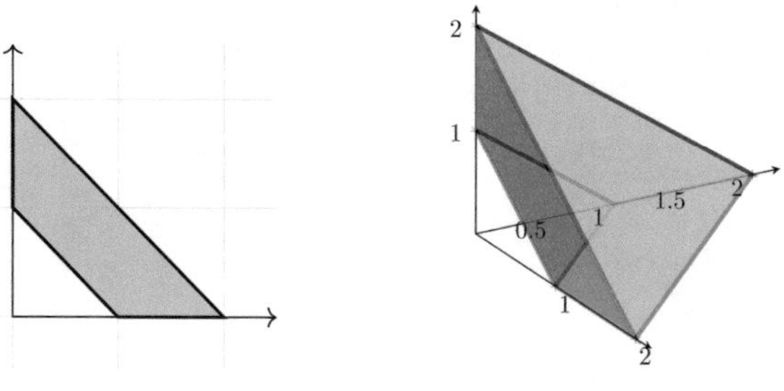

(A) $n = 2$ (B) $n = 3$

FIGURE 1. Convex hull of $\mathcal{B}$ such that $X_\mathcal{B}$ is the blow up of $\mathbb{P}^n$ at a point.

Example VI.9. If $\mathcal{A} = \{0, 2, 3\} \subset \mathbb{Z}$, then $X_\mathcal{A} = \mathrm{Cl}_{\mathbb{P}^2}(\{[1 : x^2 : x^3] : x \in \Bbbk^*\}) = V(z_0z_2^2 - z_1^3) \subset \mathbb{P}^2$. If $(x, y) = (z_1/z_0, z_2/z_0)$ are coordinates on the basic open subset $U_0 = \mathbb{P}^2 \setminus V(z_0)$ of $\mathbb{P}^2$, then $X_{\mathcal{A}} \cap U_0$ is the curve defined by $y^2 = x^3$. Note that the curve is singular at the origin (example III.92).

A toric variety of dimension n by definition contains an open subset isomorphic to $(\Bbbk^*)^n$. The following result describes a class of sets $\mathcal{A}$ such that $X_{\mathcal{A}}$ contains open subsets isomorphic to $\Bbbk^n$, i.e. $X_{\mathcal{A}}$ is a *compactification* of $\Bbbk^n$—its proof is left as exercise VI.8.

Proposition VI.10. *Let $\mathcal{A}$ be a finite subset of $\mathbb{Z}^n$.*

(1) Assume $\mathcal{A} \subset \mathbb{Z}^n_{\geq 0}$ and $\mathcal{A} \supset \{0, e_1, \ldots, e_n\}$. Show that $U_0 \cap X_{\mathcal{A}} \cong \Bbbk^n$.

(2) More generally, assume there is a vertex α of $\mathcal{A}$ such that there are precisely n edges $\mathcal{E}_1, \ldots, \mathcal{E}_n$ of $\operatorname{conv}(\mathcal{A})$ containing α and on each $\mathcal{E}_i$, there is an element of $\mathcal{A}$ of the form $\alpha + \beta_i$, where $\beta_1, \ldots, \beta_n$ is a basis of $\mathbb{Z}^n$. Then show that $U_\alpha \cap X_{\mathcal{A}} \cong \Bbbk^n$. □

3.1. Exercises.

EXERCISE VI.7. Prove assertion (3) of proposition VI.6.

EXERCISE VI.8. Prove proposition VI.10. [Hint: the first assertion follows from proposition VI.2. For the second assertion and the definition of "vertex" read the next section and use Theorem VI.12.]

4. Structure of $X_{\mathcal{A}}$

We continue with the notation and the set up of section VI.2. In this section we study the complement of $X^0_{\mathcal{A}}$ in $X_{\mathcal{A}}$ (i.e. the subvariety of $X_{\mathcal{A}}$ "at infinity") and catch a glimpse of its beautiful combinatorial structure. Let $\mathcal{P}$ be the convex hull of $\mathcal{A}$. Then $\mathcal{P}$ is a rational polytope (proposition V.29). A *face* of $\mathcal{A}$ is by definition a set $\mathcal{B}$ of the form $\mathcal{Q} \cap \mathcal{A}$ where $\mathcal{Q}$ is a face of $\mathcal{P}$. We say that $\mathcal{B}$ is a *facet* (respectively, *vertex, edge*) of $\mathcal{A}$ if $\mathcal{Q}$ is a facet (respectively, vertex, edge) of $\mathcal{P}$, see fig. 2.

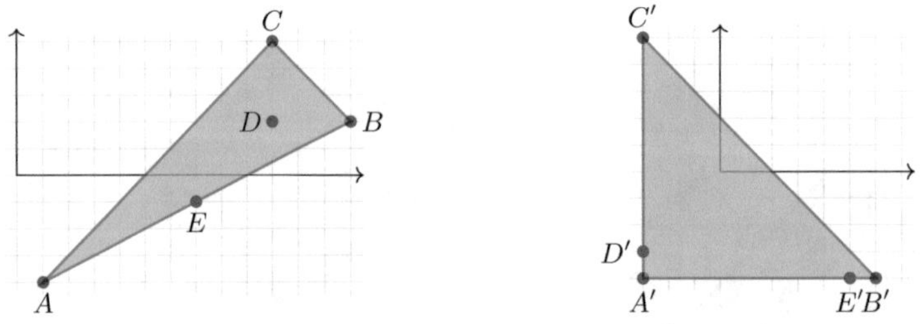

(A) $\mathcal{A} = \{A, B, C, D, E\}$
Facets: $\{A, E, B\}$, $\{B, C\}$, $\{C, A\}$
Vertices: A, B, C

(B) $\mathcal{A}' = \{A', B', C', D', E'\}$
Facets: $\{A', E', B'\}$, $\{B', C'\}$, $\{C', D', A'\}$
Vertices: A', B', C'

FIGURE 2. Faces of some planar sets.

Proposition VI.11. *Let $z \in X_{\mathcal{A}}$. Define $\mathcal{A}_z := \{\alpha \in \mathcal{A} : z \in U_\alpha\}$. Then $\mathcal{A}_z$ is a face of $\mathcal{A}$.*

PROOF. Pick $\alpha_1, \alpha_2 \in \mathcal{A}_z$. For each i, let $\mathcal{Q}_i$ be the (unique) face of $\mathcal{P}$ which contains α_i in its interior. For every $\epsilon \in (0,1)$, the convex combination $\alpha_\epsilon := \epsilon\alpha_1 + (1-\epsilon)\alpha_2$ of α_1 and α_2 belongs to the relative interior of a face $\mathcal{Q}$ of $\mathcal{P}$ containing both $\mathcal{Q}_i$ (corollary V.18). It suffices to show that $\mathcal{Q} \cap \mathcal{A} \subset \mathcal{A}_z$. Indeed, let ϵ be a rational number in $(0,1)$. Let $\beta_1, \ldots, \beta_k$ be the vertices of $\mathcal{Q}$. There are positive integers $N_1, \ldots, N_k$ such that $(\sum_{j=1}^k N_j)\alpha_\epsilon = \sum_{j=1}^k N_j\beta_j$ (corollary V.33). Write $N := \sum_{j=1}^k N_j$. Multiplying the N_j by some appropriate integer we may ensure that $N\epsilon$ is a positive integer. Then $\prod_{j=1}^k z_{\beta_j}^{N_j} = z_{\alpha_1}^{N\epsilon} z_{\alpha_2}^{N-N\epsilon}$ on $X_{\mathcal{A}}$. Since both α_i are in $\mathcal{A}_z$, it follows that $z_{\beta_j}|_z \neq 0$ for each j. Since each $\beta \in \mathcal{Q} \cap \mathcal{A}$ is a convex rational linear combination of the β_j, it follows by the same reasoning that $z_\beta|_z \neq 0$ for each $\beta \in \mathcal{Q} \cap \mathcal{A}$, as required. □

Given $\mathcal{B}' \subseteq \mathcal{B} \subseteq \mathcal{A}$, we write $\mathcal{B}' \preceq \mathcal{B}$ (respectively, $\mathcal{B}' \precneqq \mathcal{B}$) to denote that $\mathcal{B}'$ is a face (respectively, a proper face) of $\mathcal{B}$.

THEOREM VI.12. *For each face $\mathcal{B}$ of $\mathcal{A}$, define $V_{\mathcal{B}} := X_{\mathcal{A}} \setminus \bigcup_{\alpha \notin \mathcal{B}} U_\alpha$ and $O_{\mathcal{B}} := V_{\mathcal{B}} \cap \bigcap_{\beta \in \mathcal{B}} U_\beta$.*

(1) $V_{\mathcal{B}} = \bigcup_{\mathcal{B}' \preceq \mathcal{B}} O_{\mathcal{B}'}$. In particular,

$$X_{\mathcal{A}} \setminus X_{\mathcal{A}}^0 = \bigcup_{\mathcal{B} \precneqq \mathcal{A}} O_{\mathcal{B}} = \bigcup_{\mathcal{B} \precneqq \mathcal{A}} V_{\mathcal{B}} = \bigcup_{j=1}^k V_{\mathcal{A}_j}, \tag{50}$$

where $\mathcal{A}_1, \ldots, \mathcal{A}_k$ are the facets of $\mathcal{A}$.

(2) There is a one-to-one correspondence between the collection of $O_{\mathcal{B}}$ for $\mathcal{B} \preceq \mathcal{A}$ and the set of orbits of $X_{\mathcal{A}}^0$ on $X_{\mathcal{A}}$. In particular, each $V_{\mathcal{B}}$ is invariant under the action of $X_{\mathcal{A}}^0$.

(3) Each $V_{\mathcal{B}}$ is a toric variety with torus $O_{\mathcal{B}}$. More precisely, the pair $(V_{\mathcal{B}}, O_{\mathcal{B}})$ is isomorphic to $(X_{\mathcal{B}}, X_{\mathcal{B}}^0)$. The isomorphism is given by the projection map $\pi_{\mathcal{B}} : V_{\mathcal{B}} \to \mathbb{P}^{|\mathcal{B}|-1}$ which "drops" all the coordinates z_α such that $\alpha \notin \mathcal{B}$; in other words, $\pi_{\mathcal{B}}([z_\alpha : \alpha \in \mathcal{A}]) = [z_\beta : \beta \in \mathcal{B}]$.

(4) The action of $O_{\mathcal{B}}$ on $V_{\mathcal{B}}$ is compatible with the action of $X_{\mathcal{A}}^0$. More precisely, assume $y_{\mathcal{A}} \in X_{\mathcal{A}}^0$ and $y_{\mathcal{B}} \in O_{\mathcal{B}} \cong X_{\mathcal{B}}^0$ correspond to the same $x \in (\mathbb{k}^*)^n$, i.e. $y_{\mathcal{A}} = [x^\alpha : \alpha \in \mathcal{A}]$ and $y_{\mathcal{B}} = [x^\beta : \beta \in \mathcal{B}]$. Then for all $z \in V_{\mathcal{B}}$,

$$y_{\mathcal{A}} \cdot_{\mathcal{A}} z = y_{\mathcal{B}} \cdot_{\mathcal{B}} z,$$

where we write $\cdot_{\mathcal{A}}$ (respectively, $\cdot_{\mathcal{B}}$) to denote the action of $X_{\mathcal{A}}^0$ (respectively, $O_{\mathcal{B}}$) on $V_{\mathcal{B}}$.

PROOF. The first statement of assertion (1) follows from proposition VI.11, and this in turn implies the first two equalities of (50). Since $V_{\mathcal{B}'} \subseteq V_{\mathcal{B}}$ whenever $\mathcal{B}' \preceq \mathcal{B}$, and since every proper face is contained in a facet (proposition V.22), the last equality of (50) follows. We now prove the remaining assertions. Given $\mathcal{B} \preceq \mathcal{A}$, let $H_{\mathcal{B}} := V(z_\alpha : \alpha \notin \mathcal{B}) \subset \mathbb{P}^N$ be the coordinate subspace containing $V_{\mathcal{B}}$. Let $\pi_{\mathcal{B}} : \mathbb{P}^N \setminus V(z_\beta : \beta \in \mathcal{B}) \to H_{\mathcal{B}}$ be the natural projection and $Z_{\mathcal{B}}$ be the closure in $\mathbb{P}^N$ of $\pi_{\mathcal{B}}(X_{\mathcal{A}}^0)$.

Claim VI.12.1. $V_{\mathcal{B}} = Z_{\mathcal{B}}$.

PROOF. The inclusion $V_{\mathcal{B}} \subseteq Z_{\mathcal{B}}$ follows from a general property of morphisms (assertion (4) of corollary III.73). For the opposite inclusion it suffices (due to the definition of $V_{\mathcal{B}}$) to show that $X_{\mathcal{A}} \supset Z_{\mathcal{B}}$. Pick $\beta \in \mathcal{B}$. We will show that $X_{\mathcal{A}} \cap U_\beta \supseteq Z_{\mathcal{B}} \cap U_\beta$. We may assume $\beta = \alpha_0$. Write $z'_i := z_{\alpha_i}/z_{\alpha_0}$, so that $U_\beta \cong \mathbb{k}^N$ with coordinates $(z'_1, \ldots, z'_N)$. Let $f(z'_1, \ldots, z'_N) = \sum_{\gamma \in \mathbb{Z}^N} c'_\gamma z'^\gamma$ be a polynomial in $(z'_1, \ldots, z'_N)$ which vanishes on $X^0_{\mathcal{A}} \cap U_\beta$. It suffices to show that f vanishes on $Z_{\mathcal{B}} \cap U_\beta$ as well. Note that $X^0_{\mathcal{A}} \cap U_\beta = \{(x^{\alpha'_1}, \ldots, x^{\alpha'_N}) : x \in (\mathbb{k}^*)^n\}$, where $\alpha'_i := \alpha_i - \alpha_0$, $i = 1, \ldots, N$. Write $f = f' + f''$, where the monomials in f' consist solely of the z'_i such that $\alpha_i \in \mathcal{B}$ and each monomial in f'' contains at least one z'_i such that $\alpha_i \notin \mathcal{B}$. If B is the *affine hull* of $\{\alpha_i - \alpha_0 : \alpha_i \in \mathcal{B}\}$, exercise VI.9 implies that the exponent of each monomial in $f'(x^{\alpha'_1}, \ldots, x^{\alpha'_N})$ is on B, whereas the exponent of no monomial in $f''(x^{\alpha'_1}, \ldots, x^{\alpha'_N})$ is on B, in particular, the monomials that appear in $f'(x^{\alpha'_1}, \ldots, x^{\alpha'_N})$ are distinct from those appearing in $f''(x^{\alpha'_1}, \ldots, x^{\alpha'_N})$. Since $f(x^{\alpha'_1}, \ldots, x^{\alpha'_N})$ is identically zero on $(\mathbb{k}^*)^n$, it follows that $f'(x^{\alpha'_1}, \ldots, x^{\alpha'_N})$ is also identically zero on $(\mathbb{k}^*)^n$. This implies that $f \circ \pi_{\mathcal{B}}(x^{\alpha'_1}, \ldots, x^{\alpha'_N}) = f'(x^{\alpha'_1}, \ldots, x^{\alpha'_N}) = 0$ for all $x \in (\mathbb{k}^*)^n$, so that f vanishes on $\pi_{\mathcal{B}}(X^0_{\mathcal{A}})$, as required. □

It is evident that $Z_{\mathcal{B}}$ can be identified to $X_{\mathcal{B}}$ by "forgetting" the coordinates z_α for all $\alpha \notin \mathcal{B}$. Since $V_{\mathcal{B}} = Z_{\mathcal{B}}$, proposition VI.2 implies that this induces an identification of $X^0_{\mathcal{B}}$ with $O_{\mathcal{B}}$, which proves assertion (3). Assertion (4) then follows from identity (48). Since assertion (4), in particular, implies that $O_{\mathcal{B}}$ is an orbit of $X^0_{\mathcal{A}}$, this proves assertion (2) as well. □

Corollary VI.13 is an immediate corollary of theorem VI.12—we leave its proof as exercise VI.10. The second statement of corollary VI.13, in particular, implies that the complement of the torus is locally defined by a single equation on $X_{\mathcal{A}}$; in the terminology of section IV.2.6, $X_{\mathcal{P}} \setminus X^0_{\mathcal{A}}$ is the "support of a Cartier divisor" on $X_{\mathcal{A}}$.

Corollary VI.13. *If $\mathcal{V}$ is the set of vertices of $\mathcal{A}$, then $X_{\mathcal{A}} \subset \bigcup_{\alpha \in \mathcal{V}} U_\alpha$. In particular, $X_{\mathcal{A}} \setminus X^0_{\mathcal{A}} = V(\prod_{\alpha \in \mathcal{V}} z_\alpha) \cap X_{\mathcal{A}}$.* □

4.1. Exercises.

EXERCISE VI.9. Let $\mathcal{Q}$ be a face of a polyhedron $\mathcal{P} \subset \mathbb{R}^n$. Assume that the origin is in $\operatorname{aff}(\mathcal{Q})$. Let $\alpha = \sum_{j=1}^k r_j \alpha_j$, where $r_j > 0$ and $\alpha_j \in \mathcal{P}$ for each j. Show that $\alpha \in \operatorname{aff}(\mathcal{Q})$ if and only if $\alpha_j \in \mathcal{Q}$ for each j. [Hint: if $\mathcal{Q} = \operatorname{In}_\nu(\mathcal{P})$, consider the value of $\langle \nu, \cdot \rangle$ on $\mathcal{Q}$, $\mathcal{P}$ and α.]

EXERCISE VI.10. Prove corollary VI.13. [Hint: Use proposition V.19.]

5. Toric varieties from polytopes

If $\mathcal{A} \subset \mathcal{A}'$ are finite subsets of $\mathbb{Z}^n$, then the natural projection $\mathbb{P}^{|\mathcal{A}'|-1} \dashrightarrow \mathbb{P}^{|\mathcal{A}|-1}$ restricts to a rational map $\pi_{\mathcal{A},\mathcal{A}'} : X_{\mathcal{A}'} \dashrightarrow X_{\mathcal{A}}$. This map is, in general, not defined everywhere on $X_{\mathcal{A}'}$. For example, if $\mathcal{A} = \{(0,0),(1,0)\}$ and $\mathcal{A}' = \{(0,0),(1,0),(0,1)\}$, then example VI.4 implies that $X_{\mathcal{A}'} \cong \mathbb{P}^2$, $X_{\mathcal{A}} \cong \mathbb{P}^1$, and

$\pi_{\mathcal{A},\mathcal{A}'}$ maps $[x_0 : x_1 : x_2] \mapsto [x_0 : x_1]$, which is not defined at the point $[0 : 0 : 1]$. However, if in addition $\mathcal{A}$ and $\mathcal{A}'$ have the same convex hull in $\mathbb{R}^n$, then corollary VI.13 implies that $\pi_{\mathcal{A},\mathcal{A}'}$ is well defined everywhere on $X_{\mathcal{A}'}$ (exercise VI.11). If $\mathcal{P}$ is the convex hull of $\mathcal{A}$ in $\mathbb{R}^n$, this observation shows that there is a natural morphism $X_{\mathcal{P}\cap\mathbb{Z}^n} \to X_{\mathcal{A}}$, and for every positive integer k there is a natural morphism

$$X_{(k+1)\mathcal{P}\cap\mathbb{Z}^n} \to X_{(k\mathcal{P}\cap\mathbb{Z}^n)+(\mathcal{P}\cap\mathbb{Z}^n)}$$

(For subsets $\mathcal{S}$ of $\mathbb{R}^n$ and a positive integer d, in chapter V we defined $d\mathcal{S}$ as a "dilation," whereas in section VI.3 we defined it as the sum of d-copies of $\mathcal{S}$. This does not lead to any conflict for the case of convex polytopes, see remark V.2.) Proposition VI.6 implies that $X_{(k\mathcal{P}\cap\mathbb{Z}^n)+(\mathcal{P}\cap\mathbb{Z}^n)}$ is isomorphic to a subset of $X_{k\mathcal{P}\cap\mathbb{Z}^n} \times X_{\mathcal{P}\cap\mathbb{Z}^n}$, so that the projection onto the first factor induces a morphism $X_{(k\mathcal{P}\cap\mathbb{Z}^n)+(\mathcal{P}\cap\mathbb{Z}^n)} \to X_{k\mathcal{P}\cap\mathbb{Z}^n}$. Consequently, there is a sequence of morphisms

$$(51) \qquad \cdots \xrightarrow{\pi_{3,4}} X_{3\mathcal{P}\cap\mathbb{Z}^n} \xrightarrow{\pi_{2,3}} X_{2\mathcal{P}\cap\mathbb{Z}^n} \xrightarrow{\pi_{1,2}} X_{\mathcal{P}\cap\mathbb{Z}^n} \xrightarrow{\pi_1} X_{\mathcal{A}}.$$

Proposition VI.14. *For k sufficiently large, $\pi_{k,k+1} : X_{(k+1)\mathcal{P}\cap\mathbb{Z}^n} \to X_{k\mathcal{P}\cap\mathbb{Z}^n}$ is an isomorphism.*

PROOF. Let $\alpha_0, \ldots, \alpha_s$ be the vertices of $\mathcal{P}$, so that $k\alpha_j$, $j = 0, \ldots, s$, are the vertices of $k\mathcal{P}$. Corollary VI.13 implies that $X_{k\mathcal{P}\cap\mathbb{Z}^n}$ is the union of affine open sets $X_{k\mathcal{P}\cap\mathbb{Z}^n} \cap U_{k\alpha_j}$, $j = 0, \ldots, s$. For each j, proposition VI.2 implies that the coordinate ring of $X_{k\mathcal{P}\cap\mathbb{Z}^n} \cap U_{k\alpha_j}$ is the semigroup algebra $\Bbbk[S_{j,k}]$, where $S_{j,k}$ is the subsemigroup of $\mathbb{Z}^n$ generated by $\{\alpha - k\alpha_j : \alpha \in k\mathcal{P} \cap \mathbb{Z}^n\}$. Since $\alpha - k\alpha_j = (\alpha + \alpha_j) - (k+1)\alpha_j$, it follows that $S_{j,k} \subset S_{j,k+1}$, and $\pi_{k,k+1}$ is simply the morphism induced by this inclusion. Note that each $S_{j,k}$ is a subsemigroup of $\mathcal{C}_{\alpha_j} \cap \mathbb{Z}^n$, where $\mathcal{C}_{\alpha_j}$ is the *rational* convex polyhedral cone in $\mathbb{R}^n$ generated by $\{\alpha_i - \alpha_j : i = 0, \ldots, s\}$. Gordan's lemma (lemma V.34) implies that $\mathcal{C}_{\alpha_j} \cap \mathbb{Z}^n$ is finitely generated. Due to exercise VI.12 we may choose an integer K such that for each j, $S_{j,k}$ contains each of these generators for each $k > K$. For each $k > K$ and each j, it follows that $S_{j,k} = \mathcal{C}_{\alpha_j} \cap \mathbb{Z}^n$ is independent of k; consequently $X_{(k+1)\mathcal{P}\cap\mathbb{Z}^n} \cong X_{k\mathcal{P}\cap\mathbb{Z}^n}$. □

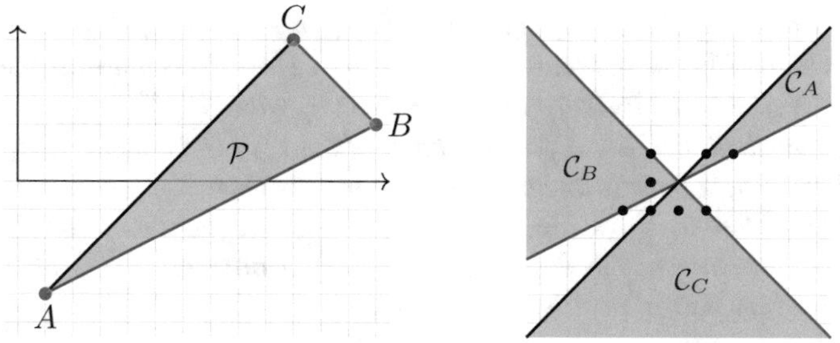

FIGURE 3. Cones of regular functions on basic open subsets of $X_{\mathcal{P}}$. The dots in each cone mark generators of the semigroup of integral points in the cone.

Given a convex integral polytope $\mathcal{P} \subseteq \mathbb{R}^n$, we write $X_{\mathcal{P}}$ for the toric variety isomorphic to $X_{k\mathcal{P}\cap\mathbb{Z}^n}$ for sufficiently large k and $X^0_{\mathcal{P}}$ for the torus $X^0_{k\mathcal{P}\cap\mathbb{Z}^n}$ of $X_{\mathcal{P}}$; we also write $\phi_{\mathcal{P}}$ for the morphism $\phi_{k\mathcal{P}\cap\mathbb{Z}^n} : (\mathbb{k}^*)^n \to X^0_{\mathcal{P}}$ defined as in (47). The arguments from the proof of proposition VI.14 show that $X_{\mathcal{P}}$ is the union of open affine subsets $U'_\alpha := X_{\mathcal{P}} \cap U_\alpha$ corresponding to vertices α of $\mathcal{P}$, and the coordinate ring of each U'_α is generated by the monomials whose exponents belong to the cone $\mathcal{C}_\alpha$ generated by $\mathcal{P} - \alpha$. See fig. 3 for the cones corresponding to the vertices of a triangle. The following proposition summarizes some basic properties of $X_{\mathcal{P}}$.

Proposition VI.15. *Let $\mathcal{A}$ be a finite subset of $\mathbb{Z}^n$ and $\mathcal{P}$ be the convex hull of $\mathcal{A}$ in $\mathbb{R}^n$. Let $G_{\mathcal{P}}$ be the subgroup of $\mathbb{Z}^n$ generated by pairwise differences of integral elements in the affine hull of $\mathcal{P}$ and $G_{\mathcal{A}}$ be the subgroup of $G_{\mathcal{P}}$ generated by the pairwise differences of elements from $\mathcal{A}$.*

(1) *If k is such that $G_{\mathcal{P}}$ is generated by $\{\alpha - \beta : \alpha, \beta \in k\mathcal{P} \cap \mathbb{Z}^n\}$, then $X_{\mathcal{P}} \cong X_{k\mathcal{P}\cap\mathbb{Z}^n}$.*
(2) *The dimension of $X_{\mathcal{P}}$ is the same as the dimension of $\mathcal{P}$. If $\dim(\mathcal{P}) = n$, then $\phi_{\mathcal{P}}$ is an isomorphism between $(\mathbb{k}^*)^n$ and $X^0_{\mathcal{P}}$, i.e. $X_{\mathcal{P}}$ is a compactification of $(\mathbb{k}^*)^n$.*
(3) *There is a natural finite-to-one morphism $\phi_{\mathcal{P},\mathcal{A}} : X_{\mathcal{P}} \to X_{\mathcal{A}}$ defined as the composition of the following maps:*

$$X_{\mathcal{P}} \cong X_{k\mathcal{P}\cap\mathbb{Z}^n} \xrightarrow{\pi_{k\mathcal{P}\cap\mathbb{Z}^n, k\mathcal{A}}} X_{k\mathcal{A}} \xrightarrow{\phi_{k\mathcal{A},\mathcal{A}}} X_{\mathcal{A}}, \tag{52}$$

where k is as in assertion (1), $\pi_{k\mathcal{P}\cap\mathbb{Z}^n,k\mathcal{A}}$ is the projection which drops coordinates corresponding to elements in $(k\mathcal{P} \cap \mathbb{Z}^n) \setminus k\mathcal{A}$, and $\phi_{k\mathcal{A},\mathcal{A}}$ is the inverse of the Veronese isomorphism of degree k (see proposition VI.6). The degree of $\phi_{\mathcal{P},\mathcal{A}}$ is the index of $G_{\mathcal{A}}$ in $G_{\mathcal{P}}$. In particular, if $G_{\mathcal{A}} = G_{\mathcal{P}}$, then $\phi_{\mathcal{P},\mathcal{A}}$ restricts to an isomorphism between $X^0_{\mathcal{P}}$ and $X^0_{\mathcal{A}}$.
(4) *There is an open cover $\{U'_\alpha\}_{\alpha\in\mathcal{P}\cap\mathbb{Z}^n}$ of $X_{\mathcal{P}}$ such that the coordinate ring of each U'_α is the semigroup algebra $\mathbb{k}[S_\alpha]$, where S_α is the semigroup of integral points in the convex polyhedral cone $\mathcal{C}_\alpha$ generated by $\{\alpha' - \alpha : \alpha' \in \mathcal{P}\}$.*
(5) *There is a natural one-to-one correspondence between the faces of $\mathcal{P}$ and the orbits of $X^0_{\mathcal{P}}$ on $X_{\mathcal{P}}$. For each face $\mathcal{Q}$ of $\mathcal{P}$, let $O_{\mathcal{Q}}$ be the corresponding orbit and $V_{\mathcal{Q}}$ be the closure of $O_{\mathcal{Q}}$ in $X_{\mathcal{P}}$. Then $V_{\mathcal{Q}}$ is naturally isomorphic to $X_{\mathcal{Q}}$, and the isomorphism identifies $O_{\mathcal{Q}}$ with $X^0_{\mathcal{Q}}$.*
(6) *In particular, $X_{\mathcal{P}} \setminus X^0_{\mathcal{P}}$ is the union of the $V_{\mathcal{Q}}$ for the facets $\mathcal{Q}$ of $\mathcal{P}$.*
(7) *Let $\mathcal{P}'$ be a convex integral polytope in $\mathbb{R}^n$. Then $X_{\mathcal{P}+\mathcal{P}'}$ is isomorphic to the closure in $X_{\mathcal{P}} \times X_{\mathcal{P}'}$ of the image of the diagonal map $(\mathbb{k}^*)^n \to X_{\mathcal{P}} \times X_{\mathcal{P}'}$ which sends $x \mapsto (\phi_{\mathcal{P}}(x), \phi_{\mathcal{P}'}(x))$. Moreover, $X_{k\mathcal{P}+l\mathcal{P}'} \cong X_{\mathcal{P}+\mathcal{P}'}$ for every pair of positive integers k, l.*

PROOF. Due to proposition VI.3 we may assume $\dim(\mathcal{P}) = n$ and $\mathcal{P}$ contains the origin. If k is as in assertion (1), then after another application of proposition VI.3 we may assume that each standard unit vector is in $k\mathcal{P} \cap \mathbb{Z}^n$, which immediately implies that $\phi_{\mathcal{P}}$ is an isomorphism and proves the first two assertions. The last assertion is a straightforward corollary of propositions VI.6 and VI.14. The remaining statements follow from propositions VI.1 and VI.2 and theorem VI.12. We leave it as an exercise to complete the proof. □

Example VI.16. Let $\mathcal{A} := \{0, 2\} \subset \mathbb{Z}$. Since $x \mapsto 2x$ maps $\{0, 1\}$ onto $\mathcal{A}$, proposition VI.3 and example VI.7 imply that $X_{\mathcal{A}} \cong \mathbb{P}^1$. If $\mathcal{P} = \operatorname{conv}(\mathcal{A}) \subset \mathbb{R}$, then $G_{\mathcal{P}} = \mathbb{Z}$ and assertion (1) of proposition VI.15 is satisfied with $k = 1$, i.e. $X_{\mathcal{P}}$ is the closure in $\mathbb{P}^2$ of the image of $\Bbbk^*$ under the map $x \mapsto [1 : x : x^2]$. Example VI.7 implies that $X_{\mathcal{P}}$ is also isomorphic to $\mathbb{P}^1$. The map $\phi_{\mathcal{P},\mathcal{A}} : X_{\mathcal{P}} \to X_{\mathcal{A}}$ is the restriction to $X_{\mathcal{P}}$ of the projection $[z_0 : z_1 : z_2] \to [z_0 : z_2]$, and therefore on the level of the tori it is simply the map $x \mapsto x^2$. Note that $\deg(\phi_{\mathcal{P},\mathcal{A}}) = 2$ is also the index of $G_{\mathcal{A}} = 2\mathbb{Z}$ in $\mathbb{Z} = G_{\mathcal{P}}$.

Example VI.17. Let $\mathcal{S}^n$ be the n-dimensional simplex in $\mathbb{R}^n$ with vertices at the origin and at the elements of the standard unit basis of $\mathbb{R}^n$, and let $\mathcal{K}^n := [0, 1]^n \subset \mathbb{R}^n$. Then $G_{\mathcal{S}^n} = G_{\mathcal{K}^n} = \mathbb{Z}^n$, and both $\mathcal{S}^n$ and $\mathcal{K}^n$ satisfy assertion (1) of proposition VI.15 with $k = 1$. Therefore it follows from examples VI.4 and VI.5 that $X_{\mathcal{S}^n} \cong \mathbb{P}^n$ and $X_{\mathcal{K}^n} \cong (\mathbb{P}^1)^n$.

Example VI.18. Consider $\mathcal{A}'$ from fig. 2b. The convex hull $\mathcal{P}'$ of $\mathcal{A}$ is a translation of $9\mathcal{S}^2$, so that propositions VI.6 and VI.15 and example VI.17 imply that $X_{\mathcal{P}'} \cong X_{\mathcal{S}^2} \cong \mathbb{P}^2$. Note that $G_{\mathcal{P}'} = G_{\mathcal{A}'} = \mathbb{Z}^2$, i.e. the map $\phi_{\mathcal{P}',\mathcal{A}'} : X_{\mathcal{P}'} \to X_{\mathcal{A}'}$ has degree one. However, it is *not* an isomorphism. Indeed, proposition VI.2 implies that the coordinate ring of $X_{\mathcal{A}'} \cap U_{\mathcal{A}'}$ is $\Bbbk[x^8, x^9, y] \cong \Bbbk[u, v, w]/\langle u^9 - v^8 \rangle$ (exercise VI.14). Therefore $X_{\mathcal{A}'} \cap U_{\mathcal{A}'}$ is isomorphic to the hypersurface in $\Bbbk^3$ defined by $u^9 - v^8 = 0$, which is *singular* at all points of the w-axis, and $\phi_{\mathcal{P}',\mathcal{A}'} : X_{\mathcal{P}'} \to X_{\mathcal{A}'}$ is a *desingularization* of $X_{\mathcal{A}'}$.

Even though $X_{\mathcal{P}}$ is by definition isomorphic to $X_{k\mathcal{P}\cap\mathbb{Z}^n}$ for $k \gg 1$, in each of the preceding examples it suffices to take $k = 1$. In general it suffices to take $k \geq \dim(\mathcal{P}) - 1$, see e.g. [CLS11, Theorem 2.2.12]. In particular, to find examples for which one needs to take $k > 1$ requires polytopes with dimensions at least 3.

5.1. Exercises.

EXERCISE VI.11. Let $\mathcal{A} \subset \mathcal{A}'$ be finite subsets of $\mathbb{Z}^n$ such that $\operatorname{conv}(\mathcal{A}) = \operatorname{conv}(\mathcal{A}')$. Show that $\pi_{\mathcal{A},\mathcal{A}'}$ is well defined everywhere on $X_{\mathcal{A}'}$. [Hint: $\mathcal{A}$ and $\mathcal{A}'$ have the same set of vertices. Use corollary VI.13.]

EXERCISE VI.12. In the notation of the proof of proposition VI.14, show that for each $\beta \in \mathcal{C}_{\alpha_j}$, there is $K \geq 0$ such that $\beta \in k\mathcal{P} - k\alpha_j$ for each $k \geq K$.

EXERCISE VI.13. Complete the proof of proposition VI.15.

EXERCISE VI.14. Show that $\Bbbk[x^8, x^9, y] \cong \Bbbk[u, v, w]/\langle u^9 - v^8 \rangle$. Prove that the hypersurface $V(u^9 - v^8)$ in $\Bbbk^3$ is singular at all points of the w-axis.

EXERCISE VI.15. In the setup of fig. 3, let S_A (respectively, S_B, S_C) be the subsemigroup of $\mathbb{Z}^n$ consisting of integral elements in $\mathcal{C}_A$ (respectively, $\mathcal{C}_B, \mathcal{C}_C$).

(1) Show that S_A is generated as a semigroup by $(1,1),(2,1)$; S_B is generated as a semigroup by $(-1,1),(-1,0),(-2,-1)$; S_C is generated as a semigroup by $(-1,-1),(0,-1),(1,-1)$.
(2) Deduce that $X_{\mathcal{P}} \cap U_A \cong \mathbb{k}^2$, $X_{\mathcal{P}} \cap U_B \cong V(uv - w^3) \subset \mathbb{k}^3$, and $X_{\mathcal{P}} \cap U_C \cong V(uv-w^2) \subset \mathbb{k}^3$. [Hint: $\mathbb{k}[X_{\mathcal{P}} \cap U_A] = \mathbb{k}[xy, x^2y]$, $\mathbb{k}[X_{\mathcal{P}} \cap U_B] = \mathbb{k}[x^{-1}y, x^{-1}, x^{-2}y^{-1}]$, $\mathbb{k}[X_{\mathcal{P}} \cap U_C] = \mathbb{k}[x^{-1}y^{-1}, y^{-1}, xy^{-1}]$.]
(3) Conclude that $X_{\mathcal{P}}$ has precisely two singular points.

EXERCISE VI.16. Let $\mathcal{P}$ be a convex integral polytope in $\mathbb{R}^n$ and α, α' be vertices of $\mathcal{P}$.

(1) Show that $x^{\alpha'-\alpha}$ is an invertible regular function on $U_\alpha \cap U_{\alpha'} \cap X_{\mathcal{P}}$.
(2) Let $\mathcal{Q}$ be the smallest face of $\mathcal{P}$ containing both α and α', and $\mathcal{Q}_\alpha$ be the cone generated by $\{\beta - \alpha : \beta \in \mathcal{Q}\}$. Show that $\alpha' - \alpha$ is in the relative interior of $\mathcal{Q}_\alpha$.
(3) Let H be the linear subspace of $\mathbb{R}^n$ spanned by $\{\beta - \alpha : \beta \in \mathcal{Q}\}$. For each $\gamma \in H \cap \mathbb{Z}^n$, show that x^γ is an invertible regular function on $U_\alpha \cap U_{\alpha'} \cap X_{\mathcal{P}}$. [Hint: Choose $\beta_1, \ldots, \beta_k \in \mathcal{Q} \cap \mathbb{Z}^n$ such that $\beta_j - \alpha$, $j = 1, \ldots, k$, generate $\mathcal{Q}_\alpha$ as a cone. Use exercise V.30 and the preceding assertions to show that for each j, $x^{\beta_j - \alpha}$ is a regular function on $U_\alpha \cap X_{\mathcal{P}}$ which does not vanish at any point of $U_\alpha \cap U_{\alpha'} \cap X_{\mathcal{P}}$.]
(4) Deduce that if $\mathcal{Q} = \mathcal{P}$, then $U_\alpha \cap U_{\alpha'} \cap X_{\mathcal{P}} = X^0_{\mathcal{P}}$.

6. Nonsingularity in codimension one on $X_{\mathcal{P}}$

Example VI.18 and exercise VI.15 above show that toric varieties $X_{\mathcal{P}}$ from polytopes might be singular. However, we will see in this section that the $X_{\mathcal{P}}$ is nonsingular outside a subvariety of dimension at most $\dim(X_{\mathcal{P}}) - 2$, i.e. $X_{\mathcal{P}}$ is "nonsingular in codimension one." Note that this is, in general, *not* true for varieties $X_{\mathcal{A}}$ (see example VI.18). We continue to use the notation of section VI.5. Let $\mathcal{Q}$ be a facet of $\mathcal{P}$. We will show that $X_{\mathcal{P}}$ is nonsingular at all points of $O_{\mathcal{Q}}$. Due to proposition VI.3 we may assume without loss of generality that $\mathcal{P}$ is *full dimensional*, i.e. $\mathcal{P}$ is an n-dimensional polytope in $\mathbb{R}^n$. Let ν be the *primitive inner normal* (see remark V.44) to $\mathcal{Q}$. Let $\mathbb{Z}^n_{\nu \geq 0} := \{\beta \in \mathbb{Z}^n : \langle \nu, \beta \rangle \geq 0\}$ and $\mathbb{Z}^n_{\nu\perp} := \{\beta \in \mathbb{Z}^n : \langle \nu, \beta \rangle = 0\}$. Choose an arbitrary element $\alpha_\nu \in \mathbb{Z}^n$ such that $\langle \nu, \alpha_\nu \rangle = 1$. We write $\mathbb{Z}_{\geq 0}\langle \alpha_\nu \rangle := \{k\alpha_\nu : k \in \mathbb{Z},\ k \geq 0\}$. The following is a straightforward implication of lemma B.57.

Lemma VI.19. *There is a basis of $\mathbb{Z}^n$ of the form $\beta_1, \ldots, \beta_n$, where $\beta_n = \alpha_\nu$ and $\beta_1, \ldots, \beta_{n-1}$ constitute a basis of $\mathbb{Z}^n_{\nu\perp}$. In particular, $\mathbb{Z}^n_{\nu\perp} \cong \mathbb{Z}^{n-1}$, and as a semigroup $\mathbb{Z}^n_{\nu \geq 0}$ is isomorphic to $\mathbb{Z}^n_{\nu\perp} + \mathbb{Z}_{\geq 0}\langle \alpha_\nu \rangle \cong \mathbb{Z}^{n-1} \times \mathbb{Z}_{\geq 0}$.* □

Proposition VI.20. *Let $U_{\mathcal{Q}} := X^0_{\mathcal{P}} \cup O_{\mathcal{Q}}$.*

(1) $U_{\mathcal{Q}}$ is an open affine neighborhood of $O_{\mathcal{Q}}$ in $X_{\mathcal{P}}$.
(2) $\mathbb{k}[U_{\mathcal{Q}}] = \mathbb{k}[x^\beta : \beta \in \mathbb{Z}^n_{\nu \geq 0}] \cong \mathbb{k}[\mathbb{Z}^{n-1} \times \mathbb{Z}_{\geq 0}]$. In particular, $U_{\mathcal{Q}} \cong (\mathbb{k}^*)^{n-1} \times \mathbb{k}$.

(3) $O_{\mathcal{Q}} = V(x^{\alpha_\nu}) \subset U_{\mathcal{Q}}$ *and* $\Bbbk[O_{\mathcal{Q}}] \cong \Bbbk[U_{\mathcal{Q}}]/\langle x^{\alpha_\nu}\rangle \cong \Bbbk[\mathbb{Z}^n_{\nu^\perp}]$. *In particular, the embedding* $O_{\mathcal{Q}} \hookrightarrow U_{\mathcal{Q}}$ *is isomorphic to the embedding* $(\Bbbk^*)^{n-1} \times \{0\} \hookrightarrow (\Bbbk^*)^{n-1} \times \Bbbk$.

PROOF. Let $\mathcal{A} := k\mathcal{P} \cap \mathbb{Z}^n$, where k is large enough so that $X_{\mathcal{P}} \cong X_{\mathcal{A}}$ and there is an integral element α_0 in the relative interior of $k\mathcal{Q}$. Since $\mathcal{Q}$ is the smallest face of $\mathcal{P}$ containing α_0 (proposition V.14), proposition VI.11 implies that $U_{\mathcal{Q}} = X_{\mathcal{P}} \cap U_{\alpha_0}$, which proves assertion (1). Let $\beta_1, \dots, \beta_n = \alpha_\nu$ be a basis of $\mathbb{Z}^n$ as in lemma VI.19. Choosing a large enough k we can also ensure that $\alpha_0 + \beta_j \in k\mathcal{P} \cap \mathbb{Z}^n$ for each $j = 1, \dots, n$ (exercise V.34). It is then straightforward to check that the semigroup generated by $\mathcal{A} - \alpha_0$ is precisely $\mathbb{Z}^n_{\nu \geq 0}$. Assertions (2) and (3) then follow from proposition VI.15 and lemma VI.19 in a straightforward manner—we leave the proof as an exercise. □

Corollary VI.21. (Nonsingularity of $X_{\mathcal{P}}$ in codimension one)**.** *The set* $\mathrm{Sing}(X_{\mathcal{P}})$ *of singular points of* $X_{\mathcal{P}}$ *is contained in* $\bigcup_i O_{\mathcal{Q}'_i}$, *where* $\mathcal{Q}'_i$ *are faces of* $\mathcal{P}$ *of dimension* $\leq n-2$. *In particular,* $\dim(\mathrm{Sing}(X_{\mathcal{P}})) \leq n-2$.

PROOF. $X_{\mathcal{P}} \setminus \bigcup_i O_{\mathcal{Q}'_i}$ is the union of $U_{\mathcal{Q}}$ over all facets $\mathcal{Q}$ of $\mathcal{P}$, which is nonsingular due to proposition VI.20. Since $\dim(\bigcup_i O_{\mathcal{Q}'_i}) = n-2$, the result follows. □

Let ν and α_ν be as in proposition VI.20. Given $g = \sum_\beta c_\beta x^\beta \in \Bbbk[x_1, x_1^{-1}, \dots, x_n, x_n^{-1}]$, the *support* $\mathrm{Supp}(g)$ of g is the set of all $\beta \in \mathbb{Z}^n$ such that $c_\beta \neq 0$, and the *Newton polytope* $\mathrm{NP}(g)$ of g is the convex hull of $\mathrm{Supp}(g)$. Let $m := \min_{\mathrm{Supp}(g)}(\nu) = \min_{\mathrm{NP}(g)}(\nu)$. Choose an arbitrary isomorphism $\psi_\nu : \mathbb{Z}^n_{\nu^\perp} \cong \mathbb{Z}^{n-1}$. Define $T_{\alpha_\nu}(g) := x^{-m\alpha_\nu} g$ and $\mathrm{In}'_{\alpha_\nu,\psi_\nu}(g) := \sum_{\langle \nu,\beta\rangle = m} c_\beta x^{\psi_\nu(\beta - m\alpha_\nu)}$ (the "T" in $T_{\alpha_\nu}(\cdot)$ is supposed to imply "translation," and "In" in $\mathrm{In}'_{\alpha_\nu,\psi_\nu}(\cdot)$ is to suggest "initial form"). See fig. 4 for an example with $\mathcal{P}$ from fig. 3. The following result is an immediate corollary of proposition VI.20. Its proof is left as an exercise.

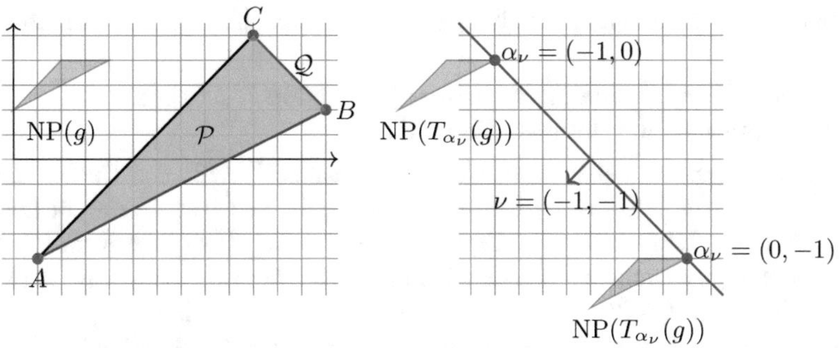

FIGURE 4. Different choices for $T_{\alpha_\nu}(g)$ when $g = 2y^2 - 3x^2y^3 + 7x^4y^4 - 6x^2y^4$.

Corollary VI.22. *(1) $T_{\alpha_\nu}(g)$ is a regular function on $U_\mathcal{Q}$ for each Laurent polynomial g.*

(2) The correspondence $T_{\alpha_\nu}(g)|_{O_\mathcal{Q}} \mapsto \mathrm{In}'_{\alpha_\nu,\psi_\nu}(g)$ induces an isomorphism $\psi_\nu^* : (\mathbb{k}^*)^{n-1} \cong O_\mathcal{Q}$.

(3) If α'_ν is another element in $\mathbb{Z}^n$ such that $\langle \nu, \alpha'_\nu \rangle = 1$, then $T_{\alpha_\nu}(g)/T_{\alpha'_\nu}(g) = x^{m(\alpha'_\nu - \alpha_\nu)}$ is invertible on $U_\mathcal{Q}$.

A basic property of the varieties $X_\mathcal{P}$ is that they are "normal," and nonsingularity in codimension one follows from normality. In this book we do not treat this notion, see [Ful93, Section 2.1] or [CLS11, Section 2.4] for an exposition of this and other fundamental properties of $X_\mathcal{P}$ including the following result (which we do not use): assume $\mathcal{P} \subset \mathbb{R}^n$ is full dimensional. Then $X_\mathcal{P}$ is nonsingular if and only if both of the following are true for every vertex α of $\mathcal{P}$:

- $\mathcal{C}_\alpha$ has precisely n edges and
- the primitive integral elements of the edges of $\mathcal{C}_\alpha$ form a basis of $\mathbb{Z}^n$.

(It then follows due to proposition VI.10 that if $X_\mathcal{P}$ is nonsingular, then it is a compactification of $\mathbb{k}^n$.)

6.1. Exercises.

EXERCISE VI.17. Complete the proof of proposition VI.20.

EXERCISE VI.18. Prove corollary VI.22.

7. Extending closed subschemes of the torus to $X_\mathcal{P}$

Recall (from section IV.2) that a *closed subscheme* V of a variety X is essentially a Zariski closed subset V' of a variety together with a *sheaf of ideals* $\mathcal{I}$ such that V' is precisely the set of zeroes of elements in $\mathcal{I}$. If $\bar{X}$ is a variety containing X as a Zariski open subset, and $\bar{V}$ is a closed subscheme of $\bar{X}$, we say that $\bar{V}$ *extends* V if the scheme-theoretic intersection $\bar{V} \cap X$ is precisely V. In this section we will study the case that $X = (\mathbb{k}^*)^n$ and $\bar{X}$ is the toric variety $X_\mathcal{P}$ corresponding to an n-dimensional convex integral polytope $\mathcal{P}$. We are specially interested in the case that

(a) both V and $\bar{V}$ are *Cartier divisors* (see section IV.2.6) and
(b) $\mathrm{Supp}(\bar{V})$ is the closure of $\mathrm{Supp}(V)$.

Each Laurent polynomial $g \in \mathbb{k}[x_1, x_1^{-1}, \ldots, x_n, x_n^{-1}]$ defines a Cartier divisor $V(g)$ on $(\mathbb{k}^*)^n$ (in fact it is not hard to see, using the fact that the ring of Laurent polynomials is a UFD, that *every* Cartier divisor on $(\mathbb{k}^*)^n$ is of the form $V(g)$ for some Laurent polynomial g—but we will not use it). There are many ways to extend $V(g)$ to $X_\mathcal{P}$, e.g. if $\mathrm{NP}(g) \subset \mathcal{P}$, then exercise VI.19 prescribes a way to extend $V(g)$ to a Cartier divisor on $X_\mathcal{P}$ which satisfies property (b) if and only if $\mathrm{NP}(g)$ intersects each facet of $\mathcal{P}$. However, we will shortly see that for some $\mathcal{P}$, there are Cartier divisors on $(\mathbb{k}^*)^n$ which can *not* be extended to $X_\mathcal{P}$ in a way to satisfy property (b). Then we will define an open subset $X^1_\mathcal{P}$ of $X_\mathcal{P}$ such that $X_\mathcal{P} \setminus X^1_\mathcal{P}$ has dimension $\leq n-2$, and every Cartier divisor on $(\mathbb{k}^*)^n$ does admit extensions to $X^1_\mathcal{P}$ satisfying both properties (a) and (b).

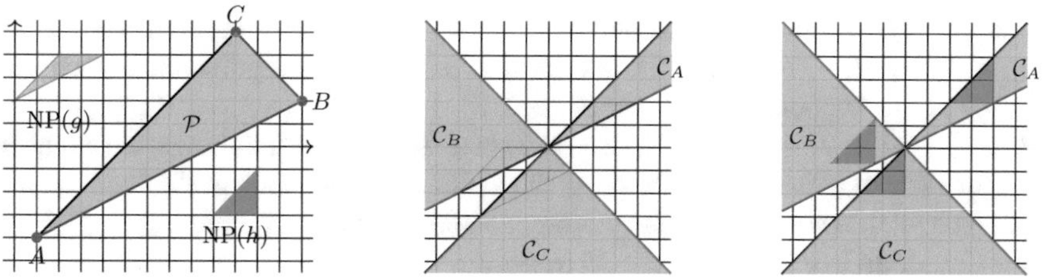

FIGURE 5. Extending Cartier divisors in dimension two.

Example VI.23. Consider $\mathcal{P}$ and g from fig. 4. Theorem VI.12 and proposition VI.15 imply that $X_{\mathcal{P}} \subset U_A \cup U_B \cup U_C$, and for each $P \in \{A, B, C\}$, the coordinate ring of $U_P \cap X_{\mathcal{P}}$ consists of Laurent polynomials supported at the cone $\mathcal{C}_P$; consider the (unique) translation of $\mathrm{NP}(g)$ which is contained in $\mathcal{C}_P$ and touches both sides of $\mathcal{C}_P$ (see the middle panel of fig. 5). More precisely, let

$$g_P := \begin{cases} 2 - 3x^2y + 7x^4y^2 - 6x^2y^2 & \text{if } P = A, \\ x^{-4}y^{-2}(2 - 3x^2y + 7x^4y^2 - 6x^2y^2) & \text{if } P = B, \\ x^{-3}y^{-3}(2 - 3x^2y + 7x^4y^2 - 6x^2y^2) & \text{if } P = C. \end{cases}$$

Then it is straightforward to check that the pair $(U_P \cap X_{\mathcal{P}}, g_P)$, $P \in \{A, B, C\}$, defines a Cartier divisor D on $X_{\mathcal{P}}$ such that D extends the Cartier divisor $V(g)$ and $\mathrm{Supp}(D)$ is the closure in $X_{\mathcal{P}}$ of $V(g) \subset (\mathbb{k}^*)^2$ (see exercise VI.20 for a more general result). On the other hand, if h is any Laurent polynomial with $\mathrm{NP}(h)$ as in the left panel of fig. 5, then the translation of $\mathrm{NP}(h)$ that is contained in $\mathcal{C}_B$ and touches both sides of $\mathcal{C}_B$ is *not* integral, see the right panel of fig. 5. It follows that if E is any Cartier divisor on $X_{\mathcal{P}}$ which extends $V(h) \subset (\mathbb{k}^*)^2$, then $\mathrm{Supp}(E)$ must contain either V_{AB} or V_{BC}, so that $\mathrm{Supp}(E)$ is larger than the closure in $X_{\mathcal{P}}$ of $V(h) \subset (\mathbb{k}^*)^2$ (exercise VI.21); in particular, E does not satisfy property (b) of the extension.

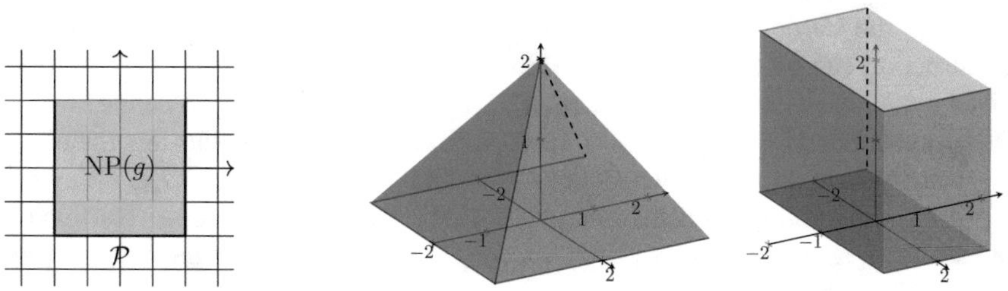

(A) Crossections of $\mathcal{P}$ and $\mathrm{NP}(g)$ (B) $\mathcal{P}$ (C) $\mathrm{NP}(g)$

FIGURE 6. Obstruction to extension of Cartier divisors in dimension ≥ 3.

Example VI.24. Let $\mathcal{P}$ be the polytope in $\mathbb{R}^3$ with vertices $(2,2,0)$, $(-2,2,0)$, $(-2,-2,0)$, $(2,-2,0)$, $(0,0,2)$, and let $g \in \Bbbk[x,y,z]$ be a polynomial such that $\mathrm{NP}(g)$ is the "cuboid" with vertices $(2,1,0)$, $(-2,1,0)$, $(-2,-1,0)$, $(2,-1,0)$, $(2,1,2)$, $(-2,1,2)$, $(-2,-1,2)$, $(2,-1,2)$ (see fig. 6). It is straightforward to check that for each $m \geq 1$, if a translation $\mathcal{Q}$ of $m\,\mathrm{NP}(g)$ is contained in $\mathcal{C}_{\mathcal{P}}$, then $\mathcal{Q}$ cannot touch all the facets of $\mathcal{C}_{\mathcal{P}}$, and therefore any extension to $X_{\mathcal{P}}$ of $V(g^m) \subset (\Bbbk^*)^3$ by a Cartier divisor fails property (b) (exercise VI.22).

In general, to find a Cartier divisor on $X_{\mathcal{P}}$ which extends $V(g) \subset (\Bbbk^*)^n$ and satisfies property (b) requires a "modification" of $X_{\mathcal{P}}$ by finding a "common refinement" of the normal fans of $\mathcal{P}$ and $\mathrm{NP}(g)$. This leads to beautiful combinatorial geometry, but we will not get into this. We will rather find an open subset $X^1_{\mathcal{P}}$ of $X_{\mathcal{P}}$ so that every Cartier divisor of $(\Bbbk^*)^n$ can be extended to a Cartier divisor on $X^1_{\mathcal{P}}$ and the extension also satisfies property (b).

7.1. The subset $X^1_{\mathcal{P}}$. Let $\mathcal{P}$ be an n-dimensional convex integral polytope in $\mathbb{R}^n$. Proposition VI.15 implies that we can identify $(\Bbbk^*)^n$ with $X^0_{\mathcal{P}}$. Denote the facets of $\mathcal{P}$ by $\mathcal{Q}_1, \ldots, \mathcal{Q}_s$, and the faces of $\mathcal{P}$ of dimension $\leq n-2$ by $\mathcal{Q}'_1, \ldots, \mathcal{Q}'_{s'}$. Define

$$X^1_{\mathcal{P}} := X^0_{\mathcal{P}} \cup \bigcup_{i=1}^{s} O_{\mathcal{Q}_i} = \bigcup_{i=1}^{s} U_{\mathcal{Q}_i} = X_{\mathcal{P}} \setminus \bigcup_{i'=1}^{s'} O_{\mathcal{Q}'_{i'}} = X_{\mathcal{P}} \setminus \bigcup_{i'=1}^{s'} V_{\mathcal{Q}'_{i'}},$$

where the equalities are implications of theorem VI.12 and propositions VI.15 and VI.20. In particular, it follows that $X^1_{\mathcal{P}}$ is Zariski open in $X_{\mathcal{P}}$, and it is the union of torus orbits in $X_{\mathcal{P}}$ of "codimension smaller than one" (this is the motivation for the "1" in the notation $X^1_{\mathcal{P}}$). Let ν_i be the primitive inner normal to $\mathcal{Q}_i$, $i = 1, \ldots, s$. Pick α_{ν_i} such that $\langle \nu_i, \alpha_{\nu_i} \rangle = 1$. Given $g \in \Bbbk[x_1, x_1^{-1}, \ldots, x_n, x_n^{-1}]$, corollary VI.22 implies that $T_{\alpha_{\nu_i}}(g)$ is a regular function on $U_{\mathcal{Q}_i}$ for each i. Since $U_{\mathcal{Q}_i} \cap U_{\mathcal{Q}_j} = X^0_{\mathcal{P}} \cong (\Bbbk^*)^n$ for $i \neq j$ (proposition VI.20) and since $T_{\alpha_{\nu_i}}(g)/T_{\alpha_{\nu_j}}(g)$ is a monomial in $x_1, \ldots, x_n$, it follows that $\{(U_{\mathcal{Q}_i}, T_{\alpha_{\nu_i}}(g))\}_i$ defines a Cartier divisor $V^1_{\mathcal{P}}(g)$ on $X^1_{\mathcal{P}}$. In exercise VI.24 you will check that $V^1_{\mathcal{P}}(g)$ is an extension to $X^1_{\mathcal{P}}$ of the Cartier divisor $V(g)$ of $(\Bbbk^*)^n$, and it also satisfies property (b). This construction can be carried out for arbitrary closed subschemes of $(\Bbbk^*)^n$. Indeed, given $g_1, \ldots, g_k \in \Bbbk[x_1, x_1^{-1}, \ldots, x_n, x_n^{-1}$, for each i, define $I_{\mathcal{Q}_i}$ to be the ideal of $\Bbbk[U_{\mathcal{Q}_i}]$ generated by $T_{\alpha_{\nu_i}}(g_j)$, $j = 1, \ldots, k$. Exercise VI.25 implies that the ideals $I_{\mathcal{Q}_i}$ can be glued over their intersection to form a *sheaf of ideals* $\mathcal{I}$ of $\mathcal{O}_{X_{\mathcal{P}}}$. We write $V^1_{\mathcal{P}}(g_1, \ldots, g_k)$ for the corresponding closed subscheme $V(\mathcal{I})$ of $X_{\mathcal{P}}$. Assertion (2) of exercise VI.25 implies that $V^1_{\mathcal{P}}(g_1, \ldots, g_k)$ does not depend on the choice of the α_{ν_i}.

Example VI.25. Exercise VI.24 implies that if $k = 1$, then $V^1_{\mathcal{P}}(g_1, \ldots, g_k)$ depends only on the ideal I of $\Bbbk[x_1, x_1^{-1}, \ldots, x_n, x_n^{-1}]$ generated by $g_1, \ldots, g_k$. We now show that this is in general false if $k > 1$. Let $\mathcal{P}$ be the triangle in $\mathbb{R}^2$ with vertices $(0,0), (1,0), (0,1)$, so that $X_{\mathcal{P}} \cong \mathbb{P}^2$ (example VI.17), and with respect to corresponding homogeneous coordinates $[z_{0,0} : z_{1,0} : z_{0,1}]$ on $\mathbb{P}^2$, $X^0_{\mathcal{P}} = \mathbb{P}^2 \setminus$

$V(z_{0,0}z_{1,0}z_{0,1})$ and $X^1_{\mathcal{P}} = X_{\mathcal{P}} \setminus \{[0:0:1],[0:1:0],[1:0:0]\}$. Note that $(x,y) := (z_{1,0}/z_{0,0}, z_{0,1}/z_{0,0})$ is a system of coordinates on $X^0_{\mathcal{P}}$. Let $g_1 := x-1$ and $g_2 := y-1$. It is straightforward to check that $V^1_{\mathcal{P}}(g_1,g_2) \cap (X^1_{\mathcal{P}} \setminus X^0_{\mathcal{P}}) = \emptyset$, and as a subscheme of $X^0_{\mathcal{P}}$, $V^1_{\mathcal{P}}(g_1,g_2) = V(g_1,g_2)$. Now let $h_1 := x-y$ and $h_2 := x+y-2+(x-y)^2$, so that the ideal generated by h_1,h_2 in $\mathbb{k}[x,x^{-1},y,y^{-1}]$ is the same as the ideal generated by g_1,g_2. It can be checked that $V^1_{\mathcal{P}}(h_1,h_2)$ contains the point $[0:1:1] \in X^1_{\mathcal{P}} \setminus X^0_{\mathcal{P}}$, so that $V^1_{\mathcal{P}}(h_1,h_2) \neq V^1_{\mathcal{P}}(g_1,g_2)$.

Let $\mathcal{Q}$ be a facet of $\mathcal{P}$ with primitive inner normal ν, and α_ν be an arbitrary element of $\mathbb{Z}^n$ such that $\langle \nu, \alpha_\nu \rangle = 1$. Proposition VI.20 implies that the ideal of $\mathbb{k}[U_{\mathcal{Q}}]$ consisting of elements vanishing on $O_{\mathcal{Q}}$ is generated by x^{α_ν}. Therefore $O_{\mathcal{Q}}$ is precisely the support of the closed *subscheme* $Z_{\alpha_\nu} := V(x^{\alpha_\nu})$ of $U_{\mathcal{Q}}$. Since $O_{\mathcal{Q}}$ is Zariski closed in $X^1_{\mathcal{P}}$ (exercise VI.23), it follows that Z_{α_ν} is in fact a closed subscheme of $X^1_{\mathcal{P}}$. We now determine the *embedded isomorphism* (see section IV.2.4) type of the "scheme-theoretic intersection" $V^1_{\mathcal{P}}(g_1,\dots,g_k) \cap Z_{\alpha_\nu}$. Let $\psi^*_\nu : (\mathbb{k}^*)^{n-1} \cong O_{\mathcal{Q}}$ be as in corollary VI.22.

Proposition VI.26. *As a closed subscheme of $O_{\mathcal{Q}}$, the scheme-theoretic intersection $V^1_{\mathcal{P}}(g_1,\dots,g_k) \cap Z_{\alpha_\nu}$ is embedded isomorphic via ψ^*_ν to the closed subscheme $V(\mathrm{In}'_{\beta_\nu,\psi_\nu}(g_1),\dots,\mathrm{In}'_{\beta_\nu,\psi_\nu}(g_k))$ of $(\mathbb{k}^*)^{n-1}$.*

PROOF. The ideal $I_{\mathcal{Q}}$ of $V^1_{\mathcal{P}}(g_1,\dots,g_k) \cap Z_{\alpha_\nu}$ is generated in $\mathbb{k}[U_{\mathcal{Q}}]$ by x^{α_ν}, $T_{\alpha_\nu}(g_1),\dots,T_{\alpha_\nu}(g_k)$. Proposition VI.20 and corollary VI.22 imply that $\mathbb{k}[U_{\mathcal{Q}}]/\langle x^{\alpha_\nu}, T_{\alpha_\nu}(g_1),\dots,T_{\alpha_\nu}(g_k)\rangle$ is isomorphic via ψ^*_ν to $\mathbb{k}[(\mathbb{k}^*)^{n-1}]/\langle \mathrm{In}'_{\beta_\nu,\psi_\nu}(g_1),\dots, \mathrm{In}'_{\beta_\nu,\psi_\nu}(g_k)\rangle$, which directly implies the result. □

7.2. Exercises.

EXERCISE VI.19. Let $\mathcal{P}$ be an n-dimensional convex integral polytope in $\mathbb{R}^n$ and g be a Laurent polynomial with $\mathrm{NP}(g) \subseteq \mathcal{P}$.

(1) Show that the collection $(U_\alpha \cap X_{\mathcal{P}}, x^{-\alpha}g)$, where α varies over the vertices of $\mathcal{P}$, defines a Cartier divisor D on $X_{\mathcal{P}}$ such that D extends the Cartier divisor $V(g)$ of $(\mathbb{k}^*)^n$. [Hint: Use exercise VI.16.]
(2) Show that the following are equivalent:
   (a) $\mathrm{Supp}(D)$ is the closure in $X_{\mathcal{P}}$ of $V(g) \subset (\mathbb{k}^*)^n$.
   (b) $\mathrm{NP}(g) \cap \mathcal{Q} \neq \emptyset$ for each facet $\mathcal{Q}$ of $\mathcal{P}$.
(3) Assume in addition that $\mathcal{A} := \mathcal{P} \cap \mathbb{Z}^n$ satisfies the hypothesis of assertion (1) of proposition VI.10 so that $U_0 \cap X_{\mathcal{P}} \cong \mathbb{k}^n$. Show that the following are equivalent:
   (a) $\mathrm{Supp}(D)$ is the closure in $X_{\mathcal{P}}$ of $V(g) \subset \mathbb{k}^n$.
   (b) $\mathrm{NP}(g) \cap \mathcal{Q} \neq \emptyset$ for each facet $\mathcal{Q}$ of $\mathcal{P}$ which is not contained in any coordinate hyperplane of $\mathbb{R}^n$.

EXERCISE VI.20. Let $\mathcal{P}$ be an n-dimensional convex integral polytope in $\mathbb{R}^n$. For each vertex α of $\mathcal{P}$, let $\mathcal{C}_\alpha$ be the cone in $\mathbb{R}^n$ generated by $\{\alpha' - \alpha : \alpha' \in \mathcal{P}\}$. Assume there is $g \in \mathbb{k}[x_1,x_1^{-1},\dots,x_n,x_n^{-1}]$ such that for each vertex α of $\mathcal{P}$, there is $\beta_\alpha \in \mathbb{Z}^n$ such that $\beta_\alpha + \mathrm{NP}(g) \subset \mathcal{C}_\alpha$ and $\beta_\alpha + \mathrm{NP}(g)$ touches every edge of $\mathcal{C}_\alpha$.

(1) Let α, α' be vertices of $\mathcal{P}$, and $\mathcal{Q}$ be the smallest face of $\mathcal{P}$ containing both α and α'. Let H be the linear subspace of $\mathbb{R}^n$ spanned by $\{\beta - \alpha : \beta \in \mathcal{Q}\}$. Show that $\beta_\alpha - \beta_{\alpha'} \in H$. [Hint: Use exercise V.32 to reduce to the case that there is an edge of $\mathcal{P}$ connecting α and α'. In that case show that $\beta_\alpha - \beta_{\alpha'}$ is on the line through the origin and $\alpha - \alpha'$.]
(2) Deduce that the collection of pairs $(U_\alpha \cap X_{\mathcal{P}}, x^{\beta_\alpha} g)$, where α varies over the vertices of $\mathcal{P}$, defines a Cartier divisor D on $X_{\mathcal{P}}$ such that D extends the Cartier divisor $V(g)$ of $(\mathbb{k}^*)^n$, and $\mathrm{Supp}(D)$ is the closure in $X_{\mathcal{P}}$ of $V(g) \subset (\mathbb{k}^*)^n$. [Hint: Use exercise VI.16.]
(3) If $\mathcal{P}$ is *simplicial*, i.e. each vertex of $\mathcal{P}$ is connected to precisely n distinct edges, then show that for every $g \in \mathbb{k}[x_1, x_1^{-1}, \ldots, x_n, x_n^{-1}]$, there is $m \geq 1$ such that $V(g^m) \subset (\mathbb{k}^*)^n$ extends to a Cartier divisor D on $X_{\mathcal{P}}$ such that $\mathrm{Supp}(D)$ is the closure in $X_{\mathcal{P}}$ of $V(g^m) \subset (\mathbb{k}^*)^n$.

EXERCISE VI.21. Let $\mathcal{P}$ and h be as in example VI.23. If D is any Cartier divisor on $X_{\mathcal{P}}$ which extends $V(h^m) \subset (\mathbb{k}^*)^2$, where m is a positive integer, then show that D satisfies property (b) if and only if m is a multiple of 3.

EXERCISE VI.22. Prove the claims made in example VI.24.

EXERCISE VI.23. Let $\mathcal{Q}$ be a facet of an n-dimensional convex integral polytope $\mathcal{P}$ in $\mathbb{R}^n$. Show that $O_{\mathcal{Q}}$ is a closed subvariety of $X^1_{\mathcal{P}}$.

EXERCISE VI.24. Let $g \in \mathbb{k}[x_1, x_1^{-1}, \ldots, x_n, x_n^{-1}]$ and $\mathcal{P}$ be a convex integral polytope of dimension n. Show that

(1) The Cartier divisor $V^1_{\mathcal{P}}(g)$ of $X^1_{\mathcal{P}}$ does not depend on the choice of the α_{ν_i}.
(2) If g/h is a monomial in $x_1, \ldots, x_n$, then $V^1_{\mathcal{P}}(g) = V^1_{\mathcal{P}}(h)$.
(3) $\mathrm{Supp}(V^1_{\mathcal{P}}(g))$ is the closure in $X^1_{\mathcal{P}}$ of $V(g) \subset (\mathbb{k}^*)^n$.

EXERCISE VI.25. Let $g_1, \ldots, g_k \in \mathbb{k}[x_1, x_1^{-1}, \ldots, x_n, x_n^{-1}]$ and $\mathcal{P}$ be a convex integral polytope of dimension n.

(1) Let $\mathcal{Q}_1, \mathcal{Q}_2$ be facets of $\mathcal{P}$, ν_i be the primitive inner normal to $\mathcal{Q}_i$, and $\alpha_{\nu_i} \in \mathbb{Z}^n$ be such that $\langle \nu_i, \alpha_{\nu_i} \rangle = 1$. Show that for each $x \in U_{\mathcal{Q}_1} \cap U_{\mathcal{Q}_2}$ the ideal of $\mathcal{O}_{x, X_{\mathcal{P}}}$ generated by $T_{\alpha_{\nu_1}}(g_1), \ldots, T_{\alpha_{\nu_1}}(g_k)$ is the same as the ideal generated by $T_{\alpha_{\nu_2}}(g_1), \ldots, T_{\alpha_{\nu_2}}(g_k)$.
(2) If g_j/h_j is a monomial in $x_1, \ldots, x_n$ for each j, then show that for each i and each $x \in U_{\mathcal{Q}_i}$, the ideal of $\mathcal{O}_{x, X_{\mathcal{P}}}$ generated by $T_{\alpha_{\nu_i}}(g_1), \ldots, T_{\alpha_{\nu_i}}(g_k)$ is the same as the ideal generated by $T_{\alpha_{\nu_i}}(h_1), \ldots, T_{\alpha_{\nu_i}}(h_k)$.

EXERCISE VI.26. Verify the claims made in example VI.25.

8. Branches of curves on the torus

8.1. Branch of a curve on a variety. Let C be a *curve*, i.e. a variety of pure dimension one. Fix a desingularization $\pi : C' \to C$ of C and a nonsingular compactification $\bar{C}'$ of C'. A *branch* of C is the germ of a point in $\bar{C}'$. Equivalently, consider the equivalence relation $\sim$ on the collection of pairs $\{(Z, z) : z \in \bar{C}'$ and Z is an open neighborhood of z in $\bar{C}'\}$ defined as follows: $(Z, z) \sim (Z', z')$

if and only if $z = z'$. Then a branch of C is a equivalence class of $\sim$. Let X be a variety containing C as a subvariety. Let $\bar{X}$ be an arbitrary projective compactification of X and $\bar{C}$ be the closure of C in $\bar{X}$. Then π extends to a map $\bar{C}' \to \bar{C}$ (corollary III.109), which we also denote by π. If $B := (Z, z)$ is a branch of C and $y := \pi(z)$, we say that y is the *center* of B on $\bar{X}$, or equivalently, B is a *branch of* C *at* y. If $y \notin X$, we say that (with respect to X) B is a *branch at infinity*, or that it is *centered at infinity*. Since $\bar{C}'$ is nonsingular, $\mathcal{O}_{z,Z}$ is a discrete valuation ring (proposition III.108), and corresponds to a *unique* discrete valuation $\operatorname{ord}_z(\cdot)$ on the field of fractions of $\mathcal{O}_{z,Z}$ (assertion 5 of proposition B.17). If f is a regular function on a neighborhood of z on X, then we write $\operatorname{ord}_z(f)$ for $\operatorname{ord}_z(\pi^*(f))$.

Example VI.27. Assume $\operatorname{char}(\mathbb{k}) \neq 2$. Let C be the curve on $\mathbb{k}^2$ given by $x^2 = y^2 - y^3$. In example IV.22 we have seen that $\pi : \mathbb{k} \to C$ given by $t \mapsto (t - t^3, 1 - t^2)$ is a desingularization of C. Let $P = (0,0) \in C$. Then $\pi^{-1}(P) = \{1, -1\}$, and $(1, \mathbb{k})$ and $(-1, \mathbb{k})$ represent the two branches of C at P (see Figure 4). Note that both these branches are centered at infinity with respect to $(\mathbb{k}^*)^2$ (since $P \notin (\mathbb{k}^*)^2$). In exercise VI.27 you are asked to compute $\operatorname{ord}_1(f|_C)$ and $\operatorname{ord}_{-1}(f|_C)$ for different $f \in \mathbb{k}[x, y]$.

8.2. Weights of a branch on the torus. Fix a system of coordinates $(x_1, \ldots, x_n)$ on $(\mathbb{k}^*)^n$. Let $B = (Z, z)$ be a branch of a curve on $(\mathbb{k}^*)^n$. The *weight* of x_j corresponding to B is $\operatorname{ord}_z(x_j|_Z)$. By ν_B we denote the element in $(\mathbb{R}^n)^*$ with coordinates $(\operatorname{ord}_z(x_1), \ldots, \operatorname{ord}_z(x_n))$ with respect to the basis dual to the standard basis of $\mathbb{R}^n$. The proof of the following result is left as an exercise.

Lemma VI.28. *(1) For each $\alpha \in \mathbb{Z}^n$, x^α restricts to a well-defined rational function on Z and* $\operatorname{ord}_z(x^\alpha) = \langle \nu_B, \alpha \rangle$.
(2) $\nu_B \neq 0$ if and only if B is centered at infinity with respect to $(\mathbb{k}^*)^n$. □

Example VI.29. Recall that the curve C from example VI.27 has two branches B_1 and B_2 at the origin. Exercise VI.27 implies that both "weight vectors" ν_{B_1} and ν_{B_2} are $(1, 1)$. Note that

(i) B_j are centered at infinity with respect to $(\mathbb{k}^*)^2$, and ν_{B_j} are nonzero.
(ii) ν_B is *not* an invariant of B, but the embedding $C \cap (\mathbb{k}^*)^2 \hookrightarrow (\mathbb{k}^*)^2$. Exercise VI.28 presents a curve $C' \cong C$ via a map that sends the origin to itself, such that the branches at the origin correspond to *different* weight vectors.

8.3. Weighted order on Laurent polynomials. Let ν be an integral element ν of $(\mathbb{R}^n)^*$. The *weighted order* corresponding to ν is an integer valued map, which by an abuse of notation we also denote by ν, on the ring of Laurent polynomials defined as follows: given a Laurent polynomial $f = \sum_\alpha c_\alpha x^\alpha$,

$$\nu(f) := \min\{\langle \nu, \alpha \rangle : c_\alpha \neq 0\}.$$

In particular, if f is the zero polynomial, then $\nu(f)$ is defined to be ∞. The *initial form* $\operatorname{In}_\nu(f)$ of f with respect to ν is the sum of all $c_\alpha x^\alpha$ such that $\langle \nu, \alpha \rangle = \nu(f)$. Given a subset $\mathcal{S}$ of $\mathbb{R}^n$, we say that f is *supported at* $\mathcal{S}$ if $\operatorname{Supp}(f) \subseteq \mathcal{S}$, and we write $\mathcal{L}(\mathcal{S})$ for the set of all Laurent polynomials supported at $\mathcal{S}$. In the case that

$\mathcal{S}\cap\mathbb{Z}^n$ is a finite set, we equip $\mathcal{L}(\mathcal{S})$ with the Zariski topology by identifying it with $\Bbbk^{|\mathcal{S}\cap\mathbb{Z}^n|}$ via the map

$$\sum_{\alpha\in\mathcal{S}\cap\mathbb{Z}^n} c_\alpha x^\alpha \mapsto (c_\alpha)_{\alpha\in\mathcal{S}\cap\mathbb{Z}^n}.$$

The result we will now prove ties these notions with those from section VI.8.2; we will encounter many of its variants in the forthcoming chapters. Let $B = (Z, z)$ be a branch of a curve on $(\Bbbk^*)^n$, and ν_B be as in section VI.8.2. By an abuse of notation, we denote by ν_B also the weighted order corresponding to ν_B.

Proposition VI.30. *Assume $\mathcal{S}\cap\mathbb{Z}^n$ has finitely many elements. There is a nonempty Zariski open subset U of $\mathcal{L}(\mathcal{S})$ such that* $\operatorname{ord}_z(f) = \nu_B(f)$ *for each $f \in U$. More precisely, define*

$$\mathcal{L}^*(\mathcal{S}) := \{f \in \mathcal{L}(\mathcal{S}) : \operatorname{ord}_z(f) = \nu_B(f)\}.$$

Then $\mathcal{L}^*(\mathcal{S})$ is a constructible subset of $\mathcal{L}(\mathcal{S})$ of dimension $|\mathcal{S}|$.

PROOF. Pick a parameter ρ of $\mathcal{O}_{z,Z}$. Then $\operatorname{ord}_z(\rho) = 1$. For each $i = 1, \ldots, n$, if $m_i := \nu_B(x_i)$, then there is a representation of the form $x_i = c_i\rho^{m_i} + h_i$ where $c_i \in \Bbbk^*$, and $h_i \in \mathcal{O}_{z,Z}$ such that $\operatorname{ord}_z(h_i) > m_i$ (assertion (2b) of proposition III.108). For each $\mathcal{A} \subseteq \mathcal{S}\cap\mathbb{Z}^n$, let $\mathcal{L}^*_B(\mathcal{A}) := \{f \in \mathcal{L}^*(\mathcal{A}) : \operatorname{Supp}(\operatorname{In}_{\nu_B}(f)) \subset \operatorname{In}_{\nu_B}(\mathcal{A})\}$. It is straightforward to check that

$$\mathcal{L}^*_B(\mathcal{A}) = \{(c_\alpha)_{\alpha\in\mathcal{S}\cap\mathbb{Z}^n} : c_\alpha = 0 \text{ if } \alpha \notin \mathcal{A}, \quad \sum_{\alpha\in\operatorname{In}_{\nu_B}(\mathcal{A})}\prod_{i=1}^n (c_i)^{\alpha_i} c_\alpha \neq 0\}$$

which implies that $\mathcal{L}^*_B(\mathcal{A})$ is a *nonempty* open subset of a closed subset of $\mathcal{L}(\mathcal{S})$. Since $\mathcal{L}^*(\mathcal{S})$ is the union of $\mathcal{L}^*_B(\mathcal{A})$ over all subsets $\mathcal{A}$ of $\mathcal{S}\cap\mathbb{Z}^n$, it follows that it is a constructible subset of $\mathcal{L}(\mathcal{S})$. The remaining parts of the proposition follows from taking $\mathcal{A} = \mathcal{S}\cap\mathbb{Z}^n$. □

8.4. Exercises.

EXERCISE VI.27. Assume $\operatorname{char}(\Bbbk) \neq 2$. Consider the desingularization $\pi : \Bbbk \to C$ from example VI.27 given by $t \mapsto (t - t^3, 1 - t^2)$. Show that

(1) $1 - t^2$ is a parameter at $\mathcal{O}_{z,\Bbbk}$ for both $z = 1$ and $z = -1$.
(2) $\operatorname{ord}_z(x|_C) = \operatorname{ord}_z(y|_C) = 1$ for both $z = 1$ and $z = -1$.
(3) $\operatorname{ord}_1((y - x)|_C) = 2$, but $\operatorname{ord}_{-1}((y - x)|_C) = 1$.

EXERCISE VI.28. Assume $\operatorname{char}(\Bbbk) \neq 2$. Let C be the curve from example VI.27, $C' := V((x + y)^2 - y^2 + y^3) \subseteq \Bbbk^2$, and $P' := (0, 0) \in C'$.

(1) The map $\phi : (x, y) \mapsto (x - y, y)$ induces an isomorphism $C \cong C'$ and maps $P \to P'$, where $P := (0, 0) \in C$.
(2) There are two branches B'_1, B'_2 of C' at the origin and the corresponding "weight vectors" $\nu_{B'_1}, \nu_{B'_2} \in (\mathbb{R}^n)^*$ are different. [Hint: Use example VI.27 and exercise VI.27.]

EXERCISE VI.29. For each of the following curves $C \subseteq \mathbb{k}^n$, show that C has a single branch B at the origin, and compute $\nu_B \in (\mathbb{R}^n)^*$:

(1) $C = V(x^2 - y^3)$. [Hint: The desingularization of C is given by $t \mapsto (t^3, t^2)$.]
(2) $C = V(y^3 - x^4, z^3 - x^5)$. [Hint: Use exercise III.91.]

EXERCISE VI.30. Prove lemma VI.28.

9. Points at infinity on toric varieties

9.1. Centers of branches at infinity on the torus. Let $\mathcal{A} := \{\alpha_0, \ldots, \alpha_N\}$ be a finite subset of $\mathbb{Z}^n$ and $B = (Z, z)$ be a branch centered at infinity on $(\mathbb{k}^*)^n$. If $\phi_{\mathcal{A}} : (\mathbb{k}^*)^n \to X^0_{\mathcal{A}}$ is the map from (47), then $\phi_{\mathcal{A}}(B)$ is a branch centered at infinity on $X^0_{\mathcal{A}}$. We now determine the torus orbit of $X_{\mathcal{A}}$ that contains the center o_B of $\phi_{\mathcal{A}}(B)$ on $X_{\mathcal{A}}$; we will see that this orbit is completely determined by the element $\nu_B \in (\mathbb{R}^n)^*$ defined in section VI.8.2. Let $\mathcal{P}$ be the convex hull of $\mathcal{A}$ and $\mathcal{B}$ be a face of $\mathcal{A}$. Then the convex hull $\mathcal{Q}$ of $\mathcal{B}$ is a face of $\mathcal{P}$. As in section V.4 we write $\Sigma_{\mathcal{P}}$ for the *normal fan* of $\mathcal{P}$, and denote the *normal cone* of $\mathcal{Q}$ by $\sigma_{\mathcal{Q}}$ and the relative interior of $\sigma_{\mathcal{Q}}$ by $\sigma^0_{\mathcal{Q}}$ (fig. 7).

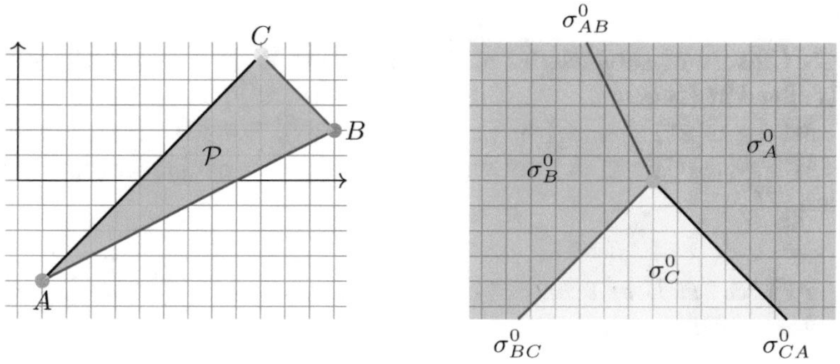

FIGURE 7. Normal fan of $\mathcal{P}$.

Proposition VI.31. *$o_B \in O_{\mathcal{B}}$ if and only if $\nu_B \in \sigma^0_{\mathcal{Q}}$. In particular, if $\mathcal{B}$ is a facet of $\mathcal{A}$, then $o_B \in O_{\mathcal{B}}$ if and only if ν_B is a positive multiple of the* primitive inner normal *to $\mathcal{Q}$.*

PROOF. Pick $\beta \in \mathcal{B}$. Theorem VI.12 implies that $x^{\alpha-\beta} = z_\alpha / z_\beta$ is a regular function on $O_{\mathcal{B}}$ for each $\alpha \in \mathcal{A}$. It also implies that $o_B \in O_{\mathcal{B}}$ if and only if the following holds: "$x^{\alpha-\beta}|_{o_B} \neq 0$ if and only if $\alpha \in \mathcal{B}$." Due to assertion (1) of lemma VI.28 the latter condition is equivalent to the condition that $\mathrm{In}_{\nu_B}(\mathcal{A}) = \mathcal{B}$, which is in turn equivalent to the condition that $\mathrm{In}_{\nu_B}(\mathcal{P}) = \mathcal{Q}$, as required. ☐

We now describe the coordinates of o_B. As in section VI.8.3, we write ν_B also for the *weighted order* on $\mathbb{k}[x_1, x_1^{-1}, \ldots, x_n, x_n^{-1}]$ induced by ν_B. Fix an arbitrary element $\rho_B \in \mathcal{O}_{z,Z}$ such that $\operatorname{ord}_z(\rho_B|_Z) = 1$. We say that ρ_B is a *parameter* of B. Define

$$(53)\qquad \begin{aligned} \operatorname{In}_B(x_j) &:= \left.\frac{x_j}{(\rho_B)^{\nu_B(x_j)}}\right|_z \in \mathbb{k}^*, \\ \operatorname{In}(B) &:= (\operatorname{In}_B(x_1), \ldots, \operatorname{In}_B(x_n)) \in (\mathbb{k}^*)^n \end{aligned}$$

That $\operatorname{In}_B(x_j)$ and $\operatorname{In}(B)$ are well defined follows from proposition III.108. Note that $\operatorname{In}(B)$ depends on the choice of ρ_B. In all cases considered in this book, whenever a branch B of a curve is considered, a corresponding parameter ρ_B is assumed to be fixed from the beginning. Let $\mathcal{B}$ be the face of $\mathcal{A}$ such that $o_B \in O_{\mathcal{B}}$ (proposition VI.31 shows that $\mathcal{B}$ is uniquely determined by ν_B). Let $\phi_{\mathcal{B}} : (\mathbb{k}^*)^n \to X^0_{\mathcal{B}}$ be the map from (47). Theorem VI.12 implies that we may think of $\phi_{\mathcal{B}}$ as a map from $(\mathbb{k}^*)^n$ to $O_{\mathcal{B}} \subset X_{\mathcal{A}}$, simply by adjoining a zero in place of each coordinate z_α on $\mathbb{P}^N$ such that $\alpha \notin \mathcal{B}$.

Proposition VI.32. $o_B = \phi_{\mathcal{B}}(\operatorname{In}(B))$. *In particular,* $\phi_{\mathcal{B}}(\operatorname{In}(B))$ *does not depend on the choice of* ρ_B *(even though* $\operatorname{In}(B)$ *does).*

PROOF. Pick $\alpha, \beta \in \mathcal{B}$. Proposition VI.31 implies that $(z_\alpha/z_\beta)|_{o_B} = x^{\alpha-\beta}|_{o_B} = (\operatorname{In}(B))^{\alpha-\beta}$. The result then follows immediately from theorem VI.12. □

9.2. Closure of subvarieties of the torus. Let W be a closed subvariety of $(\mathbb{k}^*)^n$ defined by Laurent polynomials $f_1, \ldots, f_m$ in $(\mathbb{k}^*)^n$. Let $\mathcal{A}$ be a finite subset of $\mathbb{Z}^n$ and $\phi_{\mathcal{A}} : (\mathbb{k}^*)^n \to X^0_{\mathcal{A}}$ be the map from (47). Write $\bar{W}'$ for the closure in $X_{\mathcal{A}}$ of $W' := \phi_{\mathcal{A}}(W) \subset X^0_{\mathcal{A}}$. In this section we give a partial description of the points in $\bar{W}'$.

Lemma VI.33. *Let* B *be a branch of a curve contained in* W. *Then* $\operatorname{In}(B)$ *is a common zero of* $\operatorname{In}_{\nu_B}(f_i)$, $i = 1, \ldots, m$.

PROOF. Let $B = (Z, z)$. Pick a parameter ρ_B of B. Proposition III.108 implies that for each j, $x_j/\rho_B^{\nu_B(x_j)}$ is a regular function on a neighborhood of z on Z, and it can be expressed as $\operatorname{In}_B(x_j) + g_j$, where $g_j \in \mathcal{O}_{z,Z}$, $\operatorname{ord}_z(g_j) > 0$. Consequently, for each i, $f_i/\rho_B^{\nu_B(f_i)}$ can be expressed in $\mathcal{O}_{z,Z}$ as $\operatorname{In}_{\nu_B}(f_i)(\operatorname{In}(B)) + h_i$, where $h_i \in \mathcal{O}_{z,Z}$, $\operatorname{ord}_z(h_i) > 0$. Since $f_i/\rho_B^{\nu_B(f_i)}$ maps to the zero element in $\mathcal{O}_{z,Z}$, it follows that $\operatorname{In}_{\nu_B}(f_i)(\operatorname{In}(B)) = 0$, as required. □

Let ν be an integral element of $(\mathbb{R}^n)^*$; we write ν also for the corresponding weighted order on $\mathbb{k}[x_1, x_1^{-1}, \ldots, x_n, x_n^{-1}]$ and define $W_\nu(f_1, \ldots, f_m) := V(\operatorname{In}_\nu(f_1), \ldots, \operatorname{In}_\nu(f_m)) \subset (\mathbb{k}^*)^n$. Let $\mathcal{B}$ be a face of $\mathcal{A}$. As in proposition VI.32, we regard the map $\phi_{\mathcal{B}} : (\mathbb{k}^*)^n \to X^0_{\mathcal{B}}$ from (47) as a map from $(\mathbb{k}^*)^n$ to $O_{\mathcal{B}} \subset X_{\mathcal{A}}$. As in proposition VI.31 we write $\mathcal{Q}$ for the convex hull of $\mathcal{B}$ and $\sigma^0_{\mathcal{Q}}$ for the relative interior of the corresponding cone of the normal fan of the convex hull of $\mathcal{A}$.

Corollary VI.34. $\bar{W}' \cap O_{\mathcal{B}} \subset \bigcup_{\nu \in \sigma^0_{\mathcal{Q}}} \phi_{\mathcal{B}}(W_\nu(f_1, \ldots, f_m))$.

PROOF. Let $w \in \bar{W}' \cap O_{\mathcal{B}}$. If $w \in W'$, then we must have that $\mathcal{B} = \mathcal{A}$. In that case $0 \in \sigma^0_{\mathcal{Q}}$. Since $W_0(f_1, \dots, f_m) = W$, it follows that $w \in \phi_{\mathcal{B}}(W_0(f_1, \dots, f_m)) = W'$, as required. So assume $w \in \bar{W}' \setminus W'$. Then there is an irreducible curve $C' \subset W'$ such that w is in the closure of C' (proposition III.87). Pick a branch $B = (Z, z)$ of $\phi_{\mathcal{A}}^{-1}(C)$ such that $z \mapsto w$ under the morphism induced by $\phi_{\mathcal{A}}$. Proposition VI.31 implies that $\nu_B \in \sigma^0_{\mathcal{Q}}$ and proposition VI.32 implies that $w = \phi_{\mathcal{B}}(\mathrm{In}(B))$. Since $\mathrm{In}(B) \in W_{\nu_B}(f_1, \dots, f_m)$ (lemma VI.33), the result follows. □

For each $i = 1, \dots, m$, there are only finitely many possibilities for $\mathrm{In}_\nu(f_i)$ as ν varies over $(\mathbb{R}^n)^*$. It follows that the union in the statement of corollary VI.34 can be regarded as being over a *finite* collection of $\nu \in (\mathbb{R}^n)^*$. Exercise VI.31 shows that the containment of corollary VI.34 is in general proper. However, if W is a hypersurface (i.e. $m = 1$), then exercise VI.32 shows that corollary VI.34 holds with "=" in place of $\subset$.

9.3. Exercises.

EXERCISE VI.31. Let $\mathcal{A} := \{(0,0), (1,0), (0,1)\} \subset \mathbb{R}^2$ and h_1, h_2 be as in example VI.25. Let $W := V(h_1, h_2) = \{(1,1)\} \in (\mathbb{k}^*)^2$ and $\mathcal{B} := \{(1,0), (0,1)\}$. Note that $\mathcal{Q} := \mathrm{conv}(\mathcal{B})$ is an edge of $\mathcal{P} := \mathrm{conv}(\mathcal{A})$.

(1) Show that $\bar{W}' \cap O_{\mathcal{B}} = \emptyset$.
(2) Let ν be the primitive inner normal to $\mathcal{Q}$. Show that $W_\nu(h_1, h_2) \neq \emptyset$.

EXERCISE VI.32. Let $f \in \mathbb{k}[x_1, x_1^{-1}, \dots, x_n, x_n^{-1}]$ and $W := V(f) \subset (\mathbb{k}^*)^n$. Show that in the notation of corollary VI.34, $\bar{W}' \cap O_{\mathcal{B}} = \bigcup_{\nu \in \sigma^0_{\mathcal{Q}}} \phi_{\mathcal{B}}(W_\nu(f))$ for each face $\mathcal{B}$ of $\mathcal{A}$. [Hint: Use exercise VI.24 to prove it in the case that $\mathcal{Q}$ is a facet of $\mathcal{P}$. Then use induction on $\dim(\mathcal{P})$.]

10. *Weighted projective spaces

[2] Recall that the n-dimensional projective space is the space of straight lines through the origin in $\mathbb{k}^{n+1}$. A *weighted projective space* is constructed in the same way, using *weighted rational curves* in place of straight lines. Let ω be an integral element of $(\mathbb{R}^{n+1})^*$ with coordinates $(\omega_0, \dots, \omega_n)$ with respect to the basis dual to the standard basis of $\mathbb{R}^{n+1}$. Assume each ω_j is *positive*. The corresponding weighted projective space, which we denote by $\mathbb{P}^n(\omega)$ or $\mathbb{P}^n(\omega_0, \dots, \omega_n)$, is the set of curves in $\mathbb{k}^{n+1}$ of the form $C_a := \{(a_0 t^{\omega_0}, \dots, a_n t^{\omega_n}) : t \in \mathbb{k}\}$, where $a := (a_0, \dots, a_n) \in \mathbb{k}^{n+1} \setminus \{0\}$. The *weighted homogeneous coordinates* of C_a are $[a_0 : \cdots : a_n]$. Note that the projective space $\mathbb{P}^n$ is the special case of $\mathbb{P}^n(\omega)$ for $\omega = (1, \dots, 1)$. Like $\mathbb{P}^n$, each $\mathbb{P}^n(\omega)$ can be given the structure of a complete algebraic variety. In this section we give two (equivalent) realizations of $\mathbb{P}^n(\omega)$ as a toric variety.

---

[2] The asterisk in the section name is to indicate that the material of this section is not going to be used in the proof of Bernstein's theorem. It is used for the first time in chapter VIII.

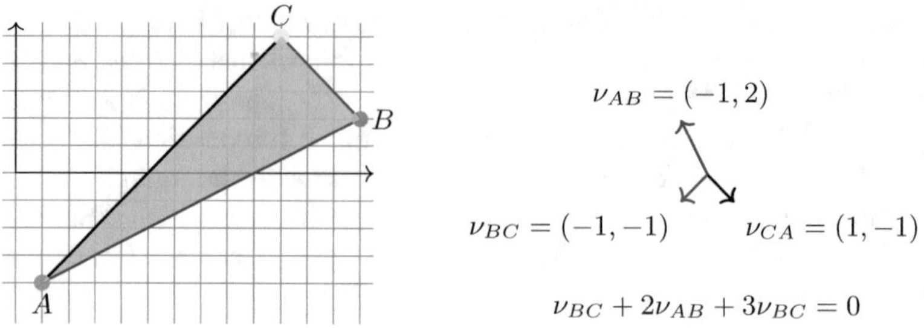

FIGURE 8. $X_{\mathcal{P}} \cong \mathbb{P}^2(1, 2, 3)$.

10.1. $\mathbb{P}^n(\omega)$ via polytopes in $\mathbb{R}^n$. Pick integral elements $\nu_0, \ldots, \nu_n$ of $(\mathbb{R}^n)^*$ such that $\nu_0, \ldots, \nu_n$ span $(\mathbb{Z}^n)^*$, and $\sum_{j=0}^n \omega_j \nu_j = 0$. Let $\mathcal{P}$ be an n-dimensional integral simplex in $\mathbb{R}^n$ such that its inner facet normals are $\nu_0, \ldots, \nu_n$; note that $\mathcal{P}$ is uniquely determined by the ν_j up to translation and scaling, see fig. 8. We will show that $\mathbb{P}^n(\omega)$ can be identified with $X_{\mathcal{P}}$.

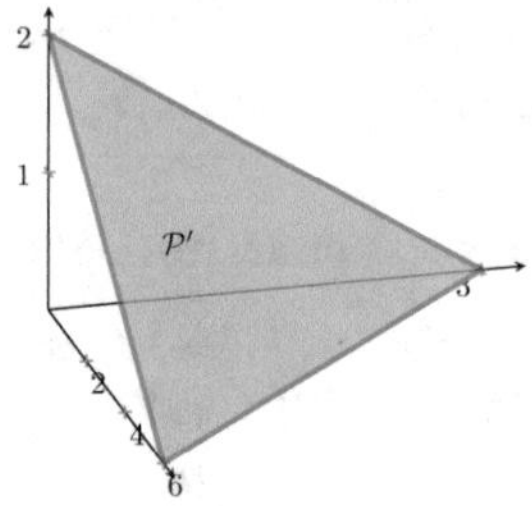

FIGURE 9. A polytope $\mathcal{P}'$ in $\mathbb{R}^3$ such that $X_{\mathcal{P}'} \cong \mathbb{P}^2(1, 2, 3)$.

10.2. $\mathbb{P}^n(\omega)$ via polytopes in $\mathbb{R}^{n+1}$. Let $p := \operatorname{lcm}(\omega_0, \ldots, \omega_n)$ and $\mathcal{P}'$ be the n-dimensional simplex in $\mathbb{R}^{n+1}$ with vertices $\beta_j := (p/\omega_j)e_j$, $j = 0, \ldots, n$, where $e_0, \ldots, e_n$ are the standard unit vectors in $\mathbb{R}^{n+1}$ (fig. 9). We will show that $X_{\mathcal{P}'} \cong X_{\mathcal{P}}$, where $\mathcal{P}$ is from section VI.10.1.

10.3. Equivalence of the constructions. Let $\mathcal{P}$ be as in section VI.10.1. For each $j = 0, \ldots, n$, let $\mathcal{Q}_j$ be the facet of $\mathcal{P}$ with inner normal ν_j, and α_j be the (unique) vertex of $\mathcal{P}$ which is not on the facet $\mathcal{Q}_j$. Pick j, $0 \leq j \leq n$. Proposition VI.15 implies that the coordinate ring of $U_{\alpha_j} \cap X_{\mathcal{P}}$ is $\Bbbk[S_j]$, where S_j is the semigroup of integral points in the polyhedral cone $\mathcal{C}_{\alpha_j} \subset \mathbb{R}^n$ generated by $\alpha_i - \alpha_j$, $i = 0, \ldots, n$. Exercise VI.33 implies that $\alpha \in S_j$ if and only if $\alpha \in \mathbb{Z}^n$ and $\langle \nu_i, \alpha \rangle \geq 0$ for each $i = 0, \ldots, \hat{j}, \ldots, n$. Consider the map $\phi : \mathbb{Z}^n \mapsto \mathbb{Z}^{n+1}$ given by $\alpha \mapsto (\langle \nu_0, \alpha \rangle, \ldots, \langle \nu_n, \alpha \rangle)$.

Claim VI.35. *ϕ induces an isomorphism between S_j and $S'_j := \{\beta = (\beta_0, \ldots, \beta_n) \in \mathbb{Z}^{n+1} : \langle \omega, \beta \rangle = 0,\ \beta_i \geq 0$ for each $i = 0, \ldots, \hat{j}, \ldots, n\}$.*

PROOF. Exercise VI.33 implies that $S_j = \phi^{-1}(S'_j)$. We now show that $S'_j = \phi(S_j)$. Indeed, let H_j and H'_j be the subgroups of $\mathbb{Z}^{n+1}$ generated respectively by $\phi(S_j)$ and S'_j. It suffices to show that $H'_j = H_j$. Since $H_j \subset H'_j$ are subgroups of $\mathbb{Z}^n$ of the same rank n, we have to show that if $k\beta \in H_j$ for some positive integer k and $\beta = (\beta_0, \dots, \beta_n) \in \mathbb{Z}^{n+1}$, then $\beta \in H_j$. Indeed, if $k\beta_j = \langle \nu_j, \alpha \rangle$ for each $j = 0, \dots, n$, then since the ν_j span $(\mathbb{Z}^n)^*$, it follows that $\alpha/k \in \mathbb{Z}^n$, and $\beta = \phi(\alpha/k) \in H_j$. Therefore $H'_j = H_j$, which proves the claim. □

As in section VI.10.2 let β_j be the vertex of $\mathcal{P}'$ on the j-th axis. Exercise VI.33 implies that S'_j is the semigroup of integral points in the polyhedral cone $\mathcal{C}'_{\beta_j} \subset \mathbb{R}^{n+1}$ generated by $\beta_i - \beta_j$, $i = 0, \dots, n$. Therefore proposition VI.15 implies that ϕ induces an isomorphism $X_{\mathcal{P}} \cong X_{\mathcal{P}'}$.

10.4. Identification with $\mathbb{P}^n(\omega)$. Let $f \in \Bbbk[x_0, \dots, x_n]$. We say that f is *weighted homogeneous* with respect to ω (or in short, ω- *homogeneous*) if ω is constant on Supp(f). If f is ω-homogeneous, then the set $V(f) := \{[a_0 : \cdots : a_n] : f(a_0, \dots, a_n) = 0\} \subset \mathbb{P}^n(\omega)$ of zeroes of f is a well-defined subset of $\mathbb{P}^n(\omega)$. As in the case of $\mathbb{P}^n$, the basic open subsets of $\mathbb{P}^n(\omega)$ are $U_j := \mathbb{P}^n(\omega) \setminus V(x_j)$, $j = 0, \dots, n$. If f is ω-homogeneous with $\omega(f)$ a multiple of ω_j, say $\omega(f) = k\omega_j$, $k \geq 0$, then $(f/x_j^k)|_{C_a}$ is constant for all $C_a \in U_j$ and therefore f/x_j^k is a well-defined function on U_j. Exercise VI.34 shows that the $\Bbbk$-algebra R_j generated by all these f/x_j^k, $k \geq 0$, is finitely generated, and if $h_1, \dots, h_s$ generate R_j as a $\Bbbk$-algebra, then they induce a bijection from U_j to an open affine subvariety of $X_{\mathcal{P}'}$ (where $\mathcal{P}'$ is as in section VI.10.2) which extends to a bijection from $\mathbb{P}^n(\omega) \to X_{\mathcal{P}'}$. We use this bijection to identify $\mathbb{P}^n(\omega)$ with $X_{\mathcal{P}'} \cong X_{\mathcal{P}}$ (where $\mathcal{P}$ is as in section VI.10.1). In particular, it follows that each U_j is an open affine subvariety of $\mathbb{P}^n(\omega)$ with coordinate ring R_j. To completely describe the identification of $\mathbb{P}^n(\omega)$ and $X_{\mathcal{P}'}$ it remains to explicitly identify points in $\mathbb{P}^n(\omega)$ with those of $X_{\mathcal{P}'}$. Given $a := [a_0 : \cdots : a_n] \in \mathbb{P}^n(\omega)$, we now compute the ideal I_a of all ω-homogeneous polynomials in $(x_0, \dots, x_n)$ that vanish at a. Let J_a be the ideal of $\Bbbk[x_0, \dots, x_n]$ generated by all the x_i such that $a_i = 0$, and all binomials of the form $a^{\alpha_2}x^{\alpha_1} - a^{\alpha_1}x^{\alpha_2}$ where $\alpha_1, \alpha_2 \in \mathbb{Z}_{\geq 0}^{n+1}$ such that $\langle \omega, \alpha_1 \rangle = \langle \omega, \alpha_2 \rangle$. It is clear that $J_a \subset I_a$. The following proposition, which you will prove in exercise VI.35, shows that the converse is also true.

Proposition VI.36. $I_a = J_a$. □

10.5. Exercises.

EXERCISE VI.33. In the notation of section VI.10.3 show that

(1) $\mathcal{C}_{\alpha_j} = \{\alpha \in \mathbb{R}^n : \langle \nu_i, \alpha \rangle \geq 0,\ i = 0, \dots, \hat{j}, \dots, n\}$.

(2) $\mathcal{C}'_{\beta_j} = \{\beta \in \mathbb{R}^{n+1} : \langle \omega, \beta \rangle = 0,\ \beta_i \geq 0,\ i = 0, \dots, \hat{j}, \dots, n\}$.

[Hint: Use corollary V.23.]

EXERCISE VI.34. For each $j = 0, \dots, n$, let R_j be as in section VI.10.4 and S'_j be as in section VI.10.3.

(1) Show that $R_j = \Bbbk[S'_j]$ for each j.

(2) Let $U'_j := U_{\beta_j} \cap X_{\mathcal{P}'}$ be the affine open subset of $X_{\mathcal{P}'}$ (where $\mathcal{P}'$ is as in section VI.10.2) with coordinate ring $\mathbb{k}[S'_j]$. If $h_1, \dots, h_s$ generate R_j as a $\mathbb{k}$-algebra, then show that the map $x \mapsto (h_1(x), \dots, h_s(x))$ induces a bijection ϕ_j from $U_j := \mathbb{P}^n(\omega) \setminus V(x_j)$ to U'_j.

(3) Show that for distinct j, the maps ϕ_j are "compatible," i.e. for each j, j', ϕ_j maps $U_j \cap U_{j'}$ bijectively onto $U'_j \cap U'_{j'}$.

EXERCISE VI.35. Prove proposition VI.36. [Hint: Use arguments analogous to those outlined in exercise VI.6.]

11. *Weighted blow up

[3]Let ν be a weighted order on $\mathbb{k}[x_1, x_1^{-1}, \dots, x_n, x_n^{-1}]$ with *positive* weights $\nu_j := \nu(x_j)$, $j = 1, \dots, n$. Identify ν with the integral element of $(\mathbb{R}^n)^*$ with coordinates $(\nu_1, \dots, \nu_n)$ with respect to the basis dual to the standard basis of $\mathbb{R}^n$. Fix a positive integer k. Let $\mathcal{B}_k := \{\alpha \in \mathbb{Z}^n_{\geq 0} : \langle \nu, \alpha \rangle = kp\}$, where $p := \operatorname{lcm}(\nu_1, \dots, \nu_n)$, and $\mathfrak{q}_k$ be the ideal of $\mathbb{k}[x_1, \dots, x_n]$ generated by $\{x^\alpha : \alpha \in \mathcal{B}_k\}$. Recall that the *blow up* $\operatorname{Bl}_{\mathfrak{q}_k}(\mathbb{k}^n)$ of $\mathbb{k}^n$ at $\mathfrak{q}_k$ is the closure in $\mathbb{k}^n \times X_{\mathcal{B}_k}$ of the graph of the map $\phi_{\mathcal{B}_k} : (\mathbb{k}^*)^n \to X_{\mathcal{B}_k}$ defined as

$$x \mapsto [x^{\alpha_0} : \cdots : x^{\alpha_N}],$$

where $\alpha_0, \dots, \alpha_N$ are the elements of $\mathcal{B}_k$. Let $\mathcal{P} := \{\alpha \in \mathbb{R}^n_{\geq 0} : \langle \nu, \alpha \rangle = p\}$. Then $\mathcal{P}$ is the $(n-1)$-dimensional simplex in $\mathbb{R}^n$ with vertices $(p/\nu_j)e_j$, $j = 1, \dots, n$, where $e_1, \dots, e_n$ form the standard basis of $\mathbb{R}^n$. Let $\mathcal{Q}$ be the n-dimensional simplex in $\mathbb{R}^n$ whose vertices are $0, e_1, \dots, e_n$, so that $\mathbb{k}^n$ can be naturally identified with the basic open set U_0 of $X_{\mathcal{Q}} \cong \mathbb{P}^n$. If k is sufficiently large, then $X_{\mathcal{B}_k} \cong X_{\mathcal{P}}$, so that assertion (7) of proposition VI.15 implies that $\operatorname{Bl}_{\mathfrak{q}_k}(\mathbb{k}^n)$ is isomorphic to the open subset of $X_{\mathcal{P}+\mathcal{Q}}$ which is the union of basic open subsets $U_{(p/\nu_j)e_j}$, $j = 1, \dots, n$ (note that each $(p/\nu_j)e_j$ is a vertex of $\mathcal{P} + \mathcal{Q}$). In particular, $\operatorname{Bl}_{\mathfrak{q}_k}(\mathbb{k}^n)$ are isomorphic for all sufficiently large k; we call the corresponding algebraic variety the *ν-weighted blow up of* $\mathbb{k}^n$ and denote it by $\operatorname{Bl}_\nu(\mathbb{k}^n)$. Let $\sigma : \operatorname{Bl}_\nu(\mathbb{k}^n) \to \mathbb{k}^n$ be the blow up map. The *exceptional divisor* on $\operatorname{Bl}_\nu(\mathbb{k}^n)$ is $E_\nu := \sigma^{-1}(0)$. Note that σ is an isomorphism on $\operatorname{Bl}_\nu(\mathbb{k}^n) \setminus E_\nu$. If V is a subvariety of $\mathbb{k}^n$, the *strict transform* of V on $\operatorname{Bl}_\nu(\mathbb{k}^n)$ is the closure in $\operatorname{Bl}_\nu(\mathbb{k}^n)$ of $\sigma^{-1}(V \setminus \{0\})$. Note that $\mathcal{P}$ is the "lower" facet of $\mathcal{P} + \mathcal{Q}$ (see fig. 10) and E_ν is precisely the subvariety $V_{\mathcal{P}}$ of $X_{\mathcal{P}+\mathcal{Q}}$ corresponding to the facet $\mathcal{P}$ of $\mathcal{P} + \mathcal{Q}$. The construction in section VI.10.2 therefore shows that $E_\nu \cong \mathbb{P}^{n-1}(\nu)$. The following proposition makes this isomorphism more explicit.

[3]The asterisk in the section name is to indicate that the material of this section is not going to be used in the proof of Bernstein's theorem. It is used for the first time in chapter IX.

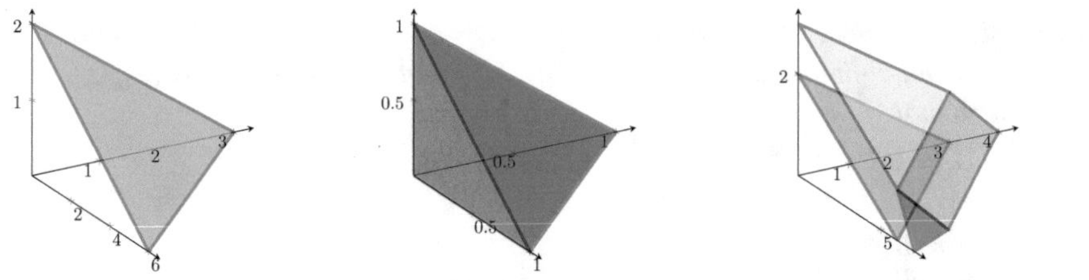

$\mathcal{P}$ $\qquad$ $\mathcal{Q}$ $\qquad$ Lower and upper faces of $\mathcal{P} + \mathcal{Q}$

FIGURE 10. Construction of $\mathrm{Bl}_{(1,2,3)}(\Bbbk^3)$.

Proposition VI.37. *For each $a = (a_1, \dots, a_n) \in \Bbbk^n \setminus \{0\}$, let $C_a := \{(a_1 t^{\nu_1}, \dots, a_n t^{\nu_n}) : t \in \Bbbk\} \subset \Bbbk^n$. The strict transform of each C_a intersects E_ν at precisely one point which we denote by $[a]$. The map $[a_1 : \cdots : a_n] \to [a]$ yields an isomorphism between $\mathbb{P}^{n-1}(\nu)$ and E_ν.*

PROOF. Fix $a = (a_1, \dots, a_n) \in \Bbbk^n \setminus \{0\}$ and j such that $a_j \neq 0$. Let C'_j be the convex polyhedral cone generated by $(p/\nu_i)e_i - (p/\nu_j)e_j$, $i = 1, \dots, \hat{j}, \dots, n$, and C_j be the convex polyhedral cone generated by C'_j and e_j. Let S_j and S'_j be the semigroups of integral elements, respectively, in C_j and C'_j. Proposition VI.15 implies that $\Bbbk[U_{\alpha_j} \cap X_{\mathcal{P}+\mathcal{Q}}] = \Bbbk[S_j]$ and $\Bbbk[U_{\alpha_j} \cap V_{\mathcal{P}}] = \Bbbk[S'_j]$. Let C'_a be the strict transform of C_a in $\mathrm{Bl}_\nu(\Bbbk^n)$. It turns out that (exercise VI.36)

(i) For every $\alpha \in S_j$, either x^α identically vanishes on C_a or $\mathrm{ord}_t(x^\alpha|_{(a_1 t^{\nu_1}, \dots, a_n t^{\nu_n})})$ is nonnegative. This implies that $C'_a \cap E_\nu \subset U_{\alpha_j}$.
(ii) $S'_j = \{\alpha = (\alpha_1, \dots, \alpha_n) \in \mathbb{Z}^n : \langle \nu, \alpha \rangle = 0,\ \alpha_i \geq 0,\ i = 1, \dots, \hat{j}, \dots, n\}$. If $\alpha \in S'_j$, it follows that
   (1) either $\alpha_i > 0$ for some i such that $a_i = 0$, in which case x^α identically vanishes on C'_a,
   (2) or $\alpha_i \neq 0$ if and only if $a_i \neq 0$, in which case x^α takes the constant (nonzero) value a^α on C'_a.

These observations together with proposition VI.36 implies that $C'_a \cap E_\nu$ corresponds precisely to the point $[a_1 : \cdots : a_n] \in \mathbb{P}^{n-1}(\nu)$, as required. □

We write O_ν for the torus $O_{\mathcal{P}}$ of $E_\nu = V_{\mathcal{P}}$. The isomorphism between E_ν and $\mathbb{P}^{n-1}(\nu)$ from proposition VI.37 induces an isomorphism $O_\nu \cong \mathbb{P}^{n-1}(\nu) \setminus V(x_1 \cdots x_n) \cong (\Bbbk^*)^{n-1}$. Proposition VI.20 implies that O_ν is a nonsingular hypersurface of $\mathrm{Bl}_\nu(\Bbbk^n)$. Let $I \subset [n] := \{1, \dots, n\}$ and $\Bbbk^I := V(x_j : j \notin I)$ be the corresponding coordinate subspace of $\Bbbk^n$. We identify $\Bbbk^I$ with $\Bbbk^{|I|}$. Let ν' be a weighted order with positive weights on $\Bbbk[x_i : i \in I]$ such that $(\nu'(x_i) : i \in I)$ is proportional to $(\nu_i : i \in I)$. It follows from the definition of a weighted blow up that $\mathrm{Bl}_{\nu'}(\Bbbk^I)$ can be identified with the strict transform of $\Bbbk^I$ on $\mathrm{Bl}_\nu(\Bbbk^n)$. The following proposition compiles some properties of the embedding $\mathrm{Bl}_{\nu'}(\Bbbk^I) \hookrightarrow \mathrm{Bl}_\nu(\Bbbk^n)$, its proof is left as an exercise.

Proposition VI.38. *Assume* $\gcd(\nu_i : i \in I) = 1$. *Let* $k := |I|$. *Then there is a Zariski open neighborhood* U *of* $O_{\nu'}$ *in* $\mathrm{Bl}_\nu(\Bbbk^n)$ *and regular functions* $(z_1, \dots, z_n)$ *on* U *such that*

(1) $U \cong \Bbbk \times (\Bbbk^*)^{k-1} \times \Bbbk^{n-k}$ *with coordinates* $(z_1, \dots, z_n)$*;*
(2) $z_1, \dots, z_k$ *are monomials in* $(x_i : i \in I)$*;*
(3) $\nu(z_1) = 1$, $\nu(z_i) = 0$, $2 \le i \le n$*;*
(4) *for all* $i' \notin I$, *there is* i' *such that* $z_{i'} = x_{i'}/z_1^{\nu_{i'}}$*;*
(5) $E_\nu \cap U = V(z_1) \cong (\Bbbk^*)^{k-1} \times \Bbbk^{n-k}$*;*
(6) $O_\nu = (E_\nu \cap U) \setminus V(z_{k+1} \cdots z_n)$*;*
(7) $\mathrm{Bl}_{\nu'}(\Bbbk^I) \cap U = V(z_{k+1}, \dots, z_n) \cong \Bbbk \times (\Bbbk^*)^{k-1}$*;*
(8) $E_{\nu'} \cap U = O_{\nu'} = V(z_1, z_{k+1}, \dots, z_n) \cong (\Bbbk^*)^{k-1}$. □

11.1. Exercises.

EXERCISE VI.36. Verify observations (i) and (ii) from the proof of proposition VI.37.

EXERCISE VI.37. Prove proposition VI.38. [Hint: Choose $\beta_1, \dots, \beta_k \in \mathbb{Z}^I$ such that $\langle \nu, \beta_1 \rangle = 1$ and $(\beta_2, \dots, \beta_k)$ is a basis of $\nu^\perp \cap \mathbb{Z}^I$. Set $\beta_j := e_j - \nu_j \beta_1$, $j = k+1, \dots, n$. Show that $(\beta_1, \dots, \beta_n)$ is a basis of $\mathbb{Z}^n$. Choose $d \ge 1$ such that the relative interior of $d\mathcal{P}$ contains a point β_0 with integral coordinates and consider the cone $\mathcal{C}_{\beta_0}$ generated by $d\mathcal{P} + \mathcal{Q} - \beta_0$. Show that $\mathcal{C}_{\beta_0} \cap \mathbb{Z}^I$ is generated as a semigroup by $\beta_1, \pm\beta_2, \dots, \pm\beta_k$, and $\mathcal{C}_{\beta_0} \cap \mathbb{Z}^n$ is generated as a semigroup by $\beta_1, \pm\beta_2, \dots, \pm\beta_k, \beta_{k+1}, \dots, \beta_n$. Set $z_j := x^{\beta_j}$, $j = 1, \dots, n$.]

CHAPTER VII

Number of zeroes on the torus: BKK bound

1. Introduction

In this chapter we derive Bernstein's theorem for the number of isolated solutions of generic systems of n Laurent polynomials on the algebraic torus $(\mathbb{k}^*)^n$ over an algebraically closed field $\mathbb{k}$, apply it to derive some properties of mixed volume and discuss some open problems related to Bernstein's theorem. The term "BKK bound" in the title of this section refers to the bound from Bernstein's theorem which is also known in the literature as "BKK theorem" after D. Bernstein, A. Kushnirenko and A. Khovanski.

2. Mixed volume

The set of convex polytopes in $\mathbb{R}^n$, $n \geq 1$, is a commutative semigroup under Minkowski addition (see chapter V). The interaction between Minkowski addition and volume gives rise to the theory of *mixed volumes*. The starting point of this theory is the following result proven in section V.7.1:

THEOREM V.39. Let $\mathcal{P}_1, \ldots, \mathcal{P}_s$ be convex polytopes in $\mathbb{R}^n$. Then there are nonnegative real numbers $v_\alpha(\mathcal{P}_1, \ldots, \mathcal{P}_s)$ for all $\alpha \in \mathcal{E}_s := \{(\alpha_1, \ldots, \alpha_s) \in \mathbb{Z}^s_{\geq 0} : \alpha_1 + \cdots + \alpha_s = n\}$ such that for all $\lambda = (\lambda_1, \ldots, \lambda_s) \in \mathbb{R}^s_{\geq 0}$,

$$\mathrm{Vol}_n(\lambda_1 \mathcal{P}_1 + \cdots + \lambda_s \mathcal{P}_s) = \sum_{\alpha \in \mathcal{E}_s} v_\alpha(\mathcal{P}_1, \ldots, \mathcal{P}_s) \lambda_1^{\alpha_1} \cdots \lambda_s^{\alpha_s},$$

where Vol_n is the n-dimensional Euclidean volume.

Definition VII.1. The *mixed volume* $\mathrm{MV}(\mathcal{P}_1, \ldots, \mathcal{P}_n)$ of convex polytopes $\mathcal{P}_1$, $\ldots, \mathcal{P}_n$ in $\mathbb{R}^n$ is $v_{(1,\ldots,1)}(\mathcal{P}_1, \ldots, \mathcal{P}_n)$.

THEOREM VII.2. *Let $\mathcal{K}$ be any collection of convex polytopes in $\mathbb{R}^n$ which is invariant under Minkowski addition*[1]. *Then $\mathrm{MV} : \mathcal{K}^n \to \mathbb{R}$ is the unique function such that*

[1]For example, $\mathcal{K}$ may be the set of all convex polytopes in $\mathbb{R}^n$, or the set of convex integral polytopes in $\mathbb{R}^n$.

P. Mondal, *How Many Zeroes?*, CMS/CAIMS Books in Mathematics 2,
https://doi.org/10.1007/978-3-030-75174-6_VII

(1) $\mathrm{MV}(\mathcal{P}, \ldots, \mathcal{P}) = n! \mathrm{Vol}_n(\mathcal{P})$ *for all* $\mathcal{P} \in \mathcal{K}$,
(2) MV *is symmetric in its arguments and*
(3) MV *is* multiadditive, *i.e.*

$$\mathrm{MV}(k_1\mathcal{P}_1 + k_1'\mathcal{P}_1', \mathcal{P}_2, \ldots, \mathcal{P}_n) = k_1 \mathrm{MV}(\mathcal{P}_1, \mathcal{P}_2, \ldots, \mathcal{P}_n) + k_1' \mathrm{MV}(\mathcal{P}_1', \mathcal{P}_2, \ldots, \mathcal{P}_n)$$

for all $k_1, k_1' \in \mathbb{Z}_{\geq 0}$ *and* $\mathcal{P}_1, \ldots, \mathcal{P}_n, \mathcal{P}_1' \in \mathcal{K}$.

Moreover, MV *can be expressed in terms of the volume (we write* $[n]$ *to denote* $\{1, \ldots, n\}$*):*

$$\mathrm{MV}(\mathcal{P}_1, \ldots, \mathcal{P}_n) = \sum_{\substack{I \subseteq [n] \\ I \neq \emptyset}} (-1)^{n-|I|} \mathrm{Vol}_n\left(\sum_{i \in I} \mathcal{P}_i\right). \tag{54}$$

PROOF. This follows from combining theorem V.39, corollary B.62 and lemma B.59. □

Example VII.3. For $n = 1$, a convex polytope is simply an interval and its mixed volume is its length. For $n = 2$, if $\mathcal{P}$, $\mathcal{Q}$ are convex polygons in $\mathbb{R}^2$, then identity (54) implies (see fig. 1) that

$$\mathrm{MV}(\mathcal{P}, \mathcal{Q}) = \mathrm{Area}(\mathcal{P} + \mathcal{Q}) - \mathrm{Area}(\mathcal{P}) - \mathrm{Area}(\mathcal{Q}). \tag{55}$$

Remark VII.4. Theorem V.39 implies that the mixed volume is nonnegative, and identity (54) implies that MV is invariant under volume-preserving transformations of $\mathbb{R}^n$. Moreover, theorem V.35 coupled with identity (54) implies that mixed volume is continuous with respect to the Hausdorff distance on polytopes. In section VII.6 we use Bernstein's theorem to deduce some other basic properties of mixed volume.

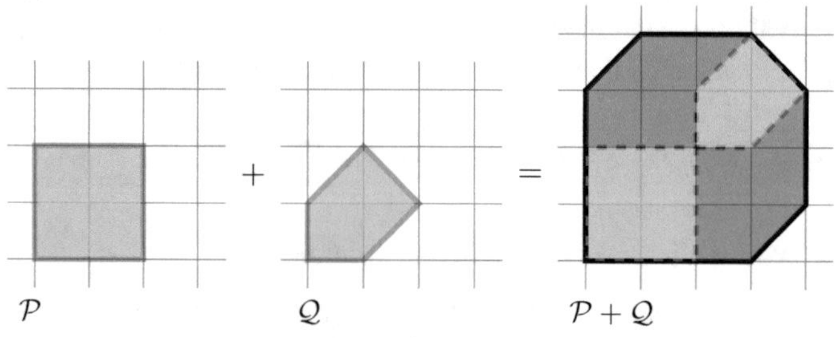

FIGURE 1. $\mathrm{MV}(\mathcal{P}, \mathcal{Q})$ is 8, which is the area of the blue part of $\mathcal{P} + \mathcal{Q}$.

3. Theorems of Kushnirenko and Bernstein

3.1. The bound. Given Laurent polynomials $f_1, \ldots, f_n$ in indeterminates $x_1, \ldots, x_n$, the number of their zeroes counted with multiplicity is

$$[f_1, \ldots, f_n]_{(\mathbb{k}^*)^n} := \sum_{a \in (\mathbb{k}^*)^n} [f_1, \ldots, f_n]_a,$$

where $[f_1, \dots, f_n]_a$ is the *intersection multiplicity* of $f_1, \dots, f_n$ at a. If $V := V(f_1, \dots, f_n)$ has non-isolated points, then $[f_1, \dots, f_n]_{(\mathbb{k}^*)^n} = \infty$ (proposition IV.28). We write $[f_1, \dots, f_n]^{iso}_{(\mathbb{k}^*)^n}$ for the sum of intersection multiplicities of $f_1, \dots, f_n$ at all the isolated points of V. Recall that the *support* $\operatorname{Supp}(f)$ of a Laurent polynomial $f = \sum_{\alpha \in \mathcal{A}} c_\alpha x^\alpha$ is the set of all $\alpha \in \mathbb{Z}^n$ such that $c_\alpha \neq 0$. Given a subset $\mathcal{S}$ of $\mathbb{R}^n$, we say that f is *supported at* $\mathcal{S}$ if $\operatorname{Supp}(f) \subseteq \mathcal{S}$, and we write $\mathcal{L}(\mathcal{S})$ for the set of all Laurent polynomials supported at $\mathcal{S}$. Given an ordered collection $\mathcal{A} := (\mathcal{A}_1, \dots \mathcal{A}_m)$ of *finite* subsets of $\mathbb{Z}^n$, we write $\mathcal{L}(\mathcal{A})$ for the set of all m-tuples $(f_1, \dots, f_m)$ of Laurent polynomials such that f_j is supported at $\mathcal{A}_j$ for each j. We say that some property holds for *generic* f_j supported at $\mathcal{A}_j$, $j = 1, \dots, m$, if it holds for all $(f_1, \dots, f_m)$ in a nonempty Zariski open subset of $\mathcal{L}(\mathcal{A}) \cong \prod_{j=1}^m \mathcal{L}(\mathcal{A}_j) \cong \mathbb{k}^{\sum_j |\mathcal{A}_j|}$.

THEOREM VII.5. (Bernstein's theorem: the bound). *Let $\mathcal{P}_j$ be the convex hull of $\mathcal{A}_j$, $j = 1, \dots, n$. If $\operatorname{Supp}(f_j) \subset \mathcal{A}_j$, $j = 1, \dots, n$, then*

$$[f_1, \dots, f_n]^{iso}_{(\mathbb{k}^*)^n} \leq \operatorname{MV}(\mathcal{P}_1, \dots, \mathcal{P}_n). \tag{56}$$

Moreover, for generic f_j supported at $\mathcal{A}_j$, $j = 1, \dots, n$, we have

$$[f_1, \dots, f_n]_{(\mathbb{k}^*)^n} = [f_1, \dots, f_n]^{iso}_{(\mathbb{k}^*)^n} = \operatorname{MV}(\mathcal{P}_1, \dots, \mathcal{P}_n). \tag{57}$$

The *Newton polytope* $\operatorname{NP}(f)$ of a Laurent polynomial f is the convex hull (in $\mathbb{R}^n$) of the support of f. Bernstein's theorem, in particular, states that the number of isolated solutions of a system of polynomials is bounded by the mixed volume of their Newton polytopes. A. Kushnirenko proved Bernstein's theorem for the case that all the Newton polytopes are identical, in which case the mixed volume equals $n!$ times the volume of any of these polytopes. D. Bernstein found theorem VII.5 while trying to understand and generalize Kushnirenko's result. Kushnirenko, however, not only gave the bound, but also gave a precise characterization of the collections of $f_1, \dots, f_n$ for which the bound is achieved. There is a natural way to understand this characterization in the case that $\mathbb{k} = \mathbb{C}$; we describe it now.

3.2. The non-degeneracy condition. Let $\mathcal{A}_1, \dots, \mathcal{A}_n$ be finite subsets of $\mathbb{Z}^n$, and $f_1, \dots, f_n$ be Laurent polynomials in $x_1, \dots, x_n$ over $\mathbb{C}$ such that $\operatorname{Supp}(f_j) = \mathcal{A}_j$, $j = 1, \dots, n$. Assume there are Laurent polynomials $g_1, \dots, g_n$ such that $\operatorname{Supp}(g_j) \subseteq \mathcal{A}_j$ for each j, and $[f_1, \dots, f_n]^{iso}_{(\mathbb{C}^*)^n} < [g_1, \dots, g_n]^{iso}_{(\mathbb{C}^*)^n}$. Write $h_j := (1-t)f_j + tg_j$, $j = 1, \dots, n$. Since the $g_j = h_j|_{t=1}$ have "more" common zeroes on $(\mathbb{C}^*)^n$ than the $f_j = h_j|_{t=0}$, intuitively we may expect that there is a curve $C(t)$ on $(\mathbb{C}^*)^n$ such that $h_j(C(t)) = 0$ and $\lim_{t \to 0} C(t)$ is *not* on $(\mathbb{C}^*)^n$, i.e. as t approaches 0, either $C(t)$ approaches one of the coordinate hyperplanes of $\mathbb{C}^n$, or $|C(t)|$ approaches infinity (see fig. 2). In any event, assuming our intuition is correct, there is a punctured neighborhood U of the origin on $\mathbb{C}$ and a parametrization $U \to C(t)$ of the form $\gamma : t \mapsto (a_1 t^{\nu_1} + \cdots, \dots, a_n t^{\nu_n} + \cdots)$, where $a = (a_1, \dots, a_n) \in (\mathbb{C}^*)^n$ and for each j, ν_j is the order (in t) of the j-th coordinate. Since $\lim_{t \to 0} C(t) \notin (\mathbb{C}^*)^n$, it follows that not all the ν_j are zero. Let ν be

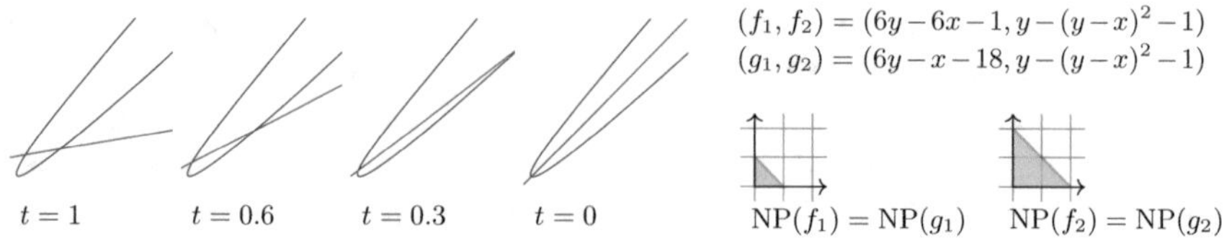

FIGURE 2. One of the common roots of $(1-t)f_j + tg_j = 0$, $j = 1, 2$, approaches infinity as $t \to 0$.

the element in $(\mathbb{R}^n)^*$ with coordinates $(\nu_1, \ldots, \nu_n)$ with respect to the basis dual to the standard basis of $\mathbb{R}^n$. Given a subset $\mathcal{S}$ of $\mathbb{R}^n$ and a Laurent polynomial $g = \sum_\alpha c_\alpha x^\alpha$ supported at $\mathcal{S}$, we write

$$\text{(58)} \qquad \mathrm{In}_{\mathcal{S},\nu}(g) := \sum_{\alpha \in \mathrm{In}_\nu(\mathcal{S})} c_\alpha x^\alpha = \begin{cases} \mathrm{In}_\nu(g) & \text{if } \mathrm{Supp}(g) \cap \mathrm{In}_\nu(\mathcal{S}) \neq \emptyset, \\ 0 & \text{otherwise,} \end{cases}$$

where $\mathrm{In}_\nu(\mathcal{S})$ is defined as in section V.1. Let $q_j := \min_\nu(\mathcal{A}_j)$. Since $\mathrm{Supp}(g_j) \subseteq \mathcal{A}_j = \mathrm{Supp}(f_j)$, it follows that

$$h_j(\gamma(t)) = t^{q_j} \mathrm{In}_{\mathcal{A}_j,\nu}(f_j)(a) + t^{q_j+1}(-\mathrm{In}_{\mathcal{A}_j,\nu}(f_j)(a) + t^\epsilon \mathrm{In}_{\mathcal{A}_j,\nu}(g_j)(a)) + \cdots$$

where ϵ is nonnegative, and the orders in t of the omitted terms are higher than q_j. Since $h_j(\gamma(t)) \equiv 0$, it follows that $\mathrm{In}_{\mathcal{A}_j,\nu}(f_j)(a) = 0$ for each $j = 1, \ldots, n$. This leads to the following definition.

Definition VII.6. Let $\mathcal{A}_1, \ldots, \mathcal{A}_m$ be finite subsets of $\mathbb{Z}^n$ and $(f_1, \ldots, f_m) \in \mathcal{L}(\mathcal{A})$. We say that $f_1, \ldots, f_m$ are $(\mathcal{A}_1, \ldots, \mathcal{A}_m)$-*non-degenerate* if they satisfy the following condition:

(ND*) for each $\nu \in (\mathbb{R}^n)^* \setminus \{0\}$, there is no common root of $\mathrm{In}_{\mathcal{A}_j,\nu}(f_j)$, $j = 1, \ldots, m$, on $(\Bbbk^*)^n$.

We say that $f_1, \ldots, f_m$ are *BKK non-degenerate* if they are $(\mathrm{Supp}(f_1), \ldots, \mathrm{Supp}(f_m))$-non-degenerate.

The preceding argument suggests that for $\Bbbk = \mathbb{C}$, $\mathcal{A}$-non-degeneracy is sufficient for the maximality of $[f_1, \ldots, f_n]^{iso}_{(\Bbbk^*)^n}$. It turns out that it is also necessary; Kushnirenko proved it in the case that the convex hulls of the $\mathcal{A}_j$ are identical and Bernstein treated the general case. Both necessity and sufficiency remain valid even if $\mathbb{C}$ is replaced by an arbitrary algebraically closed field:

THEOREM VII.7. (Bernstein's theorem: non-degeneracy condition). *Let* $\mathcal{P}_j := \mathrm{conv}(\mathcal{A}_j)$, $j = 1, \ldots, n$. *If the mixed volume of* $\mathcal{P}_1, \ldots, \mathcal{P}_n$ *is nonzero, then the bound* (56) *is satisfied with an equality if and only if* $f_1, \ldots, f_n$ *are* $(\mathcal{A}_1, \ldots, \mathcal{A}_n)$-*non-degenerate.*

Remark VII.8. Recall from section VI.8.3 that integral elements in $(\mathbb{R}^n)^*$ can be identified with *weighted orders* on the ring of Laurent polynomials. In the case that $\mathrm{NP}(f_j) = \mathrm{conv}(\mathcal{A}_j)$ for each $j = 1, \ldots, m$, condition (ND*) is equivalent to the following condition:

(ND$'^*$) for each nontrivial weighted order ν, there is no common root of $\mathrm{In}_\nu(f_j)$, $j = 1, \ldots, m$, on $(\mathbb{k}^*)^n$.

Remark VII.9. On the face of it condition (IX.24) consists of uncountably many conditions, one for each element in $(\mathbb{R}^n)^*$. However, it is equivalent to finitely many conditions. Indeed, let $\mathcal{P}_j := \mathrm{conv}(\mathcal{A}_j)$, $j = 1, \ldots, m$, and $\mathcal{P} := \mathcal{P}_1 + \cdots + \mathcal{P}_m$. For each face $\mathcal{Q}$ of $\mathcal{P}$, there are unique faces $\mathcal{Q}_j$ of $\mathcal{P}_j$ such that $\mathcal{Q} = \mathcal{Q}_1 + \cdots + \mathcal{Q}_m$ (proposition V.16). Given $f_j := \sum_\alpha c_{j,\alpha} x^\alpha$, let $f_{j,\mathcal{Q}} := \sum_{\alpha \in \mathcal{Q}_j} c_{j,\alpha} x^\alpha$ be the "component" of f_j supported at $\mathcal{Q}_j$. Then (ND*) is equivalent to the following condition:

(ND$''^*$) for each face $\mathcal{Q}$ of dimension less than n of $\mathcal{P}$, there is no common root of $f_{j,\mathcal{Q}}$, $j = 1, \ldots, m$, on $(\mathbb{k}^*)^n$.

Kushnirenko's theorem follows immediately by applying theorems VII.2, VII.5 and VII.7 to the case that each f_j is supported at the same (finite) subset of $\mathbb{Z}^n$:

Corollary VII.10. (Kushnirenko [Kou76]). *Let $f_1, \ldots, f_n$ be Laurent polynomials supported at a finite subset $\mathcal{A}$ of $\mathbb{Z}^n$ and $\mathcal{P} := \mathrm{conv}(\mathcal{A})$. Then $[f_1, \ldots, f_n]^{iso}_{(\mathbb{k}^*)^n} \leq n!\,\mathrm{Vol}_n(\mathcal{P})$. If $\mathrm{Vol}_n(\mathcal{P}) > 0$, then the bound is satisfied with equality if and only if $f_1, \ldots, f_n$ are $(\mathcal{A}, \ldots, \mathcal{A})$-non-degenerate.* □

Example VII.11. To attain the bound in Bernstein's and Kushnirenko's theorems it is *not* necessary that $\mathrm{NP}(f_j)$ be equal to $\mathcal{P}_j$ for each (or, any!) j. Indeed, consider $f_1 = 1 + x^4 + x^2y^4$ and $f_2 = xy^2 + x^3y^3 + x^6$. Newton polygons of the f_j are proper subsets of the triangle $\mathcal{P}$ with vertices $A = (0,0)$, $B = (6,0)$ and $C = (2,4)$ (see fig. 3). However, it is straightforward to check that if $\mathcal{Q}$ is any edge or vertex of $\mathcal{P}$, then either $f_{1,\mathcal{Q}}$ or $f_{2,\mathcal{Q}}$ is a monomial, so that (f_1, f_2) satisfy the non-degeneracy condition (ND$''^*$) from remark VII.9. Consequently, the number of solutions of (f_1, f_2) on $(\mathbb{k}^*)^2$ is $2\,\mathrm{Area}(\mathcal{P}) = 24$.

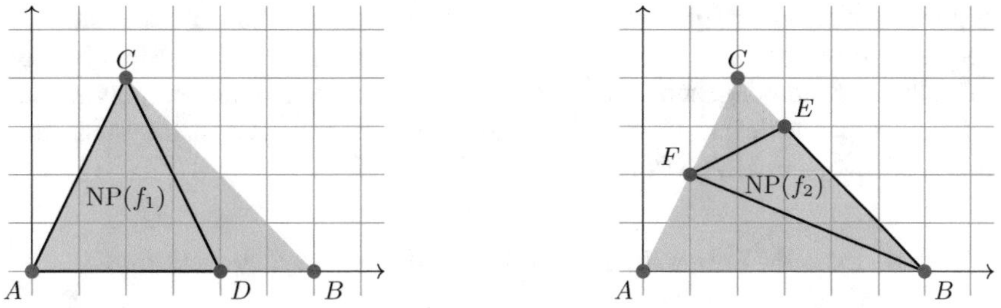

FIGURE 3. The bound in Kushnirenko's theorem is attained when $\mathcal{A}$ is the triangle ABC, $\mathrm{NP}(f_1)$ is the triangle ADC and $\mathrm{NP}(f_2)$ is the triangle BEF.

Example VII.11 shows that it is possible that $\mathrm{MV}(\mathcal{P}_1,\dots,\mathcal{P}_n) = \mathrm{MV}(\mathcal{Q}_1,\dots,\mathcal{Q}_n)$ even if each $\mathcal{P}_j$ *properly* contains $\mathcal{Q}_j$. In fact using Bernstein's theorem it is possible to precisely characterize the situations in which this happens. First we need to make a definition. Convex polytopes $\mathcal{Q}_1,\dots,\mathcal{Q}_m$ in $\mathbb{R}^n$ are said to be *dependent* if there is a nonempty subset I of $[m] := \{1,\dots,m\}$ such that $\dim(\sum_{i\in I}\mathcal{Q}_i) < |I|$; otherwise they are said to be *independent*.

THEOREM VII.12. *If $\mathcal{P}'_j$ are convex polytopes in $\mathbb{R}^n$ such that $\mathcal{P}_j \subseteq \mathcal{P}'_j$ for each j, then $\mathrm{MV}(\mathcal{P}_1,\dots,\mathcal{P}_n) \leq \mathrm{MV}(\mathcal{P}'_1,\dots,\mathcal{P}'_n)$. The inequality is strict if and only if both of the following are true:*

(1) $\mathcal{P}'_1,\dots,\mathcal{P}'_n$ are independent and
(2) there is $\nu \in (\mathbb{R}^n)^*\setminus\{0\}$ such that the collection $\{\mathrm{In}_\nu(\mathcal{P}_j) : \mathcal{P}_j \cap \mathrm{In}_\nu(\mathcal{P}'_j) \neq \emptyset\}$ of polytopes is independent.

We prove this result in corollary VII.34. Using this we can give the following alternate formulation of Bernstein's theorem in which $(\mathcal{A}_1,\dots,\mathcal{A}_n)$-non-degeneracy is replaced by a combinatorial condition plus BKK non-degeneracy; we will prove it in corollary VII.36.

THEOREM VII.13. (Bernstein's theorem—alternate version). *Let $\mathcal{A}_j$ be finite subsets of $\mathbb{Z}^n$ and f_j be Laurent polynomials supported at $\mathcal{A}_j$, $j = 1,\dots,n$. Then*

$$[f_1,\dots,f_n]^{iso}_{(\Bbbk^*)^n} \leq \mathrm{MV}(\mathrm{conv}(\mathcal{A}_1),\dots,\mathrm{conv}(\mathcal{A}_n))$$

If $\mathrm{MV}(\mathrm{conv}(\mathcal{A}_1),\dots,\mathrm{conv}(\mathcal{A}_n)) > 0$, then the bound is satisfied with an equality if and only if both of the following conditions hold:

(1) for each nontrivial weighted order ν, the collection $\{\mathrm{In}_\nu(\mathrm{NP}(f_j)) : \mathrm{In}_\nu(\mathcal{A}_j) \cap \mathrm{Supp}(f_j) \neq \emptyset\}$ of polytopes is dependent and
(2) $f_1,\dots,f_n$ are BKK non-degenerate, i.e. they satisfy (ND'^*) with $m = n$.

Remark VII.14. The Newton polytopes of polynomials in example VII.11 satisfy the property that for each nonzero $\nu \in (\mathbb{R}^n)^*$, for at least one of these polytopes the face where ν attains the minimum is a vertex. These were extensively used by A. Khovanskii (see, e.g. [Kho20]) who called these *developed* systems of polytopes. A basic reason for the usefulness of these systems is the following property (which is straightforward to check): if the collection of Newton polytopes of a system of Laurent polynomials is developed, then this system is automatically BKK non-degenerate. In particular, if $f_1,\dots,f_n$ are Laurent polynomials in $(x_1,\dots,x_n)$ such that their Newton polytopes form a developed system, then $[f_1,\dots,f_n]^{iso}_{(\Bbbk^*)^n} = \mathrm{MV}(\mathrm{NP}(f_1),\dots,\mathrm{NP}(f_n))$.

4. Proof of Bernstein-Kushnirenko non-degeneracy condition

Let $\mathcal{A} := (\mathcal{A}_1,\dots,\mathcal{A}_m)$, $m \geq 1$, be a collection of finite subsets of $\mathbb{Z}^n$. Let $\mathcal{L}(\mathcal{A}) := \prod_{j=1}^m \mathcal{L}(\mathcal{A}_j) \cong \Bbbk^{\sum_j |\mathcal{A}_j|}$ and $\mathcal{N}(\mathcal{A})$ be the set of all $\mathcal{A}$-non-degenerate $(f_1,\dots,f_m) \in \mathcal{L}(\mathcal{A})$. In section VII.4.3 we prove the following result:

THEOREM VII.15. *$\mathcal{N}(\mathcal{A})$ is a Zariski open subset of $\mathcal{L}(\mathcal{A})$. If $m \geq \min\{n, \dim(\sum_{j=1}^m \operatorname{conv}(\mathcal{A}_j)) + 1\}$, then $\mathcal{N}(\mathcal{A})$ is nonempty.*

In the case that $m = n$, define

$$(59)\quad [\mathcal{A}_1, \ldots, \mathcal{A}_n]^{iso}_{(\mathbb{k}^*)^n} := \max\{[f_1, \ldots, f_n]^{iso}_{(\mathbb{k}^*)^n} : \operatorname{Supp}(f_j) \subseteq \mathcal{A}_j,\ j = 1, \ldots, n\},$$

and let $\mathcal{M}(\mathcal{A})$ be the set of all $(f_1, \ldots, f_n) \in \mathcal{L}(\mathcal{A})$ such that $[f_1, \ldots, f_n]^{iso}_{(\mathbb{k}^*)^n} = [\mathcal{A}_1, \ldots, \mathcal{A}_n]^{iso}_{(\mathbb{k}^*)^n}$. In sections VII.4.1 and VII.4.2 we prove the following result:

THEOREM VII.16. *Assume $m = n$. Then*

$$\mathcal{M}(\mathcal{A}) = \begin{cases} \mathcal{L}(\mathcal{A}) & \text{if } [\mathcal{A}_1, \ldots, \mathcal{A}_n]^{iso}_{(\mathbb{k}^*)^n} = 0, \\ \mathcal{N}(\mathcal{A}) & \text{if } [\mathcal{A}_1, \ldots, \mathcal{A}_n]^{iso}_{(\mathbb{k}^*)^n} > 0. \end{cases}$$

In particular, $\mathcal{M}(\mathcal{A})$ is a nonempty Zariski open subset of $\mathcal{L}(\mathcal{A})$.

Theorem VII.15 is required (in section VII.4.2) for the proof of theorem VII.16. The reason to defer the proof of theorem VII.15 to section VII.4.3 is that it is somewhat more technical.

4.1. Sufficiency of Bernstein-Kushnirenko non-degeneracy condition. In this section we prove that $\mathcal{N}(\mathcal{A}) \subseteq \mathcal{M}(\mathcal{A})$ in the case that $m = n$. In his proof of this part of his theorem, Bernstein used the theory of Puiseux series of curves, which only works in characteristic zero. For arbitrary characteristics, the analogous role is played by the notion of *branches* described in section VI.8.1. Let $B = (Z, z)$ be a branch on $(\mathbb{k}^*)^n$. Let $\nu_B \in (\mathbb{R}^n)^*$ be the corresponding integral vector of weights, ρ_B be a parameter of B and $\operatorname{In}(B) \in (\mathbb{k}^*)^n$ be the corresponding n-tuple of "initial coefficients" (see section VI.9.1).

Lemma VII.17. *Let B be a branch of a curve contained in the common zero set of Laurent polynomials $f_1, \ldots, f_m$ on $(\mathbb{k}^*)^n$. Then $\operatorname{In}(B)$ is a common zero of $\operatorname{In}_{\nu_B}(f_j)$, $j = 1, \ldots, m$. In particular, if B is a branch at infinity, and $\operatorname{Supp}(f_j) \subseteq \mathcal{A}_j$ for each j, then both (ND*) and (ND′*) are violated with $\nu = \nu_B$.*

PROOF. The first assertion is a direct corollary of lemma VI.33. If B is a branch at infinity, then ν_B is a nontrivial weighted order (lemma VI.28), so that the second assertion follows from the first assertion and the second identity from (58). ☐

Corollary VII.18. *Let f_j be a Laurent polynomial supported at $\mathcal{A}_j$, $j = 1, \ldots, m$. If $f_1, \ldots, f_m$ are $(\mathcal{A}_1, \ldots, \mathcal{A}_m)$-non-degenerate, then $V(f_1, \ldots, f_m) \subset (\mathbb{k}^*)^n$ is finite.* ☐

Proposition VII.19. *Let f_j be a Laurent polynomial supported at $\mathcal{A}_j$, $j = 1, \ldots, n$. If $f_1, \ldots, f_n$ are $(\mathcal{A}_1, \ldots, \mathcal{A}_n)$-non-degenerate, then $[f_1, \ldots, f_n]^{iso}_{(\mathbb{k}^*)^n} = [\mathcal{A}_1, \ldots, \mathcal{A}_n]^{iso}_{(\mathbb{k}^*)^n}$. In particular, $\mathcal{N}(\mathcal{A}_1, \ldots, \mathcal{A}_n) \subseteq \mathcal{M}(\mathcal{A}_1, \ldots, \mathcal{A}_n)$.*

PROOF. Assume to the contrary that there are Laurent polynomials g_j supported at $\mathcal{A}_j$ such that $[g_1, \ldots, g_n]^{iso}_{(\mathbb{k}^*)^n} > [f_1, \ldots, f_n]^{iso}_{(\mathbb{k}^*)^n}$. It suffices to show that this leads to a contradiction. Define $h_j := (1-t)f_j + tg_j$, where t is a new indeterminate. Since the set of zeroes of $f_1, \dots, f_n$ on $(\mathbb{k}^*)^n$ is finite (corollary VII.18), assertion (5) of theorem IV.32 implies that the set of zeroes of $h_1, \dots, h_n$ in $(\mathbb{k}^*)^n \times \mathbb{k}$ contains a curve which has a branch $B = (Z, z)$ at infinity (with respect to $(\mathbb{k}^*)^n \times \mathbb{k}$) at $t = 0$ and $B \not\subseteq (\mathbb{k}^*)^n \times \{0\}$. Then $t|_B \not\equiv 0$ and B determines a well-defined weighted order ν_B on $\mathbb{k}[x_1, x_1^{-1}, \dots, x_n, x_n^{-1}, t]$ such that $\nu_B(t) > 0$. Let ν be the restriction of ν_B to $\mathbb{k}[x_1, x_1^{-1}, \dots, x_n, x_n^{-1}]$. Since $\nu_B(t) > 0$ and $\operatorname{Supp}(g_j) \subset \mathcal{A}_j$, it follows from (58) that

$$\operatorname{In}_{\mathcal{A}_j,\nu}(f_j) = \begin{cases} \operatorname{In}_\nu(f_j) = \operatorname{In}_{\nu_B}(h_j) & \text{if } \operatorname{Supp}(f_j) \cap \operatorname{In}_\nu(\mathcal{A}_j) \neq \emptyset, \\ 0 & \text{otherwise.} \end{cases}$$

Therefore lemma VII.17 implies that $\operatorname{In}_{\mathcal{A}_j,\nu}(f_j)$, $j = 1, \dots, n$, have a common zero on $(\mathbb{k}^*)^n$. Since B is centered at infinity with respect to $(\mathbb{k}^*)^n \times \mathbb{k}$ and $\nu_B(t) > 0$, it follows that ν is a nontrivial weighted order on $\mathbb{k}[x_1, x_1^{-1}, \dots, x_n, x_n^{-1}]$. This contradicts the $\mathcal{A}$-non-degeneracy of $f_1, \dots, f_n$, as desired. □

4.2. Necessity of Bernstein-Kushnirenko non-degeneracy condition. In this section we finish (using theorem VII.15) the proof of theorem VII.16. Assume $[\mathcal{A}_1, \dots, \mathcal{A}_n]^{iso}_{(\mathbb{k}^*)^n} > 0$ and $(f_1, \dots, f_n) \in \mathcal{L}(\mathcal{A})$ is a $\mathcal{A}$-*degenerate system* in $\mathcal{L}$, i.e. there is a nontrivial weighted order ν on $\mathbb{k}[x_1, x_1^{-1}, \dots, x_n, x_n^{-1}]$ such that $\operatorname{In}_{\mathcal{A}_1,\nu}(f_1), \dots, \operatorname{In}_{\mathcal{A}_n,\nu}(f_n)$ have a common zero on $(\mathbb{k}^*)^n$. It suffices to show that

$$[f_1, \dots, f_n]^{iso}_{(\mathbb{k}^*)^n} < [\mathcal{A}_1, \dots, \mathcal{A}_n]^{iso}_{(\mathbb{k}^*)^n}.$$

We show it in *two* ways.

Claim VII.20. *Given any point $a \in (\mathbb{k}^*)^n$, there is $(g_1, \dots, g_n) \in \mathcal{L}(\mathcal{A})$ such that*

(1) $\operatorname{In}_{\mathcal{A}_j,\nu}(g_j)(a) \neq 0$ for each $j = 1, \dots, n$ and

(2) there is an isolated zero b of $(g_1, \dots, g_n)$ on $(\mathbb{k}^*)^n$ such that $f_j(b) \neq 0$ for each $j = 1, \dots, n$.

PROOF. Due to theorem VII.15 we can pick $\mathcal{A}$-non-degenerate $(h_1, \dots, h_n) \in \mathcal{L}(\mathcal{A})$ which automatically satisfies the first property. For each j, let α_j be an arbitrary element of $\mathcal{A}_j$. For each $\epsilon := (\epsilon_1, \dots, \epsilon_n) \in (\mathbb{k}^*)^n$, let $h_{\epsilon_j,j} := h_j - \epsilon_j x^{\alpha_j}$. Since $\mathcal{N}(\mathcal{A})$ is Zariski open (theorem VII.15), it follows that $(h_{\epsilon_1,1}, \dots, h_{\epsilon_n,n})$ is in $\mathcal{N}(\mathcal{A})$ for generic $\epsilon \in (\mathbb{k}^*)^n$, and therefore $[h_{\epsilon_j,1}, \dots, h_{\epsilon_j,n}]_{(\mathbb{k}^*)^n} = [\mathcal{A}_1, \dots, \mathcal{A}_n]^{iso}_{(\mathbb{k}^*)^n} > 0$ (proposition VII.19), in particular, the map $\Phi : (\mathbb{k}^*)^n \to (\mathbb{k}^*)^n$ defined by $x \mapsto (x^{-\alpha_1}h_1, \dots, x^{-\alpha_n}h_n)$ is dominant. Therefore it suffices to take $g_j := h_{\epsilon_j,j}$ for generic $(\epsilon_1, \dots, \epsilon_n) \in (\mathbb{k}^*)^n \setminus \Phi(V(f_1 \cdots f_n))$. □

4.2.1. *Bernstein's proof of necessity of non-degeneracy (Bernstein's trick).* After a linear change of coordinates of $\mathbb{Z}^n$ (which corresponds to a monomial change of coordinates of $(\mathbb{k}^*)^n$) and translating each $\mathcal{A}_j$ by an element in $\mathbb{Z}^n$ (which corresponds to multiplying each f_j by a monomial) if necessary, we may arrange that $\nu = (0, \dots, 0, 1)$ and $\operatorname{In}_\nu(\mathcal{A}_j) = \mathcal{A}_j \cap (\mathbb{Z}^{n-1} \times \{0\})$, $j = 1, \dots, n$. For each $(g_1, \dots, g_n) \in \mathcal{L}(\mathcal{A})$, it follows that

$$\text{In}_{\mathcal{A}_j,\nu}(g_j) = g_j|_{x_n=0}. \tag{60}$$

If $(a_1, \ldots, a_n)$ is a common zero of the $\text{In}_{\mathcal{A}_j,\nu}(f_j)$ on $(\mathbb{k}^*)^n$, then it follows that $a' := (a_1, \ldots, a_{n-1}, 0)$ is also a common zero of the $\text{In}_{\mathcal{A}_j,\nu}(f_j)$ on $\mathbb{k}^n$. Let $g_1, \ldots, g_n$ and b be as in claim VII.20. Identity (60) implies that $g_j(a') \neq 0$ for each $j = 1, \ldots, n$. Take any map $c : \mathbb{k} \to \mathbb{k}^n$ such that $c(0) = a'$ and $c(1) = b$ (e.g. we may take $c(t) = (1-t)a' + tb$) and set $h_j(x,t) := g_j(c(t))f_j(x) - f_j(c(t))g_j(x)$, $j = 1, \ldots, n$. Then each h_j vanishes on the curve $C' := \{(c(t), t) : t \in \mathbb{k}\} \subset \mathbb{k}^{n+1}$. Since $(b, 1)$ is an isolated zero of $h_1(x, 1), \ldots, h_n(x, 1)$, and since $(a', 0) \in C'$ is a "point at infinity" with respect to $(\mathbb{k}^*)^n \times \mathbb{k}$, assertion (5) of theorem IV.32 implies that $[h_1(x, 0), \ldots, h_n(x, 0)]^{iso}_{(\mathbb{k}^*)^n} < [h_1(x, \epsilon), \ldots, h_n(x, \epsilon)]^{iso}_{(\mathbb{k}^*)^n}$ for generic $\epsilon \in \mathbb{k}$. Since each $h_j(x, \epsilon)$ is supported at $\mathcal{A}_j$, it follows that $[f_1, \ldots, f_n]^{iso}_{(\mathbb{k}^*)^n} < [\mathcal{A}_1, \ldots, \mathcal{A}_n]^{iso}_{(\mathbb{k}^*)^n}$, as required. □

4.2.2. *A modified version of Bernstein's trick.* Here we give an alternative proof of the necessity of Bernstein-Kushnirenko non-degeneracy by adapting Bernstein's trick to produce a curve C' as in his original proof *without* changing the coordinates on $(\mathbb{k}^*)^n$; this will be useful later, e.g. in the proofs of the weighted Bézout theorem (section VIII.3) and the extension of the BKK bound to the affine space (section X.7.3). Let $a = (a_1, \ldots, a_n)$ be a common zero of the $\text{In}_\nu(f_j)$ on $(\mathbb{k}^*)^n$. As in the original proof, let $(g_1, \ldots, g_n) \in \mathcal{L}(\mathcal{A})$ and $b = (b_1, \ldots, b_n) \in (\mathbb{k}^*)^n$ be as in claim VII.20. Fix integers $\nu'_j > \nu_j := \nu(x_j)$. Let C be the rational curve on $\mathbb{k}^n$ parameterized by $c(t) := (c_1(t), \ldots, c_n(t)) : \mathbb{k} \to \mathbb{k}^n$ given by

$$c_j(t) := a_j t^{\nu_j} + (b_j - a_j)t^{\nu'_j},\ j = 1, \ldots, n. \tag{61}$$

Note that $c(1) = b \in (\mathbb{k}^*)^n$, so that $c(t) \in (\mathbb{k}^*)^n$ for all but finitely many values of $\mathbb{k}$. Let $m_j := \min_{\mathcal{A}_j}(\nu)$, $j = 1, \ldots, n$. Then $t^{-m_j} f_j(c(t))$ and $t^{-m_j} g_j(c(t))$ are well-defined rational functions in t. The following claim, which follows from a straightforward computation, implies that $t = 0$ is *not* a pole of $t^{-m_j} f_j(c(t))$ or $t^{-m_j} g_j(c(t))$.

Claim VII.21. *Fix j, $1 \leq j \leq n$. If p_j is a Laurent polynomial supported at $\mathcal{A}_j$ then $t^{-m_j} p_j(c(t)) \in \mathbb{k}[[t]]$. The following identity holds in $\mathbb{k}[[t]]$:*

$$t^{-m_j} p_j(c(t)) = \text{In}_{\mathcal{A}_j,\nu}(p_j)(a) + t q_j(t)$$

for some $q_j(t) \in \mathbb{k}[[t]]$. □

Define

$$h_j := t^{-m_j} f_j(c(t)) g_j - t^{-m_j} g_j(c(t)) f_j. \tag{62}$$

Let T be the complement in $\mathbb{k}$ of all the poles of $\prod t^{-2m_j} f_j(c(t)) g_j(c(t))$. Then both 0 and 1 are in T. Moreover, for each j,

- $h_j(x, 1) = f_j(b) g_j(x)$ (since $f_j(b) \neq 0 = g_j(b)$) and

- $h_j(x,0) = -\mathrm{In}_{\mathcal{A}_j,\nu}(g_j)(a)f_j(x)$ (this follows from claim VII.21 since $\mathrm{In}_{\mathcal{A}_j,\nu}(f_j)(a) = 0 \neq \mathrm{In}_{\mathcal{A}_j,\nu}(g_j)(a)$).

In particular, $h_j(x,0)$ and $h_j(x,1)$ are (nonzero) constant multiples of, respectively, f_j and g_j. Each h_j vanishes on the curve $C' := \{(c(t),t) : t \in T\} \subset \mathbb{k}^{n+1}$. Note that $C' \cap ((\mathbb{k}^*)^n \times \{1\})$ contains $(b,1)$ which is an isolated zero of $h_1(x,1),\dots,h_n(x,1)$. On the other hand, since ν is nontrivial, it follows that C' has a "point at infinity" at $t=0$ with respect to $(\mathbb{k}^*)^n \times T$. Therefore assertion (5) of theorem IV.32 implies that $[h_1(x,0),\dots,h_n(x,0)]^{iso}_{(\mathbb{k}^*)^n} < [h_1(x,\epsilon),\dots, h_n(x,\epsilon)]^{iso}_{(\mathbb{k}^*)^n}$ for generic $\epsilon \in T$. Since each such $h_j(x,\epsilon)$ is supported at $\mathcal{A}_j$, it follows that $[f_1,\dots,f_n]^{iso}_{(\mathbb{k}^*)^n} < [\mathcal{A}_1,\dots,\mathcal{A}_n]^{iso}_{(\mathbb{k}^*)^n}$, as required. □

4.3. The set of non-degenerate systems. In this section we prove theorem VII.15. We start with some notation. Let $J \subseteq [m] := \{1,\dots,m\}$ and $\mathcal{L}_J(\mathcal{A}) := \prod_{j\in J}\mathcal{L}(\mathcal{A}_j)$. Write $c_{j,\alpha}$ for the coefficient of x^α for each $j \in J$, so that $(c_{j,\alpha} : j \in J,\ \alpha \in \mathcal{A}_j)$ are the coordinates on $\mathcal{L}_J(\mathcal{A})$. For $f = (c_{j,\alpha})_{j,\alpha} \in \mathcal{L}_J(\mathcal{A})$, we write f_j for the corresponding element in $\mathcal{L}(\mathcal{A}_j)$, i.e. $f_j = \sum_{\alpha\in\mathcal{A}_j} c_{j,\alpha}x^\alpha$. Let $\mathcal{A}_J := (\mathcal{A}_j : j \in J)$ and $\mathcal{N}_J(\mathcal{A})$ be the set of all $f \in \mathcal{L}_J(\mathcal{A})$ which are $\mathcal{A}_J$-non-degenerate. We will at first show that $\mathcal{N}_J(\mathcal{A})$ is Zariski open in $\mathcal{L}_J(\mathcal{A})$. If $\mathcal{B} = (\mathcal{B}_j : j \in J)$ is an ordered tuple of (finite) sets, we say that $\mathcal{B}$ is a *face* of $\mathcal{A}_J$, and write $\mathcal{B} \preceq \mathcal{A}_J$, if it satisfies the following property:

"there is $\nu \in (\mathbb{R}^n)^*$ such that $\mathcal{B}_j = \mathrm{In}_\nu(\mathcal{A}_j)$ for each $j \in J$."

For each $\mathcal{B} = (\mathcal{B}_j : j \in J) \preceq \mathcal{A}_J$ and $f = (c_{j,\alpha})_{j,\alpha} \in \mathcal{L}_J(\mathcal{A})$, we define $f_{j,\mathcal{B}_j} := \sum_{\alpha\in\mathcal{B}_j} c_{j,\alpha}x^\alpha$. Let $\mathcal{D}_{J,\mathcal{B}}(\mathcal{A})$ be the set of all $f \in \mathcal{L}_J$ such that there is a common root of $f_{j,\mathcal{B}_j}$, $j \in J$, on $(\mathbb{k}^*)^n$, so that

$$\mathcal{L}_J(\mathcal{A}) \setminus \mathcal{N}_J(\mathcal{A}) = \bigcup_{\substack{\mathcal{B}\preceq\mathcal{A}_J \\ \dim(\mathcal{P}_{J,\mathcal{B}})<n}} \mathcal{D}_{J,\mathcal{B}}(\mathcal{A}) \tag{63}$$

where $\mathcal{P}_{J,\mathcal{B}} = \sum_{j\in J}\mathrm{conv}(\mathcal{B}_j)$. Let $\mathcal{L}'_J(\mathcal{A}) := \mathcal{L}_J(\mathcal{A}) \times (\mathbb{k}^*)^n$ and $\bar{\mathcal{L}}'_J(\mathcal{A}) := \mathcal{L}_J(\mathcal{A}) \times \mathbb{P}^n$. Let $\mathcal{D}'_{J,\mathcal{B}}(\mathcal{A}) \subset \mathcal{L}'_J(\mathcal{A})$ be the collection of all (f,a), where $f \in \mathcal{D}_{J,\mathcal{B}}(\mathcal{A})$ and $a \in (\mathbb{k}^*)^n$ are such that $f_{j,\mathcal{B}_j}(a) = 0$ for each $j \in J$; let $\bar{\mathcal{D}}'_{J,\mathcal{B}}(\mathcal{A})$ be the closure of $\mathcal{D}'_{J,\mathcal{B}}(\mathcal{A})$ in $\bar{\mathcal{L}}'_J(\mathcal{A})$. Let $\pi_J : \bar{\mathcal{L}}'_J(\mathcal{A}) \to \mathcal{L}_J(\mathcal{A})$ be the natural projection and $\bar{\mathcal{D}}_{J,\mathcal{B}}(\mathcal{A}) := \pi_J(\bar{\mathcal{D}}'_{J,\mathcal{B}}(\mathcal{A}))$.

Claim VII.22. *Let $\mathcal{B} \preceq \mathcal{A}_J$. Then $\bar{\mathcal{D}}_{J,\mathcal{B}}(\mathcal{A}) \subset \bigcup_{\mathcal{B}'\preceq\mathcal{B}}\mathcal{D}_{J,\mathcal{B}'}(\mathcal{A})$.*

PROOF. Let $f^0 = (c^0_{j,\alpha})_{j,\alpha} \in \bar{\mathcal{D}}_{J,\mathcal{B}}(\mathcal{A}) \setminus \mathcal{D}_{J,\mathcal{B}}(\mathcal{A})$. Pick $a^0 \in \mathbb{P}^n$ such that $(f^0,a^0) \in \bar{\mathcal{D}}'_{J,\mathcal{B}}(\mathcal{A})$. We can find an irreducible curve C on $\bar{\mathcal{D}}'_{J,\mathcal{B}}(\mathcal{A})$ such that $(f^0,a^0) \in C$ and $C \cap \mathcal{D}'_{J,\mathcal{B}}(\mathcal{A})$ is nonempty and open in C (proposition III.87). Let $B = (Z,z)$ be a branch of C at (f^0,a^0). For each $j = 1,\dots,n$, and each $\alpha \in \mathcal{A}_j$, we write $\bar{x}_j, \bar{c}_{j,\alpha}$, respectively, for the restrictions of $x_j, c_{j,\alpha}$ to C. Then for each $j \in J$, $\bar{F}_j := \sum_{\alpha\in\mathcal{B}_j}\bar{c}_{j,\alpha}\bar{x}^\alpha$ is identically zero on C. Since $C' \cap \mathcal{D}'_{J,\mathcal{B}}(\mathcal{A}) \neq \emptyset$, it follows that no $\bar{x}_j$ is identically zero on C. Let ν'_B be the element in $(\mathbb{R}^n)^*$ with coordinates $(\nu_B(\bar{x}_1),\dots,\nu_B(\bar{x}_n))$ with respect to the basis dual to the standard basis. Fix $j \in J$. Let $\mathcal{B}'_j := \operatorname{In}_{\nu'_B}(\mathcal{B}_j)$ and $m_j := \min_{\mathcal{B}_J}(\nu'_B) = \langle \nu'_B, \beta \rangle$ for any $\beta \in \mathcal{B}'_j$. Then for each $\alpha \in \mathcal{B}_j$, since $\operatorname{ord}_z(\bar{c}_{j,\alpha}) \geq 0$, it follows that

$$\operatorname{ord}_z(\bar{c}_{j,\alpha}\bar{x}^\alpha) = \operatorname{ord}_z(\bar{c}_{j,\alpha}) + \operatorname{ord}_z(\bar{x}^\alpha) \geq \operatorname{ord}_z(\bar{x}^\alpha) = \langle \nu'_B, \alpha \rangle \geq m_j.$$

Moreover, $\operatorname{ord}_z(\bar{c}_{j,\alpha}\bar{x}^\alpha) = m_j$ if and only if $\operatorname{ord}_z(\bar{c}_{j,\alpha}) = 0$ and $\alpha \in \mathcal{B}'_j$, i.e. if and only if $c^0_{j,\alpha} \neq 0$ and $\alpha \in \mathcal{B}'_j$, i.e. if and only if $\alpha \in \operatorname{Supp}(f^0_{j,\mathcal{B}'_j})$. If ρ_B is a parameter at B, it follows that $\bar{F}_j$ can be expanded in $\Bbbk((\rho_B))$ as

$$\sum_{\alpha \in \mathcal{B}'_j} c^0_{j,\alpha} \prod_{i=1}^{n} (a^0_i)^{\alpha_i} (\rho_B)^{m_j} + \cdots$$

where $a^0_i = \operatorname{In}_B(\bar{x}_i)$, $i = 1, \ldots, n$, and the omitted terms have higher order in ρ_B. Since $\bar{F}_j$ is identically zero on C, it follows that $(a^0_1, \ldots, a^0_n)$ is a zero of $\sum_{\alpha \in \mathcal{B}'_j} c^0_{j,\alpha} x^\alpha = f^0_{j,\mathcal{B}'_j}$ for each $j \in J$. Therefore $f^0 = (f^0_j : j \in J)$ is $\mathcal{A}_J$-degenerate and is an element of $\mathcal{D}_{J,\mathcal{B}'}(\mathcal{A})$, as required. □

Corollary VII.23. *For each $\mathcal{B} \preceq \mathcal{A}_J$, the set $\bigcup_{\mathcal{B}' \preceq \mathcal{B}} \mathcal{D}_{J,\mathcal{B}'}(\mathcal{A})$ is Zariski closed in $\mathcal{L}_J(\mathcal{A})$. Consequently $\mathcal{N}_J(\mathcal{A})$ is Zariski open in $\mathcal{L}_J(\mathcal{A})$.*

PROOF. Since $\mathbb{P}^n$ is *complete* (see section III.9), it follows that $\bar{\mathcal{D}}_{J,\mathcal{B}}(\mathcal{A}) := \pi_J(\bar{\mathcal{D}}'_{J,\mathcal{B}}(\mathcal{A}))$ is Zariski closed in $\mathcal{L}_J(\mathcal{A})$. Since the relation $\preceq$ is transitive, claim VII.22 therefore implies the first assertion. The second assertion then follows from eq. (63) (and the transitivity of $\preceq$). □

Let $\mathcal{P}_j := \operatorname{conv}(\mathcal{A}_j)$, $j = 1, \ldots, n$, and $\mathcal{P}_J := \sum_{j \in J} \mathcal{P}_j$. For each $\mathcal{B} = (\mathcal{B}_j : j \in J) \preceq \mathcal{A}_J$, define $\mathcal{P}_{J,\mathcal{B}} := \sum_{j \in J} \operatorname{conv}(\mathcal{B}_j)$. Note that $\mathcal{P}_{J,\mathcal{B}}$ is a face of $\mathcal{P}_J$.

Claim VII.24. *Let $\mathcal{B} \preceq \mathcal{A}_J$ be such that $\dim(\mathcal{P}_{J,\mathcal{B}}) < |J|$. Then $\bar{\mathcal{D}}_{J,\mathcal{B}}(\mathcal{A}) \subsetneq \mathcal{L}_J(\mathcal{A})$.*

PROOF. We proceed by induction on $|J|$. If $|J| = 1$, then the assumption $\dim(\mathcal{P}_{J,\mathcal{B}}) < |J|$ is valid only if $\mathcal{P}_{J,\mathcal{B}}$ is a vertex of $\mathcal{P}_J$, and therefore $\mathcal{B}_J$ is a vertex of $\mathcal{A}_j$ for each j. In that case $f_{j,\mathcal{B}_j}$ is a monomial for each $f \in \mathcal{L}_J(\mathcal{A})$, so that $\bar{\mathcal{D}}_{J,\mathcal{B}}(\mathcal{A}) = \emptyset$. Now assume $|J| \geq 2$. Let $d := \dim(\mathcal{P}_{J,\mathcal{B}})$. Since $d < |J|$, there is $J' \subset J$ such that $|J'| = |J| - 1$ and $\dim(\mathcal{P}_{J',\mathcal{B}_{J'}}) = d$, where $\mathcal{B}_{J'} := (\mathcal{B}_j : j \in J')$. Define

$$\mathcal{N}_{J',\mathcal{B}}(\mathcal{A}) := \mathcal{L}_{J'}(\mathcal{A}) \setminus \bigcup_{\substack{\mathcal{B}' \preceq \mathcal{B}_{J'} \\ \dim(\mathcal{P}_{J',\mathcal{B}'}) < d}} \bar{\mathcal{D}}_{J',\mathcal{B}'}(\mathcal{A}).$$

For each $\bar{\mathcal{D}}_{J',\mathcal{B}'}(\mathcal{A})$ appearing in the above union, we have $\dim(\mathcal{P}_{J',\mathcal{B}'}) < d \leq |J'|$, so that the inductive hypothesis implies that $\bar{\mathcal{D}}_{J',\mathcal{B}'}(\mathcal{A})$ is a proper Zariski closed subset of $\mathcal{L}_{J'}(\mathcal{A}')$. Since $\mathcal{L}_{J'}(\mathcal{A})$ is irreducible[2], it follows that $\mathcal{N}_{J',\mathcal{B}}(\mathcal{A})$ is a nonempty Zariski open subset of $\mathcal{L}_{J'}(\mathcal{A})$.

---

[2]It is straightforward to check that $\mathcal{L}_{J'}(\mathcal{A}) \cong \Bbbk^p$ for some $p \geq 0$.

After an integral change of coordinates of $\mathbb{Z}^n$ (i.e. a monomial change of coordinates of $(\Bbbk^*)^n$) we may arrange that the affine hull of $\mathcal{P}_{J,\mathcal{B}}$ is of the form $H+\alpha$ where $\alpha \in \mathbb{Z}^n$ and H is the coordinate hyperplane of $\mathbb{R}^n$ spanned by the first d unit vectors. Pick $(f_j : j \in J') \in \mathcal{N}_{J',\mathcal{B}}(\mathcal{A})$. Let $\mathcal{B}' \preceq \mathcal{B}$. For each $j \in J'$, $f_{j,\mathcal{B}'_j}$ is of the form $x^{\alpha_j} g_{j,\mathcal{B}'_j}(x_1,\dots,x_d)$ for some $\alpha_j \in \mathbb{Z}^n$ and Laurent polynomial $g_{j,\mathcal{B}'_j}$ in $(x_1,\dots,x_d)$. Since for each $j \in J'$, the support of $f_{j,\mathcal{B}'_j}$ is a translation of the support of $g_{j,\mathcal{B}'_j}$, it follows from the definition of $\mathcal{N}_{J',\mathcal{B}}(\mathcal{A})$ that $(g_{j,\mathcal{B}'_j} : j \in J')$ is a BKK non-degenerate system of Laurent polynomials in $(x_1,\dots,x_d)$. Corollary VII.18 implies that $W_{\mathcal{B}'} := V(g_{j,\mathcal{B}'_j} : j \in J')$ is a finite set of points in $(\Bbbk^*)^d$. If j_0 is the unique element of $J \setminus J'$, then a generic $f_{j_0} \in \mathcal{L}(\mathcal{A}_{j_0})$ will satisfy the following: for each $\mathcal{B}' \preceq \mathcal{B}$, $f_{j_0,\mathcal{B}'_{j_0}}$ takes a nonzero value at each point of $W_{\mathcal{B}'}$. It follows that $f := (f_j : j \in J) \notin \bigcup_{\mathcal{B}' \preceq \mathcal{B}} \mathcal{D}_{J,\mathcal{B}'}(\mathcal{A})$. Claim VII.22 now implies that $f \notin \bar{\mathcal{D}}_{J,\mathcal{B}}(\mathcal{A})$, as required. □

PROOF OF THEOREM VII.15. Apply corollary VII.23 with $J = [m] := \{1,\dots, m\}$ to see that $\mathcal{N}(\mathcal{A})$ is Zariski open in $\mathcal{L}(\mathcal{A})$. If $\dim(\sum_{j=1}^m \mathcal{P}_j) < m$, then an application of claim VII.24 with $J = [m]$ shows that $\bar{\mathcal{D}}_{J,\mathcal{B}}(\mathcal{A})$ is a proper Zariski closed subset of $\mathcal{L}(\mathcal{A})$ for each face $\mathcal{B}$ of $\mathcal{A} = (\mathcal{A}_1,\dots,\mathcal{A}_m)$. Since $\mathcal{L}(\mathcal{A})$ is irreducible, it follows that $\mathcal{L}(\mathcal{A}) \setminus \bigcup_{\mathcal{B} \preceq \mathcal{A}} \bar{\mathcal{D}}_{J,\mathcal{B}}(\mathcal{A})$ is nonempty. Identity (63) then implies that $\mathcal{N}(\mathcal{A})$ is nonempty. If, on the other hand, $m \geq n$, an application of claim VII.24 with $J = [m]$ shows that $\bar{\mathcal{D}}_{J,\mathcal{B}}(\mathcal{A})$ is a proper Zariski closed subset of $\mathcal{L}(\mathcal{A})$ for each face $\mathcal{B}$ of $\mathcal{A}$ such that $\dim(\mathcal{P}_{[m],\mathcal{B}}) \leq n-1$, and it follows similarly that $\mathcal{N}(\mathcal{A})$ is nonempty. □

4.4. Properly non-degenerate systems. We now introduce a notion of non-degeneracy to be used in section VII.5. Let $\mathcal{A} := (\mathcal{A}_1,\dots,\mathcal{A}_m)$ and $(f_1,\dots,f_m) \in \mathcal{L}(\mathcal{A})$. We say that $f_1,\dots,f_m$ are *properly $\mathcal{A}$-non-degenerate* if for every $J \subseteq [m]$ and *every*[3] weighted order ν on $\Bbbk[x_1,x_1^{-1},\dots,x_n,x_n^{-1}]$ such that $\dim(\mathrm{In}_\nu(\sum_{j\in J}\mathrm{conv}(\mathcal{A}_j))) < |J|$, the Laurent polynomials $\mathrm{In}_{\mathcal{A}_j,\nu}(f_j)$, $j \in J$, have no common zero in $(\Bbbk^*)^n$. If $m \geq \min\{n, \dim(\sum_j \mathrm{conv}(\mathcal{A}_j)) + 1\}$, then a properly $\mathcal{A}$-non-degenerate system is also $\mathcal{A}$-non-degenerate; otherwise, there may be properly $\mathcal{A}$-non-degenerate systems which are not $\mathcal{A}$-non-degenerate.

Proposition VII.25. *The collection $\tilde{\mathcal{N}}(\mathcal{A})$ of properly $\mathcal{A}$-non-degenerate systems is a nonempty Zariski open subset of $\mathcal{L}(\mathcal{A})$. If $m \geq \min\{n, \dim(\sum_{j=1}^m \mathrm{conv}(\mathcal{A}_j)) + 1\}$, then $\tilde{\mathcal{N}}(\mathcal{A}) \subset \mathcal{N}(\mathcal{A})$.*

PROOF. Indeed, for each $J \subseteq [m]$, let $\tilde{\pi}_J$ be the natural projection from $\mathcal{L}(\mathcal{A})$ onto $\mathcal{L}_J(\mathcal{A})$. It is straightforward to check that in the notation of section VII.4.3,

$$\tilde{\mathcal{N}}(\mathcal{A}) = \mathcal{L}(\mathcal{A}) \setminus \bigcup_{\substack{J\subseteq[m]\\ \dim(\mathcal{P}_J)<|J|}} \bigcup_{\mathcal{B}\preceq\mathcal{A}_J} \tilde{\pi}_J^{-1}(\mathcal{D}_{J,\mathcal{B}}(\mathcal{A})).$$

[3]Unlike $\mathcal{A}$-non-degeneracy, ν is allowed to be the trivial weighted order $(0,\dots,0)$.

Claim VII.22 implies that $\bigcup_{\mathcal{B}\preceq\mathcal{A}_J}\tilde{\pi}_J^{-1}(\mathcal{D}_{J,\mathcal{B}}(\mathcal{A})) = \bigcup_{\mathcal{B}\preceq\mathcal{A}_J}\tilde{\pi}_J^{-1}(\bar{\mathcal{D}}_{J,\mathcal{B}}(\mathcal{A}))$, so that $\tilde{\mathcal{N}}(\mathcal{A})$ is Zariski open in $\mathcal{L}(\mathcal{A})$. Claim VII.24 implies that $\tilde{\mathcal{N}}(\mathcal{A})$ is nonempty. □

5. Proof of the BKK bound

For each $j = 1, \dots, n$, let $\mathcal{A}_j$ be a finite subset of $\mathbb{Z}^n$ and $\mathcal{P}_j$ be the convex hull of $\mathcal{A}_j$ in $\mathbb{R}^n$. In this section we show that

$$[\mathcal{A}_1, \dots, \mathcal{A}_n]^{iso}_{(\mathbb{k}^*)^n} = \mathrm{MV}(\mathcal{P}_1, \dots, \mathcal{P}_n). \tag{64}$$

In section VII.5.1 we prove two relevant results from the theory of toric varieties, which we use in section VII.5.2 to prove (64). Throughout this section we follow the convention of section VI.8.3 to identify weighted orders on the ring of Laurent polynomials in $(x_1, \dots, x_n)$ with integral elements of $(\mathbb{R}^n)^*$.

5.1. Toric propositions. Let ν be a primitive integral element in $(\mathbb{R}^n)^*$. For each Laurent polynomial g in $(x_1, \dots, x_n)$, define $T_{\alpha_\nu}(g)$ and $\mathrm{In}'_{\alpha_\nu,\psi_\nu}(g)$ as in corollary VI.22, where $\alpha_\nu \in \mathbb{Z}^n$ is such that $\langle \nu, \alpha_\nu \rangle = 1$ and $\psi_\nu : \mathbb{Z}^n_{\nu^\perp} \cong \mathbb{Z}^{n-1}$ is an isomorphism (of abelian groups). Let $\mathcal{P}$ be an n-dimensional convex integral polytope in $\mathbb{R}^n$ which has a facet $\mathcal{Q}$ with primitive inner normal ν. Let $X_{\mathcal{P}}$ be the toric variety corresponding to $\mathcal{P}$ from section VI.5, and given Laurent polynomials $g_1, \dots, g_k$, let $V^1_{\mathcal{P}}(g_1, \dots, g_k)$ be the extension from section VI.7.1 of the closed subscheme $V(g_1, \dots, g_k)$ of $(\mathbb{k}^*)^n$ to a closed subscheme of the Zariski open subset $X^1_{\mathcal{P}}$ of $X_{\mathcal{P}}$. Recall that a *possibly non-reduced curve* is a pure dimension one closed subscheme of a variety.

Proposition VII.26. *Let $f_1, \dots, f_n$ be Laurent polynomials in $(x_1, \dots, x_n)$.*

(1) *If $\mathrm{In}'_{\alpha_\nu,\psi_\nu}(f_2), \dots, \mathrm{In}'_{\alpha_\nu,\psi_\nu}(f_n)$ have no common zero on $(\mathbb{k}^*)^{n-1}$, then there is a Zariski open subset U' of $X_{\mathcal{P}}$ containing $O_{\mathcal{Q}}$ such that the support of $V^1_{\mathcal{P}}(f_2, \dots, f_n) \cap U'$ is empty.*

(2) *If the number of common zeroes of $\mathrm{In}'_{\alpha_\nu,\psi_\nu}(f_2), \dots, \mathrm{In}'_{\alpha_\nu,\psi_\nu}(f_n)$ on $(\mathbb{k}^*)^{n-1}$ is nonzero and finite, then there is a Zariski open subset U' of $X_{\mathcal{P}}$ containing $O_{\mathcal{Q}}$ such that $C' := V^1_{\mathcal{P}}(f_2, \dots, f_n) \cap U'$ is a possibly non-reduced curve and every irreducible component of C' intersects $(\mathbb{k}^*)^n$.*

(3) *If in addition to the assumptions of assertion* (2) $\mathrm{In}'_{\alpha_\nu,\psi_\nu}(f_1)$ *does* not *vanish at any of the common zeroes of $\mathrm{In}'_{\alpha_\nu,\psi_\nu}(f_2), \dots, \mathrm{In}'_{\alpha_\nu,\psi_\nu}(f_n)$ on $(\mathbb{k}^*)^{n-1}$, then f_1 restricts to a nonzero rational function on C' which can be represented in $\mathcal{O}_{a,C'}$ as a quotient of non zero-divisors for every $a \in C'$, and*

$$\sum_{a \in C' \cap O_{\mathcal{Q}}} \mathrm{ord}_a(f_1|_{C'}) = \nu(f_1)[\mathrm{In}'_{\alpha_\nu,\psi_\nu}(f_2), \dots, \mathrm{In}'_{\alpha_\nu,\psi_\nu}(f_n)]_{(\mathbb{k}^*)^{n-1}}.$$

PROOF. If $\mathrm{In}'_{\alpha_\nu,\psi_\nu}(f_2), \dots, \mathrm{In}'_{\alpha_\nu,\psi_\nu}(f_n)$ have no common zero on $(\mathbb{k}^*)^{n-1}$, then proposition VI.26 implies that $V^1_{\mathcal{P}}(f_2, \dots, f_n) \cap O_{\mathcal{Q}} = \emptyset$, so that part (1) holds with $U' := X^1_{\mathcal{P}} \setminus V^1_{\mathcal{P}}(f_2, \dots, f_n)$. Now assume the number of common zeroes of $\mathrm{In}'_{\alpha_\nu,\psi_\nu}(f_2),\dots,\mathrm{In}'_{\alpha_\nu,\psi_\nu}(f_n)$ on $(\mathbb{k}^*)^{n-1}$ is nonzero and finite. Since by definition $T_{\alpha_\nu}(f_j) = x^{-\nu(f_j)\alpha_\nu} f_j$, it follows (e.g. due to proposition VI.26) that $T_{\alpha_\nu}(f_2),\dots,T_{\alpha_\nu}(f_n), x^{\alpha_\nu}$ are n regular functions on the n-dimensional variety $U_Q := X^0_{\mathcal{P}} \cup O_Q$ such that their common zero set Z is finite and nonempty. Theorem III.80 then implies that $V^1_{\mathcal{P}}(f_2,\dots,f_n)$ has pure dimension one near each point of Z. Since Z is precisely the set of points in $V^1_{\mathcal{P}}(f_2,\dots,f_n) \cap O_Q$, this implies assertion (2) is satisfied with some $U' \subset U_Q$. Since U_Q is nonsingular, and the set of zeroes of x^{α_ν} on U_Q is precisely O_Q (proposition VI.20), and since O_Q does not contain any irreducible component of C', lemma IV.29 implies that $x^{\alpha_\nu}|_{C'}$ is a non zero-divisor in $\mathcal{O}_{a,C'}$ for each $a \in C'$. Now fix $a \in C' \cap O_Q$. Since a corresponds to a common zero of $\mathrm{In}'_{\alpha_\nu,\psi_\nu}(f_2),\dots,\mathrm{In}'_{\alpha_\nu,\psi_\nu}(f_n)$ on $(\mathbb{k}^*)^{n-1}$ (proposition VI.26), under the assumption of assertion (3), $T_{\alpha_\nu}(f_1)$ is a regular function near a which does *not* vanish at a. Therefore $T_{\alpha_\nu}(f_1)|_{C'}$ is a unit in $\mathcal{O}_{a,C'}$, and $f_1 = T_{\alpha_\nu}(f_1)(x^{\alpha_\nu})^{\nu(f_1)}$ is the quotient of two non zero-divisors in $\mathcal{O}_{a,C'}$. Proposition IV.21 then implies that $\mathrm{ord}_a(f_1|_{C'}) = \mathrm{ord}_a(T_{\alpha_\nu}(f_1)|_{C'}) + \nu(f_1)\,\mathrm{ord}_a(x^{\alpha_\nu}|_{C'}) = \nu(f_1)\,\mathrm{ord}_a(x^{\alpha_\nu}|_{C'})$. On the other hand, assertion (4) of proposition IV.28 implies that

$$\mathrm{ord}_a(x^{\alpha_\nu}|_{C'}) = [x^{\alpha_\nu}, T_{\alpha_\nu}(f_2),\dots,T_{\alpha_\nu}(f_n)]_a = \dim_{\mathbb{k}}(\mathcal{O}_{a,X_{\mathcal{P}}}/\langle x^{\alpha_\nu}, T_{\alpha_\nu}(f_2),\dots,T_{\alpha_\nu}(f_n)\rangle).$$

Proposition VI.26 then implies that

$$\begin{aligned}\mathrm{ord}_a(x^{\alpha_\nu}|_{C'}) &= \dim_{\mathbb{k}}(\mathcal{O}_{(\psi_\nu)_*(a),(\mathbb{k}^*)^{n-1}}/\langle \mathrm{In}'_{\alpha_\nu,\psi_\nu}(f_2),\dots,\mathrm{In}'_{\alpha_\nu,\psi_\nu}(f_n)\rangle)\\ &= [\mathrm{In}'_{\alpha_\nu,\psi_\nu}(f_2),\dots,\mathrm{In}'_{\alpha_\nu,\psi_\nu}(f_n)]_{(\psi_\nu)_*(a)},\end{aligned}$$

where $(\psi_\nu)_* : O_Q \cong (\mathbb{k}^*)^{n-1}$ is the inverse of the isomorphism $\psi_\nu^* : (\mathbb{k}^*)^{n-1} \cong O_Q$ from corollaryVI.22. assertion (3) now follows immediately. ☐

Corollary VII.27. *Let $f_1,\dots,f_n$ be Laurent polynomials in $(x_1,\dots,x_n)$ and $\mathcal{R}$ be the sum of Newton polytopes of $f_2,\dots,f_n$.*

1. *(1) Assume that $\mathrm{In}'_{\alpha_\nu,\psi_\nu}(f_2),\dots,\mathrm{In}'_{\alpha_\nu,\psi_\nu}(f_n)$ have no common zero on $(\mathbb{k}^*)^{n-1}$ for every primitive integral $\nu \in (\mathbb{R}^n)^*$ such that* $\dim(\mathrm{In}_\nu(\mathcal{R})) < n-1$. *Then $V(f_2,\dots,f_n) \cap (\mathbb{k}^*)^n$ is either empty or a curve.*
2. *(2) Assume in addition that $\mathrm{In}'_{\alpha_\nu,\psi_\nu}(f_1),\dots,\mathrm{In}'_{\alpha_\nu,\psi_\nu}(f_n)$ have no common zero on $(\mathbb{k}^*)^{n-1}$ for every primitive integral $\nu \in (\mathbb{R}^n)^*$ such that* dim $(\mathrm{In}_\nu(\mathcal{R})) = n-1$. *Then*

$$[f_1,\dots,f_n]_{(\mathbb{k}^*)^n} = [f_1,\dots,f_n]^{iso}_{(\mathbb{k}^*)^n} = -\sum_\nu \nu(f_1)[\mathrm{In}'_{\alpha_\nu,\psi_\nu}(f_2),\dots,\mathrm{In}'_{\alpha_\nu,\psi_\nu}(f_n)]_{(\mathbb{k}^*)^{n-1}}, \tag{65}$$

where the sum is over all primitive integral $\nu \in (\mathbb{R}^n)^*$ such that dim $(\mathrm{In}_\nu(\mathcal{R})) = n-1$.

PROOF. Let C' be the set of common zeroes of $f_2,\dots,f_n$ on $(\mathbb{k}^*)^n$. If $\dim(\mathcal{R}) < n-1$, then the assumption of assertion (1) implies that $C' = \emptyset$ and all three sides of (65) are zero. Therefore assume that $\dim \mathcal{R}$ is $n-1$ or n. Let $\mathcal{P}$ be an n-dimensional convex integral polytope in $\mathbb{R}^n$ which satisfy the following property:

(66) for every $\nu \in (\mathbb{R}^n)^*$, if ν is an inner normal to a face of $\mathcal{R}$ of dimension $n-1$, then ν is also an inner normal to a facet of $\mathcal{P}$.

For example, if $\dim(\mathcal{R}) = n$, then we can simply take $\mathcal{P} = \mathcal{R}$, and if $\dim(\mathcal{R}) = n-1$, we can take $\mathcal{P}$ to be any (convex integral) polytope which has two facets parallel to $\mathcal{R}$. Let $X_{\mathcal{P}}$ be the toric variety corresponding to $\mathcal{P}$. Recall that we can identify $(\mathbb{k}^*)^n$ with the torus $X^0_{\mathcal{P}}$ of $X_{\mathcal{P}}$ (assertion (2) of proposition VI.15). Let $\bar{C}'$ be the closure of C' in $X_{\mathcal{P}}$. Fix a primitive integral element $\nu \in (\mathbb{R}^n)^*$. Let $\mathcal{Q} := \operatorname{In}_\nu(\mathcal{P})$ and $O_{\mathcal{Q}}$ be the corresponding torus orbit on $X_{\mathcal{P}}$.

Claim VII.27.1. *Let* $\mathcal{Q}' := \operatorname{In}_\nu(\mathcal{R})$.

(1) If $\dim(\mathcal{Q}') < n-1$, then $V(\operatorname{In}'_{\alpha_\nu,\psi_\nu}(f_2), \ldots, \operatorname{In}'_{\alpha_\nu,\psi_\nu}(f_n)) \cap (\mathbb{k}^*)^{n-1} = \emptyset$, and $\bar{C}' \cap O_{\mathcal{Q}} = \emptyset$.
(2) If $\dim(\mathcal{Q}') = n-1$, then $\mathcal{Q}$ is a facet of $\mathcal{P}$, and $V(\operatorname{In}'_{\alpha_\nu,\psi_\nu}(f_2), \ldots, \operatorname{In}'_{\alpha_\nu,\psi_\nu}(f_n)) \cap (\mathbb{k}^*)^{n-1}$ is finite.

PROOF. If $\dim(\mathcal{Q}') < n-1$, then the hypothesis of the corollary implies that the set of common zeroes of $\operatorname{In}'_{\alpha_\nu,\psi_\nu}(f_2), \ldots, \operatorname{In}'_{\alpha_\nu,\psi_\nu}(f_n)$ on $(\mathbb{k}^*)^n$ is empty. Assertion (1) of the claim then follows from corollary VI.34. On the other hand, if $\dim(\mathcal{Q}') = n-1$, then property (66) implies that $\mathcal{Q}$ is a facet of $\mathcal{P}$, and the hypothesis of the corollary implies that $\operatorname{In}'_{\alpha_\nu,\psi_\nu}(f_2), \ldots, \operatorname{In}'_{\alpha_\nu,\psi_\nu}(f_n)$ are BKK non-degenerate on $(\mathbb{k}^*)^{n-1}$. Corollary VII.18 then implies that $V(\operatorname{In}'_{\alpha_\nu,\psi_\nu}(f_2), \ldots, \operatorname{In}'_{\alpha_\nu,\psi_\nu}(f_n)) \cap (\mathbb{k}^*)^{n-1}$ is finite, as required. □

Let $C := V^1_{\mathcal{P}}(f_2, \ldots, f_n)$. Recall that C is an extension of the closed subscheme of $\mathbb{k}^n$ determined by $f_2, \ldots, f_n$ to the open subset of $X_{\mathcal{P}}$ obtained by removing $O_{\mathcal{Q}}$ for all face $\mathcal{Q}$ of $\mathcal{P}$ with dimension $\leq n-2$. Assertion (1) of claim VII.27.1 implies that $\bar{C}' \subset \operatorname{Supp}(C)$. If $\operatorname{In}'_{\alpha_\nu,\psi_\nu}(f_2), \ldots, \operatorname{In}'_{\alpha_\nu,\psi_\nu}(f_n)$ have no common zero on $(\mathbb{k}^*)^{n-1}$ for each primitive integral $\nu \in (\mathbb{R}^n)^*$, then assertion (1) of proposition VII.26 implies that $\bar{C}' \cap X_{\mathcal{P},\infty} = \emptyset$, where $X_{\mathcal{P},\infty} := X_{\mathcal{P}} \setminus (\mathbb{k}^*)^n$. On the other hand $\bar{C}'$ is a *complete* variety if it is nonempty, so that it *cannot* be completely contained in the affine variety $(\mathbb{k}^*)^n$ (example III.59). It follows that $C' = \emptyset$ and all sides of (65) are zero. So assume there is primitive integral $\nu \in (\mathbb{R}^n)^*$ such that $V(\operatorname{In}'_{\alpha_\nu,\psi_\nu}(f_2), \ldots, \operatorname{In}'_{\alpha_\nu,\psi_\nu}(f_n)) \cap (\mathbb{k}^*)^{n-1} \neq \emptyset$. Then assertion (2) of claim VII.27.1 and proposition VII.26 imply that C has dimension one near $X_{\mathcal{P},\infty}$, and $\operatorname{Supp}(C) \subset \bar{C}'$. It follows that $\operatorname{Supp}(C) = \bar{C}'$, so that C is a *possibly non-reduced curve* such that $\operatorname{Supp}(C)$ is *projective*. Since $C \cap (\mathbb{k}^*)^n$ is the closed subscheme of $(\mathbb{k}^*)^n$ defined by $f_2, \ldots, f_n$, proposition IV.28 implies that

$$[f_1, \ldots, f_n]_{(\mathbb{k}^*)^n} = \sum_{a \in C \cap (\mathbb{k}^*)^n} \operatorname{ord}_a(f_1|_C).$$

Since $\mathrm{Supp}(C)$ is projective, corollary IV.25, proposition VII.26 and claim VII.27.1 then imply that

$$\begin{aligned}[f_1, \dots, f_n]_{(\Bbbk^*)^n} &= -\sum_{a \in C \cap X_{\mathcal{P},\infty}} \mathrm{ord}_a(f_1|_C) \\ &= -\sum_{\nu} \nu(f_1)[\mathrm{In}'_{\alpha_\nu,\psi_\nu}(f_2), \dots, \mathrm{In}'_{\alpha_\nu,\psi_\nu}(f_n)]_{(\Bbbk^*)^{n-1}}\end{aligned}$$

as required. □

5.2. Proof of identity (64). Define

(67) $$[\mathcal{P}_1, \dots, \mathcal{P}_n]^{iso}_{(\Bbbk^*)^n} := \max\{[f_1, \dots, f_n]^{iso}_{(\Bbbk^*)^n} : \mathrm{Supp}(f_j) \subseteq \mathcal{P}_j,\ j = 1, \dots, n\}.$$

Let $\mathcal{A}'_j := \mathcal{P}_j \cap \mathbb{Z}^n \supseteq \mathcal{A}_j$, $j = 1, \dots, n$. It follows from the definition that $[\mathcal{P}_1, \dots, \mathcal{P}_n]^{iso}_{(\Bbbk^*)^n} = [\mathcal{A}'_1, \dots, \mathcal{A}'_n]^{iso}_{(\Bbbk^*)^n} \geq [\mathcal{A}_1, \dots, \mathcal{A}_n]^{iso}_{(\Bbbk^*)^n}$. However, due to theorem VII.15 we may pick BKK non-degenerate Laurent polynomials $f_1, \dots, f_n$ such that $\mathrm{Supp}(f_j) \subseteq \mathcal{A}_j \subseteq \mathcal{A}'_j$ and $\mathrm{NP}(f_j) = \mathcal{P}_j$ for each j, and then proposition VII.19 implies that

$$[\mathcal{A}_1, \dots, \mathcal{A}_n]^{iso}_{(\Bbbk^*)^n} = [f_1, \dots, f_n]^{iso}_{(\Bbbk^*)^n} = [\mathcal{A}'_1, \dots, \mathcal{A}'_n]^{iso}_{(\Bbbk^*)^n} = [\mathcal{P}_1, \dots, \mathcal{P}_n]^{iso}_{(\Bbbk^*)^n}$$

In order to prove (64) it therefore suffices to show that

(68) $$[\mathcal{P}_1, \dots, \mathcal{P}_n]^{iso}_{(\Bbbk^*)^n} = \mathrm{MV}(\mathcal{P}_1, \dots, \mathcal{P}_n).$$

Claim VII.28. *Let $\mathcal{K}$ be the set of convex integral polytopes in $\mathbb{R}^n$ regarded as a semigroup under Minkowski addition. The function $\mathcal{K}^n \to \mathbb{R}$ given by $(\mathcal{Q}_1, \dots, \mathcal{Q}_n) \mapsto [\mathcal{Q}_1, \dots, \mathcal{Q}_n]^{iso}_{(\Bbbk^*)^n}$ is symmetric and multiadditive.*

PROOF. The symmetry is evident, so we prove the multiadditivity. Pick $\mathcal{Q}_1, \dots, \mathcal{Q}_n, \mathcal{Q}'_1 \in \mathcal{K}$. Theorem VII.15 implies that we may choose Laurent polynomials $g_1, \dots, g_n, g'_1$ such that $\mathrm{NP}(g_j) = \mathcal{Q}_j$, $j = 1, \dots, n$, $\mathrm{NP}(g'_1) = \mathcal{Q}'_1$, and both $g_1, \dots, g_n$ and $g'_1, g_2, \dots, g_n$ are BKK non-degenerate. But then $g_1 g'_1, g_2, \dots, g_n$ are also BKK non-degenerate, in particular, $g_1 g'_1, g_2, \dots, g_n$ only have isolated zeroes on $(\Bbbk^*)^n$ (corollary VII.18). It follows that

$$\begin{aligned}&[\mathcal{Q}_1 + \mathcal{Q}'_1, \mathcal{Q}_2, \dots, \mathcal{Q}_n]^{iso}_{(\Bbbk^*)^n} \\ &\quad = [g_1 g'_1, g_2, \dots, g_n]^{iso}_{(\Bbbk^*)^n} \text{ (proposition VII.19)} \\ &\quad = [g_1, \dots, g_n]^{iso}_{(\Bbbk^*)^n} + [g'_1, g_2, \dots, g_n]^{iso}_{(\Bbbk^*)^n} \text{ (proposition IV.28, assertion (5))} \\ &\quad = [\mathcal{Q}_1, \dots, \mathcal{Q}_n]^{iso}_{(\Bbbk^*)^n} + [\mathcal{Q}'_1, \mathcal{Q}_2, \dots, \mathcal{Q}_n]^{iso}_{(\Bbbk^*)^n} \text{ (proposition VII.19)}\end{aligned}$$

as required. □

Due to theorem VII.2, and claim VII.28, in order to prove (68) it suffices to show that $[\mathcal{P}, \dots, \mathcal{P}]^{iso}_{(\Bbbk^*)^n} = n!\,\mathrm{Vol}_n(\mathcal{P})$ for each convex integral polytope $\mathcal{P}$ in

$\mathbb{R}^n$. We proceed by induction on n. It is clearly true for $n = 1$, so assume it is true for $n-1$. Pick a convex integral polytope $\mathcal{P}$ in $\mathbb{R}^n$. Let $f_1, \ldots, f_n$ be *properly* $\mathcal{A}$-non-degenerate Laurent polynomials (see section VII.4.4) such that the Newton polytope of each f_j is $\mathcal{P}$. In particular $f_1, \ldots, f_n$ are BKK non-degenerate (proposition VII.25). Therefore corollary VII.18, proposition VII.19 imply that

$$[\mathcal{P}, \ldots, \mathcal{P}]^{iso}_{(\mathbb{k}^*)^n} = [f_1, \ldots, f_n]_{(\mathbb{k}^*)^n}.$$

Since the f_j are properly non-degenerate, they satisfy the hypothesis of corollary VII.27, and identity (65) implies that

$$[f_1, \ldots, f_n]_{(\mathbb{k}^*)^n} = -\sum_{\nu} \nu(f_1)[\mathrm{In}'_{\alpha_\nu,\psi_\nu}(f_2), \ldots, \mathrm{In}'_{\alpha_\nu,\psi_\nu}(f_n)]_{(\mathbb{k}^*)^{n-1}},$$

where the sum is over all primitive integral $\nu \in (\mathbb{R}^n)^*$. Fix one such ν. The proper $\mathcal{A}$-non-degeneracy of the f_j implies that $\mathrm{In}'_{\alpha_\nu,\psi_\nu}(f_2), \ldots, \mathrm{In}'_{\alpha_\nu,\psi_\nu}(f_n)$ are BKK non-degenerate on $(\mathbb{k}^*)^{n-1}$. Moreover, the Newton polytope of each $\mathrm{In}'_{\alpha_\nu,\psi_\nu}(f_j)$ is $\mathcal{P}_\nu$, which is the convex hull in $\mathbb{R}^{n-1}$ of $\{\psi_\nu(\beta - m_\nu\alpha_\nu) : \beta \in \mathrm{In}_\nu(\mathcal{P}) \cap \mathbb{Z}^n\}$, where $m_\nu := \min_{\mathcal{P}}(\nu)$. The inductive hypothesis implies that

$$[\mathrm{In}'_{\alpha_\nu,\psi_\nu}(f_2), \ldots, \mathrm{In}'_{\alpha_\nu,\psi_\nu}(f_n)]_{(\mathbb{k}^*)^{n-1}} = (n-1)!\,\mathrm{Vol}_{n-1}(\mathcal{P}_\nu).$$

It follows from the definition of $\mathrm{Vol}'_\nu(\cdot)$ (see definition V.42) that $\mathrm{Vol}_{n-1}(\mathcal{P}_\nu) = \mathrm{Vol}'_\nu(\mathrm{In}_\nu(\mathcal{P}))$. It follows that

$$\begin{aligned}
[f_1, \ldots, f_n]_{(\mathbb{k}^*)^n} &= -(n-1)! \sum_{\nu} \min_{\mathcal{P}}(\nu)\,\mathrm{Vol}'_\nu(\mathrm{In}_\nu(\mathcal{P})) \\
&= (n-1)! \sum_{\nu} \max_{\mathcal{P}}(\nu)\,\mathrm{Vol}'_\nu(\mathrm{ld}_\nu(\mathcal{P})) = n!\,\mathrm{Vol}_n(\mathcal{P}),
\end{aligned}$$

where the last equality uses corollary V.43. Therefore, $[\mathcal{P}, \ldots, \mathcal{P}]^{iso}_{(\mathbb{k}^*)^n} = n!\,\mathrm{Vol}_n(\mathcal{P})$, as required. □

6. Applications of Bernstein's theorem to convex geometry

In this section we use Bernstein's theorem to deduce some properties of mixed volume. In particular we characterize the conditions for mixed volume (of n convex polytopes in $\mathbb{R}^n$) being zero (theorem VII.33) and the conditions under which it is "strictly monotonic" (corollary VII.34). As an application back to algebraic geometry, we prove the alternate version of Bernstein's theorem (corollary VII.36). Throughout this section $\mathcal{P}_1, \ldots, \mathcal{P}_n$ denote convex polytopes in $\mathbb{R}^n$, $n \geq 1$.

Proposition VII.29. (Monotonicity of mixed volume). *If $\mathcal{P}'_j$ are convex polytopes in $\mathbb{R}^n$ such that $\mathcal{P}_j \subseteq \mathcal{P}'_j$ for each j, then* $\mathrm{MV}(\mathcal{P}_1, \ldots, \mathcal{P}_n) \leq \mathrm{MV}(\mathcal{P}'_1, \ldots, \mathcal{P}'_n)$.

PROOF. Theorem VII.5 implies that it holds for rational polytopes. The general case then follows from the observation that every polytope can be approximated arbitrarily closely (with respect to the Hausdorff distance) by rational polytopes (corollary V.31), and the mixed volume is continuous with respect to the Hausdorff distance (remark VII.4). □

Let ν be a nonzero integral element of $(\mathbb{R}^n)^*$ and $\mathbb{R}^n_{\nu\perp} := \{\alpha \in \mathbb{R}^n : \langle \nu, \alpha \rangle = 0\}$. Choose an affine transformation $\psi_\nu : \mathbb{R}^n \to \mathbb{R}^n$ such that ψ_ν restricts to an automorphism of $\mathbb{Z}^n$ and maps $\mathbb{R}^n_{\nu\perp} \cap \mathbb{Z}^n$ onto $\mathbb{Z}^{n-1} \times \{0\}$. If $\mathcal{Q}_1, \ldots, \mathcal{Q}_{n-1}$ are rational polytopes in $\mathbb{R}^n$ such that each $\mathcal{Q}_j$ is a translate of some polytope $\mathcal{Q}'_j \subset \mathbb{R}^n_{\nu\perp}$, then we define

$$\mathrm{MV}'_\nu(\mathcal{Q}_1, \ldots, \mathcal{Q}_{n-1}) := \mathrm{MV}(\psi_\nu(\mathcal{Q}'_1), \ldots, \psi_\nu(\mathcal{Q}'_{n-1})), \tag{69}$$

where the mixed volume on the right-hand side is the $(n-1)$-dimensional mixed volume on $\mathbb{R}^{n-1}$. Proposition V.40, theorem VII.2 imply that MV'_ν does not depend on the choice of ψ_ν or the translations involved.

Proposition VII.30. *Assume $\mathcal{P}_1, \ldots, \mathcal{P}_n$ are rational polytopes. Then*

(1) $\mathrm{MV}(\mathcal{P}_1, \ldots, \mathcal{P}_n) = \sum_\nu \max_{\mathcal{P}_1}(\nu) \mathrm{MV}'_\nu(\mathrm{ld}_\nu(\mathcal{P}_2), \ldots, \mathrm{ld}_\nu(\mathcal{P}_n))$, *where the sum is over all primitive integral $\nu \in (\mathbb{R}^n)^*$.*

(2) Assume $\mathcal{P}_1$ is a line segment in the direction of ν, where ν is a primitive integral element in $\mathbb{R}^n$. Let $l(\mathcal{P}_1)$ be the "integer length" of $\mathcal{P}_1$ (i.e. the Euclidean length of $\mathcal{P}_1$ is $l(\mathcal{P}_1)$ times the length of ν). Identify ν with an element of $(\mathbb{R}^n)^*$ via the basis dual to the standard basis of $\mathbb{R}^n$. Let $\pi_\nu : \mathbb{R}^n \to \mathbb{R}^n_{\nu\perp}$ be a "lattice projection in the direction normal to ν" (i.e. $\pi_\nu = \psi_\nu^{-1} \circ \pi \circ \psi_\nu$, where $\pi : \mathbb{R}^n \to \mathbb{R}^{n-1} \times \{0\}$ is the projection in the first $(n-1)$-coordinates). Then $\mathrm{MV}(\mathcal{P}_1, \ldots, \mathcal{P}_n) = l(\mathcal{P}_1) \mathrm{MV}'_\nu(\pi_\nu(\mathcal{P}_2), \ldots, \pi_\nu(\mathcal{P}_n))$.

PROOF. Due to the multiadditivity of the mixed volume we may assume that each $\mathcal{P}_j$ is integral. Pick Laurent polynomials f_j with Newton polytope $\mathcal{P}_j$ such that $f_1, \ldots, f_n$ are properly non-degenerate. Then they are BKK non-degenerate on $(\mathbb{k}^*)^n$, they satisfy the hypothesis of corollary VII.27, and for each primitive integral $\nu \in (\mathbb{R}^n)^*$, $\mathrm{In}'_{\alpha_\nu,\psi_\nu}(f_2), \ldots, \mathrm{In}'_{\alpha_\nu,\psi_\nu}(f_n)$ are BKK non-degenerate on $(\mathbb{k}^*)^{n-1}$. Then corollary VII.27, theorems VII.5 and VII.7 imply assertion (1). For assertion (2), change coordinates on $\mathbb{Z}^n$ (using lemma B.57) so that $\nu = (0, \ldots, 0, 1)$. Without loss of generality we may assume that $\mathcal{P}_1$ is the line segment bounded by the origin and $(0, \ldots, 0, l)$, where $l := l(\mathcal{P}_1)$, and in addition, $f_1 = x_n^l + 1$. Then for generic $a = (a_1, \ldots, a_n) \in (\mathbb{k}^*)^n$,

$$[f_1, \ldots, f_n]_a = [x_n^l + 1 - a_1, f_2, \ldots, f_n]_a = \sum_{\epsilon^l = a_1 - 1} [f_2|_{x_n=\epsilon}, \ldots, f_n|_{x_n=\epsilon}]_{(a_2,\ldots,a_n)}.$$

Since π_ν is simply the projection in the first $(n-1)$-coordinates, for generic a_1, the Newton polytope of $f_j|_{x_n=\epsilon}$ is $\pi_\nu(\mathcal{P}_j)$ for each j. Assertion (2) therefore follows from theorem VII.5. □

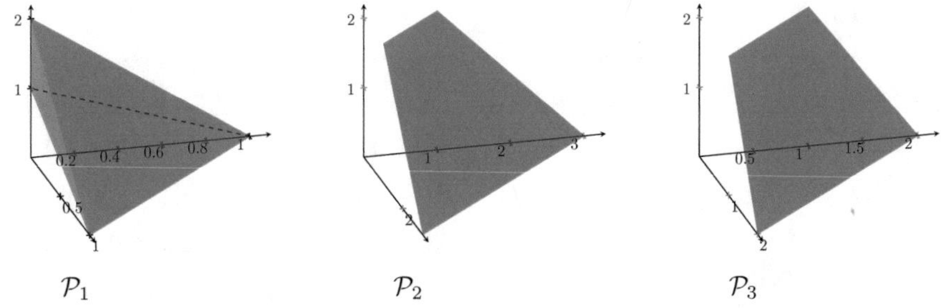

FIGURE 4. Newton polytopes of polynomials from Example VII.31.

Example VII.31. Let $f_1 = a_1x + b_1y + c_1z + d_1z^2$, $f_2 = a_2x^3 + b_2xz^2 + c_2y^3 + d_2yz^2$, $f_3 = a_3x^2 + b_3xz^2 + c_3y^2 + d_3yz^2$, where a_i, b_i, c_i, d_i's are generic elements of $\mathbb{k}$, and let $\mathcal{P}_j := \mathrm{NP}(f_j)$, $j = 1, 2, 3$ (see fig. 4). We compute $[f_1, f_2, f_3]_{(\mathbb{k}^*)^3} = \mathrm{MV}(\mathcal{P}_1, \mathcal{P}_2, \mathcal{P}_3)$ using proposition VII.30. Assertion (1) of proposition VII.30 implies that $\mathrm{MV}(\mathcal{P}_1, \mathcal{P}_2, \mathcal{P}_3) = \sum_\nu \max_{\mathcal{P}_1}(\nu)\, \mathrm{MV}'_\nu(\mathrm{ld}_\nu(\mathcal{P}_2), \mathrm{ld}_\nu(\mathcal{P}_3))$. Theorem VII.33 implies that $\mathrm{MV}'_\nu(\mathrm{ld}_\nu(\mathcal{P}_2), \mathrm{ld}_\nu(\mathcal{P}_3))$ is nonzero only if ν is one of the six outer normals of facets of $\mathcal{P}_2 + \mathcal{P}_3$ (fig. 5). When $\nu = (-1, 0, 0)$ or $(0, -1, 0)$, then $\max_{\mathcal{P}_1}(\nu) = 0$, so it suffices to consider the remaining four cases. The image of the leading faces of $\mathcal{P}_2, \mathcal{P}_3$ and $\mathcal{P}_2 + \mathcal{P}_3$ under (certain choices of) ψ_ν is given in fig. 6, and example VII.3 implies that $\mathrm{MV}'(\mathrm{ld}_\nu(\mathcal{P}_2), \mathrm{ld}_\nu(\mathcal{P}_3))$ is the area of the region shaded green inside $\psi_\nu(\mathrm{ld}_\nu(\mathcal{P}_2 + \mathcal{P}_3))$. It then follows from fig. 6 that

$\mathrm{MV}(\mathcal{P}_1, \mathcal{P}_2, \mathcal{P}_3) = \max_{\mathcal{P}_1}(1,1,1) \cdot \mathrm{Area}($  $) + \max_{\mathcal{P}_1}(-1,-1,-1) \cdot \mathrm{Area}($  $)$

$+ \max_{\mathcal{P}_1}(2,2,1) \cdot \mathrm{Area}($  $) + \max_{\mathcal{P}_1}(-2,-2,-1) \cdot \mathrm{Area}($  $)$

$$= 2 \cdot 2 - 1 \cdot 4 + 2 \cdot 3 - 1 \cdot 1$$
$$= 5$$

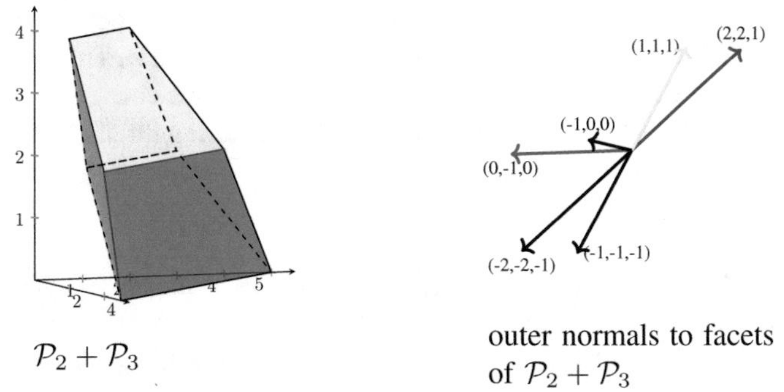

FIGURE 5. $\mathcal{P}_2 + \mathcal{P}_3$ and the outer normals to its facets.

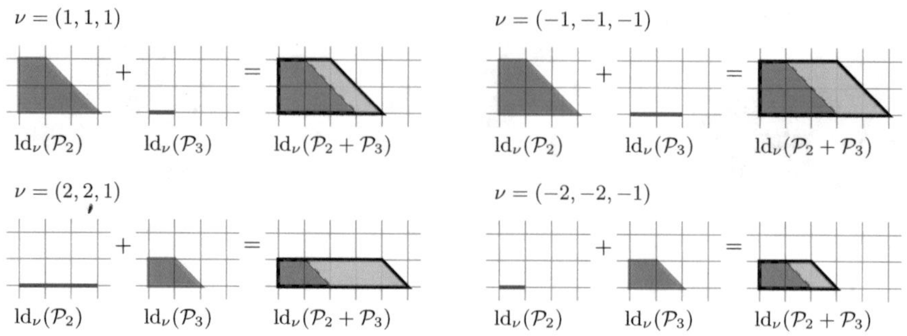

FIGURE 6. The image under ψ_ν of leading faces of $\mathcal{P}_2, \mathcal{P}_3$ and $\mathcal{P}_2 + \mathcal{P}_3$.

Definition VII.32. We say that convex polytopes $\mathcal{Q}_1, \ldots, \mathcal{Q}_m$ in $\mathbb{R}^n$ are *dependent* if there is a nonempty subset I of $[m] := \{1, \ldots, m\}$ such that $\dim(\sum_{i \in I} \mathcal{Q}_i) < |I|$; otherwise, we say that they are *independent*. In particular if $m \geq 1$ and $\mathcal{Q}_j = \emptyset$ for some j, then $\mathcal{Q}_1, \ldots, \mathcal{Q}_m$ are dependent.

THEOREM VII.33. (Minkowski). $\mathrm{MV}(\mathcal{P}_1, \ldots, \mathcal{P}_n) = 0$ *if and only if they are dependent.*

PROOF. Due to corollary V.31 and remark VII.4 and the multiadditivity of mixed volumes, it suffices to consider the case that each $\mathcal{P}_j$ is integral. At first assume there is $I \subseteq [n]$ such that $\dim(\sum_{i \in I} \mathcal{P}_i) < |I|$. A recursive application of assertion (1) of proposition VII.30 shows that $\mathrm{MV}(\mathcal{P}_1, \ldots, \mathcal{P}_n)$ can be expressed as a sum such that each summand has a multiplicative factor of the form $\mathrm{MV}'_\nu(\mathrm{ld}_\nu(\mathcal{P}'_i) : i \in I)$, where MV' denotes an $|I|$-dimensional mixed volume and $\mathcal{P}'_i$ is a face of $\mathcal{P}_i$ for each $i \in I$. Now pick BKK non-degenerate Laurent polynomials f_i, $i \in I$, such that $\mathrm{NP}(f_i) = \mathcal{P}_i$ for each i (such f_i exist due to theorem VII.15). Since $\dim(\sum_{i \in I} \mathcal{P}_i) < |I|$, it follows from the definition of BKK non-degeneracy that there is no common zero of $\mathrm{ld}_\nu(f_i)$, $i \in I$, on $(\mathbb{k}^*)^n$. Theorem VII.5 then implies that $\mathrm{MV}'_\nu(\mathrm{ld}_\nu(\mathcal{P}'_i) : i \in I) = 0$. This in turn implies that $\mathrm{MV}(\mathcal{P}_1, \ldots, \mathcal{P}_n) = 0$. Now assume that $\mathcal{P}_1, \ldots, \mathcal{P}_n$ are independent. We will show that $\mathrm{MV}(\mathcal{P}_1, \ldots, \mathcal{P}_n) > 0$. We proceed by induction on n. The case of $n = 1$ is obvious. In the general case, since $\dim(\mathcal{P}_n) \geq 1$, after a change of coordinates on $\mathbb{Z}^n$ if necessary, we may assume that it has positive length along x_n-axis. Let $\pi : \mathbb{Z}^n \to \mathbb{Z}^{n-1}$ be the projection in the first $(n-1)$-coordinates. We consider two cases:

Case 1: $\mathrm{MV}(\pi(\mathcal{P}_1), \ldots, \pi(\mathcal{P}_{n-1})) = 0$. Due to the inductive hypothesis, we may assume after a reordering of the $\mathcal{P}_j$ if necessary that there is k, $1 \leq k \leq n-1$, such that $\dim(\sum_{j=1}^k \pi(\mathcal{P}_j)) < k$. Since $\mathcal{P}_1, \ldots, \mathcal{P}_n$ are independent, it follows that $\dim(\sum_{j=1}^k \pi(\mathcal{P}_j)) = k - 1$ and $\dim(\sum_{j=1}^k \mathcal{P}_j) = k$. After a translation of one of the $\mathcal{P}_j$ if necessary, we may assume that the affine hull $\mathrm{aff}(\mathcal{P}_1 + \ldots + \mathcal{P}_k)$ of $\mathcal{P}_1 + \cdots + \mathcal{P}_k$ passes through the origin. Due to lemma B.57 we can change the basis of $\mathbb{Z}^n$ to ensure that the subgroup of $\mathbb{Z}^n$ generated by $\mathbb{Z}^n \cap \operatorname{aff}(\mathcal{P}_1 + \dots + \mathcal{P}_k)$ is $\mathbb{Z}^k \times \{(0,\dots,0)\}$. Let $\pi' : \mathbb{R}^n \to \mathbb{R}^{n-k}$ be the projection in the last $(n-k)$-coordinates. We claim that $\pi'(\mathcal{P}_{k+1}),\dots,\pi'(\mathcal{P}_n)$ are independent. Indeed, if $\dim(\sum_{j\in J} \pi'(\mathcal{P}_j)) < |J|$ for some $J \subseteq \{k+1,\dots,n\}$, then setting $J' := \{1,\dots,k\} \cup J$ will yield that $\dim(\sum_{j\in J'} \mathcal{P}_j) < |J'|$, contradicting the independence of the $\mathcal{P}_j$. Now pick generic $f_1,\dots,f_n$ such that $\mathrm{NP}(f_j) = \mathcal{P}_j$ for each j. Theorems VII.5 and VII.7 and the inductive hypothesis imply that the number of solutions Z_k of $f_1,\dots,f_k$ on $(\mathbb{k}^*)^k$ is nonzero, and for each $a = (a_1,\dots,a_k) \in Z_k$, the number of solutions of $f_{k+1},\dots,f_n$ on $\{a\} \times (\mathbb{k}^*)^{n-k}$ is nonzero. Theorem VII.5 then implies that $\mathrm{MV}(\mathcal{P}_1,\dots,\mathcal{P}_n) > 0$, as required.

Case 2: $\mathrm{MV}(\pi(\mathcal{P}_1),\dots,\pi(\mathcal{P}_{n-1})) > 0$. In this case, for each generic $\epsilon \in (\mathbb{k}^*)^n$, and for generic f_j with $\mathrm{NP}(f_j) = \mathcal{P}_j$, $j = 2,\dots,n-1$,

$$\begin{aligned}[x_n - \epsilon, f_1,\dots,f_{n-1}]_{(\mathbb{k}^*)^n} &= [f_1|_{x_n=\epsilon},\dots,f_{n-1}|_{x_n=\epsilon}]_{(\mathbb{k}^*)^{n-1}}\\ &= \mathrm{MV}(\pi(\mathcal{P}_1),\dots,\pi(\mathcal{P}_{n-1})). > 0\end{aligned}$$

Computing $[x_n - \epsilon, f_1,\dots,f_{n-1}]_{(\mathbb{k}^*)^n}$ using eq. (65) then implies due to theorem VII.5 that there is a primitive integral $\nu \in (\mathbb{R}^n)^*$ such that $\langle \nu, e_n\rangle \neq 0$ (where $e_n = (0,\dots,0,1) \in \mathbb{Z}^n$) and $\mathrm{MV}'_\nu(\mathrm{ld}_\nu(\mathcal{P}_1),\dots,\mathrm{ld}_\nu(\mathcal{P}_{n-1})) > 0$. After a translation if necessary, we may assume the origin is in the relative interior of $\mathcal{P}_n$. Then $\max_{\mathcal{P}_n}(\nu) > 0$, and therefore assertion (1) of proposition VII.30 implies that $\mathrm{MV}(\mathcal{P}_1,\dots,\mathcal{P}_n) > 0$, as required. □

Corollary VII.34. (Strict monotonicity of mixed volume, Rojas [Roj94, Corollary 9]). *If $\mathcal{P}'_j$ are convex polytopes in $\mathbb{R}^n$ such that $\mathcal{P}_j \subseteq \mathcal{P}'_j$ for each j, then $\mathrm{MV}(\mathcal{P}_1,\dots,\mathcal{P}_n) \leq \mathrm{MV}(\mathcal{P}'_1,\dots,\mathcal{P}'_n)$. The inequality is strict if and only if both of the following are true:*

(1) $\mathcal{P}'_1,\dots,\mathcal{P}'_n$ are independent and

(2) there is $\nu \in (\mathbb{R}^n)^* \setminus \{0\}$ such that the collection $\{\mathrm{In}_\nu(\mathcal{P}_j) : \mathcal{P}_j \cap \mathrm{In}_\nu(\mathcal{P}'_j) \neq \emptyset\}$ of polytopes is independent.

Remark VII.35. Recall that an empty collection of polytopes is independent. Therefore condition (2) of corollary VII.34 holds if there is $\nu \in (\mathbb{R}^n)^* \setminus \{0\}$ such that $\mathcal{P}_j \cap \mathrm{In}_\nu(\mathcal{P}'_j) = \emptyset$ for each j.

PROOF OF COROLLARY VII.34. Due to proposition VII.29 and theorem VII.33 it suffices to prove the following statement:

> if $\mathcal{P}_j \subset \mathcal{P}'_j$, $j = 1,\dots,n$, and $\mathrm{MV}(\mathcal{P}'_1,\dots,\mathcal{P}'_n) > 0$, then $\mathrm{MV}(\mathcal{P}'_1,\dots,\mathcal{P}'_n) > \mathrm{MV}(\mathcal{P}_1,\dots,\mathcal{P}_n)$ if and only if condition (2) of corollary VII.34 is true.

So assume $\mathrm{MV}(\mathcal{P}'_1,\dots,\mathcal{P}'_n) > 0$. Due to corollary V.31 and remark VII.4 and the multiadditivity of mixed volumes, we may assume in addition that all the $\mathcal{P}_j, \mathcal{P}'_j$ are integral. Then choose BKK non-degenerate Laurent polynomials f_j such that $\mathrm{NP}(f_j) = \mathcal{P}_j$, $j = 1,\dots,n$. Theorems VII.5 and VII.7 imply that

$$\mathrm{MV}(\mathcal{P}_1, \ldots, \mathcal{P}_n) = [f_1, \ldots, f_n]^{iso}_{(\Bbbk^*)^n} \leq \mathrm{MV}(\mathcal{P}'_1, \ldots, \mathcal{P}'_n)$$

and the inequality is strict if and only if there is $\nu \in (\mathbb{R}^n)^* \setminus \{0\}$ such that

(70) $\mathrm{In}_{\mathcal{P}'_j,\nu}(f_1), \ldots, \mathrm{In}_{\mathcal{P}'_n,\nu}(f_n)$ have a common zero on $(\Bbbk^*)^n$.

Since f_j are generic, then it follows from theorems VII.5 and VII.33 that condition (70) holds if and only if $\{\mathrm{In}_\nu(\mathcal{P}_j) : \mathcal{P}_j \cap \mathrm{In}_\nu(\mathcal{P}'_j) \neq \emptyset\}$ is an independent collection of polytopes. □

Corollary VII.36. (Bernstein's theorem—alternate version). *Let $\mathcal{A}_j$ be finite subsets of $\mathbb{Z}^n$ and f_j be Laurent polynomials supported at $\mathcal{A}_j$, $j = 1, \ldots, n$. Then*

$$[f_1, \ldots, f_n]^{iso}_{(\Bbbk^*)^n} \leq \mathrm{MV}(\mathrm{conv}(\mathcal{A}_1), \ldots, \mathrm{conv}(\mathcal{A}_n)).$$

If $\mathrm{MV}(\mathrm{conv}(\mathcal{A}_1), \ldots, \mathrm{conv}(\mathcal{A}_n)) > 0$, then the bound is satisfied with an equality if and only if both of the following conditions hold:

1. *for each nontrivial weighted order ν, the collection $\{\mathrm{In}_\nu(\mathrm{NP}(f_j)) : \mathrm{In}_\nu (\mathcal{A}_j) \cap \mathrm{Supp}(f_j) \neq \emptyset\}$ of polytopes is dependent and*
2. *$f_1, \ldots, f_n$ are BKK non-degenerate, i.e. they satisfy* (ND'^*) *with $m = n$.*

PROOF. The bound of corollary VII.36 follows from theorem VII.5. Theorem VII.7 implies that if $\mathrm{MV}(\mathrm{conv}(\mathcal{A}_1), \ldots, \mathrm{conv}(\mathcal{A}_n)) > 0$, then the bound holds with an equality if and only if $f_1, \ldots, f_n$ are BKK non-degenerate and $\mathrm{MV}(\mathrm{NP}(f_1), \ldots,$ $\mathrm{MV}(\mathrm{NP}(f_n) = \mathrm{MV}(\mathrm{conv}(\mathcal{A}_1), \ldots, \mathrm{conv}(\mathcal{A}_n))$. Now the result follows from corollary VII.34. □

7. Some technical results

In this section we compile a few (technical) corollaries of Bernstein's theorem that we use in later chapters.

Proposition VII.37. *Let the setup be as in theorem VII.5. Assume $\mathrm{MV}(\mathcal{P}_1, \ldots, \mathcal{P}_n)$ is nonzero. If the set of common zeroes of $f_1, \ldots, f_n$ on $(\Bbbk^*)^n$ has a positive dimensional component, then $[f_1, \ldots, f_n]^{iso}_{(\Bbbk^*)^n} < \mathrm{MV}(\mathcal{P}_1, \ldots, \mathcal{P}_n)$.*

PROOF. Since it is possible to choose a curve on such a positive dimensional component (corollary III.83), the proposition follows from theorem VII.7 and lemma VII.17. □

Corollary VII.38. *Let the notation be as in proposition VII.26. Assume $V(\mathrm{In}'_{\alpha_\nu,\psi_\nu}(f_2), \ldots, \mathrm{In}'_{\alpha_\nu,\psi_\nu}(f_n)) \cap (\Bbbk^*)^n \neq \emptyset$, so that C' is a curve. Let $\{C'_j\}_j$ be the irreducible components of C', and $\mathcal{B}'_{j,\nu}$ be the collection of all branches B of C'_j at infinity (with respect to $(\Bbbk^*)^n$) such that ν_B is proportional to ν. If*

$V(\mathrm{In}'_{\alpha_\nu,\psi_\nu}(f_1),\dots,\mathrm{In}'_{\alpha_\nu,\psi_\nu}(f_n)) \cap (\Bbbk^*)^{n-1} = \emptyset$, *then*

$$(71)\qquad \sum_{a\in C'\cap O_{\mathcal{Q}}} \operatorname{ord}_a(f_1|_{C'}) = \sum_j \sum_{(Z,z)\in\mathcal{B}'_{j,\nu}} \operatorname{ord}_z(f_1|_{C'_j})[f_2,\dots,f_n]_{C'_j} = \nu(f_1)[\mathrm{In}'_{\alpha_\nu,\psi_\nu}(f_2),\dots,\mathrm{In}'_{\alpha_\nu,\psi_\nu}(f_n)]_{(\Bbbk^*)^{n-1}},$$

where $[f_2,\dots,f_n]_{C'_j}$ *are defined as in section IV.5. If in addition* $f'_{2,\nu},\dots,f'_{n,\nu}$ *are BKK non-degenerate, then*

$$(72)\qquad \sum_{a\in C'\cap O_{\mathcal{Q}}} \operatorname{ord}_a(f_1|_{C'}) = \min_{\mathrm{NP}(f_1)}(\nu)\,\mathrm{MV}'_\nu(\mathrm{In}_\nu(\mathrm{NP}(f_2)),\dots,\mathrm{In}_\nu(\mathrm{NP}(f_n))).$$

PROOF. Theorem IV.24 implies that

$$\sum_{a\in C'\cap O_{\mathcal{Q}}} \operatorname{ord}_a(f_1|_{C'}) = \sum_j \sum_{a\in C'_j\cap O_{\mathcal{Q}}} [f_2,\dots,f_n]_{C'_j} \operatorname{ord}_a(f_1|_{C'_j}) = \sum_j \sum_{a\in C'_j\cap O_{\mathcal{Q}}} [f_2,\dots,f_n]_{C'_j} \sum_{z\in\pi_j^{-1}(a)} \operatorname{ord}_z(\pi_j^*(f_1|_{C'_j})),$$

where $\pi_j : \tilde{C}_j \to C'_j$ are desingularizations of C'_j. It follows from the definition of branches in section VI.8 that each $z \in \pi_j^{-1}(a)$, where $a \in C'_j \cap O_{\mathcal{Q}}$ corresponds to a branch $B = (Z, z)$ at infinity of C'_j. Moreover, since $\mathcal{Q}$ is a facet of $\mathcal{P}$, proposition VI.31 implies that a branch B at infinity of C'_j intersects $O_{\mathcal{Q}}$ if and only if ν_B is proportional to ν. It follows that

$$\sum_{a\in C'\cap O_{\mathcal{Q}}} \operatorname{ord}_a(f_1|_{C'}) = \sum_j \sum_{(Z,z)\in\mathcal{B}'_{j,\nu}} \operatorname{ord}_z(f_1|_{C'_j})[f_2,\dots,f_n]_{C'_j}.$$

The result now follows from proposition VII.26 and theorems VII.5, VII.7. □

8. The problem of characterizing coefficients which guarantee non-degeneracy

Let $\mathcal{A} = (\mathcal{A}_1,\dots,\mathcal{A}_n)$ be an n-tuple of finite subsets of $\mathbb{Z}^n$, and as in section VII.4, let $\mathcal{L}(\mathcal{A})$ be the collection of n-tuples $(f_1,\dots,f_n)$ of Laurent polynomials such that each f_j is supported at $\mathcal{A}_j$. Given $(f_1,\dots,f_n) \in \mathcal{L}(\mathcal{A})$, theorem VII.5 implies that if the coefficients of the f_j are generic, then

$$(73)\qquad [f_1,\dots,f_n]^{iso}_{(\Bbbk^*)^n} = \mathrm{MV}(\mathcal{P}_1,\dots,\mathcal{P}_n),$$

where $\mathcal{P}_j = \operatorname{conv}(\mathcal{A}_j)$, $j = 1,\dots,n$. On the other hand, it is straightforward to see that not *all* the coefficients of the f_j have to be generic for the equality in (73). For example, if $\mathcal{B}_j$ is the set of vertices of $\mathcal{P}_j$ and if the coefficient of x^α in each f_j is fixed for each $\alpha \in \mathcal{A}_j \setminus \mathcal{B}_j$, (73) still holds provided the coefficients of x^α in the f_j are generic for all $\alpha \in \mathcal{B}_j$. J. M. Rojas [Roj99] posed the problem of identifying all $(\mathcal{B}_1,\dots,\mathcal{B}_n)$ which have this property. The precise version of Rojas's problem for

$(\mathbb{k}^*)^n$ is as follows: let $\mathcal{B}_j \subseteq \mathcal{A}_j$, $j = 1, \dots, n$ and $\mathcal{B} := (\mathcal{B}_1, \dots, \mathcal{B}_n)$. We say that $\mathcal{B}$ *guarantees* $(\mathbb{k}^*)^n$*-maximality on* $\mathcal{L}(\mathcal{A})$ if for all choices from $\mathbb{k}$ of coefficients of x^α in f_j for all j and all $\alpha \in \mathcal{A}_j \setminus \mathcal{B}_j$, (73) holds provided the coefficients of x^β in f_k are generic for all k and all $\beta \in \mathcal{B}_k$. Then Rojas's problem[4] is to classify all $\mathcal{B}$ which guarantee $(\mathbb{k}^*)^n$-maximality on $\mathcal{L}(\mathcal{A})$.

Proposition VII.39. (Solution of Rojas's problem for $(\mathbb{k}^*)^n$). *Let* $\mathcal{Q}_j := \operatorname{conv}(\mathcal{B}_j)$, $j = 1, \dots, n$. *Then the following are equivalent:*

(1) $\mathcal{B}$ *guarantees* $(\mathbb{k}^*)^n$*-maximality on* $\mathcal{L}(\mathcal{A})$*;*
(2) $\operatorname{MV}(\mathcal{Q}_1, \dots, \mathcal{Q}_n) = \operatorname{MV}(\mathcal{P}_1, \dots, \mathcal{P}_n)$*;*
(3) one of the following holds:
  (a) $\mathcal{P}_1, \dots, \mathcal{P}_n$ *are dependent, or*
  (b) for each $\nu \in (\mathbb{R}^n)^* \setminus \{0\}$*, the collection* $\{\operatorname{In}_\nu(\mathcal{Q}_j) : \mathcal{Q}_j \cap \operatorname{In}_\nu(\mathcal{P}_j) \neq \emptyset\}$ *of polytopes is dependent.*

PROOF. It is straightforward to check that $\mathcal{B}$ guarantees $(\mathbb{k}^*)^n$-maximality on $\mathcal{L}(\mathcal{A})$ if and only if for each $(f_1, \dots, f_n) \in \mathcal{L}(\mathcal{A})$, there is $(g_1, \dots, g_n) \in \mathcal{L}(\mathcal{B})$ such that

$$[f_1 + g_1, \dots, f_n + g_n]^{iso}_{(\mathbb{k}^*)^n} = \operatorname{MV}(\mathcal{P}_1, \dots, \mathcal{P}_n)$$

The implication (1) $\Rightarrow$ (2) follows from taking $(f_1, \dots, f_n) = (0, \dots, 0)$ and applying theorem VII.5 and proposition VII.29. For the opposite implication (2) $\Rightarrow$ (1), assume $\operatorname{MV}(\mathcal{Q}_1, \dots, \mathcal{Q}_n) = \operatorname{MV}(\mathcal{P}_1, \dots, \mathcal{P}_n)$. Pick $\mathcal{B}$-non-degenerate $(g_1, \dots, g_n) \in \mathcal{L}(\mathcal{B})$ such that set $h_j := (1-t)f_j + tg_j$. Theorem VII.7 implies that $[g_1, \dots, g_n]^{iso}_{(\mathbb{k}^*)^n} \geq [h_1|_{t=\epsilon}, \dots, h_n|_{t=\epsilon}]^{iso}_{(\mathbb{k}^*)^n}$ for each $\epsilon \in \mathbb{k}$, and therefore theorem IV.32 implies that for generic $\epsilon \in \mathbb{k}$, $[h_1|_{t=\epsilon}, \dots, h_n|_{t=\epsilon}]^{iso}_{(\mathbb{k}^*)^n} = [g_1, \dots, g_n]^{iso}_{(\mathbb{k}^*)^n} = \operatorname{MV}(\mathcal{P}_1, \dots, \mathcal{P}_n)$, which implies condition (1). Finally, the equivalence of conditions (2) and (3) follows from corollary VII.34. □

We now describe a natural variant of the notion of $(\mathbb{k}^*)^n$-maximality on $\mathcal{L}(\mathcal{A})$. Consider the case that $n = 2$ and $\mathcal{A}_1 = \mathcal{A}_2 = \mathcal{P} \cap \mathbb{Z}^2$, where $\mathcal{P}$ is the bigger triangle from fig. 7. Let $\mathcal{B}_1 = \mathcal{B}_2 = \mathcal{Q} \cap \mathbb{Z}^2$, where $\mathcal{Q}$ is the smaller triangle in fig. 7. If f_1, f_2 are polynomials in two variables with Newton polytope $\mathcal{P}$, it is straightforward to check that (ND*) is satisfied if the coefficients of monomials in f_j whose exponents are in $\mathcal{B}$ are generic, and therefore the number of solutions of f_1, f_2 on $(\mathbb{k}^*)^2$ is $\operatorname{MV}(\mathcal{P}_1, \mathcal{P}_2) = 2\operatorname{Area}(\mathcal{P}) = 48$. Note however that $\operatorname{MV}(\mathcal{Q}_1, \mathcal{Q}_2) = 2\operatorname{Area}(\mathcal{Q}) = 12 < \operatorname{MV}(\mathcal{P}_1, \mathcal{P}_2)$. This motivates problem VII.40. In its statement we use the following notation: we write $\mathcal{L}^0(\mathcal{A})$ for the collection of all $(f_1, \dots, f_n) \in \mathcal{L}(\mathcal{A})$ such that $\operatorname{NP}(f_j) = \mathcal{A}_j$ for each j. Given $\mathcal{B} = (\mathcal{B}_1, \dots, \mathcal{B}_n)$ with $\mathcal{B}_j \subseteq \mathcal{A}_j$, $j = 1, \dots, n$, we denote by $\pi_\mathcal{B} : \mathcal{L}(\mathcal{A}) \to \mathcal{L}(\mathcal{B})$ be the natural projection which "forgets" the coefficients corresponding to $\alpha \notin \mathcal{B}_j$, $j = 1, \dots, n$, and we write $\mathcal{A} \setminus \mathcal{B} := (\mathcal{A}_1 \setminus \mathcal{B}_1, \dots, \mathcal{A}_n \setminus \mathcal{B}_n)$. We say that $\mathcal{B}$ *guarantees* $(\mathbb{k}^*)^n$*-maximality on* $\mathcal{L}^0(\mathcal{A})$ if for each $(h_1, \dots, h_n) \in \mathcal{L}(\mathcal{A} \setminus \mathcal{B})$, there is a nonempty Zariski open subset

[4] Rojas [Roj99] posed the problem in a more general context (instead of $(\mathbb{k}^*)^n$ he allowed for a broader class of subsets of $\mathbb{k}^n$) and presented a solution.

$\mathcal{U}$ of $\mathcal{L}(\mathcal{B})$ such that $[f_1, \ldots, f_n]^{iso}_{(\Bbbk^*)^n} = \mathrm{MV}(\mathcal{P}_1, \ldots, \mathcal{P}_n)$ for each $(f_1, \ldots, f_n) \in \mathcal{L}^0(\mathcal{A}) \cap \pi^{-1}_{\mathcal{A}\setminus\mathcal{B}}(h_1, \ldots, h_n) \cap \pi^{-1}_{\mathcal{B}}(\mathcal{U})$.

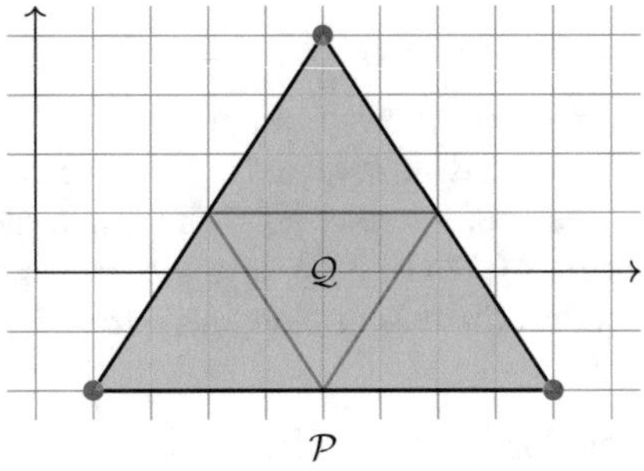

FIGURE 7. The number of solutions on $(\Bbbk^*)^2$ of polynomials with Newton polytope $\mathcal{P}$ is maximal if the coefficients of monomials from $\mathcal{Q}$ are generic.

Problem VII.40. *Classify all $\mathcal{B} = (\mathcal{B}_1, \ldots, \mathcal{B}_n)$ with $\mathcal{B}_j \subseteq \mathcal{A}_j$, $j = 1, \ldots, n$, such that $\mathcal{B}$ guarantees $(\Bbbk^*)^n$-maximality on $\mathcal{L}^0(\mathcal{A})$.*

In the remainder of this section we give some partial answers to problem VII.40. We start with a condition that guarantees $(\Bbbk^*)^n$-maximality on $\mathcal{L}^0(\mathcal{A})$. Given a weighted order ν on the ring of Laurent polynomials in $(x_1, \ldots, x_n)$ and a subset J of $[n]$, let $d_{J,\nu} := \dim(\mathrm{In}_\nu(\sum_{j\in J} \mathcal{P}_j))$ and $e_{J,\nu} := |\{j \in J : \mathrm{In}_\nu(\mathcal{P}_j) \cap \mathcal{B}_j \neq \emptyset\}|$.

Proposition VII.41. *Assume for each nontrivial weighted order ν, one of the following holds:*

1. *either $\mathrm{In}_\nu(\mathcal{P}_j)$ is a vertex of $\mathcal{P}_j$ for some j,*
2. *or there exists a nonempty subset J of $[n]$ such that $e_{J,\nu} \geq d_{J,\nu} < |J|$.*

Then $\mathcal{B}$ guarantees $(\Bbbk^*)^n$-maximality on $\mathcal{L}^0(\mathcal{A})$.

PROOF. Fix $(h_1, \ldots, h_n) \in \mathcal{L}(\mathcal{A} \setminus \mathcal{B})$. It suffices to show that for generic $(g_1, \ldots, g_n) \in \mathcal{L}(\mathcal{B})$, if $(f_1, \ldots, f_n) \in \pi^{-1}_{\mathcal{A}\setminus\mathcal{B}}(h_1, \ldots, h_n) \cap \pi^{-1}_{\mathcal{B}}(g_1, \ldots, g_n)$, then (ND*) holds provided $(f_1, \ldots, f_n) \in \mathcal{L}^0(\mathcal{A})$. Let ν be a nontrivial weighted order on the ring of Laurent polynomials. If (1) holds for some j, then $\mathrm{In}_\nu(f_j)$ is a monomial whenever $\mathrm{NP}(f_j) = \mathcal{P}_j$ and therefore is nowhere zero on $(\Bbbk^*)^n$. So without loss of generality we may assume (2) holds for some nonempty subset J of $[n]$. Let $J_\nu := \{j \in J : \mathrm{In}_\nu(\mathcal{P}_j) \cap \mathcal{B}_j \neq \emptyset\}$, so that $e_{J,\nu} = |J_\nu|$. If $e_{J,\nu} > d_{J,\nu}$, it is straightforward to check that $V(\mathrm{In}_\nu(f_j) : j \in J_\nu) \cap (\Bbbk^*)^n = \emptyset$ for generic $(g_j : j \in J_\nu)$. On the other hand, if $e_{J,\nu} = d_{J,\nu}$, then there is $j' \in J \setminus J_\nu$ (since $d_{J,\nu} < |J|$), and it is straightforward to check that $V(\mathrm{In}_\nu(f_j) : j \in J_\nu) \cap V(f_{j'}) \cap (\Bbbk^*)^n = \emptyset$ for generic $(g_j : j \in J_\nu)$. This proves the proposition. $\square$

Proposition VII.42. (Solution to problem VII.40 for $n = 2$). *Assume $n = 2$. Then $\mathcal{B} = (\mathcal{B}_1, \mathcal{B}_2)$ guarantees $(\Bbbk^*)^2$-maximality on $\mathcal{L}^0(\mathcal{A})$ if and only if for each nontrivial weighted order ν, one of the following holds:*

1. *either $\operatorname{In}_\nu(\mathcal{P}_j)$ is a vertex of $\mathcal{P}_j$ for some j,*
2. *or there is j such that $\operatorname{In}_\nu(\mathcal{P}_j) \cap \mathcal{B}_j \neq \emptyset$.*

PROOF. The $(\Leftarrow)$ direction follows from proposition VII.41. For the opposite inclusion assume there is a nontrivial weighted order ν such that for each j, $\operatorname{In}_\nu(\mathcal{P}_j)$ is an edge of $\mathcal{P}_j$ which does not intersect $\mathcal{B}_j$. Then it is clear we can pick f_j with $\operatorname{NP}(f_j) = \mathcal{P}_j$ such that $\operatorname{In}_\nu(f_1) \cap \operatorname{In}_\nu(f_j)$ have a common zero on $(\Bbbk^*)^2$ and for such (f_1, f_2), no choice of coefficients of monomials from $\mathcal{B}_j$ would make that zero disappear. □

Proposition VII.43. (Solution to problem VII.40 for $n = 3$). *Assume $n = 3$. Then $\mathcal{B}$ guarantees $(\Bbbk^*)^3$-maximality on $\mathcal{L}^0(\mathcal{A})$ if and only if for each nontrivial weighted order ν, one of the following holds:*

(1) *$\operatorname{In}_\nu(\mathcal{P}_j)$ is a vertex of $\mathcal{P}_j$ for some j;*
(2) *or there are $j_1 \neq j_2$ such that $\operatorname{In}_\nu(\mathcal{P}_{j_1} + \mathcal{P}_{j_2})$ has dimension one and $\operatorname{In}_\nu(\mathcal{P}_{j_1}) \cap \mathcal{B}_{j_1} \neq \emptyset$;*
(3) *or there are $j_1 \neq j_2$ such that $\operatorname{In}_\nu(\mathcal{P}_{j_1} + \mathcal{P}_{j_2})$ has dimension two and $\operatorname{In}_\nu(\mathcal{P}_{j_k}) \cap \mathcal{B}_{j_k} \neq \emptyset$ for each $k = 1, 2$;*
(4) *or there are $j_1 \neq j_2$ such that*
    (a) *there is no positive dimensional polytope which is a* Minkowski summand[5] *of $\operatorname{In}_\nu(\mathcal{P}_{j_k})$ for each $k = 1, 2$;*
    (b) *$\operatorname{In}_\nu(\mathcal{P}_{j_k}) \cap \mathcal{B}_{j_k} = \emptyset$ for each $k = 1, 2$ and*
    (c) *$\operatorname{In}_\nu(\mathcal{P}_{j_3}) \cap \mathcal{B}_{j_3} \neq \emptyset$, where j_3 is the single element of $\{1, 2, 3\} \setminus \{j_1, j_2\}$.*

PROOF. At first we prove the $(\Leftarrow)$ implication. Fix $(h_1, \ldots, h_n) \in \mathcal{L}(\mathcal{A} \setminus \mathcal{B})$, $(g_1, \ldots, g_n) \in \mathcal{L}(\mathcal{B})$ and $(f_1, \ldots, f_n) \in \pi_{\mathcal{A}\setminus\mathcal{B}}^{-1}(h_1, \ldots, h_n) \cap \pi_{\mathcal{B}}^{-1}(g_1, \ldots, g_n) \cap \mathcal{L}^0(\mathcal{A})$. Pick a nontrivial weighted order ν. If one of (1), (2) or (3) holds, then proposition VII.41 implies that $V(\operatorname{In}_\nu(f_1), \ldots, \operatorname{In}_\nu(f_n)) \cap (\Bbbk^*)^n$ is empty if $g_1, \ldots, g_n$ are generic. So assume (4) holds. Condition (4a) implies that $\operatorname{In}_\nu(f_{j_1})$ and $\operatorname{In}_\nu(f_{j_2})$ have no common non-invertible factor in $\Bbbk[x_1, x_1^{-1}, \ldots, x_n, x_n^{-1}]$. This implies that $V(\operatorname{In}_\nu(f_{j_1}), \operatorname{In}_\nu(f_{j_2}))$ is either empty or have codimension 2 in $(\Bbbk^*)^3$. Therefore, by choosing generic g_{j_3}, it is possible to ensure that $V(\operatorname{In}_\nu(f_{j_1}), \operatorname{In}_\nu(f_{j_2})) \cap V(\operatorname{In}_\nu(g_{j_3})) \cap (\Bbbk^*)^3 = \emptyset$, which completes the proof of $(\Leftarrow)$ implication. For the opposite implication, assume there is a nontrivial weighted order ν such that $\operatorname{In}_\nu(\mathcal{P}_j)$ is positive dimensional for each j, and one of the following holds:

(i) either $\operatorname{In}_\nu(\mathcal{P}_j) \cap \mathcal{B}_j = \emptyset$ for each j;
(ii) or there are $j_1 \neq j_2$ such that
    (1) there is a positive dimensional polytope $\mathcal{Q}$ which is a Minkowski summand of each $\operatorname{In}_\nu(\mathcal{P}_{j_k})$, $k = 1, 2$;
    (2) $\operatorname{In}_\nu(\mathcal{P}_{j_k}) \cap \mathcal{B}_{j_k} = \emptyset$ for each $k = 1, 2$ and

[5] We say that a convex polytope $\mathcal{P}$ is a *Minkowski summand* of a convex polytope $\mathcal{Q}$ if there is a convex polytope $\mathcal{R}$ such that $\mathcal{Q} = \mathcal{P} + \mathcal{R}$.

(3) $\mathrm{In}_\nu(\mathcal{P}_{j_3}) \cap \mathcal{B}_{j_3} \neq \emptyset$, where j_3 is the single element of $\{1,2,3\} \setminus \{j_1, j_2\}$.

If (i) holds then one can choose $(f_1, f_2, f_3) \in \mathcal{L}^0(\mathcal{A})$ such that $\mathrm{In}_\nu(f_j)$ have a common zero on $(\mathbb{k}^*)^n$, and the zero would be unaffected by the coefficients of monomials from the $\mathcal{B}_j$, so that $\mathcal{B}$ will not be able to ensure that (ND*) holds. On the other hand, if (ii) holds, then we can choose f_{j_1}, f_{j_2} such that $\mathrm{In}_\nu(f_{j_1})$ and $\mathrm{In}_\nu(f_{j_2})$ have a common factor g with Newton polytope $\mathcal{Q}$. Then for generic g_{j_3} and generic $f_{j_3} \in \pi_{\mathcal{B}_3}^{-1}(g_{j_3})$, Bernstein's theorem would imply that $V(\mathrm{In}_\nu(f_{j_1}), \mathrm{In}_\nu(f_{j_2}), \mathrm{In}_\nu(f_{j_3})) \cap (\mathbb{k}^*)^3 \supseteq V(g, \mathrm{In}_\nu(f_{j_3})) \cap (\mathbb{k}^*)^3 \neq \emptyset$, so that (f_1, f_2, f_3) violates (ND*), as required. □

9. Notes

All major results of this chapter are well known. A. Khovanskii in [BZ88, Section 27] gave a simple proof of Bernstein's formula in zero characteristic. The main distinction between his proof and ours is in the handling of intersection multiplicity: he counts the number of roots of systems which are *nonsingular* at the points of intersection, and then argues that every system can be deformed into such systems. We avoid this approach since in positive characteristics it would involve having as a technical overhead some versions of Bertini-type theorem every time we deal with intersection multiplicity. B. Huber and B. Sturmfels [HS95] gave a constructive proof of Bernstein's theorem (in zero characteristic) which has had a deep impact on the "homotopy continuation" method to numerically compute the solutions of polynomial systems; the techniques of their proof also give an efficient way to compute mixed volumes. Our proof that the set of BKK non-degenerate polynomials is Zariski open follows the arguments from [Oka79], and our proof that it is nonempty comes from [Mon16].

Part 3

Beyond the torus

CHAPTER VIII

Number of zeroes on the affine space I: (Weighted) Bézout theorems

In this chapter we use the results from chapter VII to prove Bézout's theorem (corollary VIII.3) and two of its classical generalizations: the weighted homogeneous version (theorem VIII.2) and the weighted multi-homogeneous version (theorem VIII.8). The weighted degrees considered in these results have the property that the weight of each variable is *positive*. In sections X.8 and X.9 we establish more general versions of these results involving arbitrary weighted degrees as special cases of the extension of Bernstein's theorem to $\mathbb{k}^n$. We continue to assume that $\mathbb{k}$ is an algebraically closed field.

1. Weighted degree

Let ω be an integral element of $(\mathbb{R}^n)^*$. The *weighted degree* corresponding to ω, which by an abuse of notation we also denote by ω, is the map $\mathbb{k}[x_1, \ldots, x_n] \to \mathbb{Z} \cup \{-\infty\}$ given by

$$f \mapsto \max_{\mathrm{Supp}(f)} (\omega).$$

Given $f = \sum_\alpha c_\alpha x^\alpha \in \mathbb{k}[x_1, \ldots, x_n]$, the *leading form* of f with respect to ω is $\mathrm{ld}_\omega(f) := \sum_{\langle \omega, \alpha \rangle = \omega(f)} c_\alpha x^\alpha$. We say that f is *weighted homogeneous* with respect to ω (or in short, *ω-homogeneous*) if $\omega(x^\alpha)$ are equal for all α such that $c_\alpha \neq 0$, or equivalently, if $\mathrm{Supp}(\mathrm{ld}_\omega(f)) = \mathrm{Supp}(f)$.

2. $\mathbb{P}^n(\omega)$ as a compactification of $\mathbb{k}^n$ when the ω_j are positive and $\omega_0 = 1$

Assume ω is an integral element of $(\mathbb{R}^{n+1})^*$ with coordinates $(\omega_0, \ldots, \omega_n)$ with respect to the basis dual to the standard basis of $\mathbb{R}^{n+1}$ such that each ω_j is *positive*. The *weighted projective space* $\mathbb{P}^n(\omega)$ corresponding to ω was constructed in section VI.10. In this section we treat the case that $\omega_0 = 1$. For each $j = 0, \ldots, n$, let $U_j := \mathbb{P}^n(\omega) \setminus V(x_j)$, be the "coordinate chart" of $\mathbb{P}^n(\omega)$ considered in section VI.10.4. Since $\omega_0 = 1$, proposition VI.10 implies that the map $(a_1, \ldots, a_n) \mapsto [1 : a_1 : \cdots : a_n]$ induces an isomorphism between $\mathbb{k}^n$ and

P. Mondal, *How Many Zeroes?*, CMS/CAIMS Books in Mathematics 2,
https://doi.org/10.1007/978-3-030-75174-6_VIII

U_0, and therefore $\mathbb{P}^n(\omega)$ is a compactification of $\Bbbk^n$. The set of points at infinity (with respect to $U_0 \cong \Bbbk^n$) on $\mathbb{P}^n(\omega)$ is $V(x_0) = \{[0 : a_1 : \cdots : a_n] : (a_1, \ldots, a_n) \in \Bbbk^n \setminus \{0\}\} \cong \mathbb{P}^{n-1}(\omega_1, \ldots, \omega_n)$. Since $\omega_0 = 1$, for each polynomial $f \in \Bbbk[x_1, \ldots, x_n]$, we can define its *weighted homogenization* with respect to ω as $\tilde{f} := x_0^{\omega(f)} f(x_1/x_0^{\omega_1}, \ldots, x_n/x_0^{\omega_n})$. It is straightforward to check that $\tilde{f}$ is ω-homogeneous and $\mathrm{ld}_\omega(\tilde{f}) = \mathrm{ld}_\omega(f)$.

Proposition VIII.1. *Identify $\Bbbk^n$ with U_0.*

(1) *Let $f \in \Bbbk[x_1, \ldots, x_n]$. Then $V(\tilde{f})$ is the Zariski closure in $\mathbb{P}^n(\omega)$ of $V(f) \subset \Bbbk^n$.*

(2) *If $f_1, \ldots, f_k \in \Bbbk[x_1, \ldots, x_n]$, then the following are equivalent:*
   (a) $\bigcap_{j=1}^k V(\tilde{f}_j) \setminus \Bbbk^n = \emptyset$.
   (b) *there is no common zero of* $\mathrm{ld}_\omega(f_1), \ldots, \mathrm{ld}_\omega(f_k)$ *on* $\Bbbk^n \setminus \{0\}$.

PROOF. At first we prove assertion (1). If f is a nonzero constant, then both $V(f)$ and $V(\tilde{f})$ are empty. So assume f is a non-constant polynomial. If $n = 1$, then $V(f)$ consists of finitely many points and it is straightforward to check that $V(f) = V(\tilde{f}) \subset \mathbb{P}^1(\omega_0, \omega_1)$. So assume $n \geq 2$. For each $j = 0, \ldots, n$, the set of points in $V(\tilde{f}) \cap U_j$ is precisely the set of zeroes on U_j of the regular function $(\tilde{f})^{\omega_j}/x_j^{\omega(f)}$. Therefore, theorem III.80 implies that each irreducible component of $\dim(V(\tilde{f}))$ has dimension $n - 1$. On the other hand,

$$V(\tilde{f}) \cap (\mathbb{P}^n \setminus \Bbbk^n) = V(\tilde{f}) \cap V(x_0) = V(\mathrm{ld}_\omega(f)) \cap V(x_0). \tag{74}$$

Since $\mathrm{ld}_\omega(f)$ is a nonzero polynomial and $V(x_0) \cong \mathbb{P}^{n-1}(\omega_1, \ldots, \omega_n)$, theorem III.80 implies that each irreducible component of $V(\tilde{f}) \setminus \Bbbk^n$ has dimension $n - 2$, and therefore cannot be an irreducible component of $V(\tilde{f})$. Consequently every irreducible component of $V(\tilde{f})$ intersects $\Bbbk^n$ and therefore contains an irreducible component of $V(f)$. Since $V(\tilde{f})$ is Zariski closed in $\mathbb{P}^n(\omega)$, this completes the proof of assertion (1). Assertion (2) follows from identity (74). □

3. Weighted Bézout theorem

We use the following notation throughout the rest of the book: for a nonnegative integer n we denote by $[n]$ the set $\{1, \ldots, n\}$; if $I \subseteq [n]$ and k is a field (in most cases k will be either $\Bbbk$ or $\mathbb{R}$), we write

$$k^I := \{(x_1, \ldots, x_n) \in k^n : x_i = 0 \text{ if } i \notin I\} \cong k^{|I|}, \tag{75}$$

$$(k^*)^I := \{\prod_{i \in I} x_i \neq 0\} \cap k^I \cong (k^*)^{|I|}, \tag{76}$$

where $k^* := k \setminus \{0\}$. Note that $k^\emptyset = (k^*)^\emptyset = \{0\}$.

THEOREM VIII.2. (Weighted Bézout theorem (theorem I.4)). *Let ω be a weighted degree on $\Bbbk[x_1, \ldots, x_n]$ with positive weights ω_j for x_j, $j = 1, \ldots, n$. Then the number of isolated solutions of polynomials $f_1, \ldots, f_n$ on $\Bbbk^n$ is bounded* *above by* $(\prod_j \omega(f_j))/(\prod_j \omega_j)$. *This bound is exact if and only if the leading weighted homogeneous forms of* $f_1, \dots, f_n$ *have no common solution other than* $(0, \dots, 0)$.

PROOF. Due to multiadditivity of intersection multiplicities (assertion (5) of proposition IV.28), we may replace each f_i by f_i^m for an appropriate positive integer m and assume that $m_{i,j} := \omega(f_i)/\omega_j$ is an integer for each i, j. Let $\mathcal{P}_i$ be the simplex in $\mathbb{R}^n$ with vertices at the origin and at $m_{i,j}e_j$, $j = 1, \dots, n$, where $e_1, \dots, e_n$ are the standard unit vectors in $\mathbb{R}^n$. Analogous to the definition of $[\mathcal{P}_1, \dots, \mathcal{P}_n]^{iso}_{(\mathbb{k}^*)^n}$ in (67), define

$$[\mathcal{P}_1, \dots, \mathcal{P}_n]^{iso}_{\mathbb{k}^n} := \max\{[p_1, \dots, p_n]^{iso}_{\mathbb{k}^n} : \operatorname{Supp}(p_j) \subseteq \mathcal{P}_j,\ j = 1, \dots, n\}.$$

It suffices to show that

(i) $[\mathcal{P}_1, \dots, \mathcal{P}_n]^{iso}_{\mathbb{k}^n} = (\prod_j \omega(f_j))/(\prod_j \omega_j)$ and
(ii) $[f_1, \dots, f_n]^{iso}_{\mathbb{k}^n} = [\mathcal{P}_1, \dots, \mathcal{P}_n]^{iso}_{\mathbb{k}^n}$ if and only if there is no common zero of $\operatorname{ld}_\omega(f_1), \dots, \operatorname{ld}_\omega(f_n)$ on $\mathbb{k}^n \setminus \{0\}$.

Assertion (i) follows from Bernstein's theorem (theorem VII.5) and the following claim.

Claim VIII.2.1. *(1)* $\operatorname{MV}(\mathcal{P}_1, \dots, \mathcal{P}_n) = (\prod_j \omega(f_j))/(\prod_j \omega_j)$.
(2) $[\mathcal{P}_1, \dots, \mathcal{P}_n]^{iso}_{\mathbb{k}^n} = [\mathcal{P}_1, \dots, \mathcal{P}_n]^{iso}_{(\mathbb{k}^*)^n}$.

PROOF. Since $\mathcal{P}_j = \frac{\omega(f_j)}{\omega(f_1)}\mathcal{P}_1$ for each $j = 1, \dots, n$, the properties of mixed volume from theorem VII.2 imply that

$$\begin{aligned}\operatorname{MV}(\mathcal{P}_1, \dots, \mathcal{P}_n) &= \operatorname{MV}(\mathcal{P}_1, \frac{\omega(f_2)}{\omega(f_1)}\mathcal{P}_1, \dots, \frac{\omega(f_n)}{\omega(f_1)}\mathcal{P}_1) = \prod_{j=2}^n \frac{\omega(f_j)}{\omega(f_1)} \operatorname{MV}(\mathcal{P}_1, \dots, \mathcal{P}_1)\\ &= \prod_{j=2}^n \frac{\omega(f_j)}{\omega(f_1)} n! \operatorname{Vol}_n(\mathcal{P}_1) = \prod_{j=2}^n \frac{\omega(f_j)}{\omega(f_1)} \prod_{j=1}^n m_{1,j} = \prod_{j=1}^n \frac{\omega(f_j)}{\omega_j}.\end{aligned}$$

This proves the first assertion of the claim. For the second assertion, let p_j be an arbitrary polynomial supported at $\mathcal{P}_j$, $j = 1, \dots, n$. Since it is clear that $[\mathcal{P}_1, \dots, \mathcal{P}_n]^{iso}_{\mathbb{k}^n} \geq [\mathcal{P}_1, \dots, \mathcal{P}_n]^{iso}_{(\mathbb{k}^*)^n}$, it suffices to show that there are p'_j supported at $\mathcal{P}_j$ such that $[p_1, \dots, p_n]^{iso}_{\mathbb{k}^n} \leq [p'_1, \dots, p'_n]^{iso}_{(\mathbb{k}^*)^n}$. Write $\mathcal{A}_j := \mathcal{P}_j \cap \mathbb{Z}^n$, $j = 1, \dots, n$. Pick $(\mathcal{A}_1, \dots, \mathcal{A}_n)$-non-degenerate $(q_1, \dots, q_n) \in \mathcal{L}(\mathcal{A}_1, \dots, \mathcal{A}_n)$, and set $r_i(x, t) := (1 - t)p_i + tq_i$. For each $\epsilon \in \mathbb{k}$, write $r_{\epsilon,i} := r_i|_{t=\epsilon}$. Note that $r_{1,i} = q_i$ and $r_{0,i} = p_i$. Pick a generic $\epsilon \in \mathbb{k}$. Due to theorem VII.15 we may assume that $(r_{\epsilon,1}, \dots, r_{\epsilon,n})$ is also $(\mathcal{A}_1, \dots, \mathcal{A}_n)$-non-degenerate. Since the intersection of an $\mathcal{A}_j$ with a coordinate subspace is a (nonempty) face of $\mathcal{A}_j$, it is then straightforward to check (e.g. using remark VII.9) from the definition of $(\mathcal{A}_1, \dots, \mathcal{A}_n)$-non-degeneracy that $r_{\epsilon,1}, \dots, r_{\epsilon,n}$ do not have any common zero on $\mathbb{k}^n \setminus (\mathbb{k}^*)^n$. Assertion (4) of theorem IV.32 then implies that $[p_1, \dots, p_n]^{iso}_{\mathbb{k}^n} \leq [r_{\epsilon,1}, \dots, r_{\epsilon,n}]^{iso}_{\mathbb{k}^n} = [r_{\epsilon,1}, \dots, r_{\epsilon,n}]^{iso}_{(\mathbb{k}^*)^n}$, as required. ☐

Now we prove assertion (ii). At first assume $[f_1, \ldots, f_n]^{iso}_{\mathbb{k}^n} < [\mathcal{P}_1, \ldots, \mathcal{P}_n]^{iso}_{\mathbb{k}^n}$. We will show that the leading weighted homogeneous forms $\mathrm{ld}_\omega(f_i)$ of f_i have a common zero on $\mathbb{k}^n \setminus \{0\}$. Let $\omega' := (1, \omega_1, \ldots, \omega_n)$. Embed $\mathbb{k}^n$ into $\mathbb{P}^n(\omega')$ via the map $(x_1, \ldots, x_n) \mapsto [1 : x_1 : \cdots : x_n]$. Set $p_i := f_i$, $i = 1, \ldots, n$. Let q_i and $r_i := (1-t)p_i + tq_i$ be as in the proof of claim VIII.2.1. Let V be the (finite) set of common zeroes of $q_1, \ldots, q_n$ on $\mathbb{k}^n$ and $C \subseteq \mathbb{k}^{n+1}$ be the union of irreducible components of $V(r_1, \ldots, r_n)$ which intersect $V \times \{1\}$. Assertion (1) of theorem IV.32 implies that C is a curve. Since $[f_1, \ldots, f_n]^{iso}_{\mathbb{k}^n} < [q_1, \ldots, q_n]^{iso}_{\mathbb{k}^n}$, assertion (5) of theorem IV.32 implies that one of the following holds:

(iii) there is a positive dimensional component of $V(f_1, \ldots, f_n) \subset \mathbb{k}^n$, or
(iv) C "has a point at infinity at $t = 0$," i.e. if $\bar{C}$ is the closure of C in $\mathbb{P}^n(\omega') \times \mathbb{P}^1$, then $\bar{C} \cap ((\mathbb{P}^n(\omega') \setminus X) \times \{0\}) \neq \emptyset$.

Denote the weighted homogeneous coordinates on $\mathbb{P}^n(\omega')$ by $[x_0 : \cdots : x_n]$. Let $\tilde{f}_i$ and $\tilde{q}_i$ be the *weighted homogenization* with respect to ω' respectively of f_i and q_i. Proposition VIII.1 implies that the closures of $V(f_i)$ and $V(q_i)$ in $\mathbb{P}^n(\omega')$ are, respectively, $V(\tilde{f}_i)$ and $V(\tilde{q}_i)$. Since $\omega(f_i) = \omega(q_i)$ for each i, it follows that the closure of $V(r_i)$ in $\mathbb{P}^n(\omega') \times \mathbb{k}$ is $V(\tilde{r}_i)$, where $\tilde{r}_i := (1-t)\tilde{f}_i + t\tilde{q}_i$. If (iii) holds, then the closure of $V(f_1, \ldots, f_n)$ in $\mathbb{P}^n(\omega')$ contains a point $a' \in \mathbb{P}^n(\omega') \setminus \mathbb{k}^n$. The weighted homogeneous coordinates of a' are of the form $[0 : a_1 : \cdots : a_n]$ with $a = (a_1, \ldots, a_n) \in \mathbb{k}^n \setminus \{0\}$. Then $\tilde{f}_i(a') = 0$, and therefore $\mathrm{ld}_\omega(f_i)(a) = 0$ for each i. On the other hand, if (iv) holds, then let $(a', 0) \in \bar{C} \cap ((\mathbb{P}^n(\omega') \setminus X) \times \{0\})$. Since $\tilde{r}_i(a', 0) = 0$ for each i, this again yields a common zero of the $\mathrm{ld}_\omega(f_i)$ on $\mathbb{k}^n \setminus \{0\}$, as required.

It remains to show the necessity of the non-degeneracy condition. Assume the leading weighted homogeneous forms $\mathrm{ld}_\omega(f_i)$ of f_i have a common solution $a \in \mathbb{k}^n \setminus \{0\}$. As in section VII.4.2, pick BKK non-degenerate $g_1, \ldots, g_n$ with a common zero $b \in \mathbb{k}^n$ such that for each i, $\mathrm{NP}(g_i) = \mathcal{P}_i$ and $\mathrm{ld}_\omega(g_i)(a) \neq 0 \neq f_i(b)$. Define a rational curve C on $\mathbb{k}^n$ via the parametrization $c(t) := (c_1(t), \ldots, c_n(t)) : \mathbb{k} \to \mathbb{k}^n$ from (61) with $\nu := -\omega$, i.e.

$$c_j(t) := a_j t^{-\omega_j} + (b_j - a_j)t^{-\omega_j + k_j},\ j = 1, \ldots, n,$$

where each k_j is a positive integer. Define $h_1, \ldots, h_n$ as in (62) with $m_j = -\omega(f_j) = -\omega(g_j)$, i.e.

$$h_j := t^{\omega(f_j)} f_j(c(t)) g_j - t^{\omega(g_j)} g_j(c(t)) f_j.$$

Note that $t^{\omega(f_j)} f_j(c(t))$ and $t^{\omega(g_j)} g_j(c(t))$ are *polynomials* in t. The same arguments as in the alternate proof of necessity of BKK non-degeneracy in section VII.4.2.2 show that $[f_1, \ldots, f_n]^{iso}_{\mathbb{k}^n} < [g_1, \ldots, g_n]^{iso}_{\mathbb{k}^n}$, so that the weighted homogeneous bound is *not* exact. □

Corollary VIII.3. (Bézout's theorem (theorem I.2)). *The number of isolated solutions of polynomials $f_1, \ldots, f_n$ on $\mathbb{k}^n$ is bounded above by $\prod_j \deg(f_j)$. This bound* *is exact if and only if the leading homogeneous forms of* $f_1, \dots, f_n$ *have no common solution other than* $(0, \dots, 0)$.

PROOF. In theorem VIII.2 take ω to be the usual degree of polynomials. □

Combining the arguments of the proof of assertion (2) of claim VIII.2.1 with theorems III.80, VII.5 and VII.33 gives the following characterization of polytopes $\mathcal{P}_1, \dots, \mathcal{P}_n$ such that $[\mathcal{P}_1, \dots, \mathcal{P}_n]^{iso}_{\mathbb{k}^n} = [\mathcal{P}_1, \dots, \mathcal{P}_n]^{iso}_{(\mathbb{k}^*)^n}$. Given a coordinate subspace H of $\mathbb{R}^n$, we write $T^H(\mathcal{P}) := \{j : \mathcal{P}_j \cap H \neq \emptyset\} \subset [n]$.

Lemma VIII.4. *Let* $\mathcal{P}_1, \dots, \mathcal{P}_n$ *be convex integral polytopes in* $\mathbb{R}^n_{\geq 0}$. *Then the following are equivalent:*

(1) $[\mathcal{P}_1, \dots, \mathcal{P}_n]^{iso}_{\mathbb{k}^n} = [\mathcal{P}_1, \dots, \mathcal{P}_n]^{iso}_{(\mathbb{k}^*)^n}$.

(2) *For each proper coordinate subspace* H *of* $\mathbb{R}^n$, *one of the following is true:*
   (a) *either there is a coordinate subspace* H' *of* $\mathbb{R}^n$ *such that* $H' \supset H$ *and* $|T^{H'}(\mathcal{P})| < \dim(H')$, *or*
   (b) $T^H(\mathcal{P})$ *is nonempty and* $\{\mathcal{P}_j \cap H : j \in T^H(\mathcal{P})\}$ *is a* dependent *collection of polytopes.* □

4. Products of weighted projective spaces

In this section we examine the closures of the hypersurfaces in a product of weighted projective spaces. This would be useful in the proof of the weighted multi-homogeneous Bézout bound (theorem VIII.8). Let $\bar{X} := \prod_{j=1}^s \mathbb{P}^{n_j}(\omega_j)$, where each ω_j is a weighted degree on $A_j := \mathbb{k}[x_{j,0}, \dots, x_{j,n_j}]$ such that the weight $\omega_{j,k}$ of $x_{j,k}$ is positive for each j, k. Let $A := \mathbb{k}[x_{j,k} : 1 \leq j \leq s,\ 0 \leq k \leq n_j]$, and for each j, let $\tilde{\omega}_j$ be the trivial extension of ω_j to A, i.e.

$$\tilde{\omega}_j(x_{k,l}) = \begin{cases} \omega_{k,l} & \text{if } k = j, \\ 0 & \text{otherwise.} \end{cases}$$

We say that a polynomial $h \in A$ is *(weighted multi-) homogeneous with respect to* $\Omega := \{\omega_1, \dots, \omega_s\}$, or in short, f is Ω*-homogeneous*, if it is $\tilde{\omega}_j$-homogeneous for each $j = 1, \dots, s$. If h is Ω-homogeneous, then $V(h) := \{a \in \bar{X} : h(a) = 0\}$ is a well-defined Zariski closed subset of $\bar{X}$.

4.1. The case that $\omega_{j,0} = 1$ for each $j = 1, \dots, s$. In this section we consider the case that $\omega_{j,0} = 1$ for each j. In this case each $\mathbb{P}^{n_j}(\omega_j)$ is a compactification of $\mathbb{k}^{n_j}$, and therefore $\bar{X}$ is a compactification of $\mathbb{k}^n$, where $n := \sum_{j=1}^s n_j$. More precisely, $\mathbb{k}^n$ can be identified with $\bar{X} \setminus V(\prod_{j=1}^s x_{j,0})$, and we may treat $B := \mathbb{k}[x_{j,k} : 1 \leq j \leq s,\ 1 \leq k \leq n_j] \subset A$ as the coordinate ring of $\mathbb{k}^n$. Let $f \in B$. Then Ω*-homogenization* $\tilde{f}$ of f is formed by substituting $x_{j,k}/x_{j,0}$ for $x_{j,k}$ in f for each j, k, and then clearing out the denominator. The following result is the analogue of assertion (1) of proposition VIII.1, and follows from the same arguments.

Proposition VIII.5. *Let $f \in B$. If $\Bbbk$ is algebraically closed, then $V(\tilde{f})$ is the Zariski closure in $\bar{X}$ of $V(f) \subset \Bbbk^n$.* □

The set of points at infinity on $\bar{X}$ is $\bar{X} \setminus \Bbbk^n = V(\prod_{j=1}^s x_{j,0}) = \bigcup_J Y_J$, where the union is over all *nonempty* subsets J of $[s] := \{1, \ldots, s\}$, and Y_J are defined as follows:

$$Y_J = V(x_{j,0} : j \in J) \setminus V(\prod_{j \notin J} x_{j,0}) \cong \prod_{j \in J} \mathbb{P}^{n_j - 1}(\omega'_j) \times \prod_{j \notin J} \Bbbk^{n_j},$$

where ω'_j are the restriction of ω_j to $\Bbbk[x_{j,1}, \ldots, x_{j,n_j}]$ for each j. Fix a nonempty subset J of $[s]$. Given $f \in B$, we would like to compute the points at infinity on $V(\tilde{f})$ which belong to Y_J. If $f = \sum_\alpha c_\alpha x^\alpha$, we write $\mathrm{ld}_{\Omega,J}(f)$ for be sum of $c_\alpha x^\alpha$ over all α such that $\tilde{\omega}_j(x^\alpha) = \tilde{\omega}_j(f)$ for each $j \in J$. In other words, $\mathrm{ld}_{\Omega,J}(f)$ is obtained from $\tilde{f}$ by substituting $x_{j,0} = 0$ for each $j \in J$ and $x_{j,0} = 1$ for each $j \notin J$.

Example VIII.6. Let $s = 3$, $n_1 = n_2 = 1$ and $n_3 = 2$ (so that $n = 4$). Let ω_1, ω_2 be the usual degree in, respectively, $x_{1,1}$ and $x_{2,1}$ coordinates, and ω_3 be the weighted degree in $(x_{3,1}, x_{3,2})$ coordinates corresponding to weights 2 for $x_{3,1}$ and 3 for $x_{3,2}$. Let $f = x_{1,1}^5 + x_{2,1}^7 + x_{1,1}^5 x_{2,1}^7 + x_{3,1}^3 + x_{1,1}^5 x_{3,2}^2$. Then $\omega_1(f) = 5$, $\omega_2(f) = 7$, $\omega_3(f) = 6$ and

$$\mathrm{ld}_{\Omega,J}(f) = \begin{cases} x_{1,1}^5 + x_{1,1}^5 x_{2,1}^7 + x_{1,1}^5 x_{3,2}^2 & \text{if } J = \{1\}, \\ x_{2,1}^7 + x_{1,1}^5 x_{2,1}^7 & \text{if } J = \{2\}, \\ x_{3,1}^3 + x_{1,1}^5 x_{3,2}^2 & \text{if } J = \{3\}, \\ x_{1,1}^5 x_{2,1}^7 & \text{if } J = \{1,2\}, \\ x_{1,1}^5 x_{3,2}^2 & \text{if } J = \{1,3\}, \\ 0 & \text{if } J = \{2,3\} \text{or } J = \{1,2,3\}. \end{cases}$$

Note the following difference from the weighted homogeneous case: if $|J| \geq 2$, then it might happen that $\mathrm{ld}_{\Omega,J}(f) = 0$ even if f is a nonzero polynomial.

Since $\mathrm{ld}_{\Omega,J}(f)$ is $\tilde{\omega}_j$-homogeneous for each $j \in J$, it defines a well-defined Zariski closed subset of Y_J. It is straightforward to check that this set is precisely the intersection of $V(\tilde{f})$ and Y_J, which is the content of the next result.

Proposition VIII.7. *Assume $\Bbbk$ is algebraically closed. If $f \in B$, then $V(\tilde{f}) \cap Y_J = V(\mathrm{ld}_{\Omega,J}(f)) \cap Y_J$. If $f_1, \ldots, f_k \in B$, then the following are equivalent*

(1) $\bigcap_{j=1}^k V(\tilde{f}_j) \setminus \Bbbk^n = \emptyset$.

(2) For each nonempty subset J of $[s]$, there is no common zero of $\mathrm{ld}_{\Omega,J}(f_1), \ldots, \mathrm{ld}_{\Omega,J}(f_k)$ on $\prod_{j \in J}(\Bbbk^{n_j} \setminus \{0\}) \times \prod_{j \notin J} \Bbbk^{n_j}$. □

5. Weighted multi-homogeneous Bézout theorem

We now generalize the weighted Bézout bound to the multi-projective setting. Let $\mathscr{I} := (I_1, \ldots, I_s)$ be an ordered partition of $[n] := \{1, \ldots, n\}$, i.e. $[n] = \bigcup_j I_j$

and $\sum_j |I_j| = n$. For each $j = 1, \ldots, s$, let ω_j be a weighted degree on $\mathbb{k}[x_k : k \in I_j]$ with positive weights $\omega_{j,k}$ for x_k, $k \in I_j$. Let $f_1, \ldots, f_n$ be polynomials on $\mathbb{k}^n$. Given $d_{i,j} := \omega_j(f_i)$, we would like to compute a (sharp) upper bound of the number of isolated solutions of $f_1, \ldots, f_n$. Let $n_j := |I_j|$ and l_j be the least common multiple of $\omega_{j,1}, \ldots, \omega_{j,n_j}$ and $\mathcal{S}_j$ be the simplex in $\mathbb{R}^{n_j}_{\geq 0}$ defined by $\{\alpha : \langle \omega_j, \alpha \rangle \leq l_j\}$. Note that $\mathrm{Supp}(f_i) \subseteq \mathcal{P}_i := \prod_{j=1}^s (d_{i,j}/l_j)\mathcal{S}_j \subset \mathbb{R}^n$. By definition $\mathrm{MV}(\mathcal{P}_1, \ldots, \mathcal{P}_n)$ is the coefficient of $\lambda_1 \cdots \lambda_n$ in the polynomial

$$\mathrm{Vol}_n(\sum_{i=1}^n \lambda_i \mathcal{P}_i) = \mathrm{Vol}_n(\sum_{i=1}^n \lambda_i \prod_{j=1}^s \frac{d_{i,j}}{l_j}\mathcal{S}_j) = \mathrm{Vol}_n(\prod_{j=1}^s(\sum_{i=1}^n \lambda_i \frac{d_{i,j}}{l_j}\mathcal{S}_j)) = \prod_{j=1}^s(\sum_{i=1}^n \lambda_i \frac{d_{i,j}}{l_j})^{n_j} \mathrm{Vol}_n(\prod_{j=1}^s \mathcal{S}_j)$$

$$= \frac{\prod_{j=1}^s \prod_{k=1}^{n_j}(l_j/\omega_{j,k})}{\prod_{j=1}^s (n_j! l_j^{n_j})} \prod_{j=1}^s (\sum_{i=1}^n \lambda_i d_{i,j})^{n_j} = \frac{\prod_{j=1}^s (\sum_{i=1}^n \lambda_i d_{i,j})^{n_j}}{\prod_{j=1}^s (n_j! \prod_{k=1}^{n_j} \omega_{j,k})}.$$

The coefficient of $\lambda_1 \cdots \lambda_n$ in $\prod_{j=1}^s (\sum_{i=1}^n \lambda_i d_{i,j})^{n_j}$ is the *permanent* of the following $n \times n$ matrix:

$$D(\mathscr{I}, \vec{d}) := \begin{pmatrix} \overbrace{d_{1,1} \quad \cdots \quad d_{1,1}}^{n_1 \text{ times}} & \cdots & \cdots & \overbrace{d_{1,s} \quad \cdots \quad d_{1,s}}^{n_s \text{ times}} \\ \vdots \qquad\qquad \vdots & & & \vdots \qquad\qquad \vdots \\ d_{n,1} \quad \cdots \quad d_{n,1} & \cdots & \cdots & d_{n,s} \quad \cdots \quad d_{n,s} \end{pmatrix}.$$

The preceding observations together with lemma VIII.4 imply that

$$(77) \quad [f_1, \ldots, f_n]^{iso}_{\mathbb{k}^n} \leq [\mathcal{P}_1, \ldots, \mathcal{P}_n]^{iso}_{\mathbb{k}^n} = \mathrm{MV}(\mathcal{P}_1, \ldots, \mathcal{P}_n) = \frac{\mathrm{perm}(D(\mathscr{I}, \vec{d}))}{(\prod_j n_j!)(\prod_{j,k} \omega_{j,k})}.$$

Note that when $s = 1$, then Ω consists of only one weighted degree ω and $\mathrm{perm}(D(\mathscr{I}, \vec{d})) = n!\omega(f_1) \cdots \omega(f_n)$, so that the bound from (77) is precisely the weighted Bézout bound $\prod_{j=1}^n (\omega(f_j)/\omega_j)$. Now we determine the condition for the attainment of this bound. For each $j = 1, \ldots, s$, fix a new indeterminate u_j. Let B'_j be the ring of polynomials (over $\mathbb{k}$) in u_j and x_k, $k \in I_j$. Note that each $X_{\mathcal{P}_i}$ is isomorphic to $\prod_{j=1}^s \mathbb{P}^{n_j}(\omega'_j)$, where ω'_j is the weighted degree on B'_j such that $\omega'_j(u_j) = 1$ and $\omega'_j(x_k) = \omega_{j,k}$, $k \in I_j$. For each $g \in \mathbb{k}[x_1, \ldots, x_n]$ and each nonempty subset J of $[s]$, define $\mathrm{ld}_{\Omega,J}(g)$ as in section VIII.4.1. Finally, note that since $\mathscr{I}$ is a partition of $[n]$, we may identify $\mathbb{k}^n$ with $\prod_{j=1}^s \mathbb{k}^{I_j}$.

THEOREM VIII.8. (Weighted multi-homogeneous Bézout theorem). *The number of isolated solutions of polynomials $f_1, \ldots, f_n$ on $\mathbb{k}^n$ is bounded by* (77). *This bound is exact if and only if the following holds: for each nonempty subset J of $[s]$, there is no common zero of* $\mathrm{ld}_{\Omega,J}(f_1), \ldots, \mathrm{ld}_{\Omega,J}(f_n)$ *on* $\prod_{j \in J}(\mathbb{k}^{I_j} \setminus \{0\}) \times \prod_{j \notin J} \mathbb{k}^{I_j}$ *(upon identification of $\mathbb{k}^n$ with $\prod_{j=1}^s \mathbb{k}^{I_j}$).*

PROOF. We have already proved that (77) holds. Let $\tilde{f}_i$ be the Ω *-homogenization* of f_i defined as in section VIII.4.1. If the bound is not exact, then it follows by the same arguments as in the proof of theorem VIII.2 that $\bigcap_{j=1}^s V(\tilde{f}_j) \setminus$

$\mathbb{k}^n \neq \emptyset$, and then proposition VIII.7 implies that the condition in the second assertion of theorem VIII.8 is violated. Now assume there is a nonempty subset J of $[s]$ such that $\mathrm{ld}_{\Omega,J}(f_1), \ldots, \mathrm{ld}_{\Omega,J}(f_n)$ have a common zero $a = (a_1, \ldots, a_n) \in \prod_{j\in J}(\mathbb{k}^{I_j} \setminus \{0\}) \times \prod_{j\notin J} \mathbb{k}^{I_j}$. Replacing each f_i by f_i^m for some appropriate positive integer m, we may assume that the vertices of each $\mathcal{P}_i$ have integer coordinates. Let $I := \{i : a_i \neq 0\} \subseteq [n]$; in other words, I is the smallest subset of $[n]$ such that $a \in (\mathbb{k}^*)^I$. Since J is nonempty, it follows that I is also nonempty; in fact $I_j \cap I \neq \emptyset$ for each $j \in J$. For each $j \in J$, let $\mathcal{T}_j$ be the facet of $\mathcal{S}_j$ determined by $\langle \omega_j, \alpha \rangle = l_j$. Fix $i \in [n]$. Define

$$\mathcal{Q}_i := \mathbb{R}^I \cap \left(\prod_{j\in J} (d_{i,j}/l_j)\mathcal{T}_j \times \prod_{j\notin J} (d_{i,j}/l_j)\mathcal{S}_j \right).$$

Then $\mathcal{Q}_i$ is a proper (nonempty) face of $\mathcal{P}_i$. Let $f_{i,\mathcal{Q}_i}$ be the component of f_i supported at $\mathcal{Q}_i$, i.e. if $f_i = \sum_\alpha c_\alpha x^\alpha$, then $f_{i,\mathcal{Q}_i} = \sum_{\alpha\in\mathcal{Q}_i} c_\alpha x^\alpha$. It is straightforward to check that $f_{i,\mathcal{Q}_i}(a) = 0$ for each i. Now choose an integral element $\nu \in (\mathbb{R}^n)^*$ such that $\mathcal{Q}_i = \mathrm{In}_\nu(\mathcal{P}_i)$ for each i [you have to check that such ν exists!]. As in section VII.4.2, pick BKK non-degenerate $g_1, \ldots, g_n$ with a common zero $b \in \mathbb{k}^n$ such that for each i, $f_i(b) \neq 0$, $\mathrm{NP}(g_i) = \mathcal{P}_i$ and $\mathrm{In}_\nu(g_i)(a) \neq 0$. Define a rational curve C on $\mathbb{k}^n$ via the parametrization from (61) and define $h_1, \ldots, h_n$ as in (62) with $m_i = \nu(g_i)$ for each i. Then $t^{-m_j} f_j(c(t))$ and $t^{-m_j} g_j(c(t))$ are polynomials in t and the same arguments as in section VII.4.2 show that $[f_1, \ldots, f_n]^{iso}_{\mathbb{k}^n} < [g_1, \ldots, g_n]^{iso}_{\mathbb{k}^n}$, as required. □

6. Notes

A version of the weighted Bézout theorem appears in [Dam99]. I. Shafarevich gave the bound for the "multi-homogeneous" case (i.e. the case that all weights are 1) of the weighted multi-homogeneous Bézout theorem in the first edition of [Sha94] in the 1970s; see also [MS87]. We could not locate any past reference for the non-degeneracy condition or the estimate for the general weighted multi-homogeneous version.

CHAPTER IX

Intersection multiplicity at the origin

1. Introduction

In this chapter, we consider the "local" version of the affine Bézout problem, i.e. the problem of estimating the intersection multiplicity of generic hypersurfaces at the origin. This computation is a crucial ingredient of the extension in chapter X of Bernstein's theorem to the affine space. Recall that the *support* of a power series $f = \sum_\alpha c_\alpha x^\alpha \in \mathbb{k}[[x_1, \ldots, x_n]]$ is $\text{Supp}(f) := \{\alpha : c_\alpha \neq 0\}$ and we say that f is *supported* at $\mathcal{A} \subset \mathbb{Z}^n$ if $\text{Supp}(f) \subset \mathcal{A}$. Now let $\mathcal{A}_1, \ldots, \mathcal{A}_n$ be (possibly infinite) subsets of $\mathbb{Z}^n_{\geq 0}$. In the case that $\mathcal{A}_j$ are finite, we saw in chapter VII that within all f_j supported at $\mathcal{A}_j$, $j = 1, \ldots, n$, $[f_1, \ldots, f_n]^{iso}_{(\mathbb{k}^*)^n}$ takes the maximum value when $f_1, \ldots, f_n$ are generic. It is possible to talk about "generic" power series supported at $\mathcal{A}_j$ even if $\mathcal{A}_j$ is infinite, and it turns out that the intersection multiplicity $[f_1, \ldots, f_n]_0$ of $f_1, \ldots, f_n$ at the origin takes the *minimum* value when f_j are generic power series supported at $\mathcal{A}_j$, $j = 1, \ldots, n$ (see theorem IX.8 for the precise statement); in this chapter, we compute this minimum and give a Bernstein-Kushnirenko-type characterization of the systems which attain the minimum.

2. Generic intersection multiplicity

Let $\mathcal{A}_j$ be a (possibly infinite) subset of $\mathbb{Z}^n_{\geq 0}$, $j = 1, \ldots, n$. Define

$$[\mathcal{A}_1, \ldots, \mathcal{A}_n]_0 := \min\{[f_1, \ldots, f_n]_0 : \forall j,\ f_j \in \mathbb{k}[[x_1, \ldots, x_n]],\ \text{Supp}(f_j) \subseteq \mathcal{A}_j\} \tag{78}$$

In this section, we motivate, state and illustrate the formula for $[\mathcal{A}_1, \ldots, \mathcal{A}_n]_0$. Its proof is given in section IX.5.

P. Mondal, *How Many Zeroes?*, CMS/CAIMS Books in Mathematics 2,
https://doi.org/10.1007/978-3-030-75174-6_IX

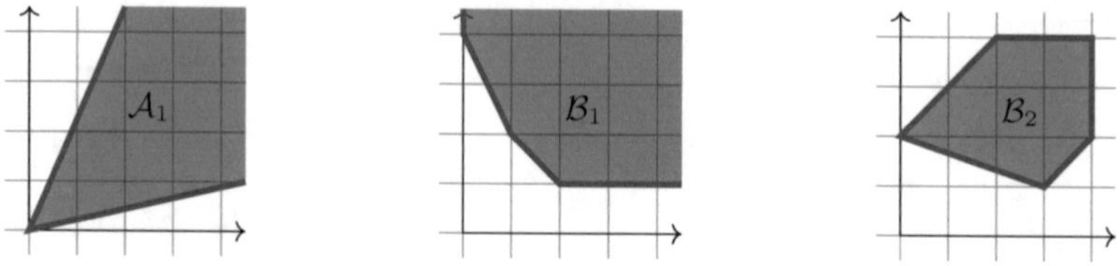

FIGURE 1. $[\mathcal{A}_1, \mathcal{A}_2]_0 = 0$ for any $\mathcal{A}_2$ since a generic f_1 supported at $\mathcal{A}_1$ has a non-zero constant term; $[\mathcal{B}_1, \mathcal{B}_2]_0 = \infty$ since every g_j supported at $\mathcal{B}_j$, $j = 1, 2$, identically vanishes on the x-axis.

2.1. Motivation for the formula. In this section, we study informally the case $\Bbbk = \mathbb{C}$ and try to motivate the formula for $[\mathcal{A}_1, \dots, \mathcal{A}_n]_0$. It is not hard to understand precisely when $[\mathcal{A}_1, \dots, \mathcal{A}_n]_0$ is zero or infinity—see fig. 1 and corollary IX.13. So consider the case that $0 < [\mathcal{A}_1, \dots, \mathcal{A}_n]_0 < \infty$, and pick f_j supported at $\mathcal{A}_j$, $j = 1, \dots, n$, such that $[f_1, \dots, f_n]_0 = [\mathcal{A}_1, \dots, \mathcal{A}_n]_0$. Then the origin is an isolated point of $f_1 = \cdots = f_n = 0$. Therefore, theorem IV.24 and proposition IV.28 imply that near the origin $f_2 = \cdots = f_n = 0$ defines a curve C, and

$$[f_1, \dots, f_n]_0 = \sum_B \operatorname{ord}_0(f_1|_B) \tag{79}$$

where the sum is over all "branches" B of C at the origin. We now try to compute this sum. So fix a branch B of C at the origin and an analytic parametrization $\gamma = (\gamma_1, \dots, \gamma_n) : U \to B$ of B, where U is a neighborhood of the origin on $\mathbb{C}$. Let H_B be the *smallest* coordinate subspace of $\mathbb{C}^n$ containing B.

2.1.1. *Case 1:* $H_B = \mathbb{C}^n$. In this case, no γ_i is identically zero so that each γ_i can be expressed as

$$\gamma_i = a_1 t^{\nu_i} + \cdots,$$

where $a_i \in \mathbb{C}^*$, t is an analytic coordinate on U and ν_i is the order (in t) of γ_i. Note that each ν_i is *positive*, since the center of B is at the origin. Let ν_B be the weighted order on $\Bbbk[x_1, \dots, x_n]$ such that $\nu_B(x_i) = \nu_i$, $i = 1, \dots, n$. Then for each j,

$$f_j(\gamma(t)) = \operatorname{In}_{\nu_B}(f_j)(a_1, \dots, a_n) t^{\nu(f_j)} + \cdots,$$

where $\operatorname{In}_{\nu_B}(\cdot)$ denotes the *initial form* with respect to ν_B, and the omitted terms have higher order in t. Since $f_j(\gamma(t)) \equiv 0$ for $j = 2, \dots, n$, we have a system of $(n-1)$ *weighted homogeneous* polynomial equations:

$$\operatorname{In}_{\nu_B}(f_j)(a_1, \dots, a_n) = 0, \ j = 2, \dots, n. \tag{80}$$

Theorem VII.5 implies that the number of solutions of (80) is the $(n-1)$-dimensional mixed volume of the Newton polytopes of $\operatorname{In}_{\nu_B}(f_j)$, $j = 2, \dots, n$. It is in fact "reasonable" to guess that in the generic case each solution of (80) corresponds to a distinct branch B of C and for each such B, the order of $f_1|_B$ at the origin is $\nu_B(f_1)$. In other words, for each weighted order ν,

$$\sum_{\substack{H_B=\mathbb{C}^n\\ \nu_B=\nu}} \operatorname{ord}_0(f_1|_B) = \nu(f_1)\operatorname{MV}(\operatorname{In}_\nu(f_2),\ldots,\operatorname{In}_\nu(f_n)) \tag{81}$$

2.1.2. *Case 2: $H_B \subsetneq \mathbb{C}^n$.* In this case, we may assume without loss of generality that H_B is the coordinate subspace spanned by $x_1,\ldots,x_k$, $k<n$; in other words $H_B = \mathbb{C}^I$ in the notation of (75) from section VIII.3, where $I := \{1,\ldots,k\}$. It follows that

$$\gamma_i = \begin{cases} a_i t^{\nu_i} + \cdots & \text{if } i = 1,\ldots,k,\\ 0 & \text{if } i = k+1,\ldots,n, \end{cases}$$

where $(a_1,\ldots,a_k) \in (\mathbb{C}^*)^k$. Since $[f_1,\ldots,f_n]_0 < \infty$ and the f_j are generic among those supported at $\mathcal{A}_j$, it follows that there are precisely $(n-k)$ elements among $f_2,\ldots,f_n$ which vanish identically on $\mathbb{C}^I$; after reordering the f_j, we may assume these are $f_{k+1},\ldots,f_n$. Since $f_j(\gamma(t)) \equiv 0$ for each $j=2,\ldots,k$, it follows that

$$\operatorname{In}_{\nu_B}(f_j|_{\mathbb{C}^I})(a_1,\ldots,a_k) = 0,\ j=2,\ldots,k, \tag{82}$$

where ν_B is the weighted order on $\mathbb{k}[x_1,\ldots,x_k]$ corresponding to weights ν_i for x_i, $i=1,\ldots,k$. As in the preceding case, the number of solutions of (82) in the generic situation is the $(k-1)$-dimensional mixed volume of the Newton polytopes of $\operatorname{In}_{\nu_B}(f_j|_{\mathbb{C}^I})$, $j=2,\ldots,k$, and each solution corresponds to a distinct branch B of C. However, each branch should be counted with proper multiplicity, and therefore this mixed volume should be multiplied by the "intersection multiplicity of $f_{k+1},\ldots,f_n$ along $\mathbb{C}^I$." It turns out (see corollary IV.34) that for generic $f_{k+1},\ldots,f_n$, this is precisely the intersection multiplicity at the origin of $f_{k+1}|_{(x_1,\ldots,x_k)=(\epsilon_1,\ldots,\epsilon_k)},\ldots,f_n|_{(x_1,\ldots,x_k)=(\epsilon_1,\ldots,\epsilon_k)}$, where $\epsilon_1,\ldots,\epsilon_k$ are generic elements from $\mathbb{C}^*$. If $\pi:\mathbb{R}^n \to \mathbb{R}^{n-k}$ is the projection onto the last $(n-k)$-coordinates, then the genericness of the ϵ_i imply that the support of $f_{k+j}|_{(x_1,\ldots,x_k)=(\epsilon_1,\ldots,\epsilon_k)}$ is precisely $\pi(\mathcal{A}_{k+j})$ for each j, and it is reasonable to guess that if the f_{k+j} and ϵ_i are generic, then

$$[f_{k+1}|_{(x_1,\ldots,x_k)=(\epsilon_1,\ldots,\epsilon_k)},\ldots,f_n|_{(x_1,\ldots,x_k)=(\epsilon_1,\ldots,\epsilon_k)}]_0 = [\pi(\mathcal{A}_{k+1}),\ldots,\pi(\mathcal{A}_n)]_0$$

It then follows as in the first case that for each weighted order ν on $\mathbb{k}[x_1,\ldots,x_k]$,

$$\begin{aligned} &\sum_{H_B=\mathbb{C}^I,\nu_B=\nu} \operatorname{ord}_0(f_1|_B)\\ &\quad= \nu(f_1|_{H_B})\operatorname{MV}(\operatorname{In}_\nu(f_2|_{H_B}),\ldots,\operatorname{In}_\nu(f_k|_{H_B}))\\ &\qquad\times [f_{k+1}|_{(x_1,\ldots,x_k)=(\epsilon_1,\ldots,\epsilon_k)},\ldots,f_n|_{(x_1,\ldots,x_k)=(\epsilon_1,\ldots,\epsilon_k)}]_0\\ &\quad= \min_{\mathcal{A}_1\cap\mathbb{R}^I}(\nu)\operatorname{MV}(\operatorname{In}_\nu(\mathcal{A}_2\cap\mathbb{R}^I),\ldots,\operatorname{In}_\nu(\mathcal{A}_k\cap\mathbb{R}^I))[\pi(\mathcal{A}_{k+1}),\ldots,\pi(\mathcal{A}_n)]_0 \end{aligned} \tag{83}$$

Therefore, $[\mathcal{A}_1, \dots, \mathcal{A}_n]_0$ should be the sum of the right-hand side of identity (83) over all appropriate I and ν. Theorem IX.1 states that this precisely the case, and the proof of theorem IX.1 in section IX.5.1 simply makes the preceding arguments rigorous.

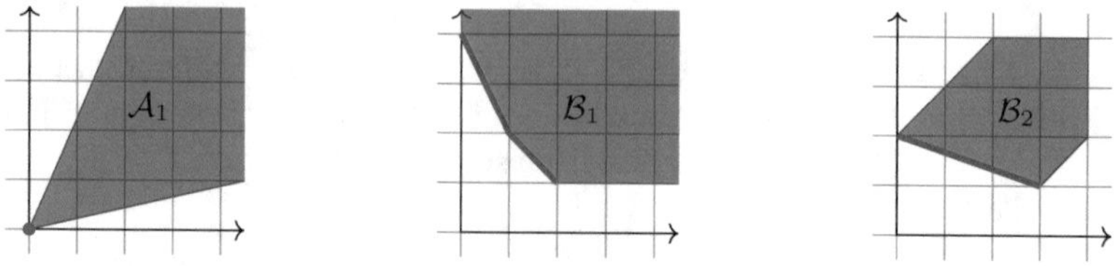

FIGURE 2. Newton diagrams (in red) of the sets from fig. 1.

2.2. Precise formulation. Let $\mathcal{A}$ be a (possibly infinite) subset of $\mathbb{Z}^n_{\geq 0}$. The convex hull of $\mathcal{A} + \mathbb{R}^n_{\geq 0}$ in $\mathbb{R}^n$ is a convex polyhedron (corollary B.48); the *Newton diagram* $\mathrm{ND}(\mathcal{A})$ of $\mathcal{A}$ is the union of the compact faces of this polyhedron (fig. 2). The Newton diagram of a power series f in $(x_1, \dots, x_n)$, denoted $\mathrm{ND}(f)$, is the Newton diagram of $\mathrm{Supp}(f)$; it is the local analogue of the Newton polytope of a polynomial. Given diagrams $\Gamma_1, \dots, \Gamma_n$ in $\mathbb{R}^n$, define

(84) $$[\Gamma_1, \dots, \Gamma_n]_0 := \min\{[f_1, \dots, f_n]_0 : \forall j,\ f_j \in \Bbbk[[x_1, \dots, x_n]],\ \mathrm{ND}(f_j) + \mathbb{R}^n_{\geq 0} \subseteq \Gamma_j + \mathbb{R}^n_{\geq 0}\}$$

We will see in theorem IX.1 below that $[\Gamma_1, \dots, \Gamma_n]_0$ can be expressed in terms of certain mixed volumes of the faces of Γ_j, and if Γ_j are Newton diagrams of $\mathcal{A}_j \subseteq \mathbb{Z}^n_{\geq 0}$, then $[\mathcal{A}_1, \dots, \mathcal{A}_n]_0 = [\Gamma_1, \dots, \Gamma_n]_0$. First, we need to introduce some notation. We write $[n] := \{1, \dots, n\}$. If $I \subseteq [n]$ and k is a field, recall that k^I is the $|I|$-dimensional coordinate subspace $\{(x_1, \dots, x_n) \in k^n : x_i = 0 \text{ if } i \notin I\}$, and $(k^*)^I = \{(x_1, \dots, x_n) \in k^I : x_i \neq 0 \text{ if } i \in I\}$. For $\mathcal{S} \subset \mathbb{R}^n$, we write $\mathcal{S}^I := \mathcal{S} \cap \mathbb{R}^I$. We denote by $\pi_I : k^n \to k^I$ the projection in the coordinates indexed by I, i.e.

(85) $$\text{the } j\text{-th coordinate of } \pi_I(x_1, \dots, x_n) := \begin{cases} x_j & \text{if } j \in I \\ 0 & \text{if } j \notin I. \end{cases}$$

Let ν be a weighted order on $\Bbbk[x_1, \dots, x_n]$ corresponding to weights ν_j for x_j, $j = 1, \dots, n$. We identify ν with the element in $(\mathbb{R}^n)^*$ with coordinates $(\nu_1, \dots, \nu_n)$ with respect to the dual basis. We say that ν is *centered at the origin* if each ν_i is positive and that ν is *primitive* if it is non-zero and the greatest common divisor of $\nu_1, \dots, \nu_n$ is 1. If ν is centered at the origin, then it also extends to a weighted order on the ring of power series in $(x_1, \dots, x_n)$. We write $\mathcal{V}_0$ for the set of weighted orders centered at the origin and $\mathcal{V}'_0$ for the primitive elements in $\mathcal{V}_0$. Given polytopes $\Gamma_1, \dots, \Gamma_n$ in $\mathbb{R}^n$, define

(86) $$[\Gamma_1, \dots, \Gamma_n]^*_0 := \sum_{\nu \in \mathcal{V}'_0} \min_{\Gamma_1}(\nu)\ \mathrm{MV}'_\nu(\mathrm{In}_\nu(\Gamma_2), \dots, \mathrm{In}_\nu(\Gamma_n))$$

where $\mathrm{MV}'_\nu(\cdot, \dots, \cdot)$ is defined as in (69).

THEOREM IX.1. ([Mon16]). *Let $\mathcal{A} := (\mathcal{A}_1, \dots, \mathcal{A}_n)$ be a collection of subsets of $\mathbb{Z}^n_{\geq 0}$ and Γ_j be the Newton diagram of $\mathcal{A}_j$, $j = 1, \dots, n$. For each $I \subset [n]$, let $T^I_{\mathcal{A}} := \{j : \mathcal{A}^I_j \neq \emptyset\}$ be the set of all indices j such that $\mathcal{A}_j$ touches the coordinate subspace $\mathbb{R}^I$ of $\mathbb{R}^n$. Define*

$$\mathscr{T}_{\mathcal{A},1} := \{I \subseteq [n] : I \neq \emptyset,\ |T^I_{\mathcal{A}}| = |I|,\ 1 \in T^I_{\mathcal{A}}\} \tag{87}$$

Then

(1) *If $0 \notin \bigcup_j \Gamma_j$ and there is $I \subset [n]$ such that $|T^I_{\mathcal{A}}| < |I|$, then*

$$[\mathcal{A}_1, \dots, \mathcal{A}_n]_0 = [\Gamma_1, \dots, \Gamma_n]_0 = \infty$$

(2) *Otherwise,*

$$[\mathcal{A}_1, \dots, \mathcal{A}_n]_0 = [\Gamma_1, \dots, \Gamma_n]_0 = \sum_{I \in \mathscr{T}_{\mathcal{A},1}} [\Gamma^I_1, \Gamma^I_{j_2}, \dots, \Gamma^I_{j_{|I|}}]^*_0 \times [\pi_{[n]\setminus I}(\Gamma_{j'_1}), \dots, \pi_{[n]\setminus I}(\Gamma_{j'_{n-|I|}})]_0 \tag{88}$$

where for each $I \in \mathscr{T}_{\mathcal{A},1}$, $j_1 = 1, j_2, \dots, j_{|I|}$ are elements of $T^I_{\mathcal{A}}$, and $j'_1, \dots, j'_{n-|I|}$ are elements of $[n] \setminus T^I_{\mathcal{A}}$.

Remark IX.2. The product of 0 and ∞, when/if it occurs in (88), is defined to be 0. Also empty intersection products and mixed volumes are defined as 1. In particular, when $n = 1$, the term $\mathrm{MV}(\mathrm{In}_\nu(\Gamma_2), \dots, \mathrm{In}_\nu(\Gamma_n))$ from (86) is defined to be 1.

Remark IX.3. (Generic intersection multiplicity is monotonic)**.** The formulae for $[\mathcal{A}_1, \dots, \mathcal{A}_n]_0$ from theorem IX.1 do not change if the $\mathcal{A}_j$ are replaced by $\mathcal{A}_j + \mathbb{R}^n_{\geq 0}$. This immediately implies that $[\cdot, \dots, \cdot]_0$ is *monotonic*, i.e. if $\mathcal{A}'_j \subseteq \mathcal{A}_j + \mathbb{R}^n_{\geq 0}$, $j = 1, \dots, n$, then $[\mathcal{A}'_1, \dots, \mathcal{A}'_n]_0 \geq [\mathcal{A}_1, \dots, \mathcal{A}_n]_0$. Precise characterization of the cases for which $[\mathcal{A}'_1, \dots, \mathcal{A}'_n]_0 > [\mathcal{A}_1, \dots, \mathcal{A}_n]_0$ is given in theorem IX.32.

It is not obvious from the outset that the term computed by (88) is invariant under the permutations of the $\mathcal{A}_j$. Some formulae which are invariant under permutations of the $\mathcal{A}_j$ are given in section IX.7. We now present an example to illustrate this invariance.

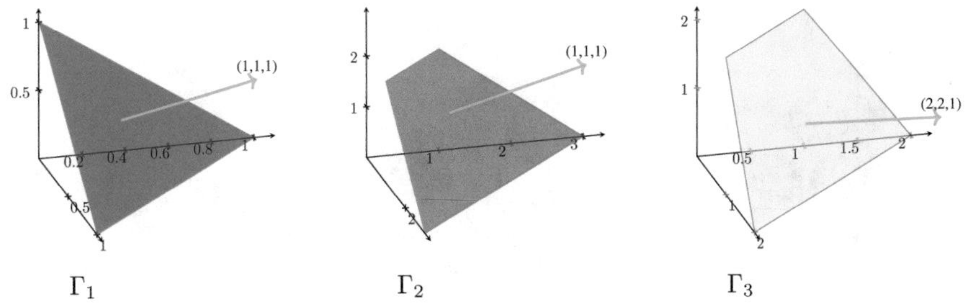

FIGURE 3. Newton diagrams of polynomials from example IX.4 and inner normals to their facets.

Example IX.4. Consider the polynomials f_1, f_2, f_3 from example VII.31. If $\mathcal{A} := (\mathrm{Supp}(f_1), \mathrm{Supp}(f_2), \mathrm{Supp}(f_3))$, then it is straightforward to check (see fig. 3) that $\mathscr{T}_{\mathcal{A},1} = \{\{1,2,3\},\{3\}\}$ so that (88) implies that

$$[f_1, f_2, f_3]_0 = [\Gamma_1, \Gamma_2, \Gamma_3]_0^* + [\pi_{\{1,2\}}(\Gamma_2), \pi_{\{1,2\}}(\Gamma_3)]_0 \, [\Gamma_1^{\{3\}}]_0^*$$

The Newton diagrams of $\pi_{\{1,2\}}(\Gamma_2)$ and $\pi_{\{1,2\}}(\Gamma_3)$ are the same diagram consisting of the line segment from $(1,0)$ to $(0,1)$, which is the Newton diagram of linear polynomials with no constant terms. It follows that $[\pi_{\{1,2\}}(\Gamma_2), \pi_{\{1,2\}}(\Gamma_3)]_0 = 1$. The Newton diagram of $\Gamma_2 + \Gamma_3$ has two facets with inner normals in $(\mathbb{R}_{>0})^3$, and these inner normals are $\nu_1 := (1,1,1)$ and $\nu_2 := (2,2,1)$ (see fig. 4). Then it follows from fig. 5 and identity (55) that

$$[f_1,f_2,f_3]_0 = \min_{\Gamma_1}(\nu_1)\,\mathrm{MV}'_{\nu_1}(\mathrm{In}_{\nu_1}(\Gamma_2), \mathrm{In}_{\nu_1}(\Gamma_3)) + \min_{\Gamma_1}(\nu_2)\,\mathrm{MV}'_{\nu_2}(\mathrm{In}_{\nu_2}(\Gamma_2), \mathrm{In}_{\nu_2}(\Gamma_3)) + 1 \cdot \mathrm{ord}_z(f_1|_{x=y=0})$$

$= \min_{\Gamma_1}(1,1,1) \cdot \mathrm{Area}($  $) + \min_{\Gamma_1}(2,2,1) \cdot \mathrm{Area}($  $) + 1 \cdot 1 = 1 \cdot 4 + 1 \cdot 1 + 1 = 6.$

On the other hand, with $\mathcal{A}' := (\mathrm{Supp}(f_3), \mathrm{Supp}(f_1), \mathrm{Supp}(f_2))$, one has $\mathscr{T}_{\mathcal{A}',1} = \{\{1,2,3\}\}$. Since the Newton diagram of $\Gamma_1 + \Gamma_2$ has only one facet and that the primitive inner normal to that facet is ν_1, we have from fig. 5 and identity (55) that

$$[f_1,f_2,f_3]_0 = [\Gamma_3, \Gamma_1, \Gamma_2]_0^* = \min_{\Gamma_3}(\nu_1)\,\mathrm{MV}'_{\nu_1}(\mathrm{In}_{\nu_1}(\Gamma_1), \mathrm{In}_{\nu_1}(\Gamma_2))$$

$= \min_{\Gamma_3}(1,1,1) \cdot \mathrm{Area}($ 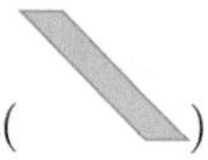 $) = 2 \cdot 3 = 6.$

Similarly, with $\mathcal{A}'' := (\mathrm{Supp}(f_2), \mathrm{Supp}(f_3), \mathrm{Supp}(f_1))$, one has $\mathscr{T}_{\mathcal{A}'',1} = \{\{1,2,3\}\}$. The Newton diagram of $\Gamma_3 + \Gamma_1$ have two facets, with inner normals ν_1 and ν_2, we have from fig. 5 and identity (55) that

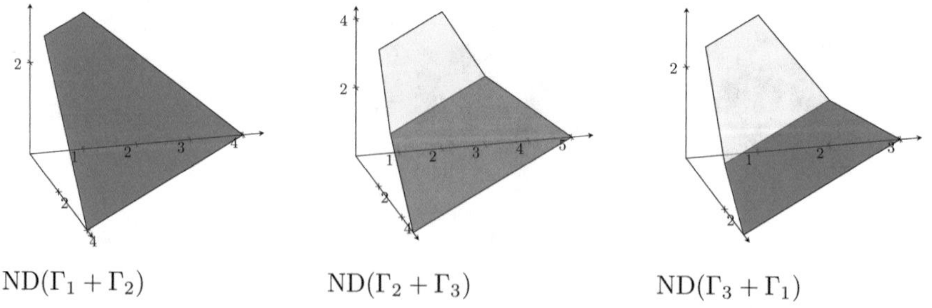

FIGURE 4. Sum of the Newton diagrams of polynomials from example IX.4.

$$
\begin{aligned}
[f_1, f_2, f_3]_0 &= [\Gamma_2, \Gamma_3, \Gamma_1]_0^* \\
&= \min_{\Gamma_2}(\nu_1)\,\mathrm{MV}'_{\nu_1}(\mathrm{In}_{\nu_1}(\Gamma_3), \mathrm{In}_{\nu_1}(\Gamma_1)) + \min_{\Gamma_2}(\nu_2)\,\mathrm{MV}'_{\nu_2}(\mathrm{In}_{\nu_2}(\Gamma_3), \mathrm{In}_{\nu_2}(\Gamma_1)) \\
&= \min_{\Gamma_2}(1,1,1) \cdot \mathrm{Area}(\;\;) + \min_{\Gamma_2}(2,2,1) \cdot \mathrm{Area}(\emptyset) = 3 \cdot 2 + 4 \cdot 0 = 6.
\end{aligned}
$$

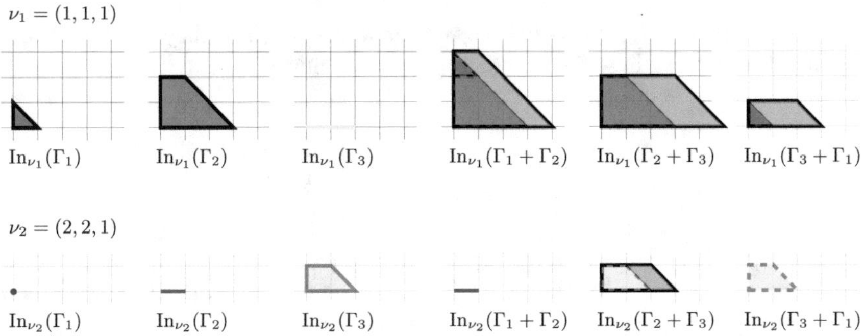

FIGURE 5. Normalized faces of the diagrams of example IX.4.

3. Characterization of minimal multiplicity systems

Given a collection $\mathcal{A} = (\mathcal{A}_1, \ldots, \mathcal{A}_m)$ of (possibly infinite) subsets of $\mathbb{Z}^n_{\geq 0}$, we write $\mathcal{L}_0(\mathcal{A}_j)$ for the space of all power series in $(x_1, \ldots, x_n)$ supported at $\mathcal{A}_j$, $j = 1, \ldots, n$, and $\mathcal{L}_0(\mathcal{A}) := \prod_{j=1}^n \mathcal{L}_0(\mathcal{A}_j)$. For the case that $m = n$, in this section, we characterize the systems $(f_1, \ldots, f_n) \in \mathcal{L}_0(\mathcal{A})$, which achieve the minimum possible intersection multiplicity at the origin. The proofs of the results of this section are given in sections IX.4, IX.6 and IX.8.

3.1. Non-degeneracy at the origin. As in the case of Bernstein's theorem, we try to guess the correct non-degeneracy condition by considering the case $\mathbb{k} = \mathbb{C}$. Also assume for simplicity that each $\mathcal{A}_j$ is *finite* so that every power series supported at $\mathcal{A}_j$ is in fact a *polynomial*. Now pick $f_j, g_j \in \mathbb{k}[x_1, \ldots, x_n]$ over $\mathbb{C}$ such that $\mathcal{A}_j = \mathrm{Supp}(f_j) \supseteq \mathrm{Supp}(g_j)$ for each j, and $[f_1, \ldots, f_n]_0 > [g_1, \ldots, g_n]_0$. Write $h_j := (1-t)f_j + tg_j$, $j = 1, \ldots, n$. Then it seems reasonable to expect that there is a curve $C(t)$ on $\mathbb{C}^n$ such that $h_j(C(t)) = 0$ and $\lim_{t \to 0} C(t) = 0$ (see fig. 6). Pick a parametrization $U \to C(t)$, where U is a neighborhood of the origin on $\mathbb{C}$ the form $\gamma : t \mapsto (\gamma_1(t), \ldots, \gamma_n(t))$, where each γ_i is a power series in t. Fix $j = 1, \ldots, n$. As in section VII.3.2, we examine the initial part of the expansion of $h_j(\gamma(t))$.

3.1.1. *Base case.* At first, we consider the case that the image of γ intersects $(\mathbb{C}^*)^n$, i.e. no γ_i is identically zero. Let $\gamma_i = a_i t^{\nu_i} + \cdots$, where $a_i \in \mathbb{C}^*$, and $\nu_i := \mathrm{ord}_t(\gamma_i)$. Let ν be the weighted order on $\mathbb{k}[x_1, \ldots, x_n]$ corresponding to weights ν_i for x_i, $i = 1, \ldots, n$. Then it follows exactly as in section VII.3.2 that $\mathrm{In}_{\mathcal{A}_j,\nu}(f_j)(a) = 0$ for each $j = 1, \ldots, n$, where $\mathrm{In}_{\mathcal{A}_j,\nu}(\cdot)$ are defined as in (58). Since $\lim_{t \to 0} \gamma(t) = 0$, it follows in addition that each ν_j is positive, i.e. ν is *centered at the origin.* This leads to the following notion.

Definition IX.5. Let $\mathcal{A} := (\mathcal{A}_1, \ldots, \mathcal{A}_m)$ be a collection of (possibly infinite) subsets of $\mathbb{Z}^n_{\geq 0}$ and $(f_1, \ldots, f_m) \in \mathcal{L}_0(\mathcal{A})$. We say that $f_1, \ldots, f_m$ are $(\mathcal{A}, *)$-*non-degenerate at the origin* if they satisfy the following condition:

$$(\mathrm{ND}_0^*) \qquad \begin{array}{l}\text{for each weighted order } \nu \text{ centered at the origin, there is} \\ \text{no common root of } \mathrm{In}_{\mathcal{A}_j,\nu}(f_j),\ j = 1, \ldots, m, \text{ on } (\mathbb{k}^*)^n.\end{array}$$

We say that $f_1, \ldots, f_m$ are $*$-*non-degenerate at the origin* if they are $(\mathcal{B}, *)$-non-degenerate at the origin with $\mathcal{B} := (\mathrm{Supp}(f_1), \ldots, \mathrm{Supp}(f_m))$.

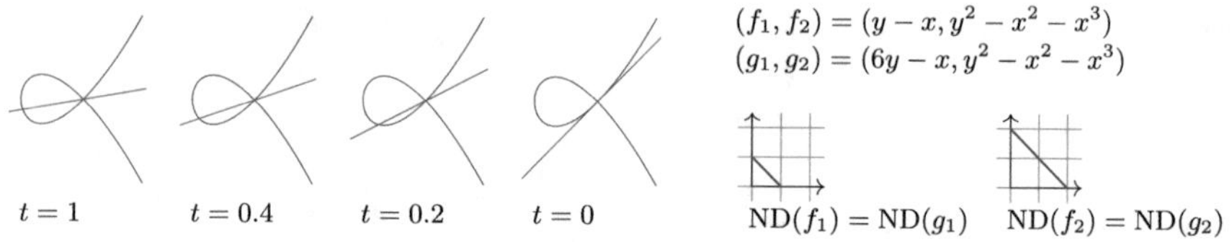

FIGURE 6. A common (non-fixed) root of $(1 - t)f_j + tg_j = 0$, $j = 1, 2$, approaches the origin as $t \to 0$.

3.1.2. *General case.* So far, we ignored the possibility that some of the γ_j can be identically zero. This happens if $\gamma(t)$ belongs to a proper coordinate subspace of $\mathbb{C}^n$. Incorporating this possibility and running the same arguments as in the first case leads to the following notion.

Definition IX.6. Let $\mathcal{A} := (\mathcal{A}_1, \ldots, \mathcal{A}_m)$ be a collection of (possibly infinite) subsets of $\mathbb{Z}^n_{\geq 0}$ and $(f_1, \ldots, f_m) \in \mathcal{L}_0(\mathcal{A})$. For each $I \subseteq [n]$ and each j, write $f_j|_{\mathbb{k}^I}$ for the power series obtained from f_j by substituting 0 for each x_k such that $k \notin I$, and write $\mathcal{A}^I := (\mathcal{A}^I_1, \ldots, \mathcal{A}^I_m) = (\mathcal{A}_1 \cap \mathbb{R}^I, \ldots, \mathcal{A}_m \cap \mathbb{R}^I)$. We say that $f_1, \ldots, f_m$ are $\mathcal{A}$-*non-degenerate at the origin* if they satisfy the following condition:

$$(\mathrm{ND}_0) \qquad \begin{array}{l} f_1|_{\mathbb{k}^I}, \ldots, f_m|_{\mathbb{k}^I} \text{ are } (\mathcal{A}^I, *)\text{-non-degenerate at} \\ \text{the origin for each nonempty subset } I \text{ of } [n] \end{array}.$$

We say that $f_1, \ldots, f_m$ are *non-degenerate at the origin* if they are $\mathcal{B}$-non-degenerate at the origin with $\mathcal{B} := (\mathrm{Supp}(f_1), \ldots, \mathrm{Supp}(f_m))$.

The preceding discussion suggests that for $\mathbb{k} = \mathbb{C}$, $\mathcal{A}$-non-degeneracy at the origin is a sufficient condition for minimum intersection multiplicity at the origin. We will see that it is in fact necessary and sufficient for all (algebraically closed) $\mathbb{k}$.

3.2. The results. Let $\mathcal{A} := (\mathcal{A}_1, \ldots, \mathcal{A}_n)$ be a collection of (possibly infinite) subsets of $\mathbb{Z}^n_{\geq 0}$. Theorem IX.8 below states the necessary and sufficient condition for the minimality of $[f_1, \ldots, f_n]_0$ for $(f_1, \ldots, f_n) \in \mathcal{L}_0(\mathcal{A})$. It also states that $[f_1, \ldots, f_n]_0$ is minimal for "generic" $(f_1, \ldots, f_n) \in \mathcal{L}_0(\mathcal{A})$. We have to be careful about the notion of "genericness" though, since the spaces $\mathcal{L}_0(\mathcal{A}_j)$ and $\mathcal{L}_0(\mathcal{A})$ are in general infinite-dimensioanl vector spaces over $\mathbb{k}$, and therefore they are *not* algebraic varieties. Let $\mathcal{A}' := (\mathcal{A}_1 \cap \mathrm{ND}(\mathcal{A}_1), \ldots, \mathcal{A}_n \cap \mathrm{ND}(\mathcal{A}_n))$. Then $\mathcal{L}_0(\mathcal{A}')$ is an algebraic variety isomorphic to $\mathbb{k}^{\sum_j |\mathcal{A}_j \cap \mathrm{ND}(\mathcal{A}_j)|}$. Let $\pi : \mathcal{L}_0(\mathcal{A}) \to \mathcal{L}_0(\mathcal{A}')$ be the natural projection which "ignores" the coefficients corresponding to exponents not in $\mathrm{ND}(\mathcal{A}_j)$, $j = 1, \ldots, n$. Write $\mathcal{M}_0(\mathcal{A})$ (respectively, $\mathcal{M}_0(\mathcal{A}')$) for the set of all $(f_1, \ldots, f_n)$ in $\mathcal{L}_0(\mathcal{A})$ (respectively, $\mathcal{L}_0(\mathcal{A}')$) with the minimum possible value for $[f_1, \ldots, f_n]_0$. We will show that $\mathcal{M}_0(\mathcal{A}')$ is a nonempty Zariski open (and therefore Zariski dense) subset of $\mathcal{L}_0(\mathcal{A}')$, and $\mathcal{M}_0(\mathcal{A}) = \pi^{-1}(\mathcal{M}'_0)$.

Remark IX.7. An *ind-variety* over a field k is a set X along with a chain of subsets $X_0 \subset X_1 \subset \cdots$ such that

(1) $X = \bigcup_i X_i$,
(2) Each X_i is an algebraic variety over k, and
(3) The inclusions $X_i \hookrightarrow X_{i+1}$ are closed embeddings of algebraic varieties.

It is not hard to see, taking arbitrary sequences of finite subsets $\mathcal{A}_{j,0} \subset \mathcal{A}_{j,1} \subset \cdots$ such that $\mathcal{A}_j = \bigcup_i \mathcal{A}_{j,i}$, that $\mathcal{L}_0(\mathcal{A})$ is an ind-variety. The notion of Zariski topology has a natural extension to the case of ind-varieties. Theorem IX.8 implies in particular that $\mathcal{M}_0(\mathcal{A})$ is a nonempty dense open subset of $\mathcal{L}_0(\mathcal{A})$ in this topology.

THEOREM IX.8. *([Mon16]).* $\mathcal{M}_0(\mathcal{A}) = \pi^{-1}(\mathcal{M}'_0)$ *and* $\mathcal{M}_0(\mathcal{A}')$ *is a nonempty Zariski open subset of* $\mathcal{L}_0(\mathcal{A}')$. *If* $[\mathcal{A}_1, \ldots, \mathcal{A}_n]_0 = \infty$, *then* $\mathcal{M}_0(\mathcal{A}) = \mathcal{L}_0(\mathcal{A})$. *Otherwise, the following are equivalent:*

(1) $(f_1, \ldots, f_n) \in \mathcal{M}_0(\mathcal{A})$
(2) $f_1, \ldots, f_n$ *are* $\mathcal{A}$*-non-degenerate at the origin.*

Theorem IX.8 is proven in section IX.4. To check $\mathcal{A}$-non-degeneracy at the origin, one needs to check $(\mathcal{A}^I, *)$-non-degeneracy for *all* (nonempty) subsets I of $[n]$. The following theorem, which we prove in section IX.6, often significantly limits the number of test cases.

THEOREM IX.9. ([Mon16]). *Let* $\mathcal{A} := (\mathcal{A}_1, \ldots, \mathcal{A}_m)$ *be a collection*[1] *of (possibly infinite) subsets of* $\mathbb{Z}^n_{\geq 0}$. *For each* $I \subseteq [n]$, *let* $\mathcal{E}^I_{\mathcal{A}} := \{j : \mathcal{A}^I_j = \emptyset\}$ *and* $\mathscr{I}_{\mathcal{A}} := \{I \subseteq [n] : I \neq \emptyset, |\mathcal{E}^I_{\mathcal{A}}| \geq n - |I|\}$. *Then for* $(f_1, \ldots, f_m) \in \mathcal{L}_0(\mathcal{A})$ *the following are equivalent:*

(1) $f_1, \ldots, f_m$ *are* $\mathcal{A}$*-non-degenerate at the origin.*
(2) $f_1|_{\mathbb{k}^I}, \ldots, f_m|_{\mathbb{k}^I}$ *are* $(\mathcal{A}^I, *)$*-non-degenerate at the origin for every* $I \in \mathscr{I}_{\mathcal{A}}$.

Remark IX.10. If $m = n$ and $0 < [\mathcal{A}_1, \ldots, \mathcal{A}_n]_0 < \infty$, then (due to corollary IX.13 below)

$$\mathscr{I}_{\mathcal{A}} = \{I \subseteq [n] : I \neq \emptyset, |\mathcal{E}^I_{\mathcal{A}}| = n - |I|\} = \{I \subseteq [n] : I \neq \emptyset, |T^I_{\mathcal{A}}| = |I|\}$$

where $T^I_{\mathcal{A}} := \{j : \mathcal{A}^I_j \neq \emptyset\}$.

Now we go back to the $m = n$ case, i.e. $\mathcal{A} := (\mathcal{A}_1, \ldots, \mathcal{A}_n)$ is a collection of subsets of $\mathbb{Z}^n$. Define $\mathscr{I}_{\mathcal{A}}$ as in theorem IX.9. Similar to the characterization of strict monotonicity of mixed volume in corollary VII.34, we give (in theorem

[1] Note that the number of subsets is m, which may be distinct from n.

IX.32) a combinatorial characterization of strict monotonicity of $[\mathcal{A}_1, \ldots, \mathcal{A}_n]_0$. As a corollary in section IX.8, we prove the following result, which says that in the same way as in the case of $(\mathbb{k}^*)^n$ (theorem VII.13), $\mathcal{A}$-non-degeneracy of a system of power series at the origin is equivalent to a combinatorial condition plus non-degeneracy at the origin with respect to their supports.

Corollary IX.11. *If* $0 < [\mathcal{A}_1, \ldots, \mathcal{A}_n]_0 < \infty$, *then the following are equivalent for* $(f_1, \ldots, f_n) \in \mathcal{L}_0(\mathcal{A})$*:*

(1) $f_1, \ldots, f_n$ *are* $\mathcal{A}$*-non-degenerate at the origin.*

(2) (a) *for each nonempty subset* I *of* $[n]$*, and each* $\nu \in (\mathbb{R}^n)^*$ *which is centered at the origin, the collection* $\{\mathrm{In}_\nu(\mathrm{ND}(f_j) \cap \mathbb{R}^I) : \mathrm{In}_\nu(\mathcal{A}_j^I) \cap \mathrm{Supp}(f_j) \neq \emptyset\}$ *of polytopes is dependent, and*

(b) $f_1, \ldots, f_n$ *are non-degenerate at the origin.*

(3) (a) *for each* $I \in \mathscr{I}_\mathcal{A}$ *and each* $\nu \in (\mathbb{R}^n)^*$ *which is centered at the origin, the collection* $\{\mathrm{In}_\nu(\mathrm{ND}(f_j) \cap \mathbb{R}^I) : \mathrm{In}_\nu(\mathcal{A}_j^I) \cap \mathrm{Supp}(f_j) \neq \emptyset\}$ *of polytopes is dependent, and*

(b) $f_1, \ldots, f_n$ *are non-degenerate at the origin.*

4. Proof of the non-degeneracy condition

In this section, we prove theorem IX.8. Let $\mathcal{A} := (\mathcal{A}_1, \ldots, \mathcal{A}_m)$, $m \geq 1$, be a collection of subsets of $\mathbb{Z}^n_{\geq 0}$. Let $\mathcal{L}_0(\mathcal{A}), \mathcal{A}'$ be as in section IX.3.2. Let $I \subseteq [n]$; define $T_\mathcal{A}^I := \{j : \mathcal{A}_j^I \neq \emptyset\}$ as in theorem IX.1. Note that $T_\mathcal{A}^I = T_{\mathcal{A}'}^I$.

Lemma IX.12. *Assume* $0 \notin \bigcup_j \mathcal{A}_j$*. Then*

(1) *If* $|T_\mathcal{A}^I| < |I|$, *then* $\dim_\mathbb{k}(\mathbb{k}[[x_i : i \in I]]/\langle f_1|_{\mathbb{k}^I}, \ldots, f_m|_{\mathbb{k}^I}\rangle) = \infty$ *for all* $(f_1, \ldots, f_m) \in \mathcal{L}_0(\mathcal{A})$.

(2) *If* $|T_\mathcal{A}^I| \geq |I|$, *then* $V(f_1, \ldots, f_m) \cap (\mathbb{k}^*)^I$ *is isolated for generic* $f_1, \ldots, f_m \in \mathcal{L}_0(\mathcal{A}')$.

PROOF. Due to Proposition IV.27 it suffices to prove the first assertion for m-tuple $(f_1, \ldots, f_m)$ of *polynomials* in $\mathcal{L}_0(\mathcal{A})$. If $|T_\mathcal{A}^I| < |I|$ then the number of f_j such that $f_j|_{\mathbb{k}^I}$ is non-zero is less than $|I|$. Since $0 \notin \mathcal{A}_j$ for any j, each $f_j|_{\mathbb{k}^I}$ is in the maximal ideal of $R_I := \mathbb{k}[[x_i : i \in I]]/\langle f_1|_{\mathbb{k}^I}, \ldots, f_m|_{\mathbb{k}^I}\rangle$. Theorem III.80 implies that the transcendence degree of R_I over $\mathbb{k}$ is positive so that $\dim_\mathbb{k}(R_I) = \infty$. The second assertion follows from Bernstein's theorem. □

Corollary IX.13. (cf. [Roj99, Lemma 2], [HJS13, Proposition 5]). *Assume* $m = n$.

(1) $[\mathcal{A}_1, \ldots, \mathcal{A}_n]_0 = 0$ *if and only if* $0 \in \bigcup_{i=1}^n \mathcal{A}_i$.

(2) $[\mathcal{A}_1, \ldots, \mathcal{A}_n]_0 = \infty$ *if and only if* $0 \notin \bigcup_{i=1}^n \mathcal{A}_i$ *and there is* $I \subseteq [n]$ *such that* $|T_\mathcal{A}^I| < |I|$. □

Let $\mathcal{N}_0(\mathcal{A})$ be the set of all $(f_1, \ldots, f_m) \in \mathcal{L}_0(\mathcal{A})$ such that $f_1, \ldots, f_m$ are $\mathcal{A}$-non-degenerate at the origin. Note that $\mathcal{N}_0(\mathcal{A}) = \pi^{-1}(\mathcal{N}_0(\mathcal{A}'))$, where $\pi : \mathcal{L}_0(\mathcal{A}) \to \mathcal{L}_0(\mathcal{A}')$ is the natural projection.

Proposition IX.14. $\mathcal{N}_0(\mathcal{A}')$ *is a Zariski open subset of* $\mathcal{L}_0(\mathcal{A}')$*. If either* $0 \in \bigcup_{i=1}^m \mathcal{A}_i$ *or* $|T_\mathcal{A}^I| \geq |I|$ *for all* $I \subseteq [n]$, *then* $\mathcal{N}_0(\mathcal{A}')$ *is nonempty.*

PROOF. For each m-tuple $\mathcal{B} = (\mathcal{B}_1, \ldots, \mathcal{B}_m)$ of subsets of $\mathbb{R}^n$ and for each $\nu \in (\mathbb{R}^n)^*$, we write $\operatorname{In}_\nu(\mathcal{B}) := (\operatorname{In}_\nu(\mathcal{B}_1), \ldots, \operatorname{In}_\nu(\mathcal{B}_m))$. If $\mathcal{B}' = \operatorname{In}_\nu(\mathcal{B})$ for some $\nu \in (\mathbb{R}^n)^*$, we say that $\mathcal{B}'$ is a *face* of $\mathcal{B}$ and write that $\mathcal{B}' \preceq \mathcal{B}$; if in addition ν is centered at the origin, we write that $\mathcal{B}' \preceq_0 \mathcal{B}$.

Claim IX.14.1. *If $\mathcal{B}' \preceq \mathcal{B} \preceq_0 \mathcal{A}'$, then $\mathcal{B}' \preceq_0 \mathcal{A}'$.*

PROOF. By assumption there is $\nu \in (\mathbb{R}^n)^*$ centered at the origin such that $\mathcal{B} = \operatorname{In}_\nu(\mathcal{A}')$. Pick $\nu' \in (\mathbb{R}^n)^*$ such that $\mathcal{B}' = \operatorname{In}_{\nu'}(\mathcal{B})$. If k is a sufficiently large positive integer, then each of the coordinates of $k\nu + \nu'$ is *positive* with respect to the basis dual to the standard basis on $\mathbb{R}^n$, and $\operatorname{In}_{k\nu+\nu'}(\mathcal{A}') = \mathcal{B}'$ so that $\mathcal{B}' \preceq_0 \mathcal{A}'$. □

Given $\mathcal{B} \preceq \mathcal{A}'$, and $f = (f_1, \ldots, f_m) \in \mathcal{L}_0(\mathcal{A}')$, define $f_{j,\mathcal{B}_j}$ as in section VII.4.3. Let $\mathcal{D}_{\mathcal{B}}(\mathcal{A}')$ be the set of all $f \in \mathcal{L}_0(\mathcal{A}')$ such that there is a common root of $f_{1,\mathcal{B}_1}, \ldots, f_{m,\mathcal{B}_m}$ on $(\mathbb{k}^*)^n$. Let $\mathcal{D}_0(\mathcal{A}') := \bigcup_{\mathcal{B} \preceq_0 \mathcal{A}'} \mathcal{D}_{\mathcal{B}}(\mathcal{A}')$. Claim IX.14.1 implies that $\mathcal{D}_0(\mathcal{A}') = \bigcup_{\mathcal{B} \preceq_0 \mathcal{A}'} \bigcup_{\mathcal{B}' \preceq \mathcal{B}} \mathcal{D}^0_{\mathcal{B}'}(\mathcal{A}')$, so that claim VII.22 implies that $\mathcal{D}_0(\mathcal{A}')$ is a Zariski closed subset of $\mathcal{L}_0(\mathcal{A}')$. Let $I \subseteq [n]$. Replacing $\mathcal{A}'$ by $\mathcal{A}'^I := (\mathcal{A}'_1 \cap \mathbb{R}^I, \ldots, \mathcal{A}'_m \cap \mathbb{R}^I)$, it follows that $\mathcal{D}_0(\mathcal{A}'^I)$ is a Zariski closed subset of $\mathcal{L}_0(\mathcal{A}'^I)$. Let $\bar{\pi}_{0,I} : \mathcal{L}_0(\mathcal{A}') \to \mathcal{L}_0(\mathcal{A}'^I)$ be the natural projection. Then $\mathcal{N}_0(\mathcal{A}') = \mathcal{L}_0(\mathcal{A}') \setminus \bigcup_{I \subseteq [n]} \bar{\pi}_{0,I}^{-1}(\mathcal{D}_0(\mathcal{A}'^I))$ is Zariski open in $\mathcal{L}_0(\mathcal{A}')$. It now remains to prove the second assertion of proposition IX.14. If $0 \in \mathcal{A}_i$ for some i, then any polynomial supported at $\mathcal{A}'_i$ with a non-zero constant term would lead to an element in $\mathcal{N}_0(\mathcal{A}')$. On the other hand, if $|T^I_{\mathcal{A}}| \geq |I|$ for every $I \subseteq [n]$, then claim VII.24 implies that $\mathcal{D}_0(\mathcal{A}'^I)$ is a proper Zariski closed subset of $\mathcal{L}_0(\mathcal{A}'^I)$ for every $I \subseteq [n]$ so that $\mathcal{N}_0(\mathcal{A}')$ is nonempty, as required. □

We now explore the relation between non-degeneracy at the origin and the intersection multiplicity at the origin. At first, we need to extend the notion of weighted orders and "initial coefficients" corresponding to branches on $(\mathbb{k}^*)^n$ to the case of branches on $\mathbb{k}^n$.

Definition IX.15. Let $B := (Z, z)$ be a branch of a curve $C \subset \mathbb{k}^n$. Identify $Z^* := Z \setminus z$ with its image on C and let $I_B := \{i : x_i|_{Z^*} \not\equiv 0\}$. Note that $\mathbb{k}^{I_B}$ is the smallest coordinate subspace of $\mathbb{k}^n$ which contains Z^*. We write ν_B for the weighted order on $\mathbb{k}[x_i, x_i^{-1} : i \in I_B]$ corresponding to the weight $\operatorname{ord}_z(x_i|_Z)$ for each $i \in I_B$. Fix a parameter ρ_B of B and define

$$\operatorname{In}_B(x_j) := \begin{cases} 0 & \text{if } j \notin I_B \\ \left.\dfrac{x_j}{(\rho_B)^{\nu_B(x_j)}}\right|_z & \text{if } j \in I_B. \end{cases} \tag{89}$$

$$\operatorname{In}(B) := (\operatorname{In}_B(x_1), \ldots, \operatorname{In}_B(x_n)) \in (\mathbb{k}^*)^{I_B}$$

Compare this definition with the case of branches on $(\mathbb{k}^*)^n$ defined in (53) in section VI.9.1. The following result is immediate from the definition.

Lemma IX.16. *If the center of B is on $(\mathbb{k}^*)^I$, then $I \subset I_B$.* □

Lemma IX.17. *Let* $(f_1, \dots, f_m) \in \mathcal{L}_0(\mathcal{A}) \cap \Bbbk[x_1, \dots, x_n]$ *and* B *be a branch of a curve contained in* $V(f_1, \dots, f_m)$. *Then* $\operatorname{In}(B) \in V(\operatorname{In}_{\nu_B}(f_1|_{\Bbbk^{I_B}}), \dots, \operatorname{In}_{\nu_B}(f_m|_{\Bbbk^{I_B}})) \cap (\Bbbk^*)^{I_B}$. *If in addition* B *is a branch at the origin, then* $f_1|_{\Bbbk^{I_B}}, \dots, f_m|_{\Bbbk^{I_B}}$ *violate condition* (ND_0^*) *(from definition IX.5) with* $\nu = \nu_B$*; in particular,* $f_1, \dots, f_m$ *violate* (ND_0) *(from definition IX.6) with* $I = I_B$.

PROOF. The first assertion is a direct corollary of lemma VI.33. If B is a branch at the origin, then ν_B is centered at the origin so that the second assertion follows from the first one. □

Corollary IX.18. *If* $f_1, \dots, f_m \in \mathcal{L}_0(\mathcal{A}) \cap \Bbbk[x_1, \dots, x_n]$ *are* $\mathcal{A}$*-non-degenerate at the origin, then the origin can not be a non-isolated point of* $V(f_1, \dots, f_m) \subset \Bbbk^n$. □

4.1. Proof of theorem IX.8. Below sometimes we work with $\Bbbk^{n+1}$ with coordinates $(x_1, \dots, x_n, t)$. In those cases, we usually denote the coordinates of elements of $\Bbbk^{n+1}$ as pairs, with the last component of the pair denoting the t-coordinate. In particular, the origin of $\Bbbk^{n+1}$ is denoted as $(0, 0)$. Take $m = n$ and define $\mathcal{M}_0(\mathcal{A})$ as in section IX.3.2.

Claim IX.19. $\mathcal{M}_0(\mathcal{A}) \supseteq \mathcal{N}_0(\mathcal{A})$.

PROOF. Let $(f_1, \dots, f_n) \in \mathcal{L}_0(\mathcal{A}) \setminus \mathcal{M}_0(\mathcal{A})$. We will show that $f_1, \dots, f_n$ are *$\mathcal{A}$-degenerate* at the origin. By our assumption, there is $(g_1, \dots, g_n) \in \mathcal{L}_0(\mathcal{A})$ such that $[g_1, \dots, g_n]_0 < [f_1, \dots, f_n]_0$. Due to proposition IV.27, we may assume all g_i and f_j are *polynomials* in $x_1, \dots, x_n$. Let t be a new indeterminate. An application of theorem IV.31 with $h_j := (1-t)f_j + tg_j$, $j = 1, \dots, n$, $X = \Bbbk^n$ and $(b_0, \epsilon_0) = (0, 1)$ implies that

(i) either the origin is a non-isolated zero of $f_1, \dots, f_n$,
(ii) or there is an irreducible component V of $V(h_1, \dots, h_n)$ in $\Bbbk^n \times \Bbbk$ containing $(0, 0)$ such that V is different from $\{0\} \times \Bbbk$.

In case (i) $f_1, \dots, f_n$ are $\mathcal{A}$-degenerate at the origin (corollary IX.18), so consider that we are in case (ii). Choose a branch B at the origin of a curve contained in $V(h_1, \dots, h_n) \subset \Bbbk^{n+1}$ which is different from $\{0\} \times \Bbbk$. Since $B \not\subset \{0\} \times \Bbbk$, it follows that $I := I_B \cap [n] \neq \emptyset$. Let ν be the restriction of ν_B to $\Bbbk[x_i : i \in I]$. Then it follows as in the proof of proposition VII.19 that for each $j = 1, \dots, n$,

$$\operatorname{In}_{\mathcal{A}_j^I, \nu}(f_j|_{\Bbbk^I}) = \begin{cases} \operatorname{In}_\nu(f_j|_{\Bbbk^I}) = \operatorname{In}_{\nu_B}(h_j|_{\Bbbk^{I_B}}) & \text{if } \operatorname{Supp}(f_j|_{\Bbbk^I}) \cap \operatorname{In}_\nu(\mathcal{A}_j^I) \neq \emptyset, \\ 0 & \text{otherwise.} \end{cases} \tag{90}$$

Lemma IX.17 implies that $h_1|_{\Bbbk^{I_B}}, \dots, h_n|_{\Bbbk^{I_B}}$ violate (ND_0^*) with $\nu = \nu_B$, and therefore identity (90) implies that $f_1, \dots, f_n$ are $\mathcal{A}$-degenerate at the origin, as desired. □

Claim IX.20. *Assume* $[\mathcal{A}_1, \dots, \mathcal{A}_n]_0 < \infty$. *Then* $\mathcal{M}_0(\mathcal{A}) \subseteq \mathcal{N}_0(\mathcal{A})$.

PROOF. If $[\mathcal{A}_1, \dots, \mathcal{A}_n]_0 = 0$, then $(f_1, \dots, f_n) \in \mathcal{M}_0(\mathcal{A})$ if and only if one of the f_j has a non-zero constant term, which immediately implies that $f_1, \dots, f_n$ are $\mathcal{A}$-non-degenerate. So assume $0 < [\mathcal{A}_1, \dots, \mathcal{A}_n]_0 < \infty$. Pick $(f_1, \dots, f_n) \in \mathcal{L}_0(\mathcal{A}) \setminus \mathcal{N}_0(\mathcal{A})$. We will show that $(f_1, \dots, f_n) \notin \mathcal{M}_0(\mathcal{A})$ following (the adapted version of) Bernstein's trick from section VII.4.2. Without loss of generality we may assume that $[f_1, \dots, f_n]_0 < \infty$, and due to proposition IV.27, we may assume in addition that each f_j is a polynomial in $(x_1, \dots, x_n)$. Since $f_1, \dots, f_n$ violate (ND_0), there is a nonempty subset I of $[n]$ and a weighted order ν centered at the origin on $\mathbb{k}[x_i : i \in I]$ such that $\text{In}_{\mathcal{A}_1^I, \nu}(f_1|_{\mathbb{k}^I}), \dots, \text{In}_{\mathcal{A}_n^I, \nu}(f_n|_{\mathbb{k}^I})$ have a common zero $a = (a_1, \dots, a_n) \in (\mathbb{k}^*)^I$. Let $T_{\mathcal{A}}^I = \{j : \mathcal{A}_j^I \neq \emptyset\}$. Since $[\mathcal{A}_1, \dots, \mathcal{A}_n]_0 < \infty$, corollary IX.13 and propositions IV.27 and IX.14 imply that there is a system $(g_1, \dots, g_n) \in \mathcal{L}_0(\mathcal{A})$ of *polynomials* in $(x_1, \dots, x_n)$ such that $\text{In}_{\mathcal{A}_j^I, \nu}(g_j|_{\mathbb{k}^I})(a) \neq 0$ for each $j \in T_{\mathcal{A}}^I$. Define $c(t) := (c_1(t), \dots, c_n(t)) : \mathbb{k} \to \mathbb{k}^I$ as follows:

$$c_i(t) := \begin{cases} a_i t^{\nu_i} & \text{if } i \in I, \\ 0 & \text{otherwise.} \end{cases}$$

For each $j \in T_{\mathcal{A}}^I$, let $m_j := \min_{\mathcal{A}_j^I}(\nu)$. Define

$$h_j := \begin{cases} t^{-m_j} g_j(c(t)) f_j - t^{-m_j} f_j(c(t)) g_j & \text{if } j \in T_{\mathcal{A}}^I, \\ f_j & \text{otherwise.} \end{cases}$$

Note that each h_j is a polynomial in $(x_1, \dots, x_n, t)$. Since $\text{In}_{\mathcal{A}_j^I, \nu}(g_j|_{\mathbb{k}^I})(a) \neq 0 = \text{In}_{\mathcal{A}_j^I, \nu}(f_j|_{\mathbb{k}^I})(a)$ for each $j \in T_{\mathcal{A}}^I$, it follows as in section VII.4.2.2 that $h_j(x, 0)$ is a non-zero constant multiple of f_j for each j. By our assumption the origin is an isolated zero of $f_1, \dots, f_n$. Since $h_1, \dots, h_n$ vanish on the curve $\{(c(t), t) : t \in \mathbb{k}\}$, theorem IV.31 implies that $[f_1, \dots, f_n]_0 > [h_1(x, \epsilon), \dots, h_n(x, \epsilon)]_0$ for generic $\epsilon \in \mathbb{k}$. Since $\text{Supp}(h_j(x, \epsilon)) \subset \mathcal{A}_j$ for each ϵ, it follows that $[f_1, \dots, f_n]_0 > [\mathcal{A}_1, \dots, \mathcal{A}_n]_0$, as required. □

Theorem IX.8 now follows from corollary IX.13, proposition IX.14 and claims IX.19 and IX.20. □

5. Proof of the bound

In this section, we prove theorem IX.1. The computation of intersection multiplicity becomes easier if a generic system satisfies a property which is stronger than (ND_0); at first, we prove that such systems exist. The proof of theorem IX.1 is then given in section IX.5.1. We start with a notation: if g is a polynomial in $(x_1, \dots, x_n)$, $a = (a_1, \dots, a_n) \in \mathbb{k}^n$ and $I \subseteq [n]$, we write g_a^I for the polynomial in $(x_i : i \in I)$ obtained from substituting $a_{i'}$ for $x_{i'}$ for each i' *not* in I.

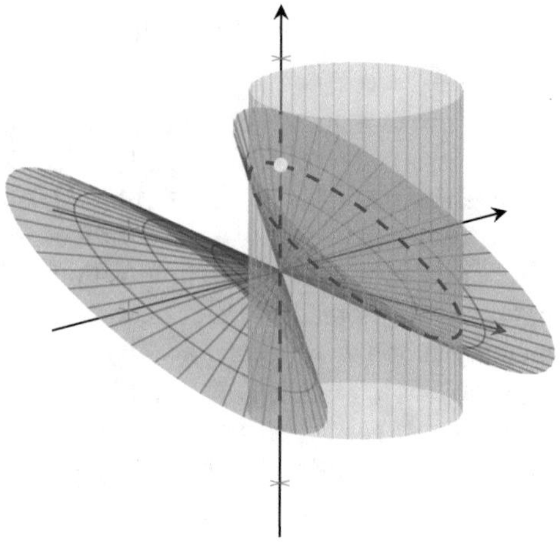

FIGURE 7. Illustration of lemma IX.21: $f := x^2+(y-1)^2-1 = 0$ and $g := z^2 + x^2 - (y+z)^2 = 0$ intersect along the z-axis and an ellipse which intersect at $(0,0,1)$. $f|_{z=1} = x^2 - 2y + y^2$ and $g|_{z=1} = x^2 - 2y - y^2$ are degenerate at the origin, since with $\nu = (1,2) \in (\mathbb{R}^n)^*$, $\mathrm{In}_\nu(f|_{z=1}) = \mathrm{In}_\nu(g|_{z=1}) = x^2 - 2y$.

Lemma IX.21. *Let $I \subseteq [n]$ and $f_1, \ldots, f_k \in \Bbbk[x_1, \ldots, x_n]$ such that $f_j|_{\Bbbk^I} = 0$ for each $j = 1, \ldots, k$. Assume $V(f_1, \ldots, f_k)$ has an irreducible component V which is not contained in $\Bbbk^I$. Then $f_{1,a}^{[n]\setminus I}, \ldots, f_{k,a}^{[n]\setminus I}$* are degenerate *at the origin for each $a \in (\Bbbk^*)^I \cap V$ (see fig. 7).*

PROOF. Let $a \in (\Bbbk^*)^I \cap V$ and B be a branch centered at a of a curve contained in V such that $B \not\subset \Bbbk^I$. Then $I_B \supsetneq I$ (lemma IX.16) and $\pi_I(\mathrm{In}(B)) = a$, where $\pi_I : \Bbbk^n \to \Bbbk^I$ is the natural projection. Let $I' := I_B \setminus I$ and ν' be the restriction of ν_B to $\Bbbk[x_{i'} : i' \in I']$. Since the center of B is on $(\Bbbk^*)^I$, it follows that $\nu_B(x_i) = 0$ for each $i \in I$. For each $j = 1, \ldots, k$, since $f_j|_{\Bbbk^I} \equiv 0$, it then follows that $\mathrm{In}_{\nu_B}(f_j|_{\Bbbk^{I_B}})(\mathrm{In}(B)) = \mathrm{In}_{\nu'}(f_{j,a}^{[n]\setminus I}|_{\Bbbk^{I'}})(a')$, where $a' := \pi_{I'}(\mathrm{In}(B)) \in (\Bbbk^*)^{I'}$. The result now follows from lemma IX.17. □

Let $f_1, \ldots, f_m$ be polynomials in $(x_1, \ldots, x_n)$ and $I \subseteq [n]$. Define

$$D_0^I(f_1, \ldots, f_m) := \{a \in (\Bbbk^*)^I : f_{1,a}^{[n]\setminus I}, \ldots, f_{m,a}^{[n]\setminus I} \text{ are degenerate at the origin}\} \tag{91}$$

The following result is immediate from theorem III.80 and lemma IX.21.

Corollary IX.22. *Let $J, J' \subseteq [m]$ and V' be an irreducible component of $V(f_{j'} : j' \in J')$ such that $V' \not\subseteq \Bbbk^I$. Assume $\dim(D_0^I(f_{j'} : j' \in J') \cap V(f_j : j \in J)) < |I| - |J|$. Then V' does* not *contain any irreducible component of $V(f_j : j \in J) \cap (\Bbbk^*)^I$.* □

Definition IX.23. Given a collection $\mathcal{A} = (\mathcal{A}_1, \dots, \mathcal{A}_m)$, $m \geq 1$, of *finite* subsets of $\mathbb{Z}^n_{\geq 0}$ and $I \subseteq [n]$, define

$$\begin{aligned} T^I_{\mathcal{A}} &:= \{j : \mathcal{A}_j \cap \mathbb{R}^I \neq \emptyset\} \subseteq [m] \\ \mathcal{T}'^I_{\mathcal{A}} &:= \{J = \{j_1, \dots, j_{n-|I|}\} \subseteq [m] \setminus T^I_{\mathcal{A}} : |J| = n - |I|, \\ &\qquad [\pi_{[n]\setminus I}(\mathcal{A}_{j_1}), \dots, \pi_{n\setminus I}(\mathcal{A}_{j_{[n]-|I|}})]_0 < \infty\} \\ \mathcal{A}^I &:= (\mathcal{A}_1 \cap \mathbb{R}^I, \dots, \mathcal{A}_m \cap \mathbb{R}^I) \end{aligned}$$

where $\pi_{[n]\setminus I} : \mathbb{R}^n \to \mathbb{R}^{[n]\setminus I}$ is defined as in (85). Due to the finiteness of the $\mathcal{A}_j$, we can identify $\mathcal{L}_0(\mathcal{A})$ with the collection $\mathcal{L}(\mathcal{A})$ of *polynomials* $(f_1, \dots, f_m)$ such that $\mathrm{Supp}(f_j) \subseteq \mathcal{A}_j$ for each j. Given $(f_1, \dots, f_m) \in \mathcal{L}(\mathcal{A})$, we say that $f_1, \dots, f_m$ are *strongly $\mathcal{A}$-non-degenerate* if for all $I \subseteq [n]$,

(a) $f_1|_{\mathbb{k}^I}, \dots, f_m|_{\mathbb{k}^I}$ are *properly $\mathcal{A}^I$-non-degenerate* (see section VII.4.4),

(b) for all $J' \in \mathcal{T}'^I_{\mathcal{A}}$, $f^{[n]\setminus I}_{j',(1,\dots,1)}$, $j' \in J'$, are *non-degenerate at the origin* and $\mathrm{NP}(f^{[n]\setminus I}_{j',(1,\dots,1)}) = \mathrm{conv}(\pi_{[n]\setminus I}(\mathcal{A}_{j'}))$ for each $j' \in J'$,

(c) for all $J \subseteq T^I_{\mathcal{A}}$ and $J' \in \mathcal{T}'^I_{\mathcal{A}}$,

$$\dim(D^I_0(f_{j'} : j' \in J') \cap V(f_j : j \in J)) < |I| - |J|, \tag{92}$$

where $D^I_0(\cdot)$ are as in (91).

Note that property (b) with $I = \emptyset$ in particular implies that

(d) $\mathrm{NP}(f_j) = \mathrm{conv}(\mathcal{A}_j)$ for each $j = 1, \dots, m$.

We write $\breve{\mathcal{N}}(\mathcal{A})$ be the collection of all $(f_1, \dots, f_m) \in \mathcal{L}(\mathcal{A})$ which are strongly $\mathcal{A}$-non-degenerate. Recall that $\mathcal{N}_0(\mathcal{A})$ stands for the collection of systems which are $\mathcal{A}$-non-degenerate at the origin. For propositions IX.24 and IX.25 below, we assume each $\mathcal{A}_j$ is finite.

Proposition IX.24. *Assume either $0 \in \bigcup_{i=1}^m \mathcal{A}_i$ or $|T^I_{\mathcal{A}}| \geq |I|$ for all $I \subseteq [n]$. Then $\breve{\mathcal{N}}(\mathcal{A}) \subseteq \mathcal{N}_0(\mathcal{A})$. In particular, if $m = n$ and $[\mathcal{A}_1, \dots, \mathcal{A}_n]_0 < \infty$, then $\breve{\mathcal{N}}(\mathcal{A}) \subseteq \mathcal{N}_0(\mathcal{A})$.*

PROOF. Follows from proposition VII.25 and corollary IX.13. □

Proposition IX.25. *$\breve{\mathcal{N}}(\mathcal{A})$ is* constructible *and it contains a nonempty Zariski open subset of $\mathcal{L}(\mathcal{A})$.*

PROOF. By propositions VII.25 and IX.14, the collection of systems that satisfy properties (a) and (b) of definition IX.23 is a nonempty Zariski open subset of $\mathcal{L}(\mathcal{A})$. Therefore, we can concentrate only on property (c). For a subset $J \subseteq [m]$, write $\mathcal{A}_J := (\mathcal{A}_j : j \in J)$. Let $I \subseteq [n]$, $J \subseteq T^I_{\mathcal{A}}$ and $J' \in \mathcal{T}'^I_{\mathcal{A}}$. Write $\mathcal{N}^I(\mathcal{A}_J, \mathcal{A}_{J'})$ for the subset of $\mathcal{L}(\mathcal{A}_J) \times \mathcal{L}(\mathcal{A}_{J'})$ consisting of all $((f_j : j \in J), (f_{j'} : j' \in J'))$ which satisfy property (92). Consider the set of maps from fig. 8, where

- $\pi_{J'}, \pi_{J,J'}, \pi_{J,I}, \pi_{J',I}$ are natural projections,
- σ is the "substitution map" which maps $((f_{j'} : j' \in J'), a) \in \mathcal{L}(\mathcal{A}_{J'}) \times (\mathbb{k}^*)^I$ to $(f^{[n]\setminus I}_{j',a} : j' \in J')$.

- $\mathcal{D}_0(\cdot)$ denotes the collection of systems which are *degenerate* at the origin, and
- $\mathcal{V}_{J,I} := \{((f_j : j \in J), a) \in \mathcal{L}(\mathcal{A}_J) \times (\mathbb{k}^*)^I : f_j(a) = 0 \text{ for each } j \in J\}$

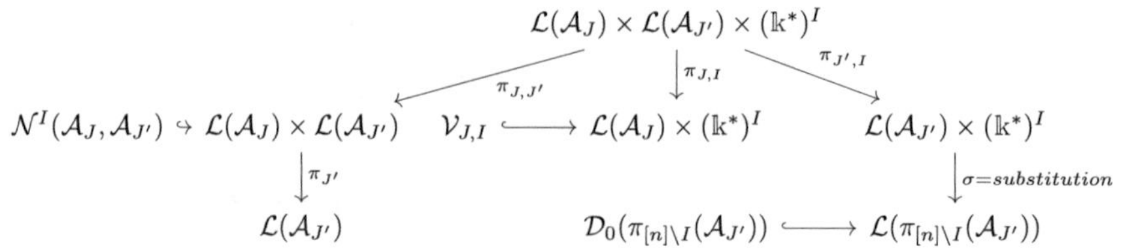

FIGURE 8. Maps from the proof of proposition IX.25.

Let $\mathcal{Z}$ be the subset of $\mathcal{L}(\mathcal{A}_J) \times \mathcal{L}(\mathcal{A}_{J'}) \times (\mathbb{k}^*)^I$ consisting of all $((f_j : j \in J), (f_{j'} : j' \in J'), a)$ such that $a \in V(f_j : j \in J)$ and $(f_{j',a}^{[n]\setminus I} : j' \in J')$ are $\pi_{[n]\setminus I}(\mathcal{A}_{J'})$-degenerate at the origin. Then $\mathcal{Z} = \pi_{J,I}^{-1}(\mathcal{V}_{J,I}) \cap (\sigma \circ \pi_{J',I})^{-1}(\mathcal{D}_0(\pi_{[n]\setminus I}(\mathcal{A}_{J'})))$. Since both $\mathcal{V}_{J,I}$ and $\mathcal{D}_0(\pi_{[n]\setminus I}(\mathcal{A}_{J'}))$ are Zariski closed (the closedness of $\mathcal{D}_0(\cdot)$ follows from proposition IX.14), it follows that $\mathcal{Z}$ is also Zariski closed. Since $\mathcal{N}^I(J, J')$ is the set of all elements in $\mathcal{L}(\mathcal{A}_J) \times \mathcal{L}(\mathcal{A}_{J'})$ whose pre-image under $\pi_{J,J'}|_{\mathcal{Z}}$ has dimension less than $|I| - |J|$, corollary III.90 implies that $\mathcal{N}^I(J, J')$ is constructible. We now show that it contains a nonempty Zariski open subset of $\mathcal{L}(\mathcal{A}_J) \times \mathcal{L}(\mathcal{A}_{J'})$. Fix any $a_0 \in (\mathbb{k}^*)^I$, and let σ_0 be the composition

$$\mathcal{L}(\mathcal{A}_{J'}) \hookrightarrow \mathcal{L}(\mathcal{A}_{J'}) \times \{a_0\} \xrightarrow{\sigma} \mathcal{L}(\pi_{[n]\setminus I}(\mathcal{A}_{J'}))$$

where the first map simply takes $(f_{j'} : j' \in J') \mapsto ((f_{j'} : j; \in J'), a_0)$. Since $J' \in \mathcal{T}_{\mathcal{A}}'^I$, proposition IX.14 implies that $\mathcal{Y}' := \sigma_0^{-1}(\mathcal{N}_0(\pi_{[n]\setminus I}(\mathcal{A}_{J'}))$ is a nonempty Zariski open subset of $\mathcal{L}(\mathcal{A}_{J'})$. Pick an arbitrary system $(f_{j'} : j' \in J') \in \mathcal{Y}'$, and let σ' be the composition

$$(\mathbb{k}^*)^I \hookrightarrow \{(f_{j'} : j' \in J')\} \times (\mathbb{k}^*)^I \xrightarrow{\sigma} \mathcal{L}(\pi_{[n]\setminus I}(\mathcal{A}_{J'}))$$

where the first map simply takes $a \mapsto ((f_{j'} : j' \in J'), a)$. Since $\sigma'(a_0) = \sigma_0(f_{j'} : j' \in J') \in \mathcal{N}_0(\pi_{[n]\setminus I}(\mathcal{A}_{J'}))$, proposition IX.14 implies that $D_0^I(f_{j'} : j' \in J') = (\sigma')^{-1}(\mathcal{D}_0(\pi_{[n]\setminus I}(\mathcal{A}_{J'}))$ is a proper Zariski closed subset of $(\mathbb{k}^*)^I$; in particular, $\dim(D_0^I(f_{j'} : j' \in J')) < |I|$. Since $J \subset T_{\mathcal{A}}^I$, lemma IX.26 below implies that there is a nonempty open subset $\mathcal{W}$ of $\mathcal{L}(\mathcal{A}_J)$ such that $((f_j : j \in J), (f_{j'} : j' \in J')) \in \mathcal{N}^I(J, J')$ for each $(f_j : j \in J) \in \mathcal{W}$. Since $(f_{j'} : j' \in J')$ was an arbitrary element from Y', exercise III.103 implies that $\mathcal{N}^I(J, J')$ contains a nonempty Zariski open subset of $\mathcal{L}(\mathcal{A}_J) \times \mathcal{L}(\mathcal{A}_{J'})$, as required. □

Lemma IX.26. *Let W be an irreducible subvariety $(\mathbb{k}^*)^n$ and $\mathcal{B} = (\mathcal{B}_1, \ldots, \mathcal{B}_k)$ be a collection of finite nonempty subsets of $\mathbb{Z}^n$. Let $\mathcal{W} := \{(f_1, \ldots, f_k) \in \mathcal{L}(\mathcal{B}) : \dim(V(f_1, \ldots, f_k) \cap W) \leq \dim(W) - k\}$. Then $\mathcal{W}$ is a constructible subset of $\mathcal{L}(\mathcal{B})$ and it contains a nonempty Zariski open subset of $\mathcal{L}(\mathcal{B})$.*

PROOF. Let $\mathcal{W}' := \{((f_1,\dots,f_k),(x_1,\dots,x_n)) \in \mathcal{L}(\mathcal{B}) \times W : f_j(x_1,\dots, x_n) = 0 \text{ for each } j\}$ and $\pi_W : \mathcal{W}' \to W$ be the natural projection. For each $w \in W$, $\pi_W^{-1}(w)$ is a linear subspace of $\mathcal{L}(\mathcal{B})$ defined by k linearly independent linear equations, so that $\dim(\pi_W^{-1}(w)) = \dim(\mathcal{L}(\mathcal{B}) - k = \sum_j |\mathcal{B}_j| - k$. Theorem III.85 then implies that $\dim(\mathcal{W}') = \sum_j |\mathcal{B}_j| - k + \dim(W) = \dim(\mathcal{L}(\mathcal{B})) + (\dim(W) - k)$. Now the result follows from applying theorem III.85 and corollary III.90 to $\pi_{\mathcal{B}}|_{\mathcal{W}'} : \mathcal{W}' \to \mathcal{L}(\mathcal{B})$, where $\pi_{\mathcal{B}} : \mathcal{L}(\mathcal{B}) \times (\Bbbk^*)^n \to \mathcal{L}(\mathcal{B})$ is the natural projection. □

5.1. Proof of theorem IX.1. Corollary IX.13 implies that theorem IX.1 holds when $[\mathcal{A}_1,\dots,\mathcal{A}_n]_0 = 0$ or ∞. So assume $0 < [\mathcal{A}_1,\dots,\mathcal{A}_n]_0 < \infty$. Let $\mathcal{A}' := (\mathcal{A}_1 \cap \Gamma_1,\dots,\mathcal{A}_n \cap \Gamma_n)$. Pick strongly $\mathcal{A}'$-non-degenerate $(f_1,\dots,f_n) \in \mathcal{L}_0(\mathcal{A}')$. Theorem IX.8 and proposition IX.24 imply that

$$[f_1,\dots,f_n]_0 = [\mathcal{A}_1,\dots,\mathcal{A}_n]_0$$

Therefore, it suffices to show that

$$[f_1,\dots,f_n]_0 = \sum_{I \in \mathscr{T}_{\mathcal{A},1}} [\Gamma_1^I, \Gamma_{j_2}^I,\dots,\Gamma_{j_{|I|}}^I]_0^* \times [\pi_{[n]\setminus I}(\Gamma_{j'_1}),\dots,\pi_{[n]\setminus I}(\Gamma_{j'_{n-|I|}})]_0, \tag{93}$$

where for each $I \in \mathscr{T}_{\mathcal{A},1}$, $j_1 = 1, j_2,\dots,j_{|I|}$ are elements of $T^I_{\mathcal{A}}$, and $j'_1,\dots,j'_{n-|I|}$ are elements of $[n] \setminus T^I_{\mathcal{A}}$. We proceed by induction on n. It is true for $n = 1$ (see remark IX.2), so assume it is true for all dimensions smaller than n. Since $0 < [f_1,\dots,f_n]_0 < \infty$, proposition IV.28 implies that on a sufficiently small Zariski open neighborhood U of the origin in $\Bbbk^n$, the closed subscheme of U defined by $f_2,\dots,f_n$ is a possibly non-reduced curve C. For each $I \subseteq [n]$, let $\{C^I_j\}_j$ be the set of irreducible components of C such that $\Bbbk^I$ is the smallest coordinate subspace of $\Bbbk^n$ containing each C^I_j.

Claim IX.27. *Let $I \subseteq [n]$, $\mathscr{T}_{\mathcal{A},1}$ be as in theorem IX.1 and $T'^I_{\mathcal{A}}$ be as in definition IX.23.*

(1) If $\{C^I_j\}_j$ is nonempty, then $I \in \mathscr{T}_{\mathcal{A},1}$.
(2) If $I \in \mathscr{T}_{\mathcal{A},1}$, then $[n] \setminus T^I_{\mathcal{A}} \in T'^I_{\mathcal{A}}$.

PROOF. For the first assertion, pick $I \subseteq [n]$ such that $\{C^I_j\}_j$ is nonempty. Since $0 < [\mathcal{A}_1,\dots,\mathcal{A}_n]_0 < \infty$, corollary IX.13 implies that $|T^I_{\mathcal{A}}| \geq |I|$. On the other hand, if $|T^I_{\mathcal{A}} \setminus \{1\}| \geq |I|$, then the proper non-degeneracy of $f_1|_{\Bbbk^I},\dots,f_n|_{\Bbbk^I}$ (property (a) of strong $\mathcal{A}$-non-degeneracy) and lemma IX.17 implies that $\{C^I_J\}_j$ is empty, which is a contradiction. Accordingly, $|T^I_{\mathcal{A}} \setminus \{1\}| = |I| - 1$ and $|T^I_{\mathcal{A}}| = |I|$, which imply that $I \in \mathscr{T}_{\mathcal{A},1}$, as required. For the second assertion, pick $I \in \mathscr{T}_{\mathcal{A},1}$ and set $J' := [n]\setminus T^I_{\mathcal{A}}$. Since $|J'| = n - |I|$, we have to show that $[\pi_{[n]\setminus I}(\mathcal{A}_{j'_1}),\dots,\pi_{n\setminus I} (\mathcal{A}_{j'_{[n]-|I|}})]_0 < \infty$, where $j'_1,\dots,j'_{n-|I|}$ are elements of J'. Indeed, otherwise, corollary IX.13 would imply that $|T^{I'}_{\mathcal{A}}| < |I'|$ for some $I' \supsetneq I$, which would in turn imply (since by assumption $0 \notin \bigcup_j \mathcal{A}_j$) that $[\mathcal{A}_1,\dots,\mathcal{A}_n]_0 = \infty$, which is a contradiction. □

Pick $I \in \mathscr{T}_{\mathcal{A},1}$. Let $j_1 = 1, j_2, \ldots, j_{|I|}$ be the elements of $T^I_{\mathcal{A}}$ and $j'_1, \ldots, j'_{n-|I|}$ be the elements of $J' := [n] \setminus T^I_{\mathcal{A}}$. Since $J' \in \mathcal{T}'^I_{\mathcal{A}}$ (claim IX.27), property (b) of strong $\mathcal{A}$-non-degeneracy and lemma IX.21 imply that $\mathbb{k}^I$ is an irreducible component of $V(f_{j'_1}, \ldots, f_{j'_{n-|I|}})$. On the other hand, applying property (c) of strong $\mathcal{A}$-non-degeneracy with $J = \{j_2, \ldots, j_{|I|}\}$, and then using corollary IX.22 shows that no irreducible component of $V(f_{j'_1}, \ldots, f_{j'_{n-|I|}})$ other than $\mathbb{k}^I$ contains any irreducible component of $V(f_{j_2}, \ldots, f_{j_{|I|}}) \cap (\mathbb{k}^*)^I$. Therefore, claim IX.27, theorem IV.24 and propositions IV.28 and IV.33 imply that

$$[f_1, \ldots, f_n]_0 = \sum_{I \in \mathscr{T}_{\mathcal{A},1}} [f^{[n]\setminus I}_{j'_1,\epsilon}, \ldots, f^{[n]\setminus I}_{j'_{n-|I|},\epsilon}]_0 \sum_j \operatorname{ord}_0(f_1|_{C^I_j})[f_{j_2}|_{(\mathbb{k}^*)^I}, \ldots, f_{j_{|I|}}|_{(\mathbb{k}^*)^I}]_{C^I_j} \tag{94}$$

where ϵ is a generic element of $(\mathbb{k}^*)^n$ and $f^{[n]\setminus I}_{\cdot,\epsilon}$ are as in lemma IX.21. Since $J' \in \mathcal{T}'^I_{\mathcal{A}}$, property (b) of strong $\mathcal{A}$-non-degeneracy and theorem IX.8 imply that

$$\begin{aligned}[f^{[n]\setminus I}_{j'_1,\epsilon}, \ldots, f^{[n]\setminus I}_{j'_{n-|I|},\epsilon}]_0 &= [\pi_{[n]\setminus I}(\mathcal{A}_{j'_1}), \ldots, \pi_{[n]\setminus I}(\mathcal{A}_{j'_{n-|I|}})]_0 \\ &= [\pi_{[n]\setminus I}(\Gamma_{j'_1}), \ldots, \pi_{[n]\setminus I}(\Gamma_{j'_{n-|I|}})]_0\end{aligned}$$

It remains to compute the inner sum of the right-hand side of (94). Let $I \subseteq [n]$. Write $R^I := \mathbb{k}[x_i : i \in I]$. Let $\mathcal{V}^I_0$ be the set of weighted orders on R^I which are centered at the origin and $\mathcal{V}'^I_0$ be the set of primitive elements in $\mathcal{V}^I_0$. For each $\nu \in \mathcal{V}'^I_0$, let $\mathcal{B}^I_{0,j,\nu}$ be the set of all branches at the origin of C^I_j such that ν_B is proportional to ν. Theorem IV.24 implies that for each I, j,

$$\operatorname{ord}_0(f_1|_{C^I_j}) = \sum_{\nu \in \mathcal{V}'^I_0} \sum_{(Z,z) \in \mathcal{B}^I_{0,j,\nu}} \operatorname{ord}_z(f_1|_{C^I_j})$$

Therefore, it suffices to show that

$$\begin{aligned}&\sum_j \sum_{(Z,z) \in \mathcal{B}^I_{0,j,\nu}} \operatorname{ord}_z(f_1|_{C^I_j})[f_{j_2}|_{(\mathbb{k}^*)^I}, \ldots, f_{j_{|I|}}|_{(\mathbb{k}^*)^I}]_{C^I_j} \\ &\quad = \min_{\Gamma^I_1}(\nu)\, \mathrm{MV}'_\nu(\mathrm{In}_\nu(\Gamma^I_{j_2}), \ldots, \mathrm{In}_\nu(\Gamma^I_{j_{|I|}}))\end{aligned} \tag{95}$$

To see it, apply corollary VII.38 (with $n = |I|$) to $f_1|_{(\mathbb{k}^*)^I}, f_{j_2}|_{(\mathbb{k}^*)^I}, \ldots, f_{j_I}|_{(\mathbb{k}^*)^I}$. Property (a) of strong $\mathcal{A}$-non-degeneracy implies that all the assumptions of proposition VII.26 and corollary VII.38 are satisfied. Part 2 of proposition VII.26 implies that each irreducible component of the resulting curve C' comes from an irreducible component of $V(f_{j_2}|_{(\mathbb{k}^*)^I}, \ldots, f_{j_I}|_{(\mathbb{k}^*)^I}) \subset (\mathbb{k}^*)^I$ and therefore the collections $\mathcal{B}'_{j,\nu}$ from corollary VII.38 are precisely the collections $\mathcal{B}^I_{0,j,\nu}$. Corollary VII.38 then implies identity (95) and completes the proof of theorem IX.1.

6. The efficient version of the non-degeneracy condition

In this section, we prove theorem IX.9. Given $I \subseteq [n]$, we write $\mathcal{V}^I$ for the set of weighted orders on $\mathbb{k}[x_i : i \in I]$. Given $I \subseteq \tilde{I}$, we say that $\nu \in \mathcal{V}^I$ and

$\tilde{\nu} \in \mathcal{V}^{\tilde{I}}$ are *compatible* if $(\nu(x_i) : i \in I)$ and $(\tilde{\nu}(x_i) : i \in I)$ are proportional, with a *positive* constant of proportionality, and $\tilde{\nu}(x_{\tilde{i}}) > 0$ for each $\tilde{i} \in \tilde{I} \setminus I$. Theorem IX.9 follows directly from lemma IX.28 below.

Lemma IX.28. *Let* $\mathcal{A} := (\mathcal{A}_1, \ldots, \mathcal{A}_m)$ *be a collection of (possibly infinite) subsets of* $\mathbb{Z}^n_{\geq 0}$*, I be a nonempty subset of* $[n]$*, and* $\nu \in \mathcal{V}^I$ *be such that*

(96) $\quad \operatorname{In}_{\tilde{\nu}}(\mathcal{A}_j)$ *is finite for each* $\tilde{\nu} \in (\mathbb{R}^n)^*$ *which is compatible with* ν.

Let $f_1, \ldots, f_m \in \mathbb{k}[[x_1, \ldots, x_n]]$ *such that* $\operatorname{Supp}(f_j) \subseteq \mathcal{A}_j$ *for each j. Assume*

(1) $|\mathcal{E}^I_{\mathcal{A}}| < n - |I|$, *where* $\mathcal{E}^I_{\mathcal{A}} := \{j : \mathcal{A}^I_j = \emptyset\}$, *and*
(2) $\operatorname{In}_{\mathcal{A}^I_j, \nu}(f_j|_{\mathbb{k}^I})$, $j = 1, \ldots, m$, *have a common zero* $u \in (\mathbb{k}^*)^n$.

Then there exists $\tilde{I} \supsetneq I$ *and* $\tilde{\nu} \in \mathcal{V}^{\tilde{I}}$ *such that*

(3) $\tilde{\nu}$ *is compatible with* ν.
(4) $\operatorname{In}_{\mathcal{A}^{\tilde{I}}_j, \tilde{\nu}}(f_j|_{\mathbb{k}^{\tilde{I}}})$, $j = 1, \ldots, m$, *have a common zero* $\tilde{u} \in (\mathbb{k}^*)^n$ *such that* $\pi_I(\tilde{u}) = \pi_I(u)$, *where* $\pi_I : (\mathbb{k}^*)^n \to (\mathbb{k}^*)^I$ *is defined as in* (85).

PROOF. Due to (96), we may assume without any loss of generality that the support of each f_j is finite, i.e. the f_j are *polynomials* in $(x_1, \ldots, x_n)$. We may also assume that $I = \{1, \ldots, k\}$, $1 \leq k \leq n$. Let $a := \pi_I(u) \in (\mathbb{k}^*)^I$ and $(a_1, \ldots, a_n)$ be the coordinates of a. At first, consider the case that $\nu(x_i) = 0$ for each $i \in I$. Assumption (2) then says that a is a common zero of $f_1, \ldots, f_m$ on $(\mathbb{k}^*)^I$. Let $y_j := x_j - a_j$, $j = 1, \ldots, n$ so that a is the origin of $\mathbb{k}^n$ with respect to $(y_1, \ldots, y_n)$ coordinates. Choose any integral $\nu' \in (\mathbb{R}^n)^*$ with positive coordinates with respect to the dual basis, and let $\pi : \operatorname{Bl}_{\nu'}(\mathbb{k}^n) \to \mathbb{k}^n$ be the ν'-weighted blow up of $\mathbb{k}^n$ with respect to $(y_1, \ldots, y_n)$ coordinates (see section VI.11). Let $E_{\nu'}$ be the exceptional divisor of π, and W' be the strict transform of $(\mathbb{k}^*)^I$ on $\operatorname{Bl}_{\nu'}(\mathbb{k}^n)$. Since $|\mathcal{E}^I_{\mathcal{A}}| < n - |I|$, there is an irreducible component V of $V(f_j : j \in \mathcal{E}^I_{\mathcal{A}}) \cap (\mathbb{k}^*)^n$ properly containing $(\mathbb{k}^*)^I$. Then the strict transform V' of (the closure of) V properly contains W'. Pick $a' \in E_{\nu'} \cap W'$, and choose an irreducible curve $C' \subset V'$ such that $a' \in C' \not\subseteq W' \cup E_{\nu'}$, and a branch $B' = (Z', z')$ of C' centered at a'. Let $\tilde{I} := I_{B'}$ and $\tilde{\nu} := \nu_{B'} \in \mathcal{V}^{\tilde{I}}$ (definition IX.15). Since π is centered at $a \in (\mathbb{k}^*)^I$ and since $\pi(B') \not\subset \mathbb{k}^I$, it follows that $I \subsetneq \tilde{I}$, $\operatorname{In}_{B'}(x_i) = a_i$ and $\tilde{\nu}(x_i) = 0$ for each $i \in I$, and $\tilde{\nu}(x_{\tilde{i}})$ is positive for each $\tilde{i} \in \tilde{I} \setminus I$. Fix $j \in [m]$. If $j \notin \mathcal{E}^I_{\mathcal{A}}$, it follows that $\min_{\mathcal{A}^{\tilde{I}}_j}(\tilde{\nu}) = 0$ and $\operatorname{In}_{\mathcal{A}^{\tilde{I}}_j, \tilde{\nu}}(f_j|_{\mathbb{k}^{\tilde{I}}}) = f_j|_{\mathbb{k}^I} = \operatorname{In}_{\mathcal{A}^I_j, \nu}(f_j|_{\mathbb{k}^I})$. This implies that $\operatorname{In}_{\mathcal{A}^{\tilde{I}}_j, \tilde{\nu}}(f_j|_{\mathbb{k}^{\tilde{I}}})(\operatorname{In}(B')) = f_j|_{\mathbb{k}^I}(a) = 0$. On the other hand, if $j \in \mathcal{E}^I_{\mathcal{A}}$, then $\operatorname{In}_{\mathcal{A}^{\tilde{I}}_j, \tilde{\nu}}(f_j|_{\mathbb{k}^{\tilde{I}}})(\operatorname{In}(B')) = 0$ due to lemma IX.17. The lemma is therefore true in the case that ν is the trivial weighted order.

Now assume ν is not the trivial weighted order. Identify ν with the element in $(\mathbb{R}^n)^*$ with coordinates $(\nu(x_1), \ldots, \nu(x_k))$ with respect to the basis dual to the standard basis of $\mathbb{R}^n$. Choose a basis $\alpha_1, \ldots, \alpha_k$ of $\mathbb{Z}^k$ such that $\langle \nu, \alpha_j \rangle = 0$ for $j = 1, \ldots, k-1$, and $\langle \nu, \alpha_k \rangle = 1$. Then $(x^{\alpha_1}, \ldots, x^{\alpha_k}, x_{k+1}, \ldots, x_n)$ are coordinates on $X := (\mathbb{k}^*)^k \times \mathbb{k}^{n-k}$. Define

$$y_j := \begin{cases} x^{\alpha_j} - a^{\alpha_j} & \text{if } 1 \leq j \leq k-1, \\ x^{\alpha_k} & \text{if } j = k, \\ x_j & \text{if } k+1 \leq j \leq n. \end{cases}$$

Write Y for the affine space $\mathbb{k}^n$ with coordinates $(y_1, \dots, y_n)$. Choose positive integers $\nu'_1, \dots, \nu'_n$ such that $\nu'_k = 1$ and $\nu'_j \gg 1$ for $j = k+1, \dots, n$. Let ν' be the element in $(\mathbb{R}^n)^*$ with coordinates $(\nu'_1, \dots, \nu'_n)$ with respect to the basis dual to the standard basis, and $\pi : Y' \to Y$ be the ν'-weighted blow up of Y with respect to $(y_1, \dots, y_n)$ coordinates, E be the exceptional divisor of π, and W' be the strict transform on Y' of $W := V(y_{k+1}, \dots, y_n) \subset Y$. Proposition VI.38 implies that there is an affine open subset U of Y' such that

(i) $U \cong \mathbb{k} \times (\mathbb{k}^*)^{k-1} \times \mathbb{k}^{n-k}$ with respect to coordinates $(z_1, \dots, z_n)$ where $z_1, \dots, z_k$ are monomials in $(y_1, \dots, y_k)$, $z_j = y_j / z_1^{\nu'_j}$ for $j = k+1, \dots, n$, $\nu'(z_1) = 1$ and $\nu'(z_j) = 0$ for $j = 2, \dots, n$,
(ii) $U \cap E = V(z_1) \cong (\mathbb{k}^*)^{k-1} \times \mathbb{k}^{n-k}$ and
(iii) $U \cap W' = V(z_{k+1}, \dots, z_n) \cong \mathbb{k} \times (\mathbb{k}^*)^{k-1}$.

We treat X as an open subset of Y via the natural embedding. There is an irreducible component V of $V(f_j : j \in \mathcal{E}^I_{\mathcal{A}}) \cap X$ such that its closure $\bar{V}$ in Y properly contains W. The strict transform V' of $\bar{V}$ on Y' properly contains W'. Pick $a' \in U \cap W' \cap E$. Choose an irreducible curve $C' \subset V'$ such that $a' \in C'$, and $C' \not\subset E \cup W'$, and $C' \cap \pi^{-1}(X) \neq \emptyset$. Pick a branch $B' = (Z', z')$ of C' centered at a'. Since $\pi(B') \cap X \neq \emptyset$, we may treat B' as a branch (possibly at infinity) of a curve on X. Define $\nu_{B'}$ and $I_{B'}$ as in definition IX.15. Since each of $x_1, \dots, x_k$ is everywhere non-zero on X, it follows that $I_{B'} \supset \{1, \dots, k\} = I$. On the other hand, since $\pi(B') \not\subset W$, it follows that there exists $j > k$ such that $x_j|_{B'} \not\equiv 0$. It follows that $I_{B'} \supsetneq I$. We show that properties (3) and (4) are true with $\tilde{I} := I_{B'}$ and $\tilde{\nu} := \nu_{B'}$. Indeed, since $a' \in E$, for each $j = 1, \dots, n$, either $y_j|_{B'} \equiv 0$, or $\operatorname{ord}_{z'}(y_j|_{B'}) > 0$. Therefore, for each $j = 1, \dots, k-1$,

(iv) $\nu_{B'}(x^{\alpha_j}) = \operatorname{ord}_{z'}((a^{\alpha_j} + y_j)|_{B'}) = 0$, since $\operatorname{ord}_{z'}(y_j|_{B'}) > 0$.

Since $\nu_{B'}(x^{\alpha_k}) = \operatorname{ord}_{z'}(y_k|_{B'}) > 0$, it follows that $\nu_{B'}$ and ν are proportional on $\mathbb{k}[x_i : i \in I]$ with a positive constant of proportionality. Pick $j \in I_{B'} \setminus I$. Then $j > k$, and $\nu_{B'}(x_j) = \operatorname{ord}_{z'}(y_j|_{B'}) > 0$. It follows that $\nu_{B'}$ and ν are compatible. It remains to exhibit property (4). Since the center of B is on $U \cap E$, properties (i) and (ii) of U imply that $(\operatorname{ord}_{z'}(y_1|_{B'}), \dots, \operatorname{ord}_{z'}(y_k|_{B'}))$ is proportional to $(\nu'_1, \dots, \nu'_k)$. Since $\nu'(z_1) = 1$, it follows that the constant of proportionality is $q := \operatorname{ord}_{z'}(z_1|_{B'})$. Therefore, $\nu_{B'}(x^{\alpha_k}) = \operatorname{ord}_{z'}(y_k|_{B'}) = \nu'_k q = q = q\nu(x^{\alpha_k})$. Since $\nu_{B'}$ is compatible with ν, it follows that

(v) $\nu_{B'}(x_j) = q\nu(x_j)$ for $j = 1, \dots, k$.

On the other hand, since $a' \in U \cap W' \cap E$, properties (ii) and (iii) imply that $\operatorname{ord}_{z'}(z_j|_{B'}) \geq 1$ if $j > k$. It follows that

(vi) for each $j \in I_{B'} \setminus I$, $\nu_{B'}(x_j) = \operatorname{ord}_{z'}(z_j|_{B'}) + \nu'_j \operatorname{ord}_{z'}(z_1|_{B'}) > q\nu'_j$.

Let $u' := \operatorname{In}(B') \in (k^*)^{I_{B'}}$. Observation (iv) implies that $u'^{\alpha_j} = a^{\alpha_j}$ for $j = 1, \dots, k-1$. Proposition VI.1 then implies that there is $t \in \mathbb{k}^*$ such that $(a_1, \dots, a_k) = (t^{\nu(x_1)}u'_1, \dots, t^{\nu(x_k)}u'_k)$. Choose a q-th root t' of t in $\mathbb{k}$ and let $\tilde{u} = (\tilde{u}_1, \dots, \tilde{u}_n)$ be an element with coordinates

$$\tilde{u}_j := \begin{cases} t'^{\nu_{B'}(x_j)}u'_j & \text{if } j \in I_{B'}, \\ \text{arbitrary element in } \mathbb{k}^* & \text{otherwise.} \end{cases}$$

$$= \begin{cases} t^{\nu(x_j)}u'_j = a_j & \text{if } j \in I, \\ t'^{\nu_{B'}(x_j)}u'_j & \text{if } j \in I_{B'} \setminus I, \\ \text{arbitrary element in } \mathbb{k}^* & \text{otherwise.} \end{cases}$$

Note that $\pi_I(\tilde{u}) = a = \pi_I(u)$. Fix $j \in [m]$. If $j \notin \mathcal{E}^I_{\mathcal{A}}$, then (v) and (vi) imply that, choosing $\nu'_{k+1}, \dots, \nu'_n$ sufficiently large, we can ensure that $\operatorname{In}_{\mathcal{A}_j^{I_{B'}}, \nu_{B'}}(f_j|_{\mathbb{k}^{I_{B'}}}) = \operatorname{In}_{\mathcal{A}_j^I, \nu}(f_j|_{\mathbb{k}^I})$, which would imply that $\operatorname{In}_{\mathcal{A}_j^{I_{B'}}, \nu_{B'}}(f_j|_{\mathbb{k}^{I_{B'}}})(\tilde{u}) = \operatorname{In}_{\mathcal{A}_j^I, \nu}(f_j|_{\mathbb{k}^I})(a) = 0$. On the other hand, if $j \in \mathcal{E}^I_{\mathcal{A}}$, then

$$\operatorname{In}_{\mathcal{A}_j^{I_{B'}}, \nu_{B'}}(f_j|_{\mathbb{k}^{I_{B'}}})(\tilde{u}) = t'^{\min_{\mathcal{A}_j^{I_{B'}}}(\nu_{B'})} \operatorname{In}_{\mathcal{A}_j^{I_{B'}}, \nu_{B'}}(f_j|_{\mathbb{k}^{I_{B'}}})(\operatorname{In}(B')) = 0$$

due to lemma IX.17. This completes the proof of property (4). □

7. Other formulae for generic intersection multiplicity

7.1. The formula of Huber-Sturmfels and Rojas. Let t be a new indeterminate. Fix positive integers $k_1, \dots, k_n$. Note that for each $f_1, \dots, f_n \in \mathbb{k}[[x_1, \dots, x_n]]$,

$$[f_1, \dots, f_n]_0 = [t, f_1 + c_1 t^{k_1}, \dots, f_n + c_n t^{k_n}]_0$$

for any $c_1, \dots, c_n \in \mathbb{k}$. It follows that, for each collection of subsets $\mathcal{A}_1, \dots, \mathcal{A}_n$ of $\mathbb{Z}^n_{\geq 0}$,

$$[\mathcal{A}_1, \dots, \mathcal{A}_n]_0 = [\hat{\mathcal{A}}_0, \dots, \hat{\mathcal{A}}_n]_0,$$

where $\hat{\mathcal{A}}_0 := \{(1, 0, \dots, 0)\} \subset \mathbb{Z}^{n+1}_{\geq 0}$ and $\hat{\mathcal{A}}_j := \{(k_j, 0, \dots, 0)\} \cup (\{0\} \times \mathcal{A}_j) \subset \mathbb{Z}^{n+1}_{\geq 0}$ for $j = 1, \dots, n$. Let $\hat{\mathcal{A}} := (\hat{\mathcal{A}}_0, \dots, \hat{\mathcal{A}}_n)$. It follows from (87) that $\mathscr{T}_{\hat{\mathcal{A}}, 1} = \{[n+1]\}$ and therefore, if $[\mathcal{A}_1, \dots, \mathcal{A}_n]_0 < \infty$, then theorem IX.1 implies that

$$[\mathcal{A}_1, \dots, \mathcal{A}_n]_0 = [\hat{\Gamma}_0, \dots, \hat{\Gamma}_n]^*_0 = \sum_{\hat{\nu} \in \hat{\mathcal{V}}'_0} \hat{\nu}_0 \operatorname{MV}'_{\hat{\nu}}(\operatorname{In}_{\hat{\nu}}(\hat{\Gamma}_1), \dots, \operatorname{In}_{\hat{\nu}}(\hat{\Gamma}_n)) \tag{97}$$

where $\hat{\Gamma}_j$ are the Newton diagrams of $\hat{\mathcal{A}}_j$, and $\hat{\nu}$ ranges over the primitive weighted orders on $\mathbb{k}[t, x_1, \dots, x_n]$ which are centered at the origin, and $\hat{\nu}_0 := \hat{\nu}(t)$. Note that $\operatorname{MV}'_{\hat{\nu}}(\operatorname{In}_{\hat{\nu}}(\hat{\Gamma}_1), \dots, \operatorname{In}_{\hat{\nu}}(\hat{\Gamma}_n))$ is positive only if $\hat{\nu}'$ is the inner normal to a "lower" facet of $\hat{\Gamma}_1 + \cdots + \hat{\Gamma}_n$ (the designation "lower" comes from the fact that $\hat{\nu}'$ points "upward" along the t-coordinate). B. Huber and B. Sturmfels presented in [HS97] the idea of "lifting" subsets of $\mathbb{Z}^n$ to one extra dimension and summing the mixed volumes of faces corresponding to certain lower facets of the sum of the lifted bodies. J. M. Rojas [Roj99] observed that the expression in the right-hand side of (97) gives the generic intersection multiplicity at the origin. Note that unlike the formula (88) from theorem IX.1, the expression in (97) is symmetric in $\mathcal{A}_1, \ldots, \mathcal{A}_n$ (provided the k_j are chosen to be equal).

FIGURE 9. $\mathcal{A}_1$ is convenient, whereas $\mathcal{A}_2$ is not. The subdiagram volume of $\mathcal{A}_j$ is the area of the region shaded in green.

7.2. Convenient Newton diagrams. We say that a subset of $\mathbb{R}^n_{\geq 0}$ is *convenient* if it contains a point on each coordinate axis. The *subdiagram volume* $V_n^-(\mathcal{A})$ of a subset $\mathcal{A}$ of $\mathbb{R}^n_{\geq 0}$ is the n-dimensional volume of the "cone" whose base is the Newton diagram of $\mathcal{A}$ and apex is at the origin; in other words, $V_n^-(\mathcal{A})$ is the n-dimensional volume of the union of all line segments from the origin to $\mathrm{ND}(\mathcal{A})$ (fig. 9).

Proposition IX.29. *Let $\mathcal{A}_1, \ldots, \mathcal{A}_n$ be subsets of $\mathbb{Z}^n_{\geq 0}$. Let $\Gamma_j := \mathrm{ND}(\mathcal{A}_j)$, $j = 1, \ldots, n$.*

(1) If $\Gamma_2, \ldots, \Gamma_n$ are convenient, then

$$[\mathcal{A}_1, \ldots, \mathcal{A}_n]_0 = \sum_{\nu \in \mathcal{V}'_0} \min_{\Gamma_1}(\nu)\, \mathrm{MV}'_\nu(\mathrm{In}_\nu(\Gamma_2), \ldots, \mathrm{In}_\nu(\Gamma_n)) \tag{98}$$

(2) (Kushnirenko [AY83, Theorem 22.8]*) If Γ is a convenient Newton diagram, and if $\Gamma_j = \Gamma$ for each j, then*

$$[\mathcal{A}_1, \ldots, \mathcal{A}_n]_0 = n! V_n^-(\Gamma) \tag{99}$$

(3) (Ajzenberg and Yuzhakov [AY83, Theorem 22.10]*) If $\Gamma_1, \ldots, \Gamma_n$ are convenient, then*

$$[\mathcal{A}_1, \ldots, \mathcal{A}_n]_0 = \sum_{\substack{I \subseteq [n] \\ I \neq \emptyset}} (-1)^{n-|I|} V_n^-(\sum_{i \in I} \Gamma_i) \tag{100}$$

PROOF. If $\Gamma_2, \ldots, \Gamma_n$ are convenient, then $\mathscr{T}_{\mathcal{A},1} = \{[n]\}$, and (98) follows from (88). Now we prove assertion (2). Let $\{\mathcal{Q}_j\}_j$ be the facets of $\Gamma_2 + \cdots + \Gamma_n$ with inner normals in $\mathbb{Z}^n_{>0}$. Then (98) implies that

$$[\mathcal{A}_1, \ldots, \mathcal{A}_n]_0 = \sum_j \min_\Gamma(\nu_j) \operatorname{MV}'_{\nu_j}(\operatorname{In}_{\nu_j}(\Gamma), \ldots, \operatorname{In}_{\nu_j}(\Gamma))$$
$$= (n-1)! \sum_j \min_\Gamma(\nu_j) \operatorname{Vol}'_{\nu_j}(\operatorname{In}_{\nu_j}(\Gamma)),$$

where ν_j' are the inner normals to $\mathcal{Q}_j$ and $\operatorname{Vol}'_{\nu_j}$ are as in corollary V.43. Now fix j, and let $\mathcal{R}_j := \operatorname{conv}(\mathcal{Q}_j \cup \{0\})$. Then $\mathcal{Q}_j$ is a facet of $\mathcal{R}_j$ with *outer* primitive normal ν_j, and all other facets of $\mathcal{R}_j$ passes through the origin. Since $\max_{\mathcal{R}_j}(\nu_j) = \min_\Gamma(\nu_j)$, corollary V.43 implies that $\operatorname{Vol}_n(\mathcal{R}_j) = (1/n)\operatorname{Vol}'_{\nu_j}(\operatorname{In}_{\nu_j}(\Gamma))$. Since $V_n^-(\Gamma) = \sum_j \operatorname{Vol}_n(\mathcal{R}_j)$, identity (99) follows. Since $[\mathcal{A}_1, \ldots, \mathcal{A}_n]_0$ is multi-additive and symmetric in the $\mathcal{A}_j$, assertion (3) then follows from corollary B.62. □

The following is a more precise version of assertion (2) of proposition IX.29.

Proposition IX.30. *Let $\mathcal{A}_1, \ldots, \mathcal{A}_n$ be subsets of $\mathbb{Z}^n_{\geq 0}$. Let $\Gamma_j := \operatorname{ND}(\mathcal{A}_j)$, $j = 1, \ldots, n$, and $\Gamma := \operatorname{ND}(\bigcup_{j=1}^n \mathcal{A}_j)$. For each $I \subseteq [n]$, let $T^I_{\mathcal{A}}$, where $\mathcal{A} := (\mathcal{A}_1, \ldots, \mathcal{A}_n)$, be as in theorem IX.1. Then*

(1) (Kushnirenko [AY83, Theorem 22.8]*) $[\mathcal{A}_1, \ldots, \mathcal{A}_n]_0 \geq n! V_n^-(\Gamma)$.*

(2) $[\mathcal{A}_1, \ldots, \mathcal{A}_n]_0 = n! V_n^-(\Gamma)$ if and only if for each nonempty $I \subseteq [n]$, $|T^I_{\mathcal{A}}| \geq |I|$ and for each weighted order ν centered at the origin on $\mathbb{k}[x_i : i \in I]$, the collection $\{\operatorname{In}_\nu(\Gamma_j \cap \mathbb{R}^I) : j \in T^I_{\mathcal{A}},\ \Gamma_j \cap \operatorname{In}_\nu(\Gamma \cap \mathbb{R}^I) \neq \emptyset\}$ of polytopes is dependent.

PROOF. If $[\mathcal{A}_1, \ldots, \mathcal{A}_n]_0 = \infty$ then both assertions of the proposition are satisfied (for the second assertion one needs to use corollary IX.13). So assume $[\mathcal{A}_1, \ldots, \mathcal{A}_n]_0 < \infty$. Then Γ is convenient so that assertion (1) follows from assertion (2) of proposition IX.29 and the definition of generic intersection multiplicity. Regarding the second assertion, theorem IX.8 implies that $[\mathcal{A}_1, \ldots, \mathcal{A}_n]_0 = [\Gamma, \ldots, \Gamma]_0$ if and only if generic $(f_1, \ldots, f_n) \in \mathcal{L}_0(\mathcal{A})$ are $\mathcal{B}$-non-degenerate at the origin, where $\mathcal{B} := (\bigcup_{j=1}^n \mathcal{A}_j, \ldots, \bigcup_{j=1}^n \mathcal{A}_j)$. The second assertion now follows from theorem VII.33. □

7.3. Making $\mathcal{A}_j$ convenient without changing $[\mathcal{A}_1, \ldots, \mathcal{A}_n]_0$. If $[f_1, \ldots, f_n]_0 < \infty$, then the ideal generated by $f_1, \ldots, f_n$ in $\mathbb{k}[[x_1, \ldots, x_n]]$ contains all sufficiently large powers of the maximal ideal of $\mathbb{k}[[x_1, \ldots, x_n]]$. It follows that if we replace f_j by $f_j + \sum_j c_{i,j} x_j^{d_{i,j}}$, then $[f_1, \ldots, f_n]_0$ does not change for sufficiently large $d_{i,j}$. Since the Newton diagrams of the f_j become convenient after these replacements, it follows that given any set of subsets $\mathcal{A}_1, \ldots, \mathcal{A}_n$ of $\mathbb{Z}^n_{\geq 0}$ such that $[\mathcal{A}_1, \ldots, \mathcal{A}_n]_0 < \infty$, we may use (98) or (100) to compute $[\mathcal{A}_1, \ldots, \mathcal{A}_n]_0$ after adding to each Γ_j appropriate vertices on the coordinate axes. In this section, we derive a "sharp" explicit bound on the placement of these vertices which guarantees that the intersection multiplicity at the origin remains unchanged. A. Khovanskii told the author in 2017 that he also had obtained, but never published, such a bound.

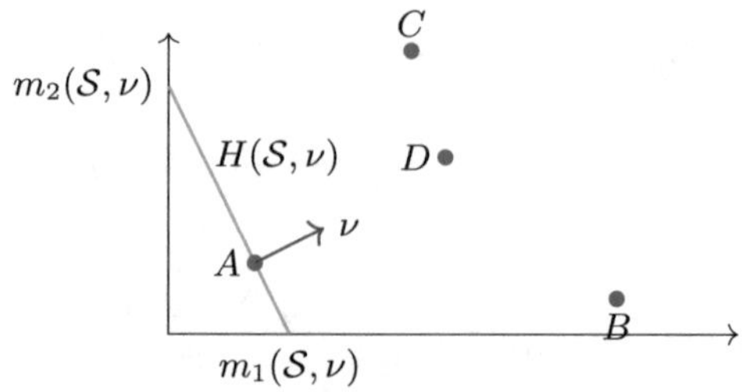

FIGURE 10. $m_i(\mathcal{S},\nu)$ for $\mathcal{S} = \{A, B, C, D\}$.

Let $\mathcal{S}$ be a compact subset of $\mathbb{R}^n$ and ν be an element of $(\mathbb{R}^n)^*$ centered at the origin. Let $H(\mathcal{S},\nu) := \{\alpha \in \mathbb{R}^n : \langle \nu, \alpha \rangle = \min_{\mathcal{S}}(\nu)\}$ be the hyperplane perpendicular to ν which contains the "face" $\mathrm{In}_\nu(\mathcal{S})$ of $\mathcal{S}$ corresponding to ν. We write $m_i(\mathcal{S},\nu)$ for the i-th coordinate of the point of the intersection of $H(\mathcal{S},\nu)$ and the i-th coordinate axis (see fig. 10). Note that

$$m_i(\mathcal{S},\nu) = \frac{\min_{\mathcal{S}}(\nu)}{\nu_i},$$

where $\nu = (\nu_1, \ldots, \nu_n)$ with respect to the coordinate dual to the standard basis of $\mathbb{R}^n$. Given a collection $\mathcal{A} = (\mathcal{A}_1, \ldots, \mathcal{A}_n)$ of subsets of $\mathbb{Z}^n_{\geq 0}$, pick $I \in \mathscr{T}_{\mathcal{A},1}$, where $\mathscr{T}_{\mathcal{A},1}$ is as in (87). Let $j_1 = 1, j_2, \ldots, j_{|I|}$ be the elements of $T^I_{\mathcal{A}}$. For each j, let Γ^I_j be the Newton diagram of $\mathcal{A}^I_j := \mathcal{A}_j \cap \mathbb{R}^I$. Let $\mathcal{V}'^I_{0,1}(\mathcal{A})$ be the set of primitive weighted orders centered at the origin on $\mathbb{k}[x_i : i \in I]$ such that the $(|I|-1)$-dimensional mixed volume of $\mathrm{In}_\nu(\Gamma^I_{j_2}), \ldots, \mathrm{In}_\nu(\Gamma^I_{j_{|I|}})$ is non-zero; recall that faces with non-zero mixed volume can be detected combinatorially (theorem VII.33). Let $e_1, \ldots, e_n$ be the standard unit vectors in $\mathbb{R}^n$. Define

(101)
$$m^I_{i,1}(\mathcal{A}) := \begin{cases} k_i & \text{if } I = \{i\},\ \Gamma^I_1 = \{k_i e_i\}, \text{ and } f_j|_{\mathbb{k}^I} = 0 \text{ for each } j > 1 & \text{(case 1)} \\ 1 & \text{else if } \mathcal{V}'^I_{0,1}(\mathcal{A}) = \emptyset & \text{(case 2)} \\ \max_{\nu \in \mathcal{V}'^I_{0,1}(\mathcal{A})} m_i(\Gamma^I_1, \nu) & \text{otherwise} & \text{(case 3)} \end{cases}$$

See fig.11 for an illustration of different cases in the definition of $m^I_{i,1}$. Define

$$m_{i,1}(\mathcal{A}) := \max_{\substack{I \in \mathscr{T}_{\mathcal{A},1} \\ i \in I}} m^I_{i,1} \tag{102}$$

Since $[n] \in \mathscr{T}_{\mathcal{A},1}$, it follows that $m^{[n]}_{i,1}(\mathcal{A})$, and therefore $m_{i,1}(\mathcal{A})$ is a well-defined nonnegative rational number for each i. Let $\mathcal{A}'_1 := \mathcal{A}_1 \cup \{m'_1 e_1, \ldots, m'_n e_n\}$, where m'_i are arbitrary integers greater than or equal to $m_{i,1}(\mathcal{A})$. Note that $\mathcal{A}'_1$ is convenient. Let $\mathcal{A}' := (\mathcal{A}'_1, \mathcal{A}_2, \ldots, \mathcal{A}_n)$.

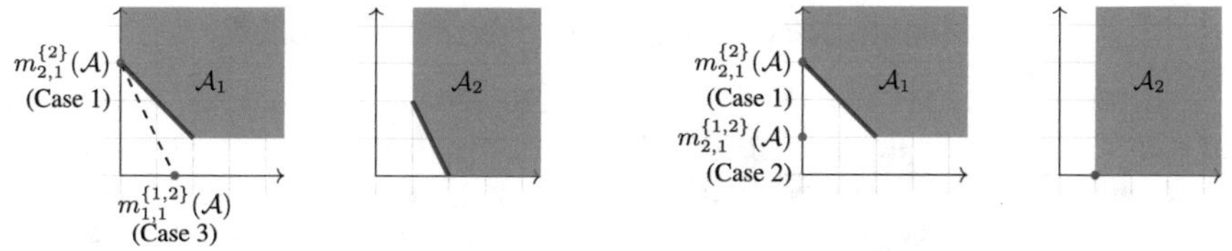

(A) Cases 1 and 3

(B) Cases 1 and 2

FIGURE 11. Different cases of (101).

Proposition IX.31. *Assume* $[\mathcal{A}_1, \dots, \mathcal{A}_n]_0 < \infty$. *Then* $[\mathcal{A}'_1, \mathcal{A}_2, \dots, \mathcal{A}_n]_0 = [\mathcal{A}_1, \dots, \mathcal{A}_n]_0$. *If in addition* $[\mathcal{A}_1, \dots, \mathcal{A}_n]_0 > 0$, *then the transformation* $\mathcal{A}_1 \mapsto \mathcal{A}'_1$ *is sharp in the following sense: if* $\mathcal{A}''_1 \supset \mathcal{A}_1 \cup \{m''_i e_i\}$ *for any* i *and any nonnegative integer* m''_i *such that* $m''_i < m_{i,1}(\mathcal{A})$, *then* $[\mathcal{A}''_1, \mathcal{A}_2, \dots, \mathcal{A}_n]_0 < [\mathcal{A}_1, \dots, \mathcal{A}_n]_0$.

PROOF. If $[\mathcal{A}_1, \dots, \mathcal{A}_n]_0 = 0$, then $0 \in \bigcup_j \mathcal{A}_j \subset \mathcal{A}'_1 \cup \bigcup_{j \geq 2} \mathcal{A}_j$ so that $[\mathcal{A}'_1, \mathcal{A}_2, \dots, \mathcal{A}_n]_0 = 0$ as well. Now assume $0 < [\mathcal{A}_1, \dots, \mathcal{A}_n]_0 < \infty$. Then $0 \notin \mathcal{A}_1$, so that $m_i(\Gamma^I_1, \nu) > 0$ for each $i \in [n]$, $I \subseteq [n]$ and each nontrivial weighted order centered at the origin on $\mathbb{k}[x_j : j \in I]$. It follows that $m_{i,1}(\mathcal{A}) > 0$ for each i, and therefore corollary IX.13 implies that $0 < [\mathcal{A}'_1, \mathcal{A}_2, \dots, \mathcal{A}_n]_0 < \infty$ and $\mathscr{T}_{\mathcal{A},1} = \mathscr{T}_{\mathcal{A}',1}$. Now pick $I \in \mathscr{T}_{\mathcal{A}',1}$. Let $j_1 = j, j_2, \dots, j_{|I|}$ be the elements of $T^I_{\mathcal{A}'}$. If Γ'^I_1 is the Newton diagram of $\mathcal{A}'_1 \cap \mathbb{R}^I$, it is straightforward to see using the definition of $[\cdot, \dots, \cdot]^*_0$ that $[\Gamma'^I_1, \Gamma^I_{j_2}, \dots, \Gamma^I_{j_{|I|}}]^*_0 = [\Gamma^I_1, \Gamma^I_{j_2}, \dots, \Gamma^I_{j_{|I|}}]^*_0$, and therefore, (88) implies that $[\mathcal{A}'_1, \mathcal{A}_2, \dots, \mathcal{A}_n]_0 = [\mathcal{A}_1, \dots, \mathcal{A}_n]_0$. On the other hand, if for some $i \in I$, Γ'^I_1 contains an element on the i-th axis with coordinates $m''_i e_i$ such that $m''_i < m_{i,1}(\Gamma^I_1, \nu)$ for some $\nu \in \mathcal{V}'^I_{0,1}(\mathcal{A})$, then $\min_{\Gamma'^I_1}(\nu) < \min_{\Gamma^I_1}(\nu)$, and it would follow that $[\Gamma'^I_1, \Gamma^I_{j_2}, \dots, \Gamma^I_{j_{|I|}}]^*_0 < [\Gamma^I_1, \Gamma^I_{j_2}, \dots, \Gamma^I_{j_{|I|}}]^*_0$, which implies the last assertion. □

It is clear that given $\mathcal{A}_1, \dots, \mathcal{A}_n$ such that $[\mathcal{A}_1, \dots, \mathcal{A}_n]_0 < \infty$, repeating the above process n times would yield a collection $\mathcal{A}'_1, \dots, \mathcal{A}'_n$ of convenient subsets of $\mathbb{Z}^n_{\geq 0}$ such that $[\mathcal{A}'_1, \dots, \mathcal{A}'_n]_0 = [\mathcal{A}_1, \dots, \mathcal{A}_n]_0$, as required. However, as fig. 12 shows, the outcome of the process is in general *not* unique: different ordering of the $\mathcal{A}_j$ might result in different $\mathcal{A}'_1, \dots, \mathcal{A}'_n$.

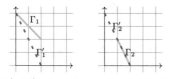

(A) $\Gamma'_1 = \Gamma'_2 =$ the line segment from $(2, 0)$ to $(0, 4)$

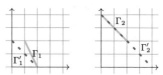

(B) Reversing the order of the $\mathcal{A}_j$ from fig. 12a changes Γ'_1 (respectively Γ'_2) to the line segment from $(2, 0)$ to $(0, 2)$ (respectively, from $(4, 0)$ to $(0, 4)$).

FIGURE 12. Dependence of $\{\mathcal{A}'_1, \mathcal{A}'_2\}$ on the ordering of $\mathcal{A}_j$: here Γ_j, Γ'_i denote respectively the Newton diagram of $\mathcal{A}_j, \mathcal{A}'_i$.

8. Monotonicity of generic intersection multiplicity

In remark IX.3, we saw that $[\cdot, \ldots, \cdot]_0$ is "monotonic" as a function on n-tuples of subsets of $\mathbb{Z}^n_{\geq 0}$. In this section, we characterize in theorem IX.32 the conditions under which it is "strictly monotonic," and as an application we prove the alternate formulation of non-degeneracy at the origin in corollary IX.11 (the counterparts of these results in the toric case are corollary VII.34 and theorem VII.13). We also state a curious implication (proposition IX.34) of theorem IX.8 that in the case the monotonicity is not strict, the intersection multiplicity is determined by the Newton diagram of the intersection.

THEOREM IX.32. *Let* $\mathcal{B}_j \subseteq \mathit{conv}(\mathcal{A}_j) + \mathbb{R}^n_{\geq 0}$, $j = 1, \ldots, n$. *Then* $[\mathcal{A}_1, \ldots, \mathcal{A}_n]_0 \leq [\mathcal{B}_1, \ldots, \mathcal{B}_n]_0$. *If* $0 < [\mathcal{A}_1, \ldots, \mathcal{A}_n]_0 < \infty$, *then the following are equivalent:*

(1) $[\mathcal{A}_1, \ldots, \mathcal{A}_n]_0 = [\mathcal{B}_1, \ldots, \mathcal{B}_n]_0$,

(2) *for each nonempty subset* I *of* $[n]$, *and each* $\nu \in (\mathbb{R}^n)^*$ *which is centered at the origin, the collection* $\{\mathrm{In}_\nu(\mathrm{ND}(\mathcal{B}^I_j)) : \mathrm{In}_\nu(\mathrm{ND}(\mathcal{A}^I_j)) \cap \mathrm{Supp}(\mathcal{B}_j) \neq \emptyset\}$ *of polytopes is dependent,*

(3) *for each* $I \in \mathscr{I}_{\mathcal{A}} := \{I \subseteq [n] : I \neq \emptyset, |\mathcal{E}^I_{\mathcal{A}}| \geq n - |I|\}$ *and each* $\nu \in (\mathbb{R}^n)^*$ *which is centered at the origin, the collection* $\{\mathrm{In}_\nu(\mathrm{ND}(\mathcal{B}^I_j)) : \mathrm{In}_\nu(\mathrm{ND}(\mathcal{A}^I_j)) \cap \mathrm{Supp}(\mathcal{B}_j) \neq \emptyset\}$ *of polytopes is dependent.*

PROOF. This follows exactly in the same way as the proof of corollary VII.34 by considering generic $(f_1, \ldots, f_n) \in \mathcal{L}_0(\mathcal{B})$, then observing that $[f_1, \ldots, f_n]_0 = [\mathcal{A}_1, \ldots, \mathcal{A}_n]_0$ if and only if $f_1, \ldots, f_n$ are $\mathcal{A}$-non-degenerate at the origin, and finally applying theorems VII.5 and VII.33 together with the genericness of $f_1, \ldots, f_n$. □

Corollary IX.33. *Corollary IX.11 holds.*

PROOF. The equivalence of assertions (1) and (2) follows exactly as in the proof of corollary VII.36.Theorem IX.9 implies that assertions (2) and (3) are equivalent. □

In proposition IX.34 below, we use the following notation: given $f = \sum_\alpha c_\alpha x^\alpha \in \mathbb{k}[[x_1, \ldots, x_n]]$ and $\mathcal{S} \subseteq \mathbb{R}^n$, we write $f_{\mathcal{S}} := \sum_{\alpha \in \mathcal{S}} c_\alpha x^\alpha$.

Proposition IX.34. *Let* $(f_1, \ldots, f_n) \in \mathcal{L}_0(\mathcal{A})$ *be such that* $[f_1, \ldots, f_n]_0 = [\mathcal{A}_1, \ldots, \mathcal{A}_n]_0$. *Then*

$$[f_{1,\mathrm{ND}(\mathcal{A}_1)}, \ldots, f_{n,\mathrm{ND}(\mathcal{A}_n)}]_0 = [f_1, \ldots, f_n]_0 = [\mathcal{A}_1, \ldots, \mathcal{A}_n]_0$$

In particular, if $\mathcal{B}_j$ *are subsets of* $\mathcal{A}_j + \mathbb{R}^n_{\geq 0}$ *such that* $[\mathcal{B}_1, \ldots, \mathcal{B}_n]_0 = [\mathcal{A}_1, \ldots, \mathcal{A}_n]_0$, *then*

$$[\mathcal{B}_1 \cap \mathrm{ND}(\mathcal{A}_1), \ldots, \mathcal{B}_n \cap \mathrm{ND}(\mathcal{A}_n)]_0 = [\mathcal{B}_1, \ldots, \mathcal{B}_n]_0 = [\mathcal{A}_1, \ldots, \mathcal{A}_n]_0$$

PROOF. Follows immediately from theorem IX.8, since $f_1, \ldots, f_n$ are non-degenerate at the origin if and only if $f_{1,\mathrm{ND}(\mathcal{A}_1)}, \ldots, f_{n,\mathrm{ND}(\mathcal{A}_n)}$ are non-degenerate at the origin. □

The requirement that $\mathcal{B}_j \subseteq \mathcal{A}_j + \mathbb{R}^n_{\geq 0}$ for each j is necessary for proposition IX.34. Indeed, if $\mathcal{A}_j, \mathcal{B}_j, j = 1, 2$, are from fig. 13, then $[\mathcal{A}_1, \mathcal{A}_2]_0 = [\mathcal{B}_1, \mathcal{B}_2]_0 = 9$, but $[\mathcal{B}_1 \cap \mathrm{ND}(\mathcal{A}_1), \mathcal{B}_2 \cap \mathrm{ND}(\mathcal{A}_2)]_0 = \infty$.

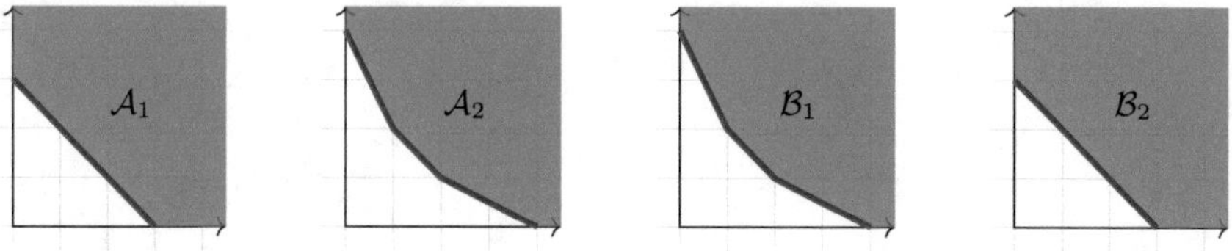

FIGURE 13. Failure of proposition IX.34 when there is j such that $\mathcal{B}_j \not\subseteq \mathcal{A}_j + \mathbb{R}^n_{\geq 0}$.

9. Notes

For convenient Newton diagrams, there is a formula for generic intersection multiplicity in terms of integer lattice points in the region bounded by the diagram and the coordinate hyperplanes (see e.g. [Est12, Theorem 5]). A. Khovanskii informed the author that he had obtained (but did not publish) a bound equivalent to (102) which reduces the computation of generic intersection multiplicity to the convenient case. Recently M. Herrero, G. Jeronimo and J. Sabia [HJS19] gave some other formulae for generic intersection mulitplicity in the general case.

CHAPTER X

Number of zeroes on the affine space II: the general case

1. Introduction

In this chapter, we compute the number of solutions on $\mathbb{k}^n$ (or more generally, on any given Zariski open subset of $\mathbb{k}^n$) of generic systems of polynomials with given supports, and give explicit BKK-type characterizations of genericness in terms of initial forms of the polynomials. As a special case, we derive generalizations of weighted (multi-homogeneous)-Bézout theorems involving arbitrary weighted degrees (i.e. weighted degrees with possibly negative or zero weights).

2. The bound

2.1. Khovanskii's formula. For polynomials $f_1, \ldots, f_n$, and any Zariski open subset U of $\mathbb{k}^n$, as in section VII.3, let $[f_1, \ldots, f_n]_U^{iso}$ be the sum of intersection multiplicities of $f_1, \ldots, f_n$ at all the isolated points of $V(f_1, \ldots, f_n) \cap U$. Given a collection $\mathcal{A} := (\mathcal{A}_1, \ldots, \mathcal{A}_n)$ of n finite subsets of $\mathbb{Z}^n_{\geq 0}$, define

$$[\mathcal{A}_1, \ldots, \mathcal{A}_n]_U^{iso} := \max\{[f_1, \ldots, f_n]_U^{iso} : \mathrm{Supp}(f_j) \subset \mathcal{A}_j,\ j = 1, \ldots, n\}.$$

In this section, we give a formula for $[\mathcal{A}_1, \ldots, \mathcal{A}_n]_U^{iso}$ in terms of (mixed volumes of) convex hulls of $\mathcal{A}_j$. For $I \subseteq [n]$, let $T^I_{\mathcal{A}} := \{j : \mathcal{A}^I_j \neq \emptyset\}$ as in theorem IX.1, and let

$$\mathscr{E}(\mathcal{A}) := \{I \subseteq [n] : \text{ there is } \tilde{I} \supseteq I \text{ such that } |T^{\tilde{I}}_{\mathcal{A}}| < |\tilde{I}|\}. \tag{103}$$

The following result, which follows immediately from theorem III.80, implies that $[\mathcal{A}_1, \ldots, \mathcal{A}_n]_U^{iso} = [\mathcal{A}_1, \ldots, \mathcal{A}_n]_{U_{\mathcal{A}}}^{iso}$, where $U_{\mathcal{A}} := U \setminus \bigcup_{i \in \mathscr{E}(\mathcal{A})} \mathbb{k}^I$.

Lemma X.1. *Let $f_1, \ldots, f_m \in \mathbb{k}[x_1, \ldots, x_n]$ and let $V := V(f_1, \ldots, f_m) \subset \mathbb{k}^n$. Given $I \subseteq [n]$, if there exists $\tilde{I} \supseteq I$ such that $|\{j : f_j|_{\mathbb{k}^{\tilde{I}}} \not\equiv 0\}| < |\tilde{I}|$, then no point of $V \cap \mathbb{k}^I$ is isolated in V.* ☐

P. Mondal, *How Many Zeroes?*, CMS/CAIMS Books in Mathematics 2,
https://doi.org/10.1007/978-3-030-75174-6_X

Define

$$\mathscr{E}(U) := \{I \subseteq [n] : \Bbbk^I \cap U = \emptyset\}, \tag{104}$$

$$\mathscr{T}(U, \mathcal{A}) := \{I \subseteq [n] : I \notin \mathscr{E}(U) \cup \mathscr{E}(\mathcal{A}),\ |T^I_{\mathcal{A}}| = |I|\}. \tag{105}$$

The following result is also a straightforward consequence of theorem III.80.

Lemma X.2. *If $I \notin \mathscr{T}(U, \mathcal{A})$, then $(\Bbbk^*)^I \cap V(f_1, \ldots, f_n) \cap U = \emptyset$ for generic $f_1, \ldots, f_n$ such that* $\mathrm{Supp}(f_j) \subseteq \mathcal{A}_j$, $j = 1, \ldots, n$. □

Remark X.3. It is possible that $\emptyset$ (i.e. the empty set) is in $\mathscr{T}(U, \mathcal{A})$; this is the case if and only if the origin is in U and $0 < [\mathcal{A}_1, \ldots, \mathcal{A}_n]_0 < \infty$ (see example X.5).

THEOREM X.4. (Khovanskii[1]). *Let $\mathcal{P}_j$ be the convex hull of $\mathcal{A}_j$, $j = 1, \ldots, n$. For $I \subseteq [n]$, let $\mathcal{P}^I_j := \mathcal{P}_j \cap \mathbb{R}^I$, and let $\pi_I : \mathbb{R}^n \to \mathbb{R}^I$ be the natural projection (as in (85)). Then*

$$[\mathcal{A}_1, \ldots, \mathcal{A}_n]^{iso}_U = \sum_{I \in \mathscr{T}(U, \mathcal{A})} \mathrm{MV}(\mathcal{P}^I_{j_1}, \ldots, \mathcal{P}^I_{j_{|I|}}) \times [\pi_{[n]\setminus I}(\mathcal{P}_{j'_1}), \ldots, \pi_{[n]\setminus I}(\mathcal{P}_{j'_{n-|I|}})]_0 \tag{106}$$

where for each $I \in \mathscr{T}(U, \mathcal{A})$, $j_1, \ldots, j_{|I|}$ are the elements of $T^I_{\mathcal{A}}$, and $j'_1, \ldots, j'_{n-|I|}$ are the elements of $[n] \setminus T^I_{\mathcal{A}}$, and $[\cdot, \ldots, \cdot]_0$ is defined as in (78).

The interpretation of the right-hand side of (106) is straightforward—for each $I \in \mathscr{T}(U, \mathcal{A})$, the corresponding summand counts with multiplicity the number of solutions on $(\Bbbk^*)^I \cap U$ of generic systems supported at $\mathcal{A}_1, \ldots, \mathcal{A}_n$. In the next section, we present another formula which sometimes is more efficient, since it involves summing over elements from a proper subset of $\mathscr{T}(U, \mathcal{A})$.

Example X.5. Let $\mathcal{A}_j$ be the support of f_j from examples VII.31 and IX.4, and U be a nonempty Zariski open subset of $\Bbbk^3$. Then $\mathscr{E}(\mathcal{A}) = \emptyset$, and

$$\mathscr{T}(U, \mathcal{A}) = \begin{cases} \{\{1,2,3\}, \{3\}, \emptyset\} & \text{if } 0 \in U \text{ (Case 1)}, \\ \{\{1,2,3\}, \{3\}\} & \text{if } 0 \notin U, \text{ but } U \text{ contains a point on the } z\text{-axis (Case 2)}, \\ \{\{1,2,3\}\} & \text{otherwise (Case 3)}. \end{cases}$$

In Case 3, identity (106) and example VII.31 imply that

$$[\mathcal{A}_1, \mathcal{A}_2, \mathcal{A}_3]^{iso}_U = \mathrm{MV}(\mathcal{P}_1, \mathcal{P}_2, \mathcal{P}_3) = 5$$

Since the projections of $\mathcal{P}_2$ and $\mathcal{P}_3$ onto the (x, y)-plane have nontrivial linear part, and $\mathcal{P}^{\{3\}}_1$ has integer length 1 (see fig. 1), identity (X.4) implies that in Case 2,

$$[\mathcal{A}_1, \mathcal{A}_2, \mathcal{A}_3]^{iso}_U = 5 + \mathrm{MV}(\mathcal{P}^{\{3\}}_1) \times [\pi_{\{1,2\}}(\mathcal{P}_2), \pi_{\{1,2\}}(\mathcal{P}_3)]_0 = 5 + 1 \cdot 1 = 6$$

[1]A. Khovanskii described this unpublished formula to the author during the Askoldfest, 2017

Finally, in Case 1, identity (X.4) and the computation from example IX.4 imply that

$$[\mathcal{A}_1, \mathcal{A}_2, \mathcal{A}_3]_U^{iso} = 6 + [\mathcal{P}_1, \mathcal{P}_2, \mathcal{P}_3]_0 = 6 + 6 = 12$$

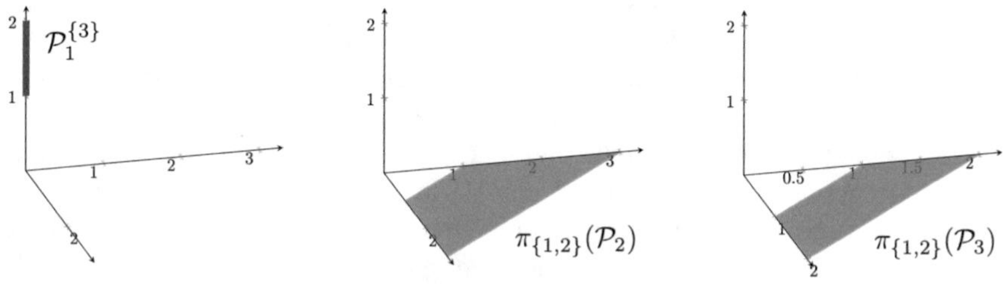

FIGURE 1. Computing $\mathrm{MV}(\mathcal{P}_1^{\{3\}}) \times [\pi_{\{1,2\}}(\mathcal{P}_2), \pi_{\{1,2\}}(\mathcal{P}_3)]_0$.

2.2. A formula in the same spirit as the formula for generic intersection multiplicity. If ν is a weighted order on $\mathbb{k}[x_1, \ldots, x_n]$, we say that ν is *centered at infinity* if $\nu(x_i) < 0$ for some $i \in [n]$. Given $I \subseteq [n]$, we say that ν is *centered at* $\mathbb{k}^I$ if $\nu(x_i) \geq 0$ for each $i \in I$ and $\nu(x_{i'}) > 0$ for each $i' \in [n] \setminus I$. Given a collection $\mathscr{S}$ of subsets of $[n]$, we denote by $\mathbb{k}^n_{\mathscr{S}}$ the complement in $\mathbb{k}^n$ of the coordinate subspaces $\mathbb{k}^I$ for all $I \in \mathscr{S}$, i.e.

$$\mathbb{k}^n_{\mathscr{S}} := \mathbb{k}^n \setminus \bigcup_{I \in \mathscr{S}} \mathbb{k}^I \tag{107}$$

We write $\mathcal{V}_{\mathscr{S}}$ for the union, over all $I \in \mathcal{S}$, of the sets of weighted orders centered at $\mathbb{k}^I$, and $\mathcal{V}_\infty$ for the set of weighted orders centered at infinity; the collection of primitive elements in $\mathcal{V}_{\mathscr{S}}$ and $\mathcal{V}_\infty$ are denoted, respectively, as $\mathcal{V}'_{\mathscr{S}}$ and $\mathcal{V}'_\infty$.

Example X.6. Taking $\mathscr{S} = \emptyset$ gives $\mathbb{k}^n_{\mathscr{S}} = \mathbb{k}^n$ and $\mathcal{V}_{\mathscr{S}} = \emptyset$. If we take $\mathscr{S} = \{\emptyset\}$, then $\mathbb{k}^n_{\mathscr{S}} = \mathbb{k}^n \setminus \{0\}$ and $\mathcal{V}_{\mathscr{S}}$ is the set $\mathcal{V}_0$ of weighted orders *centered at the origin* (see section IX.2.2). If $\mathscr{S}$ is the set of all subsets of $[n]$ consisting of $n-1$ elements, then $\mathbb{k}^n_{\mathscr{S}} = (\mathbb{k}^*)^n$ and $\mathcal{V}_{\mathscr{S}}$ is the set of all non-zero weighted orders which are not centered at infinity.

Let $\mathcal{A} := (\mathcal{A}_1, \ldots, \mathcal{A}_n)$ be a collection of subsets of $\mathbb{Z}^n_{\geq 0}$ and $\mathcal{P}_j$ be the convex hull in $\mathbb{R}^n$ of $\mathcal{A}_j$, $j = 1, \ldots, n$. Given a collection $\mathscr{S}$ of subsets of $[n]$, define

$$[\mathcal{P}_1, \ldots, \mathcal{P}_n]^*_{\mathscr{S}} := -\sum_{\nu \in \mathcal{V}'_{\mathscr{S}} \cup \mathcal{V}'_\infty} \min_{\mathcal{P}_1}(\nu)\ \mathrm{MV}'_\nu(\mathrm{In}_\nu(\mathcal{P}_2), \ldots, \mathrm{In}_\nu(\mathcal{P}_n)), \tag{108}$$

where $\mathrm{MV}'_\nu(\cdot, \ldots, \cdot)$ is defined as in (69).

THEOREM X.7. ([Mon16]). *Let U be a Zariski open subset of $\mathbb{k}^n$. We continue to use the notation of theorem X.4. Define $\mathscr{T}_1(U, \mathcal{A}) := \{I \in \mathscr{T}(U, \mathcal{A}) : 1 \in T^I_{\mathcal{A}}\}$, and for each $I \subseteq [n]$, set $\mathscr{E}^I(U, \mathcal{A}) := \{J \subseteq I : J \in \mathscr{E}(U) \cup \mathscr{E}(\mathcal{A})\}$. Then*

$$\begin{aligned}[\mathcal{A}_1, \ldots, \mathcal{A}_n]_U^{iso} = &\sum_{I \in \mathscr{T}_1(U, \mathcal{A})} [\mathcal{P}_1^I, \mathcal{P}_{j_2}^I, \ldots, \mathcal{P}_{j_{|I|}}^I]^*_{\mathscr{E}^I(U,\mathcal{A})} \\ &\times [\pi_{[n]\setminus I}(\mathcal{P}_{j'_1}), \ldots, \pi_{[n]\setminus I}(\mathcal{P}_{j'_{n-|I|}})]_0\end{aligned} \tag{109}$$

where for each $I \in \mathscr{T}_1(U, \mathcal{A})$, $j_1 = 1, j_2, \ldots, j_{|I|}$ *are the elements of* $T^I_{\mathcal{A}}$, *and* $j'_1, \ldots, j'_{n-|I|}$ *are the elements of* $[n] \setminus T^I_{\mathcal{A}}$.

There is an obvious analogy between formula (109) and the formula (88) for intersection multiplicity at the origin. The interpretation of the terms on the right-hand side of (109) is also analogous to the interpretation of the corresponding terms of (88) described in section IX.2.1; in particular, for each $I \in \mathscr{T}_1(U, \mathcal{A})$, the corresponding summand on the right-hand side of (109) is the sum, for generic $f_1, \ldots, f_n$ supported, respectively, at $\mathcal{A}_1, \ldots, \mathcal{A}_n$, of the *negative* of orders of f_1 along the branches (counted with appropriate multiplicities) of the curve determined by $f_2 = \cdots = f_n = 0$ which lie on $\mathbb{k}^I$ and are centered either at infinity or at $\mathbb{k}^{I'}$ for some $I' \in \mathscr{E}^I(U, \mathcal{A})$. As in the case of (88), and unlike (106), the symmetry of the right-hand side of (109) with respect to permutations of the $\mathcal{P}_j$ is not at all obvious. We use (109) to derive the symmetric formula of Huber and Sturmfels [HS97] and Rojas [Roj99] in section X.4.1.

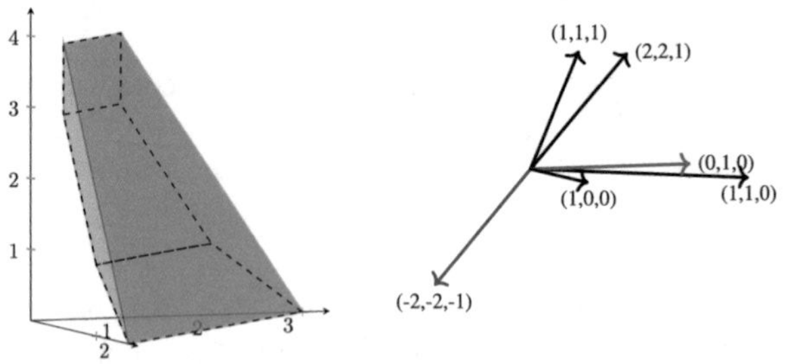

$\mathcal{P}_1 + \mathcal{P}_3$ inner normals to facets of $\mathcal{P}_1 + \mathcal{P}_3$

FIGURE 2. $\mathcal{P}_1 + \mathcal{P}_3$ and the inner normals to its facets.

Example X.8. We continue with $\mathcal{A}_1, \mathcal{A}_2, \mathcal{A}_3$ from example X.5, and compute $[\mathcal{A}_1, \mathcal{A}_2, \mathcal{A}_3]^{iso}_U$ using theorem X.7 for nonempty Zariski open subsets U of $\mathbb{k}^3$. It is straightforward to check that

$$\mathscr{T}_1(U, \mathcal{A}) = \begin{cases} \{\{1,2,3\},\{3\}\} & \text{if } U \text{ contains a point on the } z\text{-axis (Case 1 or 2 of example X.5),} \\ \{\{1,2,3\}\} & \text{otherwise (Case 3 of example X.5).} \end{cases}$$

On the other hand, if we change the order of the $\mathcal{A}_j$, or equivalently, consider the collection $\mathcal{A}' := (\mathcal{A}_2, \mathcal{A}_1, \mathcal{A}_3)$, then one checks that $\mathscr{T}_1(U, \mathcal{A}') = \{\{1,2,3\}\}$ for any (nonempty) U, and we apply identity (109) to $\mathcal{A}'$ to reduce computation. In particular, we have

$$\begin{aligned} [\mathcal{A}_1, \mathcal{A}_2, \mathcal{A}_3]^{iso}_U &= [\mathcal{A}_2, \mathcal{A}_1 \mathcal{A}_3]^{iso}_U = [\mathcal{P}_2, \mathcal{P}_1, \mathcal{P}_3]^*_{\mathscr{E}(U,\mathcal{A})} \\ &= - \sum_{\nu \in \mathcal{V}'_{\mathscr{E}(U,\mathcal{A})} \cup \mathcal{V}'_\infty} \min_{\mathcal{P}_2}(\nu) \ \mathrm{MV}'_\nu(\mathrm{In}_\nu(\mathcal{P}_1), \mathrm{In}_\nu(\mathcal{P}_3)) \end{aligned}$$

The inner normals to the facets of $\mathcal{P}_1 + \mathcal{P}_2$ are listed in fig. 2. The only element in $\mathcal{V}'_\infty$ is $(-2,-2,-1)$. If the origin is in U, then $\mathscr{E}(U) = \emptyset$, and therefore it follows from example VII.3 and fig. 3 that

$$[\mathcal{A}_1, \mathcal{A}_2, \mathcal{A}_3]_U^{iso} = -\min_{\mathcal{P}_2}(-2,-2,-1) \cdot \text{Area}($$ 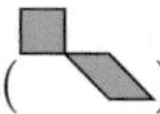 $$) = 6 \cdot 2 = 12$$

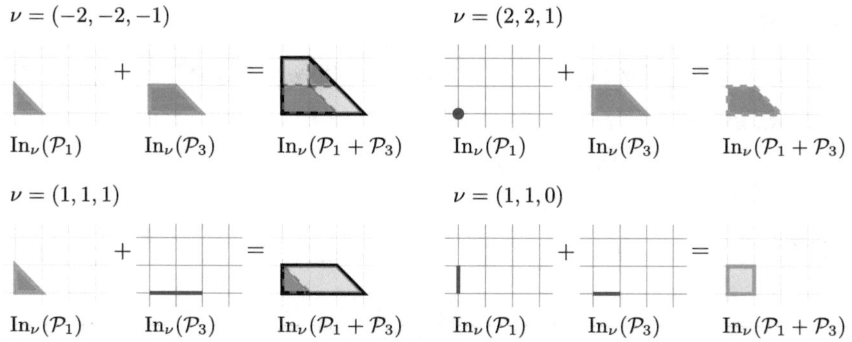

FIGURE 3. The image under ψ_ν of initial faces of $\mathcal{P}_1, \mathcal{P}_3$ and $\mathcal{P}_1 + \mathcal{P}_3$.

If the origin is not in U, but U contains some other points of the z-axis, then for $I = \{1,2,3\}$, the set $\mathscr{E}^I(U, \mathcal{A}')$ contains the emptyset, but not $\{3\}$. It follows that $\mathcal{V}'_{\mathscr{E}^I(U,\mathcal{A}')}$ does not contain $(1,1,0)$, but it contains each (primitive) weighted order centered at the origin (see example X.6); in particular, it contains $(2,2,1)$ and $(1,1,1)$. Since $\min_{\mathcal{P}_2}(\nu) = 0$ when $\nu = (1,0,0)$ or $(0,1,0)$, it does not matter if these two elements are in $\mathcal{V}'_{\mathscr{E}^I(U,\mathcal{A}')}$. It then follows from fig. 3 that

$$[\mathcal{A}_1, \mathcal{A}_2, \mathcal{A}_3]_U^{iso} = 12 - \min_{\mathcal{P}_2}(2,2,1) \cdot \text{Area}(\emptyset) - \min_{\mathcal{P}_2}(1,1,1) \cdot \text{Area}($$  $$) = 12 - 4 \cdot 0 - 3 \cdot 2 = 6$$

If U does not intersect the z-axis, then $(1,1,0)$ is also an element of $\mathcal{V}'_{\mathscr{E}^I(U,\mathcal{A}')}$, and it follows that

$$[\mathcal{A}_1, \mathcal{A}_2, \mathcal{A}_3]_U^{iso} = 6 - \min_{\mathcal{P}_2}(1,1,0) \cdot \text{Area}($$ $$) = 6 - 1 \cdot 1 = 5$$

The computations therefore agree with those from example X.5. Note that formulae (106) and (109) resolve the cases in the opposite order!

3. Derivation of the formuale for the bound

In this section, we prove theorems X.4 and X.7. Throughout this section $\mathcal{A} := (\mathcal{A}_1, \dots, \mathcal{A}_n)$ denotes a collection of finite subsets of $\mathbb{Z}^n_{\geq 0}$, and $\mathcal{P}_j$ denotes the convex hull of $\mathcal{A}_j$ in $\mathbb{R}^n$, $j = 1, \dots, n$. As in the preceding chapters, we write $\mathcal{L}(\mathcal{A})$ for the space of n-tuples of polynomials supported at $\mathcal{A}$, and as in definition IX.23, we write $\check{\mathcal{N}}(\mathcal{A})$ be the collection of all strongly $\mathcal{A}$-non-degenerate $(f_1, \dots, f_n) \in \mathcal{L}(\mathcal{A})$. Given a Zariski open subset U of $\Bbbk^n$, write $\check{\mathcal{N}}(U, \mathcal{A})$ for all

$(f_1, \dots, f_n) \in \check{\mathcal{N}}(\mathcal{A})$ such that for all $I \notin \mathscr{E}(U) \cup \mathscr{E}(\mathcal{A})$,

$$(V(f_1, \dots, f_n) \cap (\mathbb{k}^*)^I) \setminus U = \emptyset \tag{110}$$

Proposition IX.25 and lemma IX.26 imply that $\check{\mathcal{N}}(U, \mathcal{A})$ contains a nonempty Zariski open subset of $\mathcal{L}(\mathcal{A})$. Assertion (4) of theorem IV.32 therefore implies that in order to prove theorems X.4 and X.7, it suffices to show that $[f_1, \dots, f_n]_U$ equals the quantities from the right-hand sides of (106) and (109) for all $(f_1, \dots, f_n) \in \check{\mathcal{N}}(U, \mathcal{A})$.

3.1. Proof of theorem X.4. Theorem X.4 follows from the following result.

Proposition X.9. *Let $(f_1, \dots, f_n) \in \check{\mathcal{N}}(U, \mathcal{A})$ and $I \subseteq [n] \setminus (\mathscr{E}(U) \cup \mathscr{E}(\mathcal{A}))$.*

(1) If $I \notin \mathscr{T}(U, \mathcal{A})$, then $V(f_1, \dots, f_n) \cap (\mathbb{k}^*)^I = \emptyset$.

(2) If $I \in \mathscr{T}(U, \mathcal{A})$, then

$$\sum_{a \in (\mathbb{k}^*)^I} [f_1, \dots, f_n]_a = \mathrm{MV}(\mathcal{P}^I_{j_1}, \dots, \mathcal{P}^I_{j_{|I|}}) \times [\pi_{[n]\setminus I}(\mathcal{P}_{j'_1}), \dots, \pi_{[n]\setminus I}(\mathcal{P}_{j'_{n-|I|}})]_0 \tag{111}$$

where $j_1, \dots, j_{|I|}$ are elements of $T^I_{\mathcal{A}}$, and $j'_1, \dots, j'_{n-|I|}$ are elements of $[n] \setminus T^I_{\mathcal{A}}$.

PROOF. If $I \notin \mathscr{T}(U, \mathcal{A}) \cup \mathscr{E}(\mathcal{A})$, then $|T^I_{\mathcal{A}}| > |I|$, and property (a) of strongly $\mathcal{A}$-non-degenerate systems imply that $V(f_1, \dots, f_n) \cap (\mathbb{k}^*)^I = \emptyset$, which proves assertion (1). Now pick $I \in \mathscr{T}(U, \mathcal{A})$ and $a \in V(f_1, \dots, f_n) \cap (\mathbb{k}^*)^I$. Since proper non-degeneracy implies BKK non-degeneracy when dimension of the ambient affine space is equal to the number of polynomials, properties (a) and (d) of strongly $\mathcal{A}$-non-degenerate systems imply that $(f_j|_{\mathbb{k}^I} : j \in T^I_{\mathcal{A}})$ are $(\mathcal{A}^I_j : j \in T^I_{\mathcal{A}})$-non-degenerate, and corollary VII.18 implies that $V(f_1, \dots, f_n) \cap (\mathbb{k}^*)^I$ is finite. If $j \in T^I_{\mathcal{A}}$, theorem III.80 implies that $C := V(f_i : i \neq j) \cap (\mathbb{k}^*)^I$ is purely one-dimensional near a. Property (c) of strongly $\mathcal{A}$-non-degenerate systems and corollary IX.22 imply that $\mathbb{k}^I$ is an irreducible component of $V' := V(f_{j'} : j' \notin T^I_{\mathcal{A}})$, and no irreducible component of V' other than $\mathbb{k}^I$ contains any irreducible component of C. Corollary IV.34 then implies that

$$[f_1, \dots, f_n]_a = [f_{j_1}|_{\mathbb{k}^I}, \dots, f_{j_{|I|}}|_{\mathbb{k}^I}]_a \times [f_{j'_1,\epsilon}, \dots, f_{j'_{n-|I|},\epsilon}]_0 \tag{112}$$

for generic $\epsilon = (\epsilon_1, \dots, \epsilon_n) \in (\mathbb{k}^*)^I$, where $f_{j'_k,\epsilon}$ are formed from $f_{j'_k}$ by substituting ϵ_i for x_i for all $i \in I$. Due to property (110), assertion (2) follows by summing (112) over all $a \in (\mathbb{k}^*)^I$ due to theorems VII.7 and IX.8 (after using $(\mathcal{A}^I_{j_1}, \dots, \mathcal{A}^I_{j_{|I|}})$-non-degeneracy of $f_{j_1}|_{\mathbb{k}^I}, \dots, f_{j_{|I|}}|_{\mathbb{k}^I}$ and non-degeneracy at the origin of $f_{j'_1,\epsilon}, \dots, f_{j'_{n-|I|},\epsilon}$ for generic ϵ). ☐

3.2. Proof of theorem X.7. Let $(f_1, \dots, f_n) \in \check{\mathcal{N}}(U, \mathcal{A})$ and C be the closed subscheme of $U' := U \setminus \bigcup_{I \in \mathscr{E}(\mathcal{A})} \mathbb{k}^I$ defined by $f_2, \dots, f_n$. For each $I \subseteq [n]$ such that $U' \cap \mathbb{k}^I \neq \emptyset$, let $\{C^I_j\}_j$ be the set of irreducible components of C such that $\mathbb{k}^I$ is the smallest coordinate subspace of $\mathbb{k}^n$ containing each C^I_j.

Claim X.10. *Let $\mathscr{T}_1(U, \mathcal{A})$ be as in theorem X.7, $I \subseteq [n]$ such that $U' \cap \Bbbk^I \neq \emptyset$, and $T'^I_{\mathcal{A}}$ be as in definition IX.23.*

(1) If $\{C^I_j\}_j$ is nonempty, then $I \in \mathscr{T}_1(U, \mathcal{A})$.
(2) If $I \in \mathscr{T}_1(U, \mathcal{A})$, then $[n] \setminus T^I_{\mathcal{A}} \in \mathcal{T}'^I_{\mathcal{A}}$.

PROOF. For the first assertion, pick $I \subseteq [n]$ such that $\{C^I_j\}_j$ is nonempty. Since $I \notin \mathscr{E}(\mathcal{A})$, it follows that $|T^I_{\mathcal{A}}| \geq |I|$. On the other hand, if $|T^I_{\mathcal{A}} \setminus \{1\}| \geq |I|$, then property (a) of strong $\mathcal{A}$-non-degeneracy and lemma IX.17 implies that each $\{C^I_J\}_j$ is a point, which contradicts theorem III.80. Accordingly $|T^I_{\mathcal{A}} \setminus \{1\}| = |I| - 1$ and $|T^I_{\mathcal{A}}| = |I|$, which imply that $I \in \mathscr{T}_1(U, \mathcal{A})$, as required. The second assertion follows from the definition of $\mathscr{E}(\mathcal{A})$ and corollary IX.13. □

For each $I \in \mathscr{T}_1(U, \mathcal{A})$, property (a) of strong $\mathcal{A}$-non-degeneracy and assertion (1) of corollary VII.27 imply that either $\{C^I_j\}_j$ is empty, or each C^I_j has dimension one. It then follows from claim X.10 that C is a curve, and therefore lemma IV.29, proposition IV.28, and theorem IV.24 imply that

$$[f_1, \ldots, f_n]_U = \sum_{a \in C} \operatorname{ord}_a(f_1|_C) = \sum_{I,j,a} [f_2, \ldots, f_n]_{C^I_j} \operatorname{ord}_a(f_1|_{C^I_j})$$

where $[f_2, \ldots, f_n]_{C^I_j}$ are defined as in section IV.5. Pick $I \in \mathscr{T}_1(U, \mathcal{A})$. Let $j_1 = 1, j_2, \ldots, j_{|I|}$ be the elements of $T^I_{\mathcal{A}}$, and $j'_1, \ldots, j'_{n-|I|}$ be the elements of $[n] \setminus T^I_{\mathcal{A}}$. Property (c) of strongly $\mathcal{A}$-non-degenerate systems and corollary IX.22 imply that $\Bbbk^I$ is an irreducible component of $V' := V(f_{j'_1}, \ldots, f_{j'_{n-|I|}})$, and if $\{C^I_j\}_j$ is nonempty, then no irreducible component of V' other than $\Bbbk^I$ contains any irreducible component of C. Proposition IV.33 then implies that

$$[f_1, \ldots, f_n]_U = \sum_{I,j,a} \operatorname{ord}_a(f_1|_{C^I_j})[f_{j_2}|_{\Bbbk^I}, \ldots, f_{j_{|I|}}|_{\Bbbk^I}]_{C^I_j} \times [f_{j'_1,\epsilon}, \ldots, f_{j'_{n-|I|},\epsilon}]_0$$

for generic $\epsilon = (\epsilon_1, \ldots, \epsilon_n) \in (\Bbbk^*)^I$, where $f_{j'_k,\epsilon}$ are formed from $f_{j'_k}$ by substituting ϵ_i for x_i for all $i \in I$. Properties (b), (d) of strongly $\mathcal{A}$-non-degenerate systems and theorem IX.8 imply that $[f_{j'_1,\epsilon}, \ldots, f_{j'_{n-|I|},\epsilon}]_0 = [\pi_{[n]\setminus I}(\mathcal{P}_{j'_1}), \ldots, \pi_{[n]\setminus I}(\mathcal{P}_{j'_{n-|I|}})]_0$ for generic $\epsilon \in \Bbbk^I$. In order to prove theorem X.7 therefore it suffices to show that for each $I \in \mathscr{T}(U, \mathcal{A})$,

$$\sum_{j,a} \operatorname{ord}_a(f_1|_{C^I_j})[f_{j_2}|_{\Bbbk^I}, \ldots, f_{j_{|I|}}|_{\Bbbk^I}]_{C^I_j} = [\mathcal{P}^I_1, \mathcal{P}^I_{j_2}, \ldots, \mathcal{P}^I_{j_{|I|}}]^*_{\mathscr{E}^I(U,\mathcal{A})}$$

Identify $\Bbbk^I$ with $\Bbbk^k$, where $k := |I|$. Proposition VII.26 and claim VII.27.1 imply that we can find a k-dimensional convex rational polytope $\mathcal{P}$ such that

- $\Bbbk^I \hookrightarrow X_{\mathcal{P}}$,
- $V(f_{j_2}|_{\Bbbk^I}, \ldots, f_{j_k}|_{\Bbbk^I})$ extends to a complete curve $\bar{C}^I$ on $X_{\mathcal{P}}$, and
- f_1 restricts to a non-zero rational function on $\bar{C}^I$ which is representable near every point of $\bar{C}^I$ as a quotient of non zero-divisors.

Corollary IV.25 implies that

$$\sum_j \sum_{a\in U'} \operatorname{ord}_a(f_1|_{C_j^I})[f_{j_2}|_{\mathbb{k}^I},\dots,f_{j_k}|_{\mathbb{k}^I}]_{C_j^I} = -\sum_j \sum_{a\in \bar{C}_j^I\setminus U'} \operatorname{ord}_a(f_1|_{C_j^I})[f_{j_2}|_{\mathbb{k}^I},\dots,f_{j_k}|_{\mathbb{k}^I}]_{C_j^I}$$

where $\bar{C}_j^I$ are the closures of C_j^I in $X_{\mathcal{P}}$. Property (110) of the f_j implies that $a \in \bar{C}_j^I \setminus U'$ if and only if either $a \in \mathbb{k}^{I'}$ for some $I' \in \mathscr{E}(U) \cup \mathscr{E}(\mathcal{A})$, or (the germ at) a is a branch at infinity of C_j^I. Corollary VII.38 then implies that

$$\begin{aligned}\sum_j \sum_{a\in U'} &\operatorname{ord}_a(f_1|_{C_j^I})[f_{j_2}|_{\mathbb{k}^I},\dots,f_{j_k}|_{\mathbb{k}^I}]_{C_j^I}\\ &= -\sum_{\nu\in\mathcal{V}'^I_{\mathscr{E}^I(U,\mathcal{A})}\cup\mathcal{V}'^I_\infty} \min_{\mathcal{P}_1^I}(\nu)\ \operatorname{MV}'_\nu(\operatorname{In}_\nu(\mathcal{P}^I_{j_2}),\dots,\operatorname{In}_\nu(\mathcal{P}^I_{j_k}))\end{aligned}$$

where $\mathcal{V}'^I_\infty$ (respectively, $\mathcal{V}'^I_{\mathscr{E}^I(U,\mathcal{A})}$) is the set of primitive weighted orders on $\mathbb{k}[x_i : i \in I]$ which are centered at infinity (respectively, at $\mathbb{k}^{I'}$ for some $I' \in \mathscr{E}^I(U,\mathcal{A})$). Since the right-hand side of the preceding identity is precisely $[\mathcal{P}_1^I,\mathcal{P}_{j_2}^I,\dots,\mathcal{P}_{j_{|I|}}^I]^*_{\mathscr{E}^I(U,\mathcal{A})}$, this completes the proof of theorem X.7.

4. Other formulae for the bound

Throughout this section we continue to use $\mathcal{A}$ to denote a collection $(\mathcal{A}_1,\dots,\mathcal{A}_n)$ of finite subsets of $\mathbb{Z}^n_{\geq 0}$ and $\mathcal{P}_j$ to denote the convex hull of $\mathcal{A}_j$, $j = 1,\dots,n$.

4.1. The formula of Huber-Sturmfels and Rojas. Let t be a new indeterminate. Fix positive integers $k_1,\dots,k_n$. Note that for each $f_1,\dots,f_n \in \mathbb{k}[x_1,\dots,x_n]$, and each Zariski open subset U of $\mathbb{k}^n$,

$$[f_1,\dots,f_n]^{iso}_U = [t, f_1 + c_1t^{k_1},\dots,f_n + c_nt^{k_n}]^{iso}_{U\times\mathbb{k}} \tag{113}$$

for any $c_1,\dots,c_n \in \mathbb{k}$. It follows that

$$[\mathcal{A}_1,\dots,\mathcal{A}_n]^{iso}_U = [\hat{\mathcal{A}}_0,\dots,\hat{\mathcal{A}}_n]^{iso}_{\hat{U}}$$

where $\hat{U} := U \times \mathbb{k}$, $\hat{\mathcal{A}}_0 := \{(1,0,\dots,0)\} \subset \mathbb{Z}^{n+1}_{\geq 0}$ and $\hat{\mathcal{A}}_j := \{(k_j,0,\dots,0)\} \cup (\{0\}\times\mathcal{A}_j) \subset \mathbb{Z}^{n+1}_{\geq 0}$ for $j = 1,\dots,n$. Let $\hat{\mathcal{A}} := (\hat{\mathcal{A}}_0,\dots,\hat{\mathcal{A}}_n)$. It is straightforward to check that $\mathscr{T}_1(\hat{U},\hat{\mathcal{A}}) = \{[n+1]\}$ so that theorem X.7 implies that

$$\begin{aligned}[\mathcal{A}_1,\dots,\mathcal{A}_n]^{iso}_U &= [\hat{\mathcal{P}}_0,\dots,\hat{\mathcal{P}}_n]^*_{\mathscr{E}(\hat{U})\cup\mathscr{E}(\hat{\mathcal{A}})}\\ &= -\sum_{\hat{\nu}\in\hat{\mathcal{V}}'_{\mathscr{E}(\hat{U})\cup\mathscr{E}(\hat{\mathcal{A}})}\cup\hat{\mathcal{V}}'_\infty} \hat{\nu}_0\operatorname{MV}'_{\hat{\nu}}(\operatorname{In}_{\hat{\nu}}(\hat{\mathcal{P}}_1),\dots,\operatorname{In}_{\hat{\nu}}(\hat{\mathcal{P}}_n))\end{aligned}$$

where $\hat{\mathcal{P}}_j$ are the convex hulls of $\hat{\mathcal{A}}_j$, and $\hat{\nu}$ ranges over the collection $\hat{\mathcal{V}}'_{\mathscr{E}(\hat{U})\cup\mathscr{E}(\hat{\mathcal{A}})} \cup \hat{\mathcal{V}}'_\infty$ of all primitive weighted orders on $\mathbb{k}[t,x_1,\dots,x_n]$ which are either centered at infinity or centered at $\mathbb{k}^I$ for some $I \in \mathscr{E}(\hat{U}) \cup \mathscr{E}(\hat{\mathcal{A}})$, and $\hat{\nu}_0 := \hat{\nu}(t)$. Now, since $\dim(\hat{\mathcal{P}}_0) = 0$, either theorem VII.5 or theorem VII.33 implies that $\operatorname{MV}(\hat{\mathcal{P}}_0,\dots,$

$\hat{\mathcal{P}}_n) = 0$. Therefore, assertion (1) of proposition VII.30 implies that

$$[\mathcal{A}_1, \ldots, \mathcal{A}_n]_U^{iso} = \sum_{\hat{\nu} \notin \hat{\mathcal{V}}'_{\mathscr{E}(\hat{U}) \cup \mathscr{E}(\hat{\mathcal{A}})} \cup \hat{\mathcal{V}}'_\infty} \hat{\nu}_0 \operatorname{MV}'_{\hat{\nu}}(\operatorname{In}_{\hat{\nu}}(\hat{\mathcal{P}}_1), \ldots, \operatorname{In}_{\hat{\nu}}(\hat{\mathcal{P}}_n)) \tag{114}$$

If ν is a primitive weighted order on $\mathbb{k}[t, x_1, \ldots, x_n]$, then $\nu \notin \hat{\mathcal{V}}'_{\mathscr{E}(\hat{U}) \cup \mathscr{E}(\hat{\mathcal{A}})} \cup \hat{\mathcal{V}}'_\infty$ if and only if all the following hold:

(i) ν is nonnegative, and
(ii) for each $I \in \mathscr{E}(U) \cup \mathscr{E}(\mathcal{A})$, there is $i' \notin I$ such that $\nu(x_i) = 0$.

Let $\hat{\mathcal{V}}'(U, \mathcal{A})$ be the set of all primitive weighted orders $\hat{\nu}$ on $\mathbb{k}[t, x_1, \ldots, x_n]$ which satisfy properties (i), (ii) and in addition satisfy the following:

(iii) $\hat{\nu}_0 := \hat{\nu}(t)$ is positive.

Since a summand on the right-hand side of (114) has a non-zero contribution only if $\hat{\nu}_0$ is positive, it follows that for any Zariski open subset U of $\mathbb{k}^n$,

$$[\mathcal{A}_1, \ldots, \mathcal{A}_n]_U^{iso} = \sum_{\hat{\nu} \in \hat{\mathcal{V}}'(U, \mathcal{A})} \hat{\nu}_0 \operatorname{MV}'_{\hat{\nu}}(\operatorname{In}_{\hat{\nu}}(\hat{\mathcal{P}}_1), \ldots, \operatorname{In}_{\hat{\nu}}(\hat{\mathcal{P}}_n)) \tag{115}$$

In the case that $\mathbb{k} = \mathbb{C}$, $\mathscr{E}(\mathcal{A}) = \emptyset$, and $U = \mathbb{k}^n \setminus V(\prod_{j \in J} x_j) \cong (\mathbb{k}^*)^{|J|} \times \mathbb{k}^{n-|J|}$ for some $J \subseteq [n]$, formula (115) appeared in [HS97]. In this case $\hat{\mathcal{V}}'(U, \mathcal{A})$ consists of all primitive nonnegative weighted orders $\hat{\nu}$ on $\mathbb{k}[t, x_1, \ldots, x_n]$ such that $\hat{\nu}_0$ is positive, and $\hat{\nu}(x_j) = 0$ for each $j \in J$. The sum on the right-hand side of (115) in this case was termed in [HS97] as the *I-stable mixed volume* (where $I := [n] \setminus J$) of $\mathcal{A}_1, \ldots, \mathcal{A}_n$. J. M. Rojas [Roj99] observed that the formula of [HS97] works over all algebraically closed fields.

4.2. Estimates in terms of single mixed volumes. If U is nonempty, identity (106) implies that $[\mathcal{A}_1, \ldots, \mathcal{A}_n]_U^{iso} \geq \operatorname{MV}(\mathcal{P}_1, \ldots, \mathcal{P}_n)$. On the other hand, since $\mathcal{A}_j \subseteq \mathcal{A}'_j := \mathcal{A}_j \cup \{0\}$, it trivially follows that $[\mathcal{A}_1, \ldots, \mathcal{A}_n]_U^{iso} \leq [\mathcal{A}'_1, \ldots, \mathcal{A}'_n]_U^{iso} = \operatorname{MV}(\mathcal{P}'_1, \ldots, \mathcal{P}'_n)$, where $\mathcal{P}'_j$ are the convex hull of $\mathcal{A}'_j$, and the last equality follows from (106). It follows that for nonempty Zariski open subsets U of $\mathbb{k}^n$,

$$\begin{aligned} \operatorname{MV}(\operatorname{conv}(\mathcal{A}_1), \ldots, \operatorname{conv}(\mathcal{A}_n)) &\leq [\mathcal{A}_1, \ldots, \mathcal{A}_n]_U^{iso} \\ &\leq \operatorname{MV}(\operatorname{conv}(\mathcal{A}_1 \cup \{0\}) \ldots, \operatorname{conv}(\mathcal{A}_n \cup \{0\})) \end{aligned} \tag{116}$$

The upper bound in (116) is due to T. Y. Li and X. Wang [LW96]. We now examine when these bounds are exact. The lower bound is easier to handle; the following result follows directly from theorems VII.33 and X.4.

Proposition X.11. *Let U be a nonempty Zariski open subset of $\mathbb{k}^n$. Then the following are equivalent:*

(1) The first inequality in (116) holds with equality.
(2) For each $I \in \mathscr{T}(U, \mathcal{A}) \setminus \{[n]\}$, $\mathcal{P}_{j_1} \cap \mathbb{R}^I, \ldots, \mathcal{P}_{j_{|I|}} \cap \mathbb{R}^I$ are dependent, *where $j_1, \ldots, j_{|I|}$ are the elements of $T^I_{\mathcal{A}}$.* ☐

Remark X.12. Since an empty collection of convex polytopes is by definition *independent*, condition (2) of proposition X.11 implies in particular that $\emptyset \notin \mathscr{T}(U, \mathcal{A})$, which in turn implies that either U does not contain the origin, or $[\mathcal{A}_1, \dots, \mathcal{A}_n]_0$ is zero or ∞ (remark X.3).

Following A. Khovanskii [Kho78], We say that $\mathcal{A} = (\mathcal{A}_1, \dots, \mathcal{A}_n)$ is *regularly attached to the coordinate cross* if for each proper subset I of $[n]$, the set of nonempty elements of $\{\mathcal{P}_j \cap \mathbb{R}^I : j = 1, \dots, n\}$ is dependent; in particular this implies (taking $I = \emptyset$) that the origin belongs to at least one of the $\mathcal{A}_j$. The following is immediate from proposition X.11.

Corollary X.13. (Khovanskii [Kho78]). *If U is nonempty and $\mathcal{A}$ is regularly attached to the coordinate cross, then $[\mathcal{A}_1, \dots, \mathcal{A}_n]_U^{iso} = \mathrm{MV}(\mathcal{P}_1, \dots, \mathcal{P}_n)$.* □

Now let $M := \mathrm{MV}(\mathrm{conv}(\mathcal{A}_1 \cup \{0\}) \dots, \mathrm{conv}(\mathcal{A}_n \cup \{0\}))$ be the upper bound from (116). Consider as in (113) the system

$$f_j + t = 0,\ j = 1, \dots, n, \tag{117}$$

where f_j are generic polynomials supported at $\mathcal{A}_j$, with $\mathrm{NP}(f_j) = \mathrm{conv}(\mathcal{A}_j) = \mathcal{P}_j$. For generic $t \neq 0$, the corresponding system has precisely M isolated solutions, all of which are on $(\mathbb{k}^*)^n$. Therefore, the number of solutions of the system at $t = 0$ is also M if and only if there is no curve of solutions of the system (117) on $(\mathbb{k}^*)^n$ that escapes U or becomes non-isolated as t approaches 0. Theorems VII.5 and VII.33 and proposition VII.26 imply that such a curve exists if and only if there is a weighted order $\hat{\nu}$ on $\mathbb{k}[x_1, \dots, x_n, t]$ such that

(i) $\hat{\nu}(t) > 0$,
(ii) the restriction of $\hat{\nu}$ to $\mathbb{k}[x_1, \dots, x_n]$ is either centered at infinity or at $\mathbb{k}^I$ for some $I \in \mathscr{E}(U) \cup \mathscr{E}(\mathcal{A})$, and
(iii) $\mathrm{In}_{\hat{\nu}}(\hat{\mathcal{P}}_1), \dots, \mathrm{In}_{\hat{\nu}}(\hat{\mathcal{P}}_n)$ are *independent*, where $\hat{\mathcal{P}}_j := \mathrm{NP}(f_j + t) \subset \mathbb{R}^{n+1}_{\geq 0}$.

Let ν be the restriction of $\hat{\nu}$ to $\mathbb{k}[x_1, \dots, x_n]$. Let $m_\nu := \max\{\min_{\mathcal{P}_j}(\nu) : j = 1, \dots, n\}$. Since $\dim(\mathrm{In}_{\hat{\nu}}(\sum_j \hat{\mathcal{P}}_j)) \leq \dim(\mathrm{In}_\nu(\sum_j \mathcal{P}_j)) + 1$, and since $\hat{\nu}(t) > 0$, it follows from the definition of dependence of polytopes that $\mathrm{In}_{\hat{\nu}}(\hat{\mathcal{P}}_1), \dots, \mathrm{In}_{\hat{\nu}}(\hat{\mathcal{P}}_n)$ are independent if and only if

(iv) $m_\nu > 0$, and
(v) for all $J \subseteq [n]$,

$$\dim(\mathrm{In}_\nu(\sum_{j \in J} \mathcal{P}_j)) \geq \begin{cases} |J| & \text{if } \min_{\mathcal{P}_j}(\nu) < m_\nu \text{ for each } j \in J, \\ |J| - 1 & \text{if there is } j \in J \text{ such that } \min_{\mathcal{P}_j}(\nu) = m_\nu. \end{cases} \tag{118}$$

Taking $J = [n]$ in (118) implies in particular that $\dim(\mathrm{In}_\nu(\sum_j \mathcal{P}_j)) = n - 1$. These observations imply the following result.

Proposition X.14. *Let $\mathcal{P} := \sum_j \mathcal{P}_j$. The second inequality in (116) holds with equality if and only if*

(1) *either* $\dim(\mathcal{P}) \leq n-2$, *or*
(2) $\dim(\mathcal{P}) \geq n-1$, *and for each face of dimension* $(n-1)$ *of* $\mathcal{P}$ *such that its primitive inner normal*[2] ν *is centered at infinity or at* $\Bbbk^I$ *for some* $I \in \mathscr{E}(U) \cup \mathscr{E}(\mathcal{A})$, *at least one of the conditions (iv) and (v) fails for* ν. □

As a corollary we get a situation where *both* of the bounds of (116) are exact.

Corollary X.15. *Assume each* $\mathcal{A}_j$ *is* convenient *and* U *is a nonempty Zariski open subset of* $\Bbbk^n$. *Assume in addition that at least one of the following holds:*

(1) *either* U *does not contain the origin,*
(2) *or at least one of the* $\mathcal{A}_j$ *contains the origin.*

Then

$$[\mathcal{A}_1,\dots,\mathcal{A}_n]_U^{iso} = \mathrm{MV}(\mathrm{conv}(\mathcal{A}_1),\dots,\mathrm{conv}(\mathcal{A}_n)) = \mathrm{MV}(\mathrm{conv}(\mathcal{A}_1 \cup \{0\})\dots,\mathrm{conv}(\mathcal{A}_n \cup \{0\}))$$

PROOF. $\mathcal{A}_j$ satisfy the hypotheses of both propositions X.11 and X.14. □

5. Examples motivating the non-degeneracy conditions

In this section we give some examples to motivate the necessary and sufficient conditions for the equality $[f_1,\dots,f_n]_{\Bbbk^n}^{iso} = [\mathcal{A}_1,\dots,\mathcal{A}_n]_{\Bbbk^n}^{iso}$, where $\mathcal{A} := (\mathcal{A}_1,\dots,\mathcal{A}_n)$ is a given collection of finite subsets of $\mathbb{Z}_{\geq 0}^n$, and $(f_1,\dots,f_n) \in \mathcal{L}(\mathcal{A})$. We consider the case $U = \Bbbk^n$; in a sense this is the most important case, and it already captures the essence of the general case. We also assume for simplicity that $\mathscr{E}(\mathcal{A}) = \emptyset$, which ensures in particular that for generic $(f_1,\dots,f_n) \in \mathcal{L}(\mathcal{A})$, all points in the set $V(f_1,\dots,f_n)$ of common zeroes of $f_1,\dots,f_n$ in $\Bbbk^n$ are isolated. In this scenario, if we apply the intuitive reasoning from section VII.3.2 that motivated the non-degeneracy condition for Bernstein's theorem, we are led to the following condition:

(ND$_\infty$) $\quad f_1|_{\Bbbk^I},\dots,f_n|_{\Bbbk^I}$ are $\mathcal{A}^I$*-non-degenerate at infinity* for each $I \subseteq [n]$.

where $\mathcal{A}^I := (\mathcal{A}_1^I,\dots,\mathcal{A}_n^I) = (\mathcal{A}_1 \cap \mathbb{R}^I,\dots,\mathcal{A}_n \cap \mathbb{R}^I)$, and *non-degeneracy at infinity* is defined as follows: given a collection $\mathcal{B} := (\mathcal{B}_1,\dots,\mathcal{B}_m)$ of finite subsets of $\mathbb{Z}_{\geq 0}^n$ and $g_j \in \Bbbk[x_1,\dots,x_n]$, such that $\mathrm{Supp}(g_j) \subseteq \mathcal{B}_j$, $j = 1,\dots,m$, we say that $g_1,\dots,g_m$ are $\mathcal{B}$- non-degenerate at infinity if for each weighted order ν *centered at infinity* (see section X.2.2), there is no common root of $\mathrm{In}_{\mathcal{B}_j,\nu}(g_j)$, $j = 1,\dots,m$, on $(\Bbbk^*)^n$. We now present a series of examples which illustrate how condition (ND$_\infty$) falls short of characterizing the correct non-degeneracy condition, and which also suggest the ways to amend it. In all these examples $\mathcal{P}_j$ would denote the convex hull of $\mathcal{A}_j$, $j = 1,\dots,n$.

[2]If $\dim(\mathcal{P}) = n-1$, then both of the primitive normals to $\mathcal{P}$ are considered to be inner.

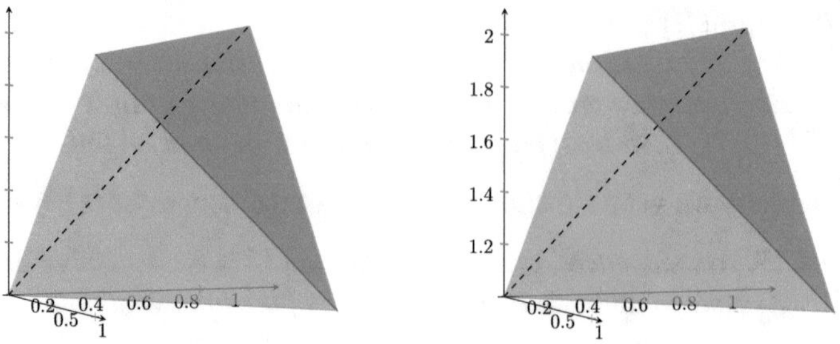

$\mathcal{P}_1 = \mathcal{P}_2$ $\qquad$ $\mathcal{P}_3 = \mathcal{P}_1 + \{(0,0,1)\}$

FIGURE 4. Newton polytopes of example X.16.

Example X.16. Let $\mathcal{A}_1 = \mathcal{A}_2 = \{(0,0,0),(1,1,0),(0,1,1),(1,0,1)\} \subset \mathbb{Z}^3_{\geq 0}$ and $\mathcal{A}_3 = \mathcal{A}_1 + \{(0,0,1)\}$. Then $\mathcal{P}_1 = \mathcal{P}_2$ is a tetrahedron and $\mathcal{P}_3$ is a translation of $\mathcal{P}_1$, and therefore (fig. 4)

$$\mathrm{MV}(\mathcal{P}_1, \mathcal{P}_2, \mathcal{P}_3) = 3!\,\mathrm{Vol}(\mathcal{P}_1) = 2$$

If f_j are polynomials such that $\mathrm{NP}(f_j) = \mathcal{P}_j$, then

$$\begin{aligned} f_1 &= a_1 + b_1 x_1 x_2 + c_1 x_2 x_3 + d_1 x_3 x_1 \\ f_2 &= a_2 + b_2 x_1 x_2 + c_2 x_2 x_3 + d_2 x_3 x_1 \\ f_3 &= x_3(a_3 + b_3 x_1 x_2 + c_3 x_2 x_3 + d_3 x_3 x_1) \end{aligned}$$

where $a_j, b_j, c_j, d_j \in \mathbb{k}^*$. We write V for the set of common zeroes of $f_1, \ldots, f_3$ on $\mathbb{k}^3$, and V^* for $V \cap (\mathbb{k}^*)^3$.

(a) If all a_j, b_j, c_j, d_j are generic, then it is straightforward to check directly that $V = V^*$. Therefore, theorem VII.5 implies that $[\mathcal{A}_1, \mathcal{A}_2, \mathcal{A}_3]^{iso}_{\mathbb{k}^3} = [f_1, f_2, f_3]^{iso}_{\mathbb{k}^3} = \mathrm{MV}(\mathcal{P}_1, \mathcal{P}_2, \mathcal{P}_3) = 2$.

(b) If $a_1 = a_2$, $b_1 = b_2$, and the remaining coefficients are generic, then $V = V^* \cup C$, where $C := \{x_3 = a_1 + b_1 x_1 x_2 = 0\}$ is a positive dimensional component of $V(f_1, f_2, f_3)$. However, f_1, f_2, f_3 still satisfy (ND*), and theorem VII.7 implies that $[f_1, f_2, f_3]^{iso}_{\mathbb{k}^3} = 2 = [\mathcal{A}_1, \mathcal{A}_2, \mathcal{A}_3]^{iso}_{\mathbb{k}^3}$.

(c) If $a_1 = a_2 = a_3$, $b_1 = b_2 = b_3$, and the rest of the coefficients are generic, then again $V = V^* \cup C$. However, (ND*) fails for the weighted order ν corresponding to weights $(-1, 1, 2)$ for (x, y, z), and theorem VII.7 implies that $[f_1, f_2, f_3]^{iso}_{\mathbb{k}^3} < 2 = [\mathcal{A}_1, \mathcal{A}_2, \mathcal{A}_3]^{iso}_{\mathbb{k}^3}$. (It is straightforward to verify directly that in this case $V^* = \emptyset$ and $[f_1, f_2, f_3]^{iso}_{\mathbb{k}^3} = 0$.)

Part (b) of example X.16 shows that it is possible that $V(f_1, \ldots, f_n)$ has a positive dimensional component on $\mathbb{k}^n$, but still $[f_1, \ldots, f_n]^{iso}_{\mathbb{k}^n} = [\mathcal{A}_1, \ldots, \mathcal{A}_n]^{iso}_{\mathbb{k}^n}$ (where $\mathcal{A}_j = \mathrm{Supp}(f_j)$). (This does not happen in the case of $(\mathbb{k}^*)^n$, see proposition VII.37.) Moreover, since the intersection of the curve C with the "torus" of the (x_1, x_2)-plane is nonempty, in part (b) of example X.16, condition (ND$_\infty$) is violated for $I = \{1, 2\}$. However, note that the intersections of $\mathcal{P}_1$ and $\mathcal{P}_2$ with the (x_1, x_2)-plane (in $\mathbb{R}^n$) are *dependent* in the terminology of definition VII.32. Moreover, in part (c) of example X.16, where $[f_1, f_2, f_3]^{iso}_{\mathbb{k}^3} < [\mathcal{A}_1, \mathcal{A}_2, \mathcal{A}_3]^{iso}_{\mathbb{k}^3}$, condition (ND_∞) is violated with $I = \{1, 2, 3\}$, and the corresponding polytopes are *independent*. This motivates the following definition.

Definition X.17. An ordered collection $\mathcal{B} = (\mathcal{B}_1, \dots, \mathcal{B}_m)$, $m \geq 1$, of collections of finite subsets of $\mathbb{R}^n$ is called $\mathbb{R}^I$*-dependent* if there is a nonempty subset J of $[m]$ such that $\mathcal{B}^I_j := \mathcal{B}_j \cap \mathbb{R}^I$ is nonempty for each $j \in J$, and the collection $\{\mathrm{conv}(\mathcal{B}^I_j) : j \in J\}$ of convex polytopes is dependent (definition VII.32); otherwise we say that $\mathcal{B}$ is $\mathbb{R}^I$*-independent*.

Example X.16 suggests that

(i) Condition (ND_∞) should be checked only for those $I \subseteq [n]$ such that $\mathcal{A}$ is $\mathbb{R}^I$-independent.

This, however, is not enough, as the next example shows.

Example X.18. Consider the following system of polynomials (fig. 5):

$$f_1 = a_1 + b_1 x_1 + c_1 x_2 x_3 + d_1 x_3 x_1$$
$$f_2 = a_2 + b_2 x_1 + c_2 x_2 x_3 + d_2 x_3 x_1$$
$$f_3 = x_3(a_3 + b_3 x_2 + c_3 x_2 x_3 + d_3 x_3 x_1)$$

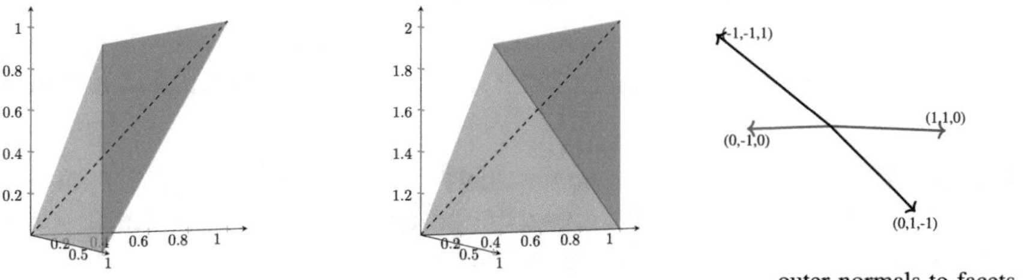

$\mathcal{P}_1 = \mathcal{P}_2$ $\mathcal{P}_3$ outer normals to facets of $\mathcal{P}_1$

FIGURE 5. Newton polytopes of example X.18.

where $a_j, b_j, c_j, d_j \in \mathbb{k}^*$, and $\mathcal{A}_j := \mathrm{Supp}(f_j)$, $j = 1, 2, 3$. We continue to write $V := V(f_1, f_2, f_3)$ and $V^* := V \cap (\mathbb{k}^*)^3$. When all the coefficients are generic, then it is straightforward to check that $V = V^*$ so that theorem VII.5 implies that

$$[f_1, f_2, f_3]^{iso}_{\mathbb{k}^3} = [\mathcal{A}_1, \mathcal{A}_2, \mathcal{A}_3]^{iso}_{\mathbb{k}^3} = \mathrm{MV}(\mathcal{P}_1, \mathcal{P}_2, \mathcal{P}_3)$$

We compute $\mathrm{MV}(\mathcal{P}_1, \mathcal{P}_2, \mathcal{P}_3)$ using proposition VII.30. If ν is the primitive outer normal to any of the facets of $\mathcal{P}_1 = \mathcal{P}_2$, it is straightforward to check that the image of the corresponding facet under the map ψ_ν from the definition of $\mathrm{MV}'_\nu(\cdot, \dots, \cdot)$ is (up to a translation) the triangle $\mathcal{T}$ with vertices $(0,0), (0,1), (1,0)$, and therefore $\mathrm{MV}'_\nu(\mathrm{ld}_\nu(\mathcal{P}_1), \mathrm{ld}_\nu(\mathcal{P}_2)) = 2!\,\mathrm{Area}(\mathcal{T}) = 1$. It follows that

$$\begin{aligned}[]
[\mathcal{A}_1, \mathcal{A}_2, \mathcal{A}_3]^{iso}_{\Bbbk^3} &= \mathrm{MV}(\mathcal{P}_1, \mathcal{P}_2, \mathcal{P}_3) \\
&= \max_{\mathcal{P}_3}(0,-1,0) + \max_{\mathcal{P}_3}(0,1,-1) + \max_{\mathcal{P}_3}(1,1,0) + \max_{\mathcal{P}_3}(-1,-1,1) \\
&= 0 + 0 + 1 + 1 \\
&= 2
\end{aligned}$$

If $a_1 = a_2$ and $b_1 = b_2$ and the other coefficients are generic, then it is straightforward to check directly that $V^* = \emptyset$ and V is the curve $\{x_3 = a_1 + b_1 x_1 = 0\}$ so that $[f_1, f_2, f_3]^{iso}_{\Bbbk^3} = 0$. However, the only $I \subseteq \{1,2,3\}$ such that $\mathcal{A}$ is $\mathbb{R}^I$-independent is $I = \{1,2,3\}$, and it is straightforward to check that $f_j|_{\Bbbk^I} = f_j$ are in fact $\mathcal{A}$-non-degenerate at infinity. In particular, f_1, f_2, f_3 satisfy condition (i), but $[f_1, f_2, f_3]^{iso}_{\Bbbk^3} < [\mathcal{A}_1, \mathcal{A}_2, \mathcal{A}_3]^{iso}_{\Bbbk^3}$.

Given $I' \subseteq I \subseteq [n]$ and a weighted order ν on $\Bbbk[x_i : i \in I]$, we say that ν is centered at $(\Bbbk^*)^{I'}$ if $\nu(x_i) = 0$ for all $i \in I'$ and $\nu(x_i) > 0$ for all $i \in I \setminus I'$. It is straightforward to check that in example X.18, the only non-zero weighted orders ν such that $\mathrm{In}_\nu(f_j)$, $j = 1,2,3$, have a common zero on $(\Bbbk^*)^3$ are of the form $(0,0,\epsilon)$ for $\epsilon > 0$, i.e. they are centered at $(\Bbbk^*)^{I'}$ with $I' := \{1,2\}$. It turns out that $\mathcal{A}$ is *hereditarily* $\mathbb{R}^{I'}$*-dependent* (see section X.6) and therefore example X.18 suggests that for $[f_1, \ldots, f_n]^{iso}_{\Bbbk^n}$ to be equal to $[\mathcal{A}_1, \ldots, \mathcal{A}_n]^{iso}_{\Bbbk^n}$, the following condition needs to be satisfied in addition to (i):

(ii) For each $I \subseteq [n]$ such that $\mathcal{A}$ is *not* hereditarily $\mathbb{R}^I$-dependent, $\mathrm{In}_{\mathcal{A}^I_j,\nu}(f_j|_{\Bbbk^I})$ do not have any common zero on $(\Bbbk^*)^n$ for all weighted orders ν on $\Bbbk[x_i : i \in I]$ which are centered at $(\Bbbk^*)^{I'}$ for some $I' \subsetneq I$ such that $\mathcal{A}$ is hereditarily $\mathbb{R}^{I'}$-dependent.

Example X.19. Let $f_1, \ldots, f_4$ be polynomials in $(x_1, \ldots, x_4)$ such that f_1 and f_2 are polynomials in (x_1, x_2) with non-zero constant terms, the (two-dimensional) mixed volume of $\mathrm{NP}(f_1)$ and $\mathrm{NP}(f_2)$ is non-zero, and

$$f_j = x_3 f_{j,1}(x_1, x_2) + x_4 f_{j,2}(x_1, x_2) \in \Bbbk[x_1, x_2, x_3, x_4],\ j = 3, 4,$$

where $f_{3,1}, f_{3,2}, f_{4,1}, f_{4,2}$ are non-zero polynomials in (x, y) such that $\mathrm{NP}(f_{3,1})$ and $\mathrm{NP}(f_{4,1})$ are positive dimensional. Let $\mathcal{A}_j$ be the support of f_j and $\mathcal{P}_j$ be the Newton polytope of f_j, $j = 1, \ldots, 4$. The only $I \subseteq \{1,2,3,4\}$ such that the supports of f_j are $\mathbb{R}^I$-independent is $I = \{1,2\}$. It follows that for generic coefficients, all the common zeroes of $f_1, \ldots, f_4$ are isolated and contained in the (x_1, x_2)-plane. Moreover, theorem X.4 implies that

$$[\mathcal{A}_1, \mathcal{A}_2, \mathcal{A}_3, \mathcal{A}_4]^{iso}_{\Bbbk^4} = \mathrm{MV}(\mathcal{P}_1^{\{1,2\}}, \mathcal{P}_2^{\{1,2\}}) \times [\pi_{\{3,4\}}(\mathcal{P}_3), \pi_{\{3,4\}}(\mathcal{P}_4)]_0 = \mathrm{MV}(\mathcal{P}_1, \mathcal{P}_2)$$

Now fix BKK non-degenerate f_1, f_2, and a common zero $z = (z_1, z_2)$ of f_1, f_2 on $(\Bbbk^*)^2$. Take $f_{3,1}$ and $f_{4,1}$ such that $f_{3,1}(z) = f_{4,1}(z) = 0$. Then $\{(z_1, z_2, t, 0) : t \in \Bbbk\} \subseteq V(f_1, f_2, f_3, f_4)$ so that $(z_1, z_2, 0, 0)$ is no longer an isolated point of

$V(f_1, f_2, f_3, f_4)$ (even though it is an isolated point of $V(f_1, f_2, f_3, f_4) \cap \mathbb{k}^{\{1,2\}}$). It follows that $[f_1, f_2, f_3, f_4]^{iso}_{\mathbb{k}^4} < \mathrm{MV}(\mathcal{P}_1, \mathcal{P}_2) = [\mathcal{A}_1, \mathcal{A}_2, \mathcal{A}_3, \mathcal{A}_4]^{iso}_{\mathbb{k}^4}$. However, $f_1, \ldots, f_4$ satisfy both conditions (i) and (ii).

Example X.19 leads us to another condition that needs to be satisfied in addition to (i) and (ii) for $[f_1, \ldots, f_n]^{iso}_{\mathbb{k}^n}$ to be equal to $[\mathcal{A}_1, \ldots, \mathcal{A}_n]^{iso}_{\mathbb{k}^n}$:

(iii) For each $I \subseteq [n]$ such that $\mathcal{A}$ is hereditarily $\mathbb{R}^I$-dependent, $\mathrm{In}_{\mathcal{A}_j^I, \nu}(f_j|_{\mathbb{k}^I})$ do not have any common zero on $(\mathbb{k}^*)^n$ for all weighted orders ν on $\mathbb{k}[x_i : i \in I]$ which are centered at $(\mathbb{k}^*)^{I'}$ for some $I' \subsetneq I$ such that $\mathcal{A}$ is not hereditarily $\mathbb{R}^{I'}$-dependent.

Note that the coordinate subspaces in condition (iii) are in a sense "dual" to those in condition (ii).

6. Non-degeneracy conditions

Let $\mathcal{A} := (\mathcal{A}_1, \ldots, \mathcal{A}_n)$ be a collection of finite subsets of $\mathbb{Z}^n_{\geq 0}$ and $I \subseteq [n]$. We say that $\mathcal{A}$ is *hereditarily* $\mathbb{R}^I$*-dependent* if $\mathcal{A}$ is $\mathbb{R}^I$-dependent (see definition X.17) and there is $I' \supseteq I$ such that

(a) $\mathcal{A}$ is $\mathbb{R}^{I'}$-dependent,
(b) $|T^{I'}_{\mathcal{A}}| = |I'|$,
(c) $|T^{\tilde{I}}_{\mathcal{A}}| > |\tilde{I}|$ for each $\tilde{I}$ such that $I \subseteq \tilde{I} \subsetneq I'$.

Remark X.20. If $I = \emptyset$, then $\mathbb{R}^I$ is the origin, and $\mathcal{A}$ is $\mathbb{R}^I$-dependent if and only if the origin belongs to some $\mathcal{A}_j$. However, even if $\mathcal{A}$ is $\mathbb{R}^I$-dependent, it might not be hereditarily $\mathbb{R}^I$-dependent, see fig. 6.

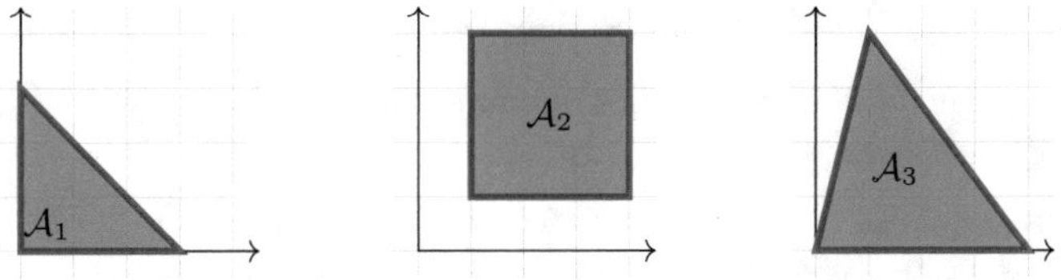

FIGURE 6. Both $(\mathcal{A}_1, \mathcal{A}_2)$ and $(\mathcal{A}_2, \mathcal{A}_3)$ are $\mathbb{R}^I$-dependent for $I = \emptyset$. However, only the latter pair is hereditarily $\mathbb{R}^I$-dependent (with $I' = \{2\}$).

The relevance of "hereditary dependence" to affine Bézout problem is given by proposition X.21 below: it states that if $\mathcal{A}$ is hereditarily $\mathbb{R}^I$-dependent, then for all $(f_1, \ldots, f_n) \in \mathcal{L}(\mathcal{A})$, either $V^I := V(f_1, \ldots, f_n) \cap (\mathbb{k}^*)^I$ is empty, or all points of V^I are non-isolated in (the possibly larger set) $V(f_1, \ldots, f_n)$ (however, points of V^I might be isolated in V^I itself!). This is not necessarily true if $\mathcal{A}$ is simply $\mathbb{R}^I$-dependent, e.g. the system $(x + y - 1, 2x - y - 2)$ (over a field of characteristic not equal to two) has an isolated zero on the coordinate subspace $y = 0$.

Proposition X.21. *If $\mathcal{A}$ is hereditarily $\mathbb{R}^I$-dependent, then $V(f_1, \ldots, f_n)$ has no isolated point on $(\mathbb{k}^*)^I$ for each $(f_1, \ldots, f_n) \in \mathcal{L}(\mathcal{A})$.*

We prove proposition X.21 in section X.7.1. Now we introduce the correct non-degeneracy condition for (arbitrary open subsets of) the affine space. Let U be a Zariski open subset of $\mathbb{k}^n$. Let $\mathscr{E}(\mathcal{A})$ and $\mathscr{E}(U)$ be as in section X.2.1. Define

$$\mathscr{D}(U, \mathcal{A}) := \{I \subseteq [n] : I \notin \mathscr{E}(U) \cup \mathscr{E}(\mathcal{A}),\ \mathcal{A} \text{ is hereditarily } \mathbb{R}^I\text{-dependent}\} \tag{119}$$

$$\mathscr{I}(U, \mathcal{A}) := \{I \subseteq [n] : I \notin \mathscr{E}(U) \cup \mathscr{E}(\mathcal{A}),\ \mathcal{A} \text{ is not hereditarily } \mathbb{R}^I\text{-dependent}\} \tag{120}$$

We say that polynomials $f_1, \ldots, f_n$ are *$(U, \mathcal{A})$-non-degenerate* if the following conditions are true:

(a) For each nonempty $I \in \mathscr{I}(U, \mathcal{A})$, there is no common zero of $f_1|_{\mathbb{k}^I}, \ldots, f_n|_{\mathbb{k}^I}$ on $(\mathbb{k}^*)^I \setminus U$ (note that this condition is vacuously true when $U = \mathbb{k}^n$, or more generally, when $U = \mathbb{k}^n_{\mathscr{S}}$ (see (107)) for some collection $\mathscr{S}$ of subsets of $[n]$),

(b) For each nonempty $I \in \mathscr{I}(U, \mathcal{A})$ and for each weighted order ν on $\mathbb{k}[x_i : i \in I]$ such that ν is centered at infinity or at $(\mathbb{k}^*)^{I'}$ for some $I' \in \mathscr{E}(U) \cup \mathscr{E}(\mathcal{A}) \cup \mathscr{D}(U, \mathcal{A})$, there is no common zero of $\mathrm{In}_{\mathcal{A}_1^I, \nu}(f_1|_{\mathbb{k}^I}), \ldots, \mathrm{In}_{\mathcal{A}_n^I, \nu}(f_n|_{\mathbb{k}^I})$ on $(\mathbb{k}^*)^n$.

(c) For each nonempty $I \in \mathscr{D}(U, \mathcal{A})$ and for each weighted order ν on $\mathbb{k}[x_i : i \in I]$ such that ν is centered at $(\mathbb{k}^*)^{I'}$ for some $I' \in \mathscr{I}(U, \mathcal{A})$, there is no common zero of $\mathrm{In}_{\mathcal{A}_1^I, \nu}(f_1|_{\mathbb{k}^I}), \ldots, \mathrm{In}_{\mathcal{A}_n^I, \nu}(f_n|_{\mathbb{k}^I})$ on $(\mathbb{k}^*)^n$.

We prove the following result in section X.7.3.

THEOREM X.22. ([Mon16]). *Assume $[\mathcal{A}_1, \ldots, \mathcal{A}_n]_U^{iso} > 0$. Then for each $(f_1, \ldots, f_n) \in \mathcal{L}(\mathcal{A})$, the following are equivalent:*

(1) $[f_1, \ldots, f_n]_U^{iso} = [\mathcal{A}_1, \ldots, \mathcal{A}_n]_U^{iso}$,
(2) $f_1, \ldots, f_n$ are $(U, \mathcal{A})$-non-degenerate.

The collection $\mathcal{N}(U, \mathcal{A})$ of all $(U, \mathcal{A})$-non-degenerate $(f_1, \ldots, f_n) \in \mathcal{L}(\mathcal{A})$ is a nonempty Zariski open subset of $\mathcal{L}(\mathcal{A})$.

In both examples X.16 and X.18, $\mathscr{D}(\mathbb{k}^3, \mathcal{A})$ is the set $\{\emptyset, \{1\}, \{2\}, \{1, 2\}\}$ of all subsets of $\{1, 2\}$, and $\mathscr{I}(\mathbb{k}^3, \mathcal{A})$ is the set of remaining subsets of $\{1, 2, 3\}$. It is straightforward to check that in cases (a) and (b) of example X.16, f_1, f_2, f_3 are $(\mathbb{k}^3, \mathcal{A})$-non-degenerate, but in case (c), condition (b) of $(\mathbb{k}^3, \mathcal{A})$-non-degeneracy fails for $I = \{1, 2, 3\}$ and $\nu = (-1, 1, 2)$, which is centered at infinity. In example X.18, condition (b) of $(\mathbb{k}^3, \mathcal{A})$-non-degeneracy fails for $I = \{1, 2, 3\}$ and $\nu = (0, 0, 1)$, which is centered at $(\mathbb{k}^*)^{I'}$, where $I' := \{1, 2\}$. In the scenario of example X.19, $\mathscr{I}(\mathbb{k}^4, \mathcal{A})$ is the collection of all subsets of $\{1, 2\}$, and $\mathscr{D}(\mathbb{k}^4, \mathcal{A})$ is the set of remaining subsets of $\{1, 2, 3, 4\}$, and condition (c) of $(\mathbb{k}^4, \mathcal{A})$-non-degeneracy fails for $I = \{1, 2, 3, 4\}$ and $\nu = (0, 0, 1, 2)$, which is centered at $(\mathbb{k}^*)^{I'}$, where $I' := \{1, 2\}$.

6.1. A more efficient formulation of $(U, \mathcal{A})$-non-degeneracy. In this section we describe a criterion equivalent to $(U, \mathcal{A})$-non-degeneracy, but which involves checking fewer conditions. Recall that for $I \subseteq [n]$, we write $T_{\mathcal{A}}^I := \{j \in [n] :$

$\mathcal{A}_j \cap \mathbb{R}^I \neq \emptyset\}$. Define

$$\begin{aligned} \mathscr{D}^*(U,\mathcal{A}) &:= \{I \in \mathscr{D}(U,\mathcal{A}) : I \neq \emptyset,\ |T^I_{\mathcal{A}}| = |I|\} \\ &= \{I \subseteq [n] : I \neq \emptyset,\ |T^I_{\mathcal{A}}| = |I|,\ \mathcal{A} \text{ is } \mathbb{R}^I\text{-dependent}\} \end{aligned} \tag{121}$$

$$\begin{aligned} \mathscr{I}^*(U,\mathcal{A}) &:= \{I \in \mathscr{I}(U,\mathcal{A}) : I \neq \emptyset,\ |T^I_{\mathcal{A}}| = |I|\} \\ &= \{I \subseteq [n] : I \neq \emptyset,\ |T^I_{\mathcal{A}}| = |I|,\ \mathcal{A} \text{ is } \mathbb{R}^I\text{-independent}\} \end{aligned} \tag{122}$$

Proposition X.23. ([Mon16]). *For polynomials $f_1, \ldots, f_n$ in $(x_1, \ldots, x_n)$, the following are equivalent:*

(1) f_i's are $(U,\mathcal{A})$-non-degenerate.
(2) (i) property (a) of $(U,\mathcal{A})$-non-degeneracy holds,
(ii) property (b) of $(U,\mathcal{A})$-non-degeneracy holds with $\mathscr{I}(U,\mathcal{A})$ replaced by $\mathscr{I}^*(U,\mathcal{A})$,
(iii) property (c) of $(U,\mathcal{A})$-non-degeneracy holds with $\mathscr{D}(U,\mathcal{A})$ replaced by $\mathscr{D}^*(U,\mathcal{A})$.

We prove proposition X.23 in section X.7.1.

Example X.24. (Warning!). In property (b) of $(U,\mathcal{A})$-non-degeneracy $\mathscr{D}(U,\mathcal{A})$ can *not* be replaced by $\mathscr{D}^*(U,\mathcal{A})$, and in property (c) of $(U,\mathcal{A})$-non-degeneracy $\mathscr{I}(U,\mathcal{A})$ can *not* be replaced by $\mathscr{I}^*(U,\mathcal{A})$. Indeed, at first consider the system

$$\begin{aligned} f_1 &= 1 + x_1 \\ f_2 &= 1 + x_1 + x_2 \\ f_3 &= 1 + x_1 + 2x_2 + ax_3x_4 \\ f_4 &= x_4(1 + x_1 + bx_2 + cx_3 + dx_4) \end{aligned}$$

for generic $a, b, c, d \in \Bbbk^*$. Let $U = \Bbbk^4$ and $\mathcal{A}_j = \operatorname{Supp}(f_j)$, $j = 1, \ldots, 4$. Then $\mathscr{D}(U,\mathcal{A})$ is the collection of all subsets of $\{1,2,3\}$, $\mathscr{I}(U,\mathcal{A})$ is the collection of all subsets of $\{1,2,3,4\}$ containing 4, $\mathscr{D}^*(U,\mathcal{A}) = \{\{1,2,3\}\}$, and $\mathscr{I}^*(U,\mathcal{A}) = \{\{1,2,3,4\}\}$. Condition (b) of $(U,\mathcal{A})$-non-degeneracy fails with $I = \{1,2,3,4\}$ and $\nu = (0,1,1,1)$ so that the center of ν is $(\Bbbk^*)^{I'}$, where $I' := \{1\} \in \mathscr{D}(U,\mathcal{A})$. However, it is straightforward to check that condition (b) would not have been violated had $\mathscr{D}(U,\mathcal{A})$ been replaced by $\mathscr{D}^*(U,\mathcal{A})$. Now consider the system

$$\begin{aligned} g_1 &= 1 + x_1 + x_2 + x_3 \\ g_2 &= 1 + x_1 + 2x_2 + 3x_3 \\ g_3 &= x_2(ax_1 + bx_2 + cx_3) + x_3(a'x_1 + b'x_2 + c'x_3) \\ g_4 &= x_4(1 + x_1) \end{aligned}$$

where a, b, c, a', b', c' are generic elements in $\Bbbk^*$. Let $\mathcal{B} := (\operatorname{Supp}(g_1), \ldots, \operatorname{Supp}(g_4))$, and $U := \Bbbk^4$. Then $\mathscr{D}(U,\mathcal{B})$ is the collection of all subsets of $\{1,2,3,4\}$ containing 4, $\mathscr{I}(U,\mathcal{B})$ is the collection of all subsets of $\{1,2,3,4\}$ not containing 4, $\mathscr{D}^*(U,\mathcal{B}) = \{\{1,2,3,4\}\}$, and $\mathscr{I}^*(U,\mathcal{B}) = \{\{1,2,3\}\}$. It is straightforward to check that g_1, g_2, g_3, g_4 violate condition (c) of $(U,\mathcal{B})$-non-degeneracy with $I = \{1,2,3,4\}$ and $\nu = (0,1,1,1)$ (so that the center of ν is $(\mathbb{k}^*)^{I'}$, where $I' := \{1\} \in \mathscr{I}(U,\mathcal{B})$). It is also straightforward to check that condition (c) would not have been violated had $\mathscr{I}(U,\mathcal{B})$ been replaced by $\mathscr{I}^*(U,\mathcal{B})$.

The following combinatorial description of strict monotonicity of $[\cdot,\dots,\cdot]_U^{iso}$ follows from theorem X.22 and proposition X.23 exactly as in the proof of theorem IX.32.

Corollary X.25. *Let $\mathcal{B}_j \subseteq \operatorname{conv}(\mathcal{A}_j)$, $j = 1,\dots,n$. Then $[\mathcal{A}_1,\dots,\mathcal{A}_n]_U^{iso} \geq [\mathcal{B}_1,\dots,\mathcal{B}_n]_U^{iso}$. If $[\mathcal{A}_1,\dots,\mathcal{A}_n]_U^{iso} > 0$, then the following are equivalent:*

(1) $[\mathcal{A}_1,\dots,\mathcal{A}_n]_U^{iso} = [\mathcal{B}_1,\dots,\mathcal{B}_n]_U^{iso}$,

(2) (a) For each nonempty $I \in \mathscr{I}(U,\mathcal{A})$ and for each weighted order ν on $\mathbb{k}[x_i : i \in I]$ such that ν is centered at infinity or at $(\mathbb{k}^*)^{I'}$ for some $I' \in \mathscr{E}(U) \cup \mathscr{E}(\mathcal{A}) \cup \mathscr{D}(U,\mathcal{A})$, the collection $\{\operatorname{In}_\nu(\operatorname{ND}(\mathcal{B}_j^I)) : \operatorname{In}_\nu(\operatorname{ND}(\mathcal{A}_j^I)) \cap \operatorname{Supp}(\mathcal{B}_j) \neq \emptyset\}$ of polytopes is dependent, and

(b) for each nonempty $I \in \mathscr{D}(U,\mathcal{A})$ and for each weighted order ν on $\mathbb{k}[x_i : i \in I]$ such that ν is centered at $(\mathbb{k}^*)^{I'}$ for some $I' \in \mathscr{I}(U,\mathcal{A})$, the collection $\{\operatorname{In}_\nu(\operatorname{ND}(\mathcal{B}_j^I)) : \operatorname{In}_\nu(\operatorname{ND}(\mathcal{A}_j^I)) \cap \operatorname{Supp}(\mathcal{B}_j) \neq \emptyset\}$ of polytopes is dependent.

(3) (a) For each nonempty $I \in \mathscr{I}^*(U,\mathcal{A})$ and for each weighted order ν on $\mathbb{k}[x_i : i \in I]$ such that ν is centered at infinity or at $(\mathbb{k}^*)^{I'}$ for some $I' \in \mathscr{E}(U) \cup \mathscr{E}(\mathcal{A}) \cup \mathscr{D}(U,\mathcal{A})$, the collection $\{\operatorname{In}_\nu(\operatorname{ND}(\mathcal{B}_j^I)) : \operatorname{In}_\nu(\operatorname{ND}(\mathcal{A}_j^I)) \cap \operatorname{Supp}(\mathcal{B}_j) \neq \emptyset\}$ of polytopes is dependent, and

(b) for each nonempty $I \in \mathscr{D}^*(U,\mathcal{A})$ and for each weighted order ν on $\mathbb{k}[x_i : i \in I]$ such that ν is centered at $(\mathbb{k}^*)^{I'}$ for some $I' \in \mathscr{I}(U,\mathcal{A})$, the collection $\{\operatorname{In}_\nu(\operatorname{ND}(\mathcal{B}_j^I)) : \operatorname{In}_\nu(\operatorname{ND}(\mathcal{A}_j^I)) \cap \operatorname{Supp}(\mathcal{B}_j) \neq \emptyset\}$ of polytopes is dependent. □

7. Proof of the non-degeneracy conditions

7.1. Reduction to the more efficient non-degeneracy criterion. In this section we prove propositions X.21 and X.23.

PROOF OF PROPOSITION X.21. Let $I' \supseteq I$ be as in the definition of hereditary dependence. By restricting all f_j's to $\mathbb{k}^{I'}$, we may assume that $I' = [n]$. Let Z^I be the set of isolated points $V(f_1,\dots,f_n)$ which are on $(\mathbb{k}^*)^I$. Assume to the contrary of the claim that $Z^I \neq \emptyset$. Let $(g_1,\dots,g_n) \in \mathcal{L}(\mathcal{A})$ be a system such that

(1) $\operatorname{NP}(g_j) = \operatorname{conv}(\mathcal{A}_j)$, $j = 1,\dots,n$, and

(2) $g_1|_{\mathbb{k}^{\tilde I}},\dots,g_n|_{\mathbb{k}^{\tilde I}}$ are *properly $\mathcal{A}^{\tilde I}$-non-degenerate* (see section VII.4.4) for each $\tilde I \subseteq [n]$, where $\mathcal{A}^{\tilde I} := (\mathcal{A}_1 \cap \mathbb{R}^{\tilde I},\dots,\mathcal{A}_n \cap \mathbb{R}^{\tilde n})$ (proposition VII.25 implies that the set of such systems is a nonempty Zariski open subset of $\mathcal{L}(\mathcal{A})$).

Let t be a new indeterminate, and $h_j := tg_j + (1-t)f_j \in \mathbb{k}[x_1,\dots,x_n,t]$, $j = 1,\dots,n$. Pick $z \in Z^I$. Then $(z,0)$ is an isolated point of $V(h_1,\dots,h_n,t)$ on $\mathbb{k}^{n+1}$, and therefore theorem III.80 implies that there is a Zariski open neighborhood U of $(z,0)$ in $\mathbb{k}^{n+1}$ such that $C := V(h_1,\dots,h_n) \cap U$ is a curve. Let C' be an irreducible component of C containing $(z, 0)$, and $\bar{C}'$ be the closure of C' in $\mathbb{P}^n \times \mathbb{P}^1$. Then $C' \not\subseteq \Bbbk^n \times \{\epsilon\}$ for any $\epsilon \in \Bbbk$, and corollary IV.26 implies that $\bar{C}'$ intersects $\mathbb{P}^n \times \{1\}$. Pick a branch B of $\bar{C}'$ centered at a point $(z', 1) \in \bar{C}' \cap (\mathbb{P}^n \times \{1\})$. Define $I_B \subseteq [n+1]$ and the weighted order ν_B on the coordinate ring of $\Bbbk^{I_B}$ as in definition IX.15. Let $\tilde{I} := I_B \cap [n]$ and $\tilde{\nu}$ be the restriction of ν_B to $\Bbbk[x_i : i \in \tilde{I}]$. Since $\mathrm{NP}(f_j) \subseteq \mathrm{NP}(g_j)$ for each j, it follows that $\tilde{\nu}(f_j|_{\Bbbk^{\tilde{I}}}) \geq \tilde{\nu}(g_j|_{\Bbbk^{\tilde{I}}})$. Since $\nu_B(t-1) > 0$, it follows that $\nu_B(t) = 0$, and therefore $\mathrm{In}_{\nu_B}(h_j|_{\Bbbk^{I_B}}) = t\,\mathrm{In}_{\tilde{\nu}}(g_j|_{\Bbbk^{\tilde{I}}})$ for each j. Lemma IX.17 then implies that $\mathrm{In}_{\tilde{\nu}}(g_j|_{\Bbbk^{\tilde{I}}})$ have a common zero in $(\Bbbk^*)^{\tilde{I}}$. On the other hand, since $\bar{C}'$ contains the point $(z, 0) \in (\Bbbk^*)^I$, it follows that $I \subseteq \tilde{I}$, and therefore it follows from the definition of hereditary dependence (applied with $I' = [n]$) that either $|T_{\mathcal{A}}^{\tilde{I}}| > |\tilde{I}|$, or $|T_{\mathcal{A}}^{\tilde{I}}| = |\tilde{I}|$ and $\mathcal{A}$ is $\mathbb{R}^I$-dependent. In any event, there is a nonempty subset J of $T_{\mathcal{A}}^{\tilde{I}}$ such that $\dim(\sum_{j \in J} \mathrm{NP}(g_j|_{\Bbbk^{\tilde{I}}})) < |J|$. Since $g_1|_{\Bbbk^{\tilde{I}}}, \ldots, g_n|_{\Bbbk^{\tilde{I}}}$ are properly $\mathcal{A}^{\tilde{I}}$-non-degenerate, it then follows by definition of proper non-degeneracy that there is no common zero of $\mathrm{In}_{\tilde{\nu}}(g_j|_{\Bbbk^{\tilde{I}}})$ on $(\Bbbk^*)^{\tilde{I}}$. This contradiction completes the proof. □

PROOF OF PROPOSITION X.23. Since $\mathscr{D}^*(U, \mathcal{A}) \subseteq \mathscr{D}(U, \mathcal{A})$ and $\mathscr{I}^*(U, \mathcal{A}) \subseteq \mathscr{I}(U, \mathcal{A})$, it suffices to show the following:

(1) property (b) of $(U, \mathcal{A})$-non-degeneracy holds if it holds with $\mathscr{I}(U, \mathcal{A})$ replaced by $\mathscr{I}^*(U, \mathcal{A})$,
(2) property (c) of $(U, \mathcal{A})$-non-degeneracy holds if it holds with $\mathscr{D}(U, \mathcal{A})$ replaced by $\mathscr{D}^*(U, \mathcal{A})$.

It follows from the definition of hereditary dependence that for every $I \in \mathscr{I}(U, \mathcal{A}) \setminus \mathscr{I}^*(U, \mathcal{A})$, there exists $I' \in \mathscr{I}^*(U, \mathcal{A})$ such that $I \subsetneq I'$ and $|T_{\mathcal{A}}^{I'}| > |I'|$ for each I' such that $I \subseteq I' \subsetneq \tilde{I}$. The same statement also holds if we replace $\mathscr{I}(U, \mathcal{A})$ and $\mathscr{I}^*(U, \mathcal{A})$, respectively, by $\mathscr{D}(U, \mathcal{A})$ and $\mathscr{D}^*(U, \mathcal{A})$. Since restricting all f_j to $\Bbbk^{\tilde{I}}$ yields a system with the same number of non-zero polynomials as the number of variables, it suffices to prove the following claim: "if there is $I \subseteq [n]$ such that $|T_{\mathcal{A}}^I| > |I|$ and $\mathrm{In}_{\mathcal{A}_1^I, \nu}(f_1|_{\Bbbk^I}), \ldots, \mathrm{In}_{\mathcal{A}_n^I, \nu}(f_n|_{\Bbbk^I})$ have a common zero on $(\Bbbk^*)^n$ for some weighted order ν on $\Bbbk[x_i : i \in I]$, then there is $\tilde{I} \supsetneq I$ and a weighted order $\tilde{\nu}$ on $\Bbbk[x_{\tilde{i}} : \tilde{i} \in \tilde{I}]$ such that $\tilde{\nu}$ is compatible with ν, and $\mathrm{In}_{\mathcal{A}_1^{\tilde{I}}, \tilde{\nu}}(f_1|_{\Bbbk^{\tilde{I}}}), \ldots, \mathrm{In}_{\mathcal{A}_n^{\tilde{I}}, \tilde{\nu}}(f_n|_{\Bbbk^{\tilde{I}}})$ have a common zero in $\tilde{u} \in (\Bbbk^*)^n$." This follows from lemma IX.28. □

7.2. Understanding condition (c) of $(U, \mathcal{A})$-non-degeneracy.

In this section we show that if condition (c) of $(U, \mathcal{A})$-non-degeneracy is violated for $(f_1, \ldots, f_n) \in \mathcal{L}(\mathcal{A})$ with I, ν and I', then the common zero on $(\Bbbk^*)^n$ of $\mathrm{In}_{\mathcal{A}_1^I, \nu}(f_1|_{\Bbbk^I}), \ldots, \mathrm{In}_{\mathcal{A}_n^I, \nu}(f_1|_{\Bbbk^I})$ corresponds to a non-isolated point of $V(f_1, \ldots, f_n)$. Recall the definition of $\pi_I : \Bbbk^n \to \Bbbk^I$ from (85).

Lemma X.26. *Let $(f_1, \ldots, f_n) \in \mathcal{L}(\mathcal{A})$, and ν be a weighted order on $\Bbbk[x_1, \ldots, x_n]$ centered at $(\Bbbk^*)^I$, $I \subseteq [n]$. Assume*

(1) $\mathrm{conv}(\mathcal{A}_1), \ldots, \mathrm{conv}(\mathcal{A}_n)$ are dependent polytopes in $\mathbb{R}^n$, and
(2) $\mathrm{In}_{\mathcal{A}_1, \nu}(f_1), \ldots, \mathrm{In}_{\mathcal{A}_n, \nu}(f_n)$ have a common zero $a \in (\Bbbk^*)^n$.

Then $\pi_I(a) \in (\mathbb{k}^*)^I$ is a non-isolated point of the set $V(f_1, \ldots, f_n)$ of common zeroes of $f_1, \ldots, f_n$ on $\mathbb{k}^n$.

Note that both conditions (1) and (2) are necessary for the conclusion of lemma X.26 to hold. For example, $f_1 = 1 + x_1, f_2 = 1 + x_1 + x_2$ satisfy condition (2), but not (1), with $I = \{1\}$, $\nu = (0, 1)$, $\mathcal{A}_j = \operatorname{Supp}(f_j)$, $j = 1, 2$, and $a = (-1, c)$ for some arbitrary $c \in \mathbb{k}^*$. In this case $\pi_I(a) = (-1, 0)$ is an *isolated* point of $V(f_1, f_2)$. On the other hand, $f_1 = 1 + x_1$, $f_2 = x_2$, and $\mathcal{A}_j = \operatorname{Supp}(f_j)$, $j = 1, 2$, satisfy, with the same I, ν and a, condition (1), but not (2), and $\pi_I(a) = (-1, 0)$ is again an isolated point of $V(f_1, f_2)$.

PROOF OF LEMMA X.26. It is immediate to check that $\pi_I(a)$ is in $V(f_1, \ldots, f_n)$; we only have to show that it is non-isolated in there. Since $\mathcal{P}_j := \operatorname{conv}(\mathcal{A}_j)$, $j = 1, \ldots, n$, are dependent, it follows that there is $J \subseteq [n]$ such that $p := \dim(\sum_{j \in J} \mathcal{P}_j) < |J|$. Let Π be the (unique) p-dimensional linear subspace of $\mathbb{R}^n$ such that $\sum_{j \in J} \mathcal{P}_j$ is contained in a translate of Π. Let $\nu_j := \nu(x_j)$, $j = 1, \ldots, n$. We identify ν with the element of $(\mathbb{R}^n)^*$ with coordinates $(\nu_1, \ldots, \nu_n)$ with respect to the basis dual to the standard basis of $\mathbb{R}^n$. Let $\Pi_0 := \{\alpha \in \Pi : \langle \nu, \alpha \rangle = 0\}$ and $r := \dim(\Pi_0)$. Choose a basis $\alpha_1, \ldots, \alpha_r$ of $\Pi_0 \cap \mathbb{Z}^n$. Let $c_i := a^{\alpha_i}$, $i = 1, \ldots, n$. Let Y be the subvariety of $(\mathbb{k}^*)^n$ determined by $x^{\alpha_i} - c_i$, $i = 1, \ldots, r$, and $\bar{Y}$ be the closure of Y in $\mathbb{k}^n$. Then Y, and therefore $\bar{Y}$, is an irreducible variety of codimension r in $\mathbb{k}^n$.

Claim X.26.1. $\pi_I(a) \in \bar{Y}$. *Moreover, if g is any Laurent polynomial in $(x_1, \ldots, x_n)$ such that* $\operatorname{Supp}(g) \subseteq \Pi_0$, *then g restricts to a constant function on Y with value $g(a)$.*

PROOF. The second assertion is obvious, so we prove the first assertion. If ν is the trivial weighted order, then π_I is the identity and therefore $\pi_I(a) = a \in Y$. Otherwise let C be the curve on $\mathbb{k}^n$ parametrized by $t \mapsto c(t) := (a_1 t^{\nu_1}, \ldots, a_n t^{\nu_n})$. Then $C \cap (\mathbb{k}^*)^n \subseteq Y$ so that $C \subseteq \bar{Y}$. Since $\pi_I(a) = c(0) \in C$, the claim is proved. □

Note that r equals either p or $p - 1$. If $r = p$, then $\Pi_0 = \Pi$ and for each $j \in J$, f_j is ν-homogeneous and is of the form $f_j = x^{\beta_j} g_j$ for some $\beta_j \in \mathbb{Z}^n$ and Laurent polynomial g_j such that $\operatorname{Supp}(g_j) \subseteq \Pi_0$. Claim X.26.1 implies that $f_j|_Y \equiv 0$ for each $j \in J$ so that $\pi_I(a) \in \bar{Y} \subseteq V(f_j : j \in J)$. It follows that at least one of the irreducible components of $V(f_j : j \in J)$ containing $\pi_I(a)$ has codimension smaller than $|J|$ in $\mathbb{k}^n$. The lemma then follows due to theorem III.80. It remains to consider the case that $r = p - 1$. Lemma B.57 implies that $\alpha_1, \ldots, \alpha_r$ can be extended to a basis $\alpha_1, \ldots, \alpha_n$ of $\mathbb{Z}^n$ such that $\alpha_1, \ldots, \alpha_{r+1}$ is a basis of Π, and $\langle \nu, \alpha_j \rangle \geq 0$ for each $j = 1, \ldots, n$; this in particular implies that $\langle \nu, \alpha_{r+1} \rangle > 0$. Let $y_i := x^{\alpha_i}$, $i = 1, \ldots, n$. Then the y_i form a system of coordinates on $(\mathbb{k}^*)^n$ and the projection onto $(y_{r+1}, \ldots, y_n)$ restricts to an isomorphism $Y \cong (\mathbb{k}^*)^{n-r}$. Pick $\beta_j := (\beta_{j,1}, \ldots, \beta_{j,n}) \in \mathbb{Z}^n$ such that $x_j = \prod_{i=1}^n y_i^{\beta_{j,i}}$, $j = 1, \ldots, n$. Then $\bar{Y}$ is the closure of the image of the map $\psi : (\mathbb{k}^*)^{n-r} \to \mathbb{k}^n$ given by

$$\psi(y_{r+1}, \ldots, y_n) := (c'_1 \prod_{i=r+1}^{n} y_i^{\beta_{1,i}}, \ldots, c'_n \prod_{i=r+1}^{n} y_i^{\beta_{n,i}}),$$

where $c'_j := \prod_{i=1}^{r} c_i^{\beta_{j,i}}$, $j = 1, \ldots, n$. Let $\bar{Y}'$ be the closure of the image of the map $\psi' : (\mathbb{k}^*)^{n-r} \to \mathbb{k}^n$ given by

$$\psi'(y_{r+1}, \ldots, y_n) := (\prod_{i=r+1}^{n} y_i^{\beta_{1,i}}, \ldots, \prod_{i=r+1}^{n} y_i^{\beta_{n,i}})$$

Then $\bar{Y}'$ is isomorphic to $\bar{Y}$ via the map $\rho : (x_1, \ldots, x_n) \mapsto (c'_1 x_1, \ldots, c'_n x_n)$. Let $\beta'_j := (\beta_{j,r+1}, \ldots, \beta_{j,n})$, $j = 1, \ldots, n$, and $\mathcal{B}' := \{\beta'_0, \ldots, \beta'_n\}$, where β'_0 is the origin in $\mathbb{Z}^{n-r}$. Let $X_{\mathcal{B}'}$ be the corresponding toric variety. In the notation of theorem VI.12 $\bar{Y}'$ is isomorphic to the affine open subset $X_{\mathcal{B}'} \cap U_{\beta'_0}$ of $X_{\mathcal{B}'}$. Since $\dim(\bar{Y}') = n - r$, it follows that the convex hull $\mathcal{P}'$ of $\mathcal{B}'$ in $\mathbb{R}^{n-r}$ has dimension $n - r$. Let $\nu'_j := \langle \nu, \alpha_j \rangle$, $j = r+1, \ldots, n$, and $\nu' \in (\mathbb{R}^{n-r})^*$ be the element with coordinates $(\nu'_{r+1}, \ldots, \nu'_n)$. Then for each $j = 1, \ldots, n$, $\langle \nu', \beta'_j \rangle = \sum_{i=r+1}^{n} \beta_{j,i} \langle \nu, \alpha_i \rangle = \langle \nu, \sum_{i=1}^{n} \beta_{j,i} \alpha_i \rangle = \nu_j \geq 0$. It follows that $\min_{\mathcal{P}'}(\nu') = 0$ (since $\langle \nu', \beta'_0 \rangle = 0$). Since $\nu'_{r+1} > 0$, there is a facet $\mathcal{Q}'$ of $\mathcal{P}'$ containing $\mathrm{In}_{\nu'}(\mathcal{P}')$ such that the first coordinate of the inner normal with respect to the dual basis is positive (this follows e.g. from corollaries V.21 and V.26 and proposition V.28). Let $\mathcal{C}' := \mathcal{B}' \cap \mathcal{Q}'$, and $V_{\mathcal{C}'}$ be the corresponding torus invariant codimension one subvariety of $X_{\mathcal{B}'}$ (see theorem VI.12). Let $Z' := \bar{Y}' \cap V_{\mathcal{C}'}$, and $Z := \rho(Z') \subseteq \bar{Y}$.

Claim X.26.2. $\pi_I(a) \in Z \subseteq V(f_j : j \in J)$.

PROOF. Let $\eta \in (\mathbb{R}^{n-r})^*$ be the primitive inner normal to $\mathcal{Q}'$, and $(\eta_{r+1}, \ldots, \eta_n)$ be the coordinates of η in the dual basis of $(\mathbb{R}^{n-r})^*$. For each $b = (b_{r+1}, \ldots, b_n) \in (\mathbb{k}^*)^{n-r}$, consider the rational curve on $\bar{Y}'$ parametrized by

$$t \mapsto c'_b(t) := \psi'(b_{r+1} t^{\eta_{r+1}}, \ldots, b_n t^{\eta_n}) = (b^{\beta'_1} t^{\eta'_1}, \ldots, b^{\beta'_n} t^{\eta'_n})$$

where $\eta'_j := \langle \eta, \beta'_j \rangle$ for each $j = 1, \ldots, n$. Since $0 = \beta_0 \in \mathrm{In}_{\nu'}(\mathcal{P}') \subset \mathcal{Q}'$, it follows that $\min_{\mathcal{P}'}(\eta) = 0$ so that each η'_j is nonnegative, and it is zero if and only if $\beta'_j \in \mathcal{Q}'$. In particular, $c'_b(0)$ is well defined, and is an element in $\mathbb{k}^n$. Moreover, theorem VI.12 implies that the set $\{c'_b(0) : b \in (\mathbb{k}^*)^{n-r}\}$ is precisely $\bar{Y}' \cap O_{\mathcal{C}'}$; in particular it is a dense Zariski open subset of Z'. Therefore, in order to prove $Z \subseteq V(f_j : j \in J)$, it suffices to show that $f_j(\rho(c'_b(0))) = 0$ for each $j \in J$. Fix $j \in J$. Write f_j as $f_j = f_{j,0} + f_{j,1} + \cdots$, where $f_{j,k}$ are ν-homogeneous polynomials with $\nu(f_{j,0}) < \nu(f_{j,1}) < \cdots$. We will show that

$$f_{j,k}(\rho(c'_b(0))) = 0 \tag{123}$$

for each k. Let $m_j := \min_\nu(\mathcal{A}_j)$. Note that $\nu(f_{j,k}) \geq m_j$ for each k. At first consider the case that $\nu(f_{j,k}) = m_j$. This implies that $k = 0$, $\mathrm{In}_\nu(\mathcal{A}_j) \cap \mathrm{Supp}(f_j)$

is nonempty, and $f_{j,k} = \mathrm{In}_{\mathcal{A}_j,\nu}(f_j)$. Since $\mathrm{In}_{\mathcal{A}_j,\nu}(f_j)(a) = 0$, in this case claim X.26.1 implies that $f_{j,k}$ is identically zero on $\rho(c'_b(t))$, $t \in \mathbb{k}$; in particular, (123) holds. Now fix k such that $\nu(f_{j,k}) > m_j$. It suffices to show that

$$\mathrm{ord}_t(f_{j,k} \circ \rho \circ c'_b(t)) > 0 \tag{124}$$

Pick $\alpha_0 \in \mathrm{In}_\nu(\mathcal{A}_j)$ and $\alpha \in \mathrm{Supp}(f_{j,k})$. Then $\alpha - \alpha_0 \in \Pi$, and $\langle \nu, \alpha - \alpha_0 \rangle = \nu(f_{j,k}) - m_j > 0$. It follows that $\alpha - \alpha_0 = \sum_{j=1}^{r+1} m_j \alpha_j$ with integers m_j such that $m_{r+1} > 0$. Since $x^{\alpha_j}|_Y \equiv c_j$ for $j = 1, \ldots, r$, and $x^{\alpha_{r+1}}|_Y \equiv y_{r+1}|_Y$, it follows that

$$\mathrm{ord}_t(x^\alpha \circ \rho \circ c'_b(t)) = \mathrm{ord}_t(x^{\alpha_0} \circ \rho \circ c'_b(t)) + m_{r+1}\eta_{r+1}$$

Since $\eta_{r+1} > 0$ (due to our choice of $\mathcal{Q}'$) and $\alpha_0 \in \mathbb{Z}^n_{\geq 0}$, (124) follows. It remains to prove that $\pi_I(a) \in Z$. Consider the rational curve C' on $\bar{Y}'$ parametrized by

$$t \mapsto \psi'(c_{r+1}t^{\nu'_{r+1}}, \ldots, c_n t^{\nu'_n}) = ((\prod_{i=r+1}^{n} c_i^{\beta_{1,i}})t^{\nu_1}, \ldots, (\prod_{i=r+1}^{n} c_i^{\beta_{n,i}})t^{\nu_n})$$

Since $\mathrm{In}_{\nu'}(\mathcal{P}')$ is a face of $\mathcal{Q}'$, proposition VI.31 implies that the center of the branch of C' at $t = 0$ is contained in Z'. Note that $\rho(C')$ is precisely the curve C from claim X.26.1. It follows from claim X.26.1 that $\pi_I(a)$ is the center of the branch of C at $t = 0$, and therefore it is on Z, as required. ☐

Since $\dim(Z) = n - r - 1 > n - |J|$, the lemma follows from claim X.26.2 and theorem III.80. ☐

Corollary X.27. *Let $I \in \mathscr{D}(U, \mathcal{A}) \setminus \{\emptyset\}$ and ν be a weighted order on $\mathbb{k}[x_i : i \in I]$ such that ν is centered at $(\mathbb{k}^*)^{I'}$ for some $I' \subseteq I$, and $\mathrm{In}_{\mathcal{A}_1^I,\nu}(f_1|_{\mathbb{k}^I}), \ldots, \mathrm{In}_{\mathcal{A}_n^I,\nu}(f_n|_{\mathbb{k}^I})$ have a common zero $a \in (\mathbb{k}^*)^n$. Then $\pi_{I'}(a) \in (\mathbb{k}^*)^{I'}$ is a non-isolated point of the zero-set $V(f_1, \ldots, f_n)$ of $f_1, \ldots, f_n$ on $\mathbb{k}^n$.*

PROOF. It is straightforward to check that $\pi_{I'}(a) \in V(f_1, \ldots, f_n)$. There is $J \supseteq I$ such that $\mathcal{A}$ is $\mathbb{R}^J$-dependent, $|T^J_{\mathcal{A}}| = |J|$, and $|T^{\tilde{I}}_{\mathcal{A}}| > |\tilde{I}|$ for each $\tilde{I}$ such that $I \subseteq \tilde{I} \subsetneq J$. Replacing the f_j by $f_j|_{\mathbb{k}^J}$ and applying lemma IX.28 reduces the corollary to the case that $I = [n]$. Then it follows from lemma X.26. ☐

7.3. Proof of theorem X.22. We divide the proof of theorem X.22 in three parts:

7.3.1. *Proof of the implication (2) ⇒ (1) from theorem X.22.* We proceed as in the proof of proposition VII.19. Pick $(f_1, \ldots, f_n), (g_1, \ldots, g_n) \in \mathcal{L}(\mathcal{A})$ such that $[g_1, \ldots, g_n]^{iso}_U > [f_1, \ldots, f_n]^{iso}_U$. We will show that $f_1, \ldots, f_n$ are $(U, \mathcal{A})$-degenerate. Define $h_j := (1-t)f_j + tg_j$, where t is a new indeterminate. The set of isolated zeroes of $g_1, \ldots, g_n$ on U is nonempty. Theorem IV.32 implies that we can find an irreducible curve C contained in the set of zeroes of $h_1, \ldots, h_n$ on $U \times \mathbb{k}$ such that

(i) C intersects $Z \times \{1\}$,

(ii) for generic $\epsilon \in \mathbb{k}$, the set $C_\epsilon := C \cap (U \times \{\epsilon\})$ is nonempty, and each point of C_ϵ is an isolated zero of $h_1|_{t=\epsilon}, \dots, h_n|_{t=\epsilon}$ on U; and

(iii) (1) either there is $(z, 0) \in C$ such that z is a non-isolated zero of $f_1, \dots, f_n$ on U,

(2) or C has a "point at infinity with respect to U" at $t = 0$, i.e. if $\bar{C}$ is the closure of C in $\mathbb{P}^n \times \mathbb{k}$, then there is $(z, 0) \in \bar{C}$ such that $z \notin U$.

Claim X.28. *Let B be a branch of $\bar{C} \subset \mathbb{P}^n \times \mathbb{k}$. Let $\tilde{I} := I_B \cap [n]$ (where $I_B \subseteq [n+1]$ is defined as in definition IX.15) and $\tilde{\nu}$ be the restriction of ν_B to $\mathbb{k}[x_i : i \in \tilde{I}]$.*

(1) $C_\epsilon \subseteq (\mathbb{k}^*)^{\tilde{I}} \times \{\epsilon\}$ for generic $\epsilon \in \mathbb{k}$.
(2) $\tilde{I} \in \mathscr{I}(U, \mathcal{A})$.
(3) Assume the center of B is $(z, 0)$ where $z \in \mathbb{P}^n$. Then
(a) $\operatorname{In}_{\mathcal{A}_j^{\tilde{I}}, \tilde{\nu}}(f_j|_{\mathbb{k}^{\tilde{I}}})$, $j = 1, \dots, n$, have a common zero on $(\mathbb{k}^*)^n$.
(b) If $z \in \mathbb{P}^n \setminus \mathbb{k}^n$, then $\tilde{I} \neq \emptyset$ and $\tilde{\nu}$ is centered at infinity. In particular, $f_1, \dots, f_n$ violate condition (b) of $(U, \mathcal{A})$-non-degeneracy with $I = \tilde{I}$ and $\nu = \tilde{\nu}$.
(c) Otherwise let J be the (unique) subset of $[n]$ such that $z \in (\mathbb{k}^*)^J$. Then either $\tilde{I} = J = \emptyset$, or $\tilde{I} \neq \emptyset$ and $\tilde{\nu}$ is centered at $(\mathbb{k}^*)^J$. In particular, if $J \in \mathscr{E}(U) \cup \mathscr{E}(\mathcal{A}) \cup \mathscr{D}(U, \mathcal{A})$, then $f_1, \dots, f_n$ violate property (b) of $(U, \mathcal{A})$-non-degeneracy with $I = \tilde{I}$, $I' = J$ and $\nu = \tilde{\nu}$.

PROOF. Assertion (1) follows from the definition of I_B. Since $(h_1|_{t=\epsilon}, \dots, h_n|_{t=\epsilon}) \in \mathcal{L}(\mathcal{A})$ for all ϵ, assertion (1), property (ii) and lemma X.1 imply that $\tilde{I} \notin \mathscr{E}(U) \cup \mathscr{E}(\mathcal{A})$, and then proposition X.21 implies that $\tilde{I} \in \mathscr{I}(U, \mathcal{A})$. This proves assertion (2). Now assume we are in the situation of assertion (3). Fix j, $1 \leq j \leq n$. Since $\nu_B(t|_{\mathbb{k}^{I_B}}) > 0$, it follows that

$$\operatorname{In}_{\mathcal{A}_j^{\tilde{I}}, \tilde{\nu}}(f_j|_{\mathbb{k}^{\tilde{I}}}) = \begin{cases} \operatorname{In}_{\tilde{\nu}}(f_j|_{\mathbb{k}^{\tilde{I}}}) = \operatorname{In}_{\nu_B}(h_j|_{\mathbb{k}^{I_B}}) & \text{if } \operatorname{Supp}(f_j) \cap \operatorname{In}_{\tilde{\nu}}(\mathcal{A}_j^{\tilde{I}}) \neq \emptyset, \\ 0 & \text{otherwise.} \end{cases}$$

Assertion (3a) now follows from lemma IX.17. If $z \in \mathbb{P}^n \setminus \mathbb{k}^n$, then there is at least one j such that $1/x_j$ is a regular function near z which vanishes at z. This j has to be in $\tilde{I}$ and $\tilde{\nu}(x_j)$ has to be negative, which proves the first statement of part (3b). The second statement then follows from assertions (2) and (3a). The first statement of part (3c) is obvious. If $J \in \mathscr{E}(U) \cup \mathscr{E}(\mathcal{A}) \cup \mathscr{D}(U, \mathcal{A})$, then assertion (2) and the first statement of part (3c) imply that $\tilde{I} \neq \emptyset$, and then the second statement follows from assertions (2) and (3a). □

Now we resume the proof of the implication (2) ⇒ (1) from theorem X.22. At first assume (iii.1) holds. Pick a branch B of C centered at $(z, 0)$ and let $\tilde{I}, J, \tilde{\nu}$ be as in part (3c) of claim X.28. Part (3c) of claim X.28 implies that $f_1, \dots, f_n$ are $(U, \mathcal{A})$-degenerate if $J \in \mathscr{E}(U) \cup \mathscr{E}(\mathcal{A}) \cup \mathscr{D}(U, \mathcal{A})$. So assume $J \in \mathscr{I}(U, \mathcal{A})$. Pick an irreducible curve C' on U such that $z \in C' \subseteq V(f_1, \dots, f_n)$. Let J' be the smallest subset of $[n]$ such that $C' \subseteq \mathbb{k}^{J'}$. Since $J' \supseteq J$ and $J \notin \mathscr{E}(U) \cup \mathscr{E}(\mathcal{A})$, it follows that $J' \notin \mathscr{E}(U) \cup \mathscr{E}(\mathcal{A})$ as well. If $J' \in \mathscr{I}(U, \mathcal{A})$, then lemma IX.17

implies that $f_1, \ldots, f_n$ violate condition (b) of $(U, \mathcal{A})$-non-degeneracy with $I = J'$ and some weighted order ν on $\mathbb{k}[x_j : j \in J']$ centered at infinity (take a branch B' of C' centered at infinity, and set $\nu = \nu_{B'}$). On the other hand, if $J' \in \mathscr{D}(U, \mathcal{A})$, then $f_1, \ldots, f_n$ violate condition (c) of $(U, \mathcal{A})$-non-degeneracy with $I = J'$ and $I' = J$ (take a branch B' of C' centered at z and take $\nu := \nu_{B'}$). This completes the proof of (2) $\Rightarrow$ (1) in the case that (iii.1) holds. Now assume we are in case (iii.2). Pick $z \in \mathbb{P}^n \setminus U$ such that $(z, 0) \in \bar{C}$. If $z \in \mathbb{P}^n \setminus \mathbb{k}^n$, then part (3b) of claim X.28 implies that $f_1, \ldots, f_n$ are $(U, \mathcal{A})$-degenerate. So assume $z \in \mathbb{k}^n \setminus U$. Define J as in part (3c) of claim X.28. If $J \in \mathscr{E}(U) \cup \mathscr{E}(\mathcal{A}) \cup \mathscr{D}(U, \mathcal{A})$, then part (3c) of claim X.28 implies that $f_1, \ldots, f_n$ are $(U, \mathcal{A})$-degenerate. So assume $J \in \mathscr{I}(U, \mathcal{A})$. But then $f_1, \ldots, f_n$ violate condition (a) of $(U, \mathcal{A})$-non-degeneracy. This completes the proof of the implication (2) $\Rightarrow$ (1).

7.3.2. *Proof of the implication (1) $\Rightarrow$ (2) from theorem X.22.* Assume $[\mathcal{A}_1, \ldots, \mathcal{A}_n]_U^{iso} > 0$ and pick an $(U, \mathcal{A})$-degenerate system $f_1, \ldots, f_n \in \mathcal{L}(\mathcal{A})$. We will show that $[\mathcal{A}_1, \ldots, \mathcal{A}_n]_U^{iso} > [f_1, \ldots, f_n]_U^{iso}$. Recall the definition of $\mathscr{I}^*(U, \mathcal{A})$ from (122).

Claim X.29. *There is $I \in \mathscr{I}^*(U, \mathcal{A})$ such that one of the following holds:*

(1) Either I is nonempty, and there is a weighted order ν on $\mathbb{k}[x_i : i \in I]$ such that $\mathrm{In}_{\mathcal{A}_j^I, \nu}(f_j|_{\mathbb{k}^I})$, $j = 1, \ldots, n$, have a common zero $a \in (\mathbb{k}^*)^n$, and one of the following holds:
   (a) ν is centered at infinity,
   (b) or ν is centered at $(\mathbb{k}^*)^{I'}$ for some $I' \subseteq I$, and $\pi_{I'}(a) \notin U$,
   (c) or ν is centered at $(\mathbb{k}^*)^{I'}$ for some $I' \subseteq I$ and $\pi_{I'}(a)$ is a non-isolated zero of $f_1, \ldots, f_n$.

(2) Or there is an isolated point a of $V(f_1, \ldots, f_n) \cap \mathbb{k}^I \cap U$ which is not *isolated in $V(f_1, \ldots, f_n) \subset \mathbb{k}^n$.*

PROOF. If $f_1, \ldots, f_n$ violate property (a) of $(U, \mathcal{A})$-non-degeneracy, then there is a common zero a' of $f_1|_{\mathbb{k}^{I'}}, \ldots, f_n|_{\mathbb{k}^{I'}}$ on $(\mathbb{k}^*)^{I'} \setminus U$ for some $I' \in \mathscr{I}(U, \mathcal{A})$. Lemma IX.28 then implies that the claim holds with case (1b) holds. If property (b) of $(U, \mathcal{A})$-non-degeneracy fails with $I \in \mathscr{I}^*(U, \mathcal{A})$, then either the claim holds with case (1a), or there is a weighted order ν on $\mathbb{k}[x_i : i \in I]$ such that $\mathrm{In}_{\mathcal{A}_j^I, \nu}(f_j|_{\mathbb{k}^I})$ have a common zero $a \in (\mathbb{k}^*)^n$, and ν is centered at $(\mathbb{k}^*)^{I'}$ for some $I' \in \mathscr{E}(U) \cup \mathscr{E}(\mathcal{A}) \cup \mathscr{D}(U, \mathcal{A})$. It is straightforward to check that $\pi_{I'}(a)$ is in the set V of common zeroes of $f_1, \ldots, f_n$ on $\mathbb{k}^n$. If $I' \in \mathscr{E}(U)$, then we are in case (1b), since $\pi_{I'}(a) \in (\mathbb{k}^*)^{I'}$ and $(\mathbb{k}^*)^{I'} \cap U = \emptyset$. If $I' \in \mathscr{E}(\mathcal{A}) \cup \mathscr{D}(U, \mathcal{A})$, then $\pi_{I'}(a)$ has to be a non-isolated point of V due to lemma X.1 and proposition X.21, which is case (1c). Due to proposition X.23 the only case left to consider is that of $f_1, \ldots, f_n$ violating property (c) of $(U, \mathcal{A})$-non-degeneracy. Then there is $J \in \mathscr{D}(U, \mathcal{A})$ and a weighted order η on $\mathbb{k}[x_j : j \in J]$ centered at $(\mathbb{k}^*)^{J'}$ for some $J' \in \mathscr{I}(U, \mathcal{A})$ such that $\mathrm{In}_{\mathcal{A}_1^J, \eta}(f_1|_{\mathbb{k}^J}), \ldots, \mathrm{In}_{\mathcal{A}_n^J, \eta}(f_n|_{\mathbb{k}^J})$ have a common zero b on $(\mathbb{k}^*)^n$. Corollary X.27 implies that $\pi_{J'}(b)$ is a non-isolated point of $V(f_1, \ldots, f_n)$. Pick the smallest subset I of $[n]$ containing J' such that $|\mathcal{T}_{\mathcal{A}}^I| = |I|$. Since $J' \notin \mathscr{E}(U) \cup \mathscr{E}(\mathcal{A})$, it follows that $I \notin \mathscr{E}(U) \cup \mathscr{E}(\mathcal{A})$, and since $\mathcal{A}$ is not hereditarily $\mathbb{R}^{J'}$-dependent, it follows that $\mathcal{A}$ is $\mathbb{R}^I$-independent; in particular,

$I \in \mathscr{I}^*(U, \mathcal{A})$. If $\pi_{J'}(b)$ is an isolated point of $V(f_1, \ldots, f_n) \cap \Bbbk^I$ (which is e.g. the case if $I = J' = \emptyset$), then case (2) holds with $a = \pi_{J'}(b)$. Otherwise picking a branch at $\pi_{J'}(b)$ of a curve contained in $V(f_1, \ldots, f_n) \cap \Bbbk^I$ and applying lemmas IX.17 and IX.28 shows that case (1c) holds. □

Claim X.30. *Assume case* (2) *of claim* X.29 *holds. Then there is* $(g_1, \ldots, g_n) \in \mathcal{L}(\mathcal{A})$ *such that* $g_j|_{\Bbbk^I} = f_j|_{\Bbbk^I}$ *for each* j, *and* a *is an isolated point of* $V(g_1, \ldots, g_n) \subset \Bbbk^n$.

PROOF. Fix $\tilde{I} \supsetneq I$. Since $|T^I_{\mathcal{A}}| = |I|$ and $I \notin \mathscr{E}(\mathcal{A})$, it follows that $T^{\tilde{I}}_{\mathcal{A}} \setminus T^I_{\mathcal{A}}$ contains at least $|\tilde{I}| - |I|$ elements. For each weighted order ν on $\Bbbk[x_i : i \in \tilde{I}]$ centered at $(\Bbbk^*)^I$, choosing generically coefficients of f_j, $j \in T^{\tilde{I}}_{\mathcal{A}} \setminus T^I_{\mathcal{A}}$, it can be ensured that $\text{In}_{\mathcal{A}^{\tilde{I}}_j, \nu}(f_j)$, $j \in T^{\tilde{I}}_{\mathcal{A}} \setminus T^I_{\mathcal{A}}$, have no common zero $\tilde{a}$ on $(\Bbbk^*)^n$ such that $\pi_I(\tilde{a}) = a$. The claim now follows due to lemma IX.17. □

At first consider case (2) of claim X.29. Pick $(g_1, \ldots, g_n)$ as in claim X.30. Apply theorem IV.32 to $X = U$ and $h_j = (1-t)f_j + tg_j$, $j = 1, \ldots, n$. It is straightforward to see that in this case $\{a\} \times \Bbbk$ is an irreducible component of the curve C from theorem IV.32 so that

$$[f_1, \ldots, f_n]^{iso}_U < [h_1|_{t=\epsilon}, \ldots, h_n|_{t=\epsilon}]^{iso}_U$$

for generic $\epsilon \in \Bbbk$. It follows that $[f_1, \ldots, f_n]^{iso}_U < [\mathcal{A}_1, \ldots, \mathcal{A}_n]^{iso}_U$, as required. Now assume there are I, ν and a as in case (1) of claim X.29. We may assume without loss of generality that $I = T^I_{\mathcal{A}} = \{1, \ldots, k\}$ for some k, $1 \leq k \leq n$.

Claim X.31. *There is* $(g_1, \ldots, g_n) \in \mathcal{L}(\mathcal{A})$ *such that*

(1) $\text{In}_{\mathcal{A}^I_j, \nu}(g_j|_{\Bbbk^I})(a) \neq 0$ *for each* $j = 1, \ldots, k$.
(2) there is a common zero b *of* $g_1, \ldots, g_k$ *on* $(\Bbbk^*)^I \cap U$ *such that*
  (a) b *is an isolated point of* $V(g_1, \ldots, g_n) \subset \Bbbk^n$, *and*
  (b) $f_j(b) \neq 0$ *for each* $j = 1, \ldots, k$.

PROOF. Let $\mathcal{L}'(\mathcal{A})$ be the collection of all $(g_1, \ldots, g_n) \in \mathcal{L}(\mathcal{A})$ such that

(i) $g_1|_{\Bbbk^J}, \ldots, g_n|_{\Bbbk^J}$ are $\mathcal{A}^J$-non-degenerate (in the sense of definition VII.6) for all $J \supset I$, and
(ii) there is no common zero of $g_1, \ldots, g_k$ on $(\Bbbk^*)^I \setminus U$.

Since $I \notin \mathscr{E}(U) \cup \mathscr{E}(\mathcal{A})$, theorem VII.15 and lemma IX.26 imply that $\mathcal{L}'(\mathcal{A})$ contains a nonempty Zariski open subset of $\mathcal{L}(\mathcal{A})$. It follows that $\mathcal{L}'_a(\mathcal{A}) := \{(g_1, \ldots, g_n) \in \mathcal{L}'(\mathcal{A}) : \text{In}_{\mathcal{A}^I_j, \nu}(g_j|_{\Bbbk^I})(a) \neq 0$ for each $j = 1, \ldots, k\}$ also contains a nonempty Zariski open subset of $\mathcal{L}(\mathcal{A})$. Since $\mathcal{A}$ is $\mathbb{R}^I$-independent, it follows that the (k-dimensional) mixed volume of $\text{conv}(\mathcal{A}_j) \cap \mathbb{R}^I$, $j = 1, \ldots, k$, is non-zero. Due to (ii), the arguments of claim VII.20 then imply that we can find $(g_1, \ldots, g_n) \in \mathcal{L}'_a(\mathcal{A})$ and a common zero b of $g_1, \ldots, g_n$ on $(\Bbbk^*)^I \cap U$ such that $f_j(b) \neq 0$ for each $j = 1, \ldots, k$. Property (i) together with lemma IX.17 then imply that b must be an isolated zero of $g_1, \ldots, g_n$ on U. □

Now we follow the process from section VII.4.2.2. Fix integers $\nu'_j > \nu_j := \nu(x_j)$. Let C be the rational curve on $\mathbb{k}^n$ parametrized by $c(t) := (c_1(t), \dots, c_n(t)) : \mathbb{k} \to \mathbb{k}^n$ given by

$$c_j(t) := \begin{cases} a_j t^{\nu_j} + (b_j - a_j)t^{\nu'_j} & \text{if } 1 \le j \le k, \\ 0 & \text{otherwise.} \end{cases}$$

Let $m_j := \min_{\mathcal{A}_j^I}(\nu)$, $1 \le j \le k$. Define

$$h_j := \begin{cases} t^{-m_j} f_j(c(t)) g_j - t^{-m_j} g_j(c(t)) f_j & \text{if } 1 \le j \le k, \\ (1-t) f_j + t g_j & \text{otherwise.} \end{cases}$$

Assertion (1) of claim X.31 implies that each $h_j|_{t=0}$ is a non-zero constant times f_j, and assertion (2) of claim X.31 implies that each $h_j|_{t=1}$ is a non-zero constant times g_j and $c(1) = b$ is an isolated zero of $g_1, \dots, g_n$ on U. The assumptions of case (1) of claim X.29 implies that the center of C at $t = 0$ is either out of U, or it is a non-isolated zero of $f_1, \dots, f_n$. Since each h_j vanishes on the curve $C' := \{(c(t), t) : t \in \mathbb{k}\} \subset \mathbb{k}^{n+1}$, assertion (5) of theorem IV.32 implies that $[f_1, \dots, f_n]_U^{iso} < [h_1|_{t=\epsilon}, \dots, h_n|_{t=\epsilon}]_{(\mathbb{k}^*)^n}^{iso}$ for generic $\epsilon \in \mathbb{k}$. It follows that $[f_1, \dots, f_n]_U^{iso} < [\mathcal{A}_1, \dots, \mathcal{A}_n]_U^{iso}$, as required.

7.3.3. *Proof that $\mathcal{N}(U, \mathcal{A})$ is a nonempty Zariski open subset of $\mathcal{L}(\mathcal{A})$.* As in section VII.4.3 we call $\mathcal{B} = (\mathcal{B}_1, \dots, \mathcal{B}_n)$ a *face* of $\mathcal{A}$ and write $\mathcal{B} \preceq \mathcal{A}$, if there is $\nu \in (\mathbb{R}^n)^*$ such that $\mathcal{B}_j = \operatorname{In}_\nu(\mathcal{A}_j)$ for each j; in that case we also write $\mathcal{B} := \operatorname{In}_\nu(\mathcal{A})$ and we say that $\mathcal{B}$ is *centered at infinity* (respectively, *centered at* $(\mathbb{k}^*)^I$ for some $I \subseteq [n]$) if ν is centered at infinity (respectively, centered at $(\mathbb{k}^*)^I$). Moreover, if g is a polynomial supported at $\mathcal{A}_j$, we write $g_{\mathcal{B}_j}$ for $\operatorname{In}_{\mathcal{A}_j,\nu}(g)$. Consider the systems of polynomials that violate either property (b) or property (c) of $(U, \mathcal{A})$-non-degeneracy: pick $I \subseteq [n]$ and a weighted order ν on $\mathbb{k}[x_i : i \in I]$. Let $\mathcal{B} := \operatorname{In}_\nu(\mathcal{A}^I)$ and $\mathcal{D}_{\mathcal{B}}^I$ be the set of all $(g_1, \dots, g_n) \in \mathcal{L}(\mathcal{A})$ such that there is a common root of $(g_j|_{\mathbb{k}^I})_{\mathcal{B}_j}$, $j = 1, \dots, n$, on $(\mathbb{k}^*)^n$. If $(f_1, \dots, f_n) \in \mathcal{L}(\mathcal{A})$ is in the closure of $\mathcal{D}_{\mathcal{B}}^I$, then claim VII.22 implies that $(f_1, \dots, f_n) \in \mathcal{D}_{\mathcal{B}'}^I$ for some $\mathcal{B}' \preceq \mathcal{B}$. Note that

(i) If $\mathcal{B}$ is centered at infinity, then $\mathcal{B}'$ is also centered at infinity.
(ii) If $\mathcal{B}$ is centered at $(\mathbb{k}^*)^{I'}$ for some $I' \in \mathscr{E}(U) \cup \mathscr{E}(\mathcal{A})$, then $\mathcal{B}'$ is also centered at $(\mathbb{k}^*)^{I''}$ for some $I'' \in \mathscr{E}(U) \cup \mathscr{E}(\mathcal{A})$.
(iii) If $\mathcal{B}$ is centered at $(\mathbb{k}^*)^{I'}$ for some $I' \subseteq [n]$, then for each $j = 1, \dots, n$,

$$f_j|_{\mathbb{k}^{I'}} = \begin{cases} f_{j,\mathcal{B}_j} & \text{if } \mathcal{B}_j \cap \mathbb{R}^{I'} \neq \emptyset, \\ 0 & \text{otherwise.} \end{cases}$$

It follows that

(1) if $I' \in \mathscr{D}(U, \mathcal{A})$ and $\mathcal{B}'$ is centered at $(\mathbb{k}^*)^{I''}$ for some $I'' \in \mathscr{I}(U, \mathcal{A})$, then $(f_1, \dots, f_n)$ violate property (c) of $(U, \mathcal{A})$-non-degeneracy with I, I' replaced, respectively, by I', I'';

(2) if $I' \in \mathscr{I}(U, \mathcal{A})$ and $\mathcal{B}'$ is centered at infinity, then $(f_1, \ldots, f_n)$ violate property (b) of $(U, \mathcal{A})$-non-degeneracy with I, I' replaced, respectively, by I', I'';

(3) if $I' \in \mathscr{I}(U, \mathcal{A})$ and $\mathcal{B}'$ is centered at $(\mathbb{k}^*)^{I''}$ for some $I'' \in \mathscr{D}(U, \mathcal{A})$, then $(f_1, \ldots, f_n)$ violate property (b) of $(U, \mathcal{A})$-non-degeneracy with I, I' replaced, respectively, by I', I'';

It follows from these observations that the set of systems which violate at least one of the properties (b) and (c) of $(U, \mathcal{A})$-non-degeneracy is Zariski closed in $\mathcal{L}(\mathcal{A})$. Now we tackle property (a). Pick $I \in \mathscr{I}(U, \mathcal{A})$ and let $\mathcal{D}_U^I$ be the set of all $(g_1, \ldots, g_n) \in \mathcal{L}(\mathcal{A})$ such that there is a common root of $g_1, \ldots, g_n$ on $(\mathbb{k}^*)^I \setminus U$. If $(f_1, \ldots, f_n) \in \mathcal{L}(\mathcal{A})$ is in the closure of $\mathcal{D}_U^I$, then the arguments from the proof of claim VII.22 imply that there is a weighted order ν on $\mathbb{k}[x_i : i \in I]$ and a common zero a of $\operatorname{In}_{\mathcal{A}_j^I, \nu}(f_j|_{\mathbb{k}^I})$, $j = 1, \ldots, n$, on $(\mathbb{k}^*)^n$, such that

(iv) either ν is centered at infinity, in which case $f_1, \ldots, f_n$ violate property (b) of $(U, \mathcal{A})$-non-degeneracy,

(v) or ν is centered at $(\mathbb{k}^*)^{I'}$ for some $I' \subseteq I$, and $\pi_{I'}(a) \in (\mathbb{k}^*)^{I'} \setminus U$; in this case $f_1, \ldots, f_n$ violate property (a) (with I replaced by I') of $(U, \mathcal{A})$-non-degeneracy if $I' \in \mathscr{I}(U, \mathcal{A})$, and property (b) of $(U, \mathcal{A})$-non-degeneracy if $I' \in \mathscr{D}(U, \mathcal{A})$.

It follows that the set of $(U, \mathcal{A})$-degenerate systems is Zariski closed in $\mathcal{L}(\mathcal{A})$, as required. □

8. Weighted Bézout theorem: general version

8.1. Weighted Bézout theorem II: all $\omega(f_j) \geq 0$. For weights $\omega = (\omega_1, \ldots, \omega_n)$ to be applicable in weighted Bézout theorem (theorem VIII.2), each ω_i has to be positive. In this section we replace this condition by a weaker one—that $\omega(f_j)$ has to be nonnegative for each j. This opens up a new possibility: if some ω_i is nonpositive, then the set of polynomials f with $\omega(f)$ bounded above by a given integer will be an infinite-dimensional vector space over $\mathbb{k}$, and the number of (isolated) solutions of n such polynomials can be arbitrarily large. Therefore to estimate number of solutions one has to bound the degree in each x_i such that $\omega_i \leq 0$. It is natural then to consider, given an integer d, and a nonnegative integer m_i for each i such that $\omega_i \leq 0$, the set of polynomials supported at the polytope

$$\mathcal{P}(\omega, d, \vec{m}) := \{\alpha \in \mathbb{R}^n : \langle \omega, \alpha \rangle \leq d,\ \alpha_i \geq 0,\ i = 1, \ldots, n,\ \alpha_i \leq m_i,\ i \in I_0 \cup I_-\}$$

where $I_- := \{i : \omega_i < 0\}$ and $I_0 := \{i : \omega_i = 0\}$ (see fig. 7). The reason for our restriction to the case of nonnegative $\omega(f_j)$ is the following observation:

Proposition X.32. *Let $d_1, \ldots, d_k \in \mathbb{Z}$ and $m_{j,i} \in \mathbb{Z}_{\geq 0}$, $j = 1, \ldots, k$, $i \in I_0 \cup I_-$. If $d_j \geq 0$ for each j, then*

$$\sum_j \mathcal{P}(\omega, d_j, \vec{m}_j) = \mathcal{P}(\omega, \sum_j d_j, \sum_j \vec{m}_j) \tag{125}$$

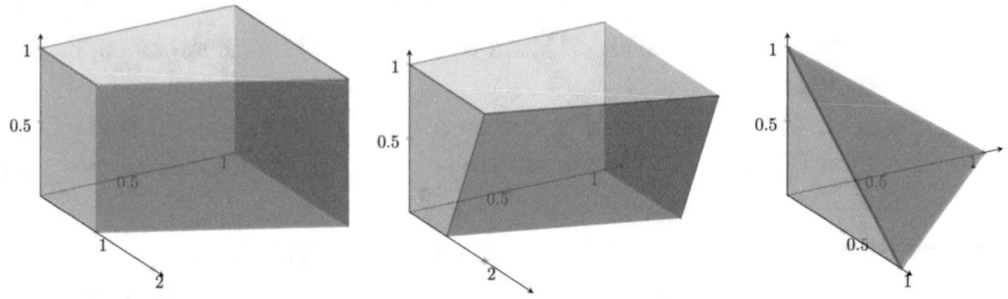

(A) $\omega = (1, -1, 0)$, $d = 1, m_2 = m_3 = 1$

(B) $\omega = (1, -1, -1)$, $d = 1, m_2 = m_3 = 1$

(C) $\omega = (1, 1, 1)$, $d = 1$

FIGURE 7. $P(\omega, d, \vec{m})$ for different $\omega, d, \vec{m}$.

PROOF. $\mathcal{P}(\omega, d_j, \vec{m}_j)$ is the product of an $(n - |I_0|)$-dimensional polytope of the same form with the I_0-dimensional box formed by the product over all $i \in I_0$ of the closed intervals from 0 to $m_{j,i}$ on x_i-axis. It is then straightforward to see that to prove (125) it suffices to prove it under the additional condition that $I_0 = \emptyset$. So assume $I_0 = \emptyset$. Let $m := \max_{i,j} m_{j,i}$ and define

$$x'_i := \begin{cases} x_i & \text{if } i \notin I_- \\ m - x_i & \text{if } i \in I_- \end{cases}$$

$$\omega'_i := |\omega_i|,\ i = 1, \ldots, n.$$

It is straightforward to check that in $(x'_1, \ldots, x'_n)$-coordinates, in the notation of exercise V.44, up to a reordering of the x_i if necessary,

$$\mathcal{P}(\omega, d_j, \vec{m}_j) = \mathcal{Q}(\vec{\omega}', d'_j, \vec{m}_j) + (m'_{j,1}, \ldots, m'_{j,n}) \tag{126}$$

where

$$d'_j := d_j + \sum_{i \in I_-} |\omega_i| m_i, \text{ and}$$

$$m'_{j,i} := \begin{cases} 0 & \text{if } i \notin I_- \\ m - m_{j,i} & \text{if } i \in I_- \end{cases}$$

It then follows from exercise V.44 that

$$\sum_j \mathcal{P}(\omega, d_j, \vec{m}_j) = \mathcal{Q}(\vec{\omega}', \sum_j d'_j, \sum_j \vec{m}_j) + \sum_j (m'_{j,1}, \ldots, m'_{j,n}) = \mathcal{P}(\omega, \sum_j d_j, \sum_j \vec{m}_j)$$

as required. □

Remark X.33. (125) may fail to hold if $d_j < 0$ for some j. Indeed, it follows from exercise V.44 and (126) that with $\omega = (-1, -1, 1)$, $m_{1,1} = m_{1,2} = 1$ and $m_{2,1} = m_{2,2} = 3$, $\mathcal{P}(\omega, 1, \vec{m}_1) + \mathcal{P}(\omega, -3, \vec{m}_1) \subsetneq \mathcal{P}(\omega, -2, \vec{m}_1 + \vec{m}_2)$.

We will now compute the mixed volume of $\mathcal{P}(\omega, d_j, \vec{m}_j)$, $j = 1, \ldots, n$. Exercise V.45 and (126) imply that

(127)

$$\operatorname{Vol}(\mathcal{P}(\omega, d, \vec{m})) = \frac{\prod_{i \in I_0} m_i}{(n - |I_0|)! \prod_{i \notin I_0} |\omega_i|} \sum_{\substack{I \subseteq I_- \\ d + \sum_{i \in I} |\omega_i| m_i > 0}} (-1)^{|I_-| - |I|} (d + \sum_{i \in I} |\omega_i| m_i)^{n - |I_0|}$$

In particular, if $d \geq 0$, then

$$\operatorname{Vol}(\mathcal{P}(\omega, d, \vec{m})) = \frac{\prod_{i \in I_0} m_i}{(n - |I_0|)! \prod_{i \notin I_0} |\omega_i|} \sum_{I \subseteq I_-} (-1)^{|I_-| - |I|} (d + \sum_{i \in I} |\omega_i| m_i)^{n - |I_0|}$$

Therefore proposition X.32 implies that for each $d_1, \ldots, d_n, \lambda_1, \ldots, \lambda_n \geq 0$,

$$\operatorname{Vol}(\sum_{j=1}^{n} \lambda_j \mathcal{P}(\omega, d_j, \vec{m}_j)) = \frac{\prod_{i \in I_0} (\sum_{j=1}^{n} \lambda_j m_{j,i})}{(n - |I_0|)! \prod_{i \notin I_0} |\omega_i|} \sum_{I \subseteq I_-} (-1)^{|I_-| - |I|} (\sum_{j=1}^{n} \lambda_j (d_j + \sum_{i \in I} |\omega_i| m_{j,i}))^{n - |I_0|}$$

The mixed volume of $\mathcal{P}(\omega, d_1, \vec{m}_1), \ldots, \mathcal{P}(\omega, d_n, \vec{m}_n)$ is the coefficient of $\lambda_1 \cdots \lambda_n$ in the right-hand side of the above expression. For each $J_0 \subseteq [n]$ such that $|J_0| = |I_0|$, the coefficient of $\prod_{j \in J_0} \lambda_j$ in $\prod_{i \in I_0} (\sum_{j=1}^{n} \lambda_j m_{j,i})$ is the *permanent* of the $|I_0| \times |I_0|$ matrix

$$D_{I_0, J_0} := \begin{pmatrix} m_{j_1, i_1} & \cdots & m_{j_1, i_k} \\ \vdots & & \vdots \\ m_{j_k, i_1} & \cdots & m_{j_k, i_k} \end{pmatrix} \tag{128}$$

where $k := |I_0| = |J_0|$ and $i_1, \ldots, i_k$ (respectively, $j_1, \ldots, j_k$) are elements of I_0 (respectively, J_0). (Note that $\operatorname{perm}(D_{I_0, J_0})$ does not depend on the ordering of the elements of I_0 or J_0. If $I_0 = J_0 = \emptyset$, then D_{I_0, J_0} is the empty matrix, and its permanent is by convention 1.) On the other hand, the coefficient of $\prod_{j \notin J_0} \lambda_j$ in $(\sum_{j=1}^{n} \lambda_j (d_j + \sum_{i \in I} |\omega_i| m_{j,i}))^{n-k}$ is $(n-k)! \prod_{j \notin J_0} (d_j + \sum_{i \in I} |\omega_i| m_{j,i})$. Combining all these together yields:

Proposition X.34. *Let $d_1, \ldots, d_n \in \mathbb{Z}$ and $m_{j,i} \in \mathbb{Z}_{\geq 0}$, $j = 1, \ldots, k$, $i \in I_0 \cup I_-$. If $d_j \geq 0$ for each j, then the mixed volume of $\mathcal{P}(\omega, d_1, \vec{m}_1), \ldots, \mathcal{P}(\omega, d_n, \vec{m}_n)$ is*

$$\sum_{\substack{J_0 \subseteq [n] \\ |J_0| = |I_0|}} \operatorname{perm}(D_{I_0, J_0}) \sum_{I \subseteq I_-} (-1)^{|I_-| - |I|} \frac{\prod_{j \in [n] \setminus J_0} (d_j + \sum_{i \in I_-} |\omega_i| m_{j,i})}{\prod_{i \in [n] \setminus I_0} |\omega_i|}$$

The following result describes the faces of $\mathcal{P}(\omega, d, \vec{m})$. Its proof is left as an exercise. Let $\eta \in (\mathbb{R}^n)^*$ with coordinates $(\eta_1, \ldots, \eta_n)$ with respect to the basis dual to the standard basis of $\mathbb{R}^n$.

Proposition X.35. *Define $M := \sup\{\eta_i / \omega_i : \omega_i > 0\}$. Assume $d \geq 0$. Then*

(1) If $M \leq 0$ (which is the case if in particular $\omega_i \leq 0$ for each i), then $\mathrm{ld}_\eta(\mathcal{P}(\omega, d, \vec{m}))$ is the set of all $(\alpha_1, \ldots, \alpha_n) \in \mathcal{P}$ such that

$$\alpha_i = \begin{cases} 0 & \text{if } \eta_i < 0, \\ m_i & \text{if } \eta_i > 0. \end{cases}$$

(2) If $M > 0$, then $\mathrm{ld}_\eta(\mathcal{P}(\omega, d, \vec{m}))$ is the set of all $(\alpha_1, \ldots, \alpha_n) \in \mathcal{P}$ such that $\sum_i \alpha_i \omega_i = d$ and

$$\alpha_i = \begin{cases} 0 & \text{if } (\omega_i = 0,\ \eta_i < 0) \text{ or } (\omega_i > 0,\ \frac{\eta_i}{\omega_i} < M) \text{ or } (\omega_i < 0,\ \frac{\eta_i}{\omega_i} > M), \\ m_i & \text{if } (\omega_i = 0,\ \eta_i > 0) \text{ or } (\omega_i < 0,\ \frac{\eta_i}{\omega_i} < M). \end{cases} \quad \square$$

We are now ready to prove the second version of the weighted Bézout theorem. Let f_j be polynomials supported at $\mathcal{P}(\omega, d_j, \vec{m}_j)$, $j = 1, \ldots, n$. For each $I \subseteq I_0 \cup I_-$, define

$$\mathcal{L}_{\vec{m}_j, I} := \{\alpha = (\alpha_1, \ldots, \alpha_n) : \alpha_i = m_{j,i} \text{ for each } i \in I\}$$

$$\mathcal{L}_{\omega, d_j, \vec{m}_j, I} := \{\alpha \in \mathcal{L}_{\vec{m}_j, I} : \sum_{i=1}^{n} \omega_i \alpha_i = d_j\}$$

Denote the coefficient of x^α in f_j by $c_{j\alpha}$, i.e. $f_j = \sum_\alpha c_{j\alpha} x^\alpha$. Write

$$\mathrm{ld}_{\vec{m}_j, I}(f_j) := \sum_{\alpha \in \mathcal{L}_{\vec{m}_j, I}} c_{j\alpha} x^\alpha,$$

$$\mathrm{ld}_{\omega, d_j, \vec{m}_j, I}(f_j) := \sum_{\alpha \in \mathcal{L}_{\omega, d_j, \vec{m}_j, I}} c_{j\alpha} x^\alpha$$

THEOREM X.36. (Weighted Bézout theorem II). *Given polynomials $f_1, \ldots, f_n$ in $(x_1, \ldots, x_n)$ and any weighted degree ω on $\Bbbk[x_1, \ldots, x_n]$, the number N of isolated solutions of $f_1, \ldots, f_n$ on $\Bbbk^n$ satisfies:*

(129)

$$N \leq \sum_{\substack{J_0 \subseteq [n] \\ |J_0| = |I_0|}} \mathrm{perm}(D_{I_0, J_0}) \sum_{I \subseteq I_-} (-1)^{|I_-| - |I|} \frac{\prod_{j \in [n] \setminus J_0} \left(\max\{\omega(f_j), 0\} + \sum_{i \in I_-} |\omega_i| \deg_{x_i}(f_j) \right)}{\prod_{i \in [n] \setminus I_0} |\omega_i|}$$

where D_{I_0, J_0} is defined as in (128) with $m_{j,i} := \deg_{x_i}(f_j)$. More generally, if $d_1, \ldots, d_n$ and $m_{j,i}$ are nonnegative *integers such that each $d_j \geq \omega(f_j)$ and $m_{j,i} \geq \deg_{x_i}(f_j)$, then*

(130)

$$N \leq \sum_{\substack{J_0 \subseteq [n] \\ |J_0| = |I_0|}} \mathrm{perm}(D_{I_0, J_0}) \sum_{I \subseteq I_-} (-1)^{|I_-| - |I|} \frac{\prod_{j \in [n] \setminus J_0} \left(d_j + \sum_{i \in I_-} |\omega_i| m_{j,i} \right)}{\prod_{i \in [n] \setminus I_0} |\omega_i|}$$

If the right-hand side of (130) is non-zero, i.e. $\mathcal{P}(\omega, d_1, \vec{m}_1), \ldots, \mathcal{P}(\omega, d_n, \vec{m}_n)$ are independent (which is the case e.g. if each d_j and each $m_{j,i}$ is positive), then (130) holds with an equality if and only if both of the following are true:

(1) *For each* nonempty *subset I of $I_0 \cup I_-$, $\operatorname{ld}_{\vec{m}_j,I}(f_j)$, $j = 1,\dots,n$, have no common zero on $\mathbb{k}^n \setminus \bigcup_{i\in I} V(x_i)$.*
(2) *For each (possibly empty) subset I of $I_0 \cup I_-$, $\operatorname{ld}_{\omega,d_j,\vec{m}_j,I}(f_j)$, $j = 1,\dots,n$, have no common zero on $\mathbb{k}^n \setminus \big(\cup_{i\in I} V(x_i)) \bigcup (\cap_{i\in I_+} V(x_i))\big)$ (where $I_+ := \{i : \omega_i > 0\}$).*

PROOF. Inequalities (129) and (130) follow directly from theorem X.4 and proposition X.34. To find the non-degeneracy conditions we apply theorem X.22 and proposition X.23 with $\mathcal{A} := (\mathcal{P}(\omega, d_1, \vec{m}_1) \cap \mathbb{Z}^n, \dots, \mathcal{P}(\omega, d_n, \vec{m}_n) \cap \mathbb{Z}^n)$ and $U := \mathbb{k}^n$. It is straightforward to check that $\mathscr{E}(\mathcal{A}) = \mathscr{E}(U) = \emptyset$, and if $\mathcal{P}(\omega, d_j, \vec{m}_j)$, $j = 1,\dots,n$, are independent then $\mathscr{D}(U, \mathcal{A}) = \emptyset$ and $\mathscr{I}^*(U, \mathcal{A}) = [n]$. Therefore theorem X.22 and proposition X.23 imply that if $\mathcal{P}(\omega, d_j, \vec{m}_j)$, $j = 1,\dots,n$, are independent (which due to theorem VII.33 and proposition X.34 is equivalent to the right-hand side of (130) being non-zero), then (130) holds with an equality if and only if the following holds:

(131) for each weighted order ν centered at infinity on $\mathbb{k}[x_1,\dots,x_n]$, there is no common zero of $\operatorname{In}_{\mathcal{A}_1,\nu}(f_1),\dots,\operatorname{In}_{\mathcal{A}_n,\nu}(f_n)$ on $(\mathbb{k}^*)^n$.

Pick a weighted order ν centered at infinity on $\mathbb{k}[x_1,\dots,x_n]$. Now apply proposition X.35 to $\eta := -\nu$. If $M := \sup\{\eta_i/\omega_i : \omega_i > 0\} \leq 0$, then $I := \{i \in I_0 \cup I_- : \eta_i > 0\}$ is nonempty and proposition X.35 implies that

$$\operatorname{In}_{\mathcal{A}_j,\nu}(f_j) = \operatorname{ld}_{\vec{m}_j,I}(f_j)|_{\{x_{i'}=0:i'\in I'\}}$$

where $I' := \{i \in [n] : \eta_i < 0\}$. Therefore (131) is equivalent to the condition that $\operatorname{ld}_{\vec{m}_j,I}(f_j)$, $j = 1,\dots,n$, have no common zero on $\Big(\mathbb{k}^n \setminus \bigcup_{i\in[n]\setminus I'} V(x_i)\Big) \cap \bigcap_{i'\in I'} V(x_{i'})$. Since it is possible for I' to be any subset of $[n] \setminus I$, taking into account all such possibilities leads to condition (1). Now consider the case that $M > 0$. In this case define $I := \{i \in I_0 : \eta_i > 0\} \cup \{i \in I_- : \eta_i/\omega_i < M\}$ and $I' := \{i \in I_0 : \eta_i < 0\} \cup \{i \in I_- : \eta_i/\omega_i > M\} \cup \{i \in I_+ : \eta_i/\omega_i < M\}$. Proposition X.35 implies that

$$\operatorname{In}_{\mathcal{A}_j,\nu}(f_j) = \operatorname{ld}_{\omega,d_j,\vec{m}_j,I}(f_j)|_{\{x_{i'}=0:i'\in I'\}}$$

It then follows as in the preceding case that to satisfy (131) for all choices of I' is equivalent to condition (2), and this completes the proof. □

Remark X.37. One does not really need the main results of this chapter to establish the bounds (129) and (130) of theorem X.36—these can be established in the same way as the proof of the classical weighted Bézout bound from section VIII.3 once the mixed volume computation from proposition X.34 is available. However, we used theorem X.22 and proposition X.23 in an essential way to establish the nondegeneracy conditions of theorem X.36.

8.2. Weighted Bézout theorem III: the general case. In this section we consider weighted Bézout theorem without any restriction on the ω_i or $\omega(f_j)$. In this generality we do not know of any compact expressions for either the mixed volume or the faces of $\mathcal{P}(\omega, d_j, \vec{m}_j)$. Therefore our version of general weighted Bézout theorem below is little more than direct application of the extension of Bernstein's theorem to the affine space.

THEOREM X.38. (Weighted Bézout theorem III). *Let $d_j \in \mathbb{Z}$ and $m_{j,i} \in \mathbb{Z}_{\geq 0}$ be such that $d_j \geq \omega(f_j)$ for each j and each $m_{j,i} \geq \deg_{x_i}(f_j)$. Given $I \subseteq I_-$, write*

$\mathcal{B}(I) := \{j : \sum_{i \in I_- \setminus I} m_{j,i}\omega_i > d_j\}$. Let $\mathscr{G} := \{I \subseteq I_- : |\mathcal{B}(\tilde{I})| \leq |\tilde{I}|$ for each $\tilde{I} \subseteq I\}$. Then the number N of isolated solutions of $f_1, \ldots, f_n$ on $\Bbbk^n$ is

$$N \leq \sum_{I \in \mathscr{G}, |\mathcal{B}(I)| = |I|} \mathrm{MV}_{n-k}(\mathcal{P}_{j'_1} \cap \mathbb{R}^{\mathcal{I}'}, \ldots, \mathcal{P}_{j'_{n-k}} \cap \mathbb{R}^{\mathcal{I}'}) \times [\pi_I(\mathcal{P}_{j_1}), \ldots, \pi_I(\mathcal{P}_{j_k})]_0 \tag{132}$$

where

- *$\mathcal{I}' := [n] \setminus I$,*
- *$\mathcal{P}_j$ are short for $\mathcal{P}(\omega, d_j, \vec{m}_j)$,*
- *$j_1, \ldots, j_k$ (respectively, $j'_1, \ldots, j'_{n-k}$) are elements of $\mathcal{B}(I)$ (respectively, $\{1, \ldots, n\} \setminus \mathcal{B}(I)$), and*
- *$\mathrm{MV}_{n-k}(\cdot, \ldots, \cdot)$ is the $(n-k)$-dimensional mixed volume.*

The bound in (132) holds with an equality if and only if $f_1, \ldots, f_n$ are $(\Bbbk^n, \mathcal{A})$-nondegenerate, where $\mathcal{A} := (\mathcal{P}_1 \cap \mathbb{Z}^n, \ldots, \mathcal{P}_n \cap \mathbb{Z}^n)$.

PROOF. Follows immediately from theorems X.4 and X.22 and proposition X.23. □

9. Weighted multi-homogeneous Bézout theorem: general version

In this section we generalize the weighted multi-homogeneous version of Bézout's theorem (theorem VIII.8) by replacing the weighted degrees with positive weights by weighted degrees from theorem X.36. As in the setting of theorem VIII.8, let $\mathscr{I} := (I_1, \ldots, I_s)$ be an ordered partition of $[n] := \{1, \ldots, n\}$, and for each $j = 1, \ldots, s$, let ω_j be a weighted degree on $\Bbbk[x_k : k \in I_j]$ with weights $\omega_{j,k}$ for x_k, $k \in I_j$. Let $I_{j,+}$ (respectively, $I_{j,0}, I_{j,-}$) be the set of all $k \in I_j$ such that $\omega_{j,k} > 0$ (respectively, $\omega_{j,k} = 0$, $\omega_{j,k} < 0$). Given nonnegative integers $d_{i,j} \geq \omega_j(f_i)$ and $m_{i,j,k} \geq \deg_{x_k}(f_i)$ for each $k \in I_{j,0} \cup I_{j,-}$, we consider the polytope

$$\mathcal{P}_i := \prod_{j=1}^{s} \mathcal{P}(\omega_j, d_{i,j}, \vec{m}_{i,j})$$

Let $n_j := |I_j|$, $j = 1, \ldots, s$. The mixed volume of $\mathcal{P}_1, \ldots, \mathcal{P}_n$ is the coefficient of $\lambda_1 \cdots \lambda_n$ in the polynomial

$$\begin{aligned}\mathrm{Vol}_n(\sum_{i=1}^n \lambda_i \mathcal{P}_i) &= \mathrm{Vol}_n(\sum_{i=1}^n \lambda_i \prod_{j=1}^s \mathcal{P}(\omega_j, d_{i,j}, \vec{m}_{i,j})) = \mathrm{Vol}_n(\prod_{j=1}^s (\sum_{i=1}^n \lambda_i \mathcal{P}(\omega_j, d_{i,j}, \vec{m}_{i,j})))\\ &= \mathrm{Vol}_n(\prod_{j=1}^s \mathcal{P}(\omega_j, \sum_{i=1}^n \lambda_i d_{i,j}, \sum_{i=1}^n \lambda_i \vec{m}_{i,j})) = \prod_{j=1}^s \mathrm{Vol}_{n_j}(\mathcal{P}(\omega_j, \sum_{i=1}^n \lambda_i d_{i,j}, \sum_{i=1}^n \lambda_i \vec{m}_{i,j}))\end{aligned}$$

where the third equality follows from proposition X.32. After a refinement of $\mathscr{I}$ if necessary, we may, and will, assume that $\omega_{j,k} \neq 0$ for each j, k, i.e. $I_j = I_{j,+} \cup I_{j,-}$ for each j. Then (127) implies that

$$\mathrm{Vol}_n(\sum_{i=1}^n \lambda_i \mathcal{P}_i) = \prod_{j=1}^s \left(\frac{1}{n_j! \prod_{k \in I_j} |\omega_{j,k}|} \sum_{I \subseteq I_{j,-}} (-1)^{|I_{j,-}| - |I|} (\sum_{i=1}^n \lambda_i (d_{i,j} + \sum_{k \in I} |\omega_{j,k}| m_{i,j,k}))^{n_j} \right)$$

Let $I_- := \bigcup_j I_{j,-}$. For each $I \subseteq I_-$, write

$$d_{I,i,j} := d_{i,j} + \sum_{k \in I \cap I_j} |\omega_{j,k}| m_{i,j,k} \tag{133}$$

Let $\Omega := \{\omega_1, \ldots, \omega_s\}$, and $D(\Omega, \vec{d}, \vec{m}, I)$ be the following $n \times n$ matrix:

$$D(\Omega, \vec{d}, \vec{m}, I) := \begin{pmatrix} \overbrace{d_{I,1,1} \cdots d_{I,1,1}}^{n_1 \text{ times}} & \cdots & \cdots & \overbrace{d_{I,1,s} \cdots d_{I,1,s}}^{n_s \text{ times}} \\ \vdots \quad\quad \vdots & & & \vdots \quad\quad \vdots \\ d_{I,n,1} \cdots d_{I,n,1} & \cdots & \cdots & d_{I,n,s} \cdots d_{I,n,s} \end{pmatrix}$$

The preceding discussion together with theorem X.4 or lemma VIII.4 imply that

$$\begin{aligned}[f_1, \ldots, f_n]^{iso}_{\mathbb{k}^n} &\leq [\mathcal{P}_1, \ldots, \mathcal{P}_n]^{iso}_{\mathbb{k}^n} = \text{MV}(\mathcal{P}_1, \ldots, \mathcal{P}_n) \\ &= \sum_{I \subseteq I_-} (-1)^{|I_-|-|I|} \frac{\text{perm}(D(\Omega, \vec{d}, \vec{m}, I))}{(\prod_j n_j!)(\prod_{j,k} \omega_{j,k})} \end{aligned} \tag{134}$$

It is straightforward to check that the bound (130) from weighted Bézout theorem II corresponds to the special case of (134) in which $n_j = 1$ for all but possibly one $j \in \{1, \ldots, s\}$. We will now see that the criterion for attainment of this bound is an amalgam of the non-degeneracy criteria of weighted multi-homogeneous Bézout theorem (theorem VIII.8) and weighted Bézout theorem II (theorem X.36). Given $I \subseteq I_-$, $J \subseteq [s]$, and $l \in [n]$, let $\mathcal{L}_{\Omega,\vec{d},\vec{m},I,J,i}$ be the set of all $\alpha = (\alpha_1, \ldots, \alpha_n) \in \mathbb{Z}^n_{\geq 0}$ such that

- $\alpha_k = m_{i,j,k}$ for each $j \in [s]$ and $k \in I_{j,-} \cap I$, and
- $\sum_{k \in I_j} \omega_{j,k}\alpha_k = d_{i,j}$ for each $j \in J$.

Given $g = \sum_\alpha c_\alpha x^\alpha \in \mathbb{k}[x_1, \ldots, x_n]$ supported at $\mathcal{P}_i$, define

$$\text{ld}_{\Omega,\vec{d},\vec{m},I,J,i}(g) := \sum_{\alpha \in \mathcal{L}_{\Omega,\vec{d},\vec{m},I,J,i}} c_\alpha x^\alpha$$

We are now ready to prove version II of the weighted multi-homogeneous Bézout theorem. But at first we recall the assumptions:

(a) $\omega_{j,k} \neq 0$ for each $j = 1, \ldots, s$, and $k \in I_j$.
(b) $d_{i,j} \geq \max\{\omega_j(f_i), 0\}$ for each $i = 1, \ldots, n$, and $j = 1, \ldots, s$.
(c) $m_{i,j,k} \geq \deg_{x_k}(f_i)$ for each $i = 1, \ldots, n$, and $j = 1, \ldots, s$, and $k \in I_{j,-}$.

THEOREM X.39. (Weighted multi-homogeneous Bézout theorem II). *Under the above assumptions the number of isolated solutions of polynomials $f_1, \ldots, f_n$ on $\mathbb{k}^n$ is bounded by* (134). *This bound is exact if and only if the following holds: for each pair I, J such that $I \subseteq I_-$, $J \subseteq [s]$, and at least one of I and J is nonempty, there is no common zero of* $\text{ld}_{\Omega,\vec{d},\vec{m},I,J,1}(f_1), \ldots, \text{ld}_{\Omega,\vec{d},\vec{m},I,J,n}(f_n)$ *on* $\mathbb{k}^n \setminus \big(\cup_{i \in I} V(x_i)) \bigcup (\cup_{j \in J} \cap_{k \in I_{j,+}} V(x_k))\big)$.

PROOF. This follows from theorems X.4 and X.22 and proposition X.23 via arguments similar to those in the proof of theorem X.36. □

10. Open problems

10.1. Systems with isolated zeroes on a given coordinate subspace. Given finite subsets $\mathcal{A}_1, \ldots, \mathcal{A}_n$ of $\mathbb{Z}^n_{\geq 0}$ and a coordinate subspace $\mathbb{k}^I$ of $\mathbb{k}^n$, it is straightforward to identify if for *generic* f_j supported at $\mathcal{A}_j$, there is any common zero of $f_1, \ldots, f_n$ on $\mathbb{k}^I$, and in case there are such points, if they are isolated in $V(f_1, \ldots,$

f_n) or not. This is the content of the next result, which is straightforward to prove from theorem VII.33 and lemma X.1.

Proposition X.40. (Zeroes of generic systems). *Let* $\mathcal{A} := (\mathcal{A}_1, \ldots, \mathcal{A}_n)$. *For all* $f = (f_1, \ldots, f_n) \in \mathcal{L}(\mathcal{A})$, *write* $V(f) := V(f_1, \ldots, f_n) \subset \mathbb{k}^n$ *and* $(V^*)^I(f) := V(f_1, \ldots, f_n) \cap (\mathbb{k}^*)^I$.

(1) The following are equivalent:
  - *(a)* $(V^*)^I(f) = \emptyset$ *for generic* $f \in \mathcal{L}(\mathcal{A})$.
  - *(b)* $\mathcal{A}$ *is* $\mathbb{R}^I$*-dependent.*

(2) The following are equivalent:
  - *(a)* $(V^*)^I(f) \neq \emptyset$ *and all points of* $(V^*)^I(f)$ *are isolated in* $V(f)$ *for generic* $f \in \mathcal{L}(\mathcal{A})$.
  - *(b)* $\mathcal{A}$ *is* $\mathbb{R}^I$*-independent and* $I \notin \mathscr{E}(\mathcal{A})$.

(3) The following are equivalent:
  - *(a)* $(V^*)^I(f) \neq \emptyset$ *and all points of* $(V^*)^I(f)$ *are non-isolated in* $V(f)$ *for generic* $f \in \mathcal{L}(\mathcal{A})$.
  - *(b)* $\mathcal{A}$ *is* $\mathbb{R}^I$*-independent and* $I \in \mathscr{E}(\mathcal{A})$. □

Problem X.41. (Existence of systems with given support and isolated zeroes on given coordinate subspace). *Characterize those* $I \subseteq [n]$ *for which there are* $f_1, \ldots, f_n$ *such that* $\mathrm{Supp}(f_j) = \mathcal{A}_j$, $j = 1, \ldots, n$, $(V^*)^I(f_1, \ldots, f_n) \neq \emptyset$ *and all points (or some points) of* $(V^*)^I(f_1, \ldots, f_n)$ *are isolated in* $V(f_1, \ldots, f_n)$.

If I is as in problem X.41, then

(a) lemma X.1 implies that $I \notin \mathscr{E}(\mathcal{A})$;
(b) theorems VII.7 and VII.33 imply that
  (1) either I is $\mathbb{R}^I$-independent and $|T^I_{\mathcal{A}}| = |I|$ (i.e. $I \in \mathscr{I}^*(U, \mathcal{A})$),
  (2) or $|T^I_{\mathcal{A}}| > |I|$ (in particular, $\mathcal{A}$ is $\mathbb{R}^I$-dependent);
(c) proposition X.21 implies that $\mathcal{A}$ is not hereditarily $\mathbb{R}^I$-dependent.

In the context of these observations problem X.41 boils down to the following problem.

Problem X.41′. *If* $I \notin \mathscr{E}(\mathcal{A})$ *and* $\mathcal{A}$ *is not hereditarily* $\mathbb{R}^I$*-dependent and* $|T^I_{\mathcal{A}}| > |I|$, *does there exist* $f_1, \ldots, f_n$ *such that* $\mathrm{Supp}(f_j) = \mathcal{A}_j$ *for each* j, $(V^*)^I(f_1, \ldots f_n) \neq \emptyset$ *and all points (or some points) of* $(V^*)^I(f_1, \ldots, f_n)$ *are isolated in* $V(f_1, \ldots, f_n)$*? If not, then characterize those* $I \subseteq [n]$ *which satisfy the hypothesis of the preceding question but fail the conclusion.*

10.2. Non-isolated zeroes and non-degeneracy. In contrast to the case of $(\mathbb{k}^*)^n$, example X.16 shows that for $f = (f_1, \ldots, f_n) \in \mathcal{L}(\mathcal{A})$, the existence of non-isolated solutions of $f_1, \ldots, f_n$ does not automatically mean that $f_1, \ldots, f_n$ are $(\mathbb{k}^n, \mathcal{A})$-non-degenerate. More precisely, part (b) of example X.16 shows that if $I \in \mathscr{D}(\mathbb{k}^n, \mathcal{A})$, then it is possible for $f_1, \ldots, f_n$ to be $(\mathbb{k}^n, \mathcal{A})$-non-degenerate even if $(V^*)^I(f)$ has non-isolated points. On the other hand, condition (b) of $(U, \mathcal{A})$-non-degeneracy implies that if $I \in \mathscr{I}(\mathbb{k}^n, \mathcal{A})$ and $(V^*)^I(f)$ has non-isolated points, then $f_1, \ldots, f_n$ are $(\mathbb{k}^n, \mathcal{A})$-degenerate. The question is if it is sufficient.

Problem X.42. *If $I \in \mathscr{D}(\Bbbk^n, \mathcal{A})$, does there exist $f = (f_1, \ldots, f_n)$ such that* $\mathrm{Supp}(f_j) = \mathcal{A}_j$ *for each j, and $(V^*)^I(f)$ is nonempty (and due to proposition X.21 necessarily positive dimensional), but $f_1, \ldots, f_n$ are $(\Bbbk^n, \mathcal{A})$-non-degenerate? If not, then characterize those $I \in \mathscr{D}(\Bbbk^n, \mathcal{A})$ for which there is no such $f \in \mathcal{L}(\mathcal{A})$.*

10.3. Simple criteria for equality of Li and Wang's bound. Since the upper bound of Li and Wang from (116) is so simple, it would be interesting to find simple criteria under which it holds with equality. Proposition X.14 gives a characterization of all such scenarios, so the question is if it can be made "more explicit" in any sense, or if there are simpler criteria (e.g. as in corollary X.15 or assertion (1) of proposition X.14) which are sufficient. One possible criterion was proposed in [RW96] and [Roj99], but that turns out to be incorrect. Indeed, both [RW96, Theorem 1] and [Roj99, Affine Point Theorem II] imply the following: if $\mathscr{E}(\mathcal{A}) = \emptyset$ and the intersection of each $\mathcal{A}_j$ with each of the n coordinate hyperplanes is nonempty, then Li and Wang's bound holds with equality. This is indeed the case for $n \leq 2$, but as the following example shows, it is false in higher dimensions.

Example X.43. Let $f_1 := ax + by + cx^2$, $f_2 := a'x + b'y + c'x^2$, $f_3 := pz^k x + q$, where $a, b, c, a', b', c', p, q$ are generic elements in $\Bbbk$ and $k \geq 1$. It is straightforward to check directly that on $\Bbbk^n$ there are precisely k solutions for $f_1 = f_2 = f_3 = 0$ and $2k$ solutions for $f_1 = f_2 = f_3 = t$ for generic $t \neq 0$ so that Li and Wang's bound fails for $\mathcal{A}_j := \mathrm{Supp}(f_j)$, $j = 1, 2, 3$. Note that both conditions (iv) and (v) from section X.4.2 hold with the weighted degree ν on $\Bbbk[x, y, z]$ corresponding to weights $x \mapsto k$, $y \mapsto k$, $z \mapsto -1$.

10.4. "Compact" formulae for general weighted and weighted multi-homogeneous versions of Bézout's theorem. There are scenarios not covered in weighted Bézout theorem II (theorem X.36) for which very similar bound exists, e.g. in the case that $|I_-| = 1$ (and no restriction that the d_j have to be nonnegative). This motivates the question: is it possible to find a formula that is more explicit than the one from weighted Bézout theorem III (theorem X.38), and which is more general than version II? That would also lead to a more general version of weighted multi-homogeneous Bézout theorem II (theorem X.39).

CHAPTER XI

Milnor number of a hypersurface at the origin

1. Introduction

The modern theory of applications of Newton polyhedra to affine Bézout problem started from A. Kushnirenko's work aimed at answering V. I. Arnold's question on *Milnor numbers* of generic singularities. In [Kou76] Kushnirenko gave a beautiful formula for a lower bound of the Milnor number at the origin in terms of volumes of the region bounded by the Newton diagram, and showed that the bound is attained in the case that the singularity is *Newton non-degenerate*. In this chapter, we show that the notion of *non-degeneracy at the origin* introduced in chapter IX can be used to derive (and generalize) Kushnirenko's result on Milnor numbers. In particular, based on non-degeneracy at the origin, we introduce a non-degeneracy criterion which generalizes Newton non-degeneracy and *inner Newton non-degeneracy*[1], the latter introduced by C. T. C. Wall [Wal99]. We show that in zero characteristic the new criterion is necessary and sufficient for the Milnor number to be the minimum, and the minimum Milnor number can be obtained by Kushnirenko's bound. In positive characteristic, this criterion is sufficient, but not necessary.

2. Milnor number

The *Milnor number* $\mu_0(f)$ of a power series f in $(x_1, \ldots, x_n)$ is the dimension over $\mathbb{k}$ of the quotient of $\mathbb{k}[[x_1, \ldots, x_n]]$ by the ideal generated by the partial derivatives of f, i.e. $\mu_0(f) = [\partial_1 f, \ldots, \partial_n f]_0$ (where $\partial_i(\cdot)$ is short for $\partial(\cdot)/\partial x_i$). It is a fundamental measure of complexity of the singularity of $V(f)$ at the origin (in the case that f is the Taylor series of a rational function, or in the case that $\mathbb{k} = \mathbb{C}$ and f is a analytic at the origin).

Proposition XI.1. *Let $f \in \mathbb{k}[x_1, \ldots, x_n]$ such that $f(0) = 0$.*

(1) $\mu_0(f) = 0$ if and only the origin is a nonsingular point of $V(f)$.
(2) If $V(f)$ has a non-isolated singularity at the origin, then $\mu_0(f) = \infty$. The converse holds if $\operatorname{char}(\mathbb{k}) = 0$.

[1]This terminology is taken from [BGM12].

P. Mondal, *How Many Zeroes?*, CMS/CAIMS Books in Mathematics 2,
https://doi.org/10.1007/978-3-030-75174-6_XI

PROOF. The first assertion is clear, so we prove the second assertion. If $V(f)$ has a non-isolated singularity at the origin, then the origin is a non-isolated point of $V(\partial_1 f, \dots, \partial_n f)$, so that $\mu_0(f) = \infty$ (proposition IV.28). Now assume $\mu_0(f) = \infty$. Then the origin is a non-isolated point of $V(\partial_1 f, \dots, \partial_n f)$ (proposition IV.28) and therefore there is an irreducible curve C containing the origin such that $C \subseteq V(\partial_1 f, \dots, \partial_n f)$ (corollary III.83). It suffices to show that $f|_C \equiv 0$ if $\operatorname{char}(\mathbb{k}) = 0$. Pick a nonsingular point $a \in C$. Then there is an isomorphism $\phi : \mathbb{k}[[t]] \cong \hat{\mathcal{O}}_{a,C}$, and $d(f \circ \phi)/dt = \sum_{j=1}^n (\partial_j f)(d(\phi_j)/dt) \equiv 0 \in \hat{\mathcal{O}}_{a,C}$. Since $\operatorname{char}(\mathbb{k}) = 0$, it follows that f is constant on C. Since $0 \in C$ and $f(0) = 0$, it follows that $C \subseteq V(f)$, as required. □

Example XI.2. The converse to assertion (XI.1) of proposition XI.1 may not be true if $p := \operatorname{char}(\mathbb{k})$ is positive; consider, e.g. the case that $n = 1$ and $f(x) = x^p$, or $n = 2$ and $f(x, y) = x^p + y^q$, where $q \geq 2$ is relatively prime to p. The latter example in particular shows that Milnor number can be infinite even for *isolated* singular points.

In the case that $\mathbb{k} = \mathbb{C}$ and the origin is an isolated singular point of $V(f)$, Milnor originally defined $\mu_0(f)$ in [Mil68, Chapter 7] in the following way: let S^{2n-1}_ϵ be the sphere of radius ϵ centered at the origin of $\mathbb{C}^n \cong \mathbb{R}^{2n}$ and $S^{2n-1} := S^{2n-1}_1$ be the unit sphere of $\mathbb{C}^n$. Given a morphism $g : \mathbb{C}^n \to \mathbb{C}^n$ such that the origin is an isolated zero of $g^{-1}(0)$, the *multiplicity* of g at the origin is the *degree of the mapping*[2] $S^{2n-1}_\epsilon \mapsto S^{2n-1}$ given by $z \mapsto g(z)/||g(z)||$ (where $|| \cdot ||$ is the Euclidean distance). Milnor defined $\mu_0(f)$ as the multiplicity at the origin of the map $z \mapsto (\partial_1 f, \dots, \partial_n f)$. Milnor showed that for all sufficiently small ϵ, if $\phi : S^{2n-1}_\epsilon \setminus V(f) \to S^1$ is the map given by $z \mapsto f(z)/||f(z)||$, then each fiber of ϕ is a smooth $(2n-2)$-dimensional real manifold with homotopy type of a "bouquet" $S^{n-1} \vee \cdots \vee S^{n-1}$ of spheres, and $\mu_0(f)$ is precisely the number of spheres in the bouquet. The fact that the multiplicity of a map $g : \mathbb{C}^n \to \mathbb{C}^n$ at the origin equals $[g_1, \dots, g_n]_0$ was left in [Mil68, Appendix B] as an exercise; a proof can be found in [AGZV85, Chapter I.5].

3. Generic Milnor number

Let $\mathcal{A}$ be a (possibly infinite) subset of $\mathbb{Z}^n_{\geq 0}$. We write $\mathcal{L}_0(\mathcal{A})$ for the set of all power series in $(x_1, \dots, x_n)$ supported at $\mathcal{A}$. For each $j = 1, \dots, n$, define

$$\partial_j \mathcal{A} := \{\alpha - e_j : \alpha \in \mathcal{A},\ \alpha - e_j \in \mathbb{Z}^n_{\geq 0},\ p \text{ does not divide } \alpha_j\} \tag{135}$$

[2]The degree of a differentiable map $\phi : M \to N$ between oriented differentiable manifolds of the same dimension, where M is compact and N is connected, is the sum of *sign* of df_x over all $x \in \phi^{-1}(y)$ for a generic $y \in N$, where df_x is the derivative map from the tangent space of M at x to the tangent space of N at y, and the sign of df_x is either 1 or -1 depending on whether df_x preserves or reverses orientation.

where e_j is the j-th standard unit vector in $\mathbb{Z}^n$ and $p := \operatorname{char}(\Bbbk)$. Note that $\partial_j \mathcal{A}$ is the support of $\partial_j g$ for *generic*[3] $g \in \mathcal{L}_0(\mathcal{A})$. Define

$$\mu_0(\mathcal{A}) := \min\{\mu_0(f) : f \in \mathcal{L}_0(\mathcal{A})\}$$

In theorem XI.3 below, we estimate $\mu_0(\mathcal{A})$ in terms of the intersection multiplicity $[\Gamma_1, \ldots, \Gamma_n]_0$ at the origin of the *Newton diagrams* Γ_j of $\partial_j \mathcal{A}$. Given $f \in \mathcal{L}_0(\mathcal{A})$, we say that f is *partially $\mathcal{A}$-non-degenerate at the origin* if the partial derivatives of f are $(\partial_1 \mathcal{A}, \ldots, \partial_n \mathcal{A})$-non-degenerate at the origin in the sense of definition IX.6, i.e. if the following property holds:

(136) for each nonempty subset I of $[n]$ and each weighted order ν centered at the origin on $\Bbbk[x_i : i \in I]$, there is no common zero of $\operatorname{In}_{\mathcal{A}_j^I, \nu}((\partial_j f)|_{\Bbbk^I}), j = 1, \ldots, n$, on $(\Bbbk^*)^n$

where $\mathcal{A}_j^I := \partial_j \mathcal{A} \cap \mathbb{R}^I$; also recall that $\operatorname{In}_{\mathcal{S},\eta}(g)$, where η is a weighted degree, $\mathcal{S} \subseteq \mathbb{R}^n$ and g is a Laurent polynomial supported at $\mathcal{S}$, is defined as follows:

$$\operatorname{In}_{\mathcal{S},\eta}(g) := \sum_{\alpha \in \operatorname{In}_\eta(\mathcal{S})} c_\alpha x^\alpha = \begin{cases} \operatorname{In}_\eta(g) & \text{if } \operatorname{Supp}(g) \cap \operatorname{In}_\eta(\mathcal{S}) \neq \emptyset, \\ 0 & \text{otherwise.} \end{cases}$$

If f is partially $\mathcal{A}$-non-degenerate at the origin for $\mathcal{A} = \operatorname{Supp}(f)$, we simply say that f is *partially non-degenerate at the origin*. In other words, f is partially non-degenerate at the origin if it satisfies the following property:

(137) for each nonempty subset I of $[n]$ and each weighted order ν centered at the origin on $\Bbbk[x_i : i \in I]$, there is no common zero of $\operatorname{In}_\nu((\partial_j f)|_{\Bbbk^I}), j = 1, \ldots, n$, on $(\Bbbk^*)^n$.

Recall that one does *not* have to check this condition for *all* nonempty subsets of $[n]$—see theorem IX.9 and remark IX.10.

THEOREM XI.3. ([Mon16]). *Let $\Gamma_j := \operatorname{ND}(\partial_j \mathcal{A})$, $j = 1, \ldots, n$. Assume $0 \notin \mathcal{A}$. Then*

(138) $$\mu_0(\mathcal{A}) \geq [\Gamma_1, \ldots, \Gamma_n]_0$$

Moreover,

(1) *The following are equivalent for all $f \in \mathcal{L}_0(\mathcal{A})$:*
   (a) *$\mu_0(f) = [\Gamma_1, \ldots, \Gamma_n]_0 < \infty$,*
   (b) *f is partially $\mathcal{A}$-non-degenerate at the origin.*
(2) *Let $\mathcal{M}_0'(\mathcal{A})$ be the set of all $f \in \mathcal{L}_0(\mathcal{A})$ which are partially $\mathcal{A}$-non-degenerate at the origin. Let $\mathcal{A}'$ be any* finite *subset of $\mathcal{A}$ such that $\partial_j \mathcal{A}' \supseteq \Gamma_j \cap \partial_j \mathcal{A}$, $j = 1, \ldots, n$. Then $\mathcal{M}_0'(\mathcal{A}')$ is a Zariski open subset of $\mathcal{L}_0(\mathcal{A}')$,*

[3]"Generic" refers to elements of a nonempty Zariski open (dense) subset of $\mathcal{L}_0(\mathcal{A})$ in the Zariski topology mentioned in remark IX.7.

and $\mathcal{M}'_0(\mathcal{A}) = \pi^{-1}(\mathcal{M}'_0(\mathcal{A}'))$, *where* $\pi : \mathcal{L}_0(\mathcal{A}) \to \mathcal{L}_0(\mathcal{A}')$ *is the natural projection.*

(3) *If* $[\Gamma_1, \ldots, \Gamma_n]_0 = \infty$, *then* $\mathcal{M}'_0(\mathcal{A}) = \emptyset$. *If* $[\Gamma_1, \ldots, \Gamma_n]_0 < \infty$, *then* $\mu_0(\mathcal{A}) = [\Gamma_1, \ldots, \Gamma_n]_0$ *if and only if* $\mathcal{M}'_0(\mathcal{A})$ *is nonempty.*

(4) *If* $\mathrm{char}(\Bbbk) = 0$ *and* $[\Gamma_1, \ldots, \Gamma_n]_0 < \infty$, *then* $\mathcal{M}'_0(\mathcal{A})$ *is nonempty and* $\mu_0(\mathcal{A}) = [\Gamma_1, \ldots, \Gamma_n]_0$.

Example XI.4. Assertion (4) of theorem XI.3 may not be true, i.e. the bound in (138) may be strict, in the case that $p := \mathrm{char}(\Bbbk) > 0$. For example, let $\mathcal{A} := \{(p+1,1),(1,p+1)\} \subset \mathbb{Z}^2_{\geq 0}$. Then $\partial_1\mathcal{A} = \{(p,1),(0,p+1)\}$ and $\partial_2\mathcal{A} = \{(p+1,0),(1,p)\}$. If $f_1 = a_{1,1}x_1^p x_2 + a_{1,2}x_2^{p+1}$ and $f_2 = a_{2,1}x_1^{p+1} + a_{2,2}x_1x_2^p$ are generic polynomials supported, respectively, at $\partial_1\mathcal{A}$ and $\partial_2\mathcal{A}$, then $[f_1, f_2]_0 = (p+1)^2$. Therefore, $[\Gamma_1, \Gamma_2]_0 = (p+1)^2$. On the other hand, if $f = ax_1^{p+1}x_2 + bx_1x_2^{p+1}$ is a generic polynomial supported at $\mathcal{A}$, then $\partial_1 f = x_2(ax_1^p + bx_2^p)$ and $\partial_2 f = x_1(ax_1^p + bx_2^p)$ so that $\mu_0(\mathcal{A}) = [x_2(ax_1^p + bx_2^p), x_1(ax_1^p + bx_2^p)]_0 = \infty > [\Gamma_1, \Gamma_2]_0$. It is straightforward to check that $\partial_1 f, \partial_2 f$ are $\mathcal{A}$-degenerate at the origin, i.e. $\mathcal{M}'_0(\mathcal{A}) = \emptyset$.

PROOF OF THEOREM XI.3. Assertions (1), (2) and (3) follow from theorem IX.8 and corollary IX.18. Therefore, it suffices to show that $\mathcal{M}'_0(\mathcal{A})$ is nonempty in the case that $\mathrm{char}(\Bbbk) = 0$. We may assume without loss of generality that $0 < \mu_0(\mathcal{A}) < \infty$. Pick any finite subset $\mathcal{A}'$ of $\mathcal{A}$ satisfying the assumptions of assertion (2). Let $I \subseteq [n]$ and ν be a weighted order on $\Bbbk[x_i : i \in I]$ centered at the origin. Denote by $\mathcal{L}^I_\nu$ the set of all $g \in \mathcal{L}_0(\mathcal{A}')$ such that

- $\mathrm{Supp}(g) = \mathcal{A}'$, and
- $V(\mathrm{In}_\nu((\partial_1 g)|_{\Bbbk^I}), \ldots, \mathrm{In}_\nu((\partial_n g)|_{\Bbbk^I})) \cap (\Bbbk^*)^n = \emptyset$

It suffices to show that $\mathcal{L}^I_\nu$ contains a nonempty Zariski open subset of $\mathcal{L}_0(\mathcal{A}')$. We may assume without loss of generality that $I = \{1, \ldots, k\}$. Take $g \in \mathcal{L}_0(\mathcal{A}')$, and express it as

$$g = g_0(x_1, \ldots, x_k) + \sum_{i=k+1}^{n} x_i g_i(x_1, \ldots, x_k) + \cdots,$$

where the omitted terms have quadratic or higher order in $(x_{k+1}, \ldots, x_n)$. Since $0 < [\Gamma_1, \ldots, \Gamma_n]_0 < \infty$, corollary IX.13 implies that $(\partial_j g)|_{\Bbbk^I} \not\equiv 0$ for $l \geq k$ values of j; denote them by $1 \leq j_1 < \cdots < j_l \leq n$. If $g_0 \equiv 0$, then each $j_i > k$, and $V(\mathrm{In}_\nu((\partial_1 g)|_{\Bbbk^I}), \ldots, \mathrm{In}_\nu((\partial_n g)|_{\Bbbk^I})) = V(\mathrm{In}_\nu(g_{j_1}), \ldots, \mathrm{In}_\nu(g_{j_l}))$. Therefore, $g \in \mathcal{L}^I_\nu$ if $g_{j_1}, \ldots, g_{j_l}$ are *BKK non-degenerate*. Since $l \geq k$, theorem VII.15 then implies that $\mathcal{L}^I_\nu$ contains a nonempty Zariski open subset of $\mathcal{L}_0(\mathcal{A}')$, as required. So assume $g_0 \not\equiv 0$. Let $h_0 := \mathrm{In}_\nu(g_0)$ and $V'(h_0) := V(\partial_1 h_0, \ldots, \partial_k h_0) \cap (\Bbbk^*)^n$. Since $V(\mathrm{In}_\nu((\partial_1 g)|_{\Bbbk^I}), \ldots, \mathrm{In}_\nu((\partial_n g)|_{\Bbbk^I})) \subseteq V'(h_0)$, it suffices to show that the set of all polynomials $h \in \mathcal{L}_0(\mathcal{A}'_0)$, where $\mathcal{A}'_0 := \mathrm{Supp}(h_0)$, such that $V'(h) = \emptyset$ contains a nonempty Zariski open subset of $\mathcal{L}_0(\mathcal{A}'_0)$. Let $Z := \{(x,h) : x \in (\Bbbk^*)^n,\ h \in \mathcal{L}_0(\mathcal{A}'_0),\ \partial_1 h(x) = \cdots = \partial_k h(x) = 0\} \subset (\Bbbk^*)^n \times \mathcal{L}_0(\mathcal{A}'_0)$, and let $\pi_1 : Z \to (\Bbbk^*)^n$ and $\pi_2 : Z \to \mathcal{L}_0(\mathcal{A}'_0)$ be the natural projections. It suffices to show that $\dim(\pi_2(Z)) < \dim \mathcal{L}_0(\mathcal{A}'_0)$. We prove this by a dimension count. Denote the elements in $\mathcal{A}'_0$ by $\alpha_i = (\alpha_{i,1}, \ldots, \alpha_{i,n})$, $i = 1, \ldots, N$, and the coefficients of x^{α_i} (in a polynomial supported at $\mathcal{A}'_0$) by a_i. Let $x \in (\mathbb{k}^*)^n$. Then $\pi_1^{-1}(x)$ is the subspace of $\mathcal{L}_0(\mathcal{A}'_0)$ defined by a system of linear equations of the form

$$\begin{pmatrix} \alpha_{1,1}x^{\alpha_1 - e_1} & \cdots & \alpha_{N,1}x^{\alpha_N - e_1} \\ \vdots & & \vdots \\ \alpha_{1,k}x^{\alpha_1 - e_k} & \cdots & \alpha_{N,k}x^{\alpha_N - e_k} \end{pmatrix} \begin{pmatrix} a_1 \\ \vdots \\ a_N \end{pmatrix} = 0 \tag{139}$$

where $e_1, \ldots, e_k$ are the standard unit vectors in $\mathbb{Z}^k$. Since $\mathbf{char}(\mathbb{k}) = \mathbf{0}$ (note: this is the only place the assumption of zero characteristic is used), the rank (as a matrix over $\mathbb{k}$) of the left-most matrix in (139) is the same as the rank (as a matrix over $\mathbb{Q}$) of

$$B := \begin{pmatrix} \alpha_{1,1} & \cdots & \alpha_{N,1} \\ \vdots & & \vdots \\ \alpha_{1,k} & \cdots & \alpha_{N,k} \end{pmatrix}$$

Since ν has positive weights, and the α_j belong to a level set (corresponding to a positive value) of ν, it follows that $\mathrm{Rank}(B) = 1 + \dim(\mathrm{NP}(h_0))$ and therefore $\dim(\pi_1^{-1}(x)) = N - 1 - \dim(\mathrm{NP}(h_0))$. Since this is independent of x, it follows that $\dim(Z) = N + n - 1 - \dim(\mathrm{NP}(h_0))$. On the other hand, if $x \in V'(h)$ for some $h \in \mathcal{L}_0(\mathcal{A}'_0)$, then since the support of each of $\partial_j h$ is contained in a translation of $\mathrm{NP}(h_0)$, it is straightforward to check that $\partial_j h(xz^\beta) = 0$ for each $j = 1, \ldots, k$, and each $z \in (\mathbb{k}^*)^n$ and each $\beta \in \mathbb{Z}^n$ which is normal (with respect to the "dot product") to $\mathrm{NP}(h_0)$. Therefore, the dimension of each fiber of π_2 is at least $n - \dim(\mathrm{NP}(h_0))$. It follows that $\dim(\pi_2(Z)) \le \dim(Z) - n + \dim(\mathrm{NP}(h_0)) = N - 1$, as required.
□

Corollary XI.5. *Let* $\Gamma_j := \mathrm{ND}(\partial_j f)$, $j = 1, \ldots, n$. *Assume* $f(0) = 0$ *and* f *is partially non-degenerate. Then the following are equivalent:*

(1) $\mu_0(f) < \infty$.
(2) $[\Gamma_1, \ldots, \Gamma_n]_0 < \infty$.
(3) $|\{j : \Gamma_j \cap \mathbb{R}^I \neq \emptyset\}| \ge |I|$ *for each* $I \subseteq [n]$.

PROOF. Combine corollary IX.13 and theorem XI.3. □

4. Classical notions of non-degeneracy

4.1. Newton non-degeneracy. For a power series f in $(x_1, \ldots, x_n)$, we write $\mathrm{j}(f)$ for the ideal generated by the partial derivatives of f. We say that f is *Newton non-degenerate* iff $\mathrm{j}(\mathrm{In}_\nu(f))$ has no zero on $(\mathbb{k}^*)^n$ for each weighted order ν on $\mathbb{k}[[x_1, \ldots, x_n]]$ centered at the origin. Newton non-degeneracy is possibly the most studied non-degeneracy property of hypersurface germs: it is a Zariski open condition (in the same sense as partial non-degeneracy at the origin) and, in the characteristic zero case, also nonempty. In this section, we discuss its relationship with partial non-degeneracy at the origin. In general, Newton non-degeneracy does *not* imply "finite determinacy," i.e. f can be Newton non-degenerate but still $V(f)$ may have a non-isolated singularity at the origin (take, e.g. $f := x_1 \cdots x_n$, $n \geq 2$). However, if the Newton diagram of f is *convenient* then Newton non-degeneracy of f implies that the origin is an isolated singularity of $V(f)$ (corollary XI.8), which can be resolved by a "toric modification." As a result, the invariants of the singularity can be computed combinatorially in terms of the diagram (see e.g. [Oka97]). We show in proposition XI.7 that for isolated singularities, Newton non-degeneracy is a special case of partial non-degeneracy at the origin. However, the following example shows that even in the case of convenient diagrams partial non-degeneracy at the origin does not imply Newton non-degeneracy.

Example XI.6. Let $f := x_1 + (x_2 + x_3)^q$, where $q \geq 2$. Then $\mathrm{ND}(f)$ is convenient and f is not Newton non-degenerate (take ν with weights $(q+1, 1, 1)$ for (x, y, z)). However, f is partially non-degenerate at the origin with $\mu_0(f) = 0$.

We now show that Newton non-degeneracy implies partial non-degeneracy at the origin. The following notation is used in its proof: let $I \subseteq I' \subseteq [n]$, $\nu \in (\mathbb{R}^I)^*$ and $\nu' \in (\mathbb{R}^{I'})^*$. We say that ν and ν' are *compatible* if the weighted order on $\Bbbk[x_i : i \in I]$ induced by ν and the weighted order on $\Bbbk[x_{i'} : i' \in I']$ induced by ν' are compatible in the sense of section IX.6.

Proposition XI.7. ([Mon16]). *Let $f \in \Bbbk[[x_1, \ldots, x_n]]$ and $\Gamma_j := \mathrm{ND}(\partial_j f)$, $j = 1, \ldots, n$. If f is Newton non-degenerate and $[\Gamma_1, \ldots, \Gamma_n]_0 < \infty$, then f is partially non-degenerate at the origin. In particular, if f is Newton non-degenerate and $\mathrm{ND}(f)$ is convenient, then f is partially non-degenerate at the origin.*

PROOF. We start with a direct proof of the second assertion since it is easier to see. Assume $\Gamma := \mathrm{ND}(f)$ is convenient and f is Newton non-degenerate. Pick a nonempty subset I of $[n]$ and a weighted order ν on $\Bbbk[x_i : i \in I]$ which is centered at the origin. Since Γ is convenient, $\Gamma \cap \mathbb{R}^I \neq \emptyset$. Therefore, we can find a weighted order ν' on $\Bbbk[x_1, \ldots, x_n]$ such that ν' is compatible with ν and $\mathrm{In}_{\nu'}(\Gamma) \subset \mathbb{R}^I$. Then $\mathrm{In}_{\nu'}(f)$ depends only on $(x_i : i \in I)$. Since f is Newton non-degenerate, it follows that $\partial_i(\mathrm{In}_{\nu'}(f))$, $i \in I$, do not have any common zero in $(\Bbbk^*)^n$. But if $i \in I$ is such that $\partial_i(\mathrm{In}_{\nu'}(f))$ is not identically zero, then $\partial_i(\mathrm{In}_{\nu'}(f)) = \partial_i(\mathrm{In}_{\nu}(f|_{\Bbbk^I})) = \mathrm{In}_{\nu}((\partial_i f)|_{\Bbbk^I})$. This implies that $\mathrm{In}_{\nu}((\partial_i f)|_{\Bbbk^I})$, $i \in I$, do not have any common zero on $(\Bbbk^*)^I$, as required for partial non-degeneracy of f at the origin.

Now we prove the first assertion. If $[\Gamma_1, \ldots, \Gamma_n]_0 = 0$, then corollary IX.13 implies that f is partially non-degenerate at the origin. So assume $0 < [\Gamma_1, \ldots, \Gamma_n]_0 < \infty$ and f is partially *degenerate* at the origin. It suffices to show that f is Newton degenerate. Pick $I \subseteq [n]$ and a *primitive* weighted order ν centered at the origin on $\Bbbk[x_i : i \in I]$ such that $\mathrm{In}_{\nu}((\partial_i f)|_{\Bbbk^I})$, $i \in I$, have a common zero $(a_1, \ldots, a_n) \in (\Bbbk^*)^n$. At first, consider the case that $f|_{\Bbbk^I} \not\equiv 0$. Then as in the convenient case pick a weighted order ν' on $\Bbbk[x_1, \ldots, x_n]$ such that ν' is compatible with ν and $\mathrm{In}_{\nu'}(f) \in \Bbbk[x_i : i \in I]$. Then for each $j = 1, \ldots, n$, $\partial_j(\mathrm{In}_{\nu'}(f))$

is either $\mathrm{In}_\nu((\partial_j f)|_{\Bbbk^I})$ or is identically zero. It follows that $(a_1, \dots, a_n)$ is a common zero of $\mathrm{j}(\mathrm{In}_{\nu'}(f))$ on $(\Bbbk^*)^n$, so that f is Newton degenerate, as required. Now assume that $f|_{\Bbbk^I} \equiv 0$. We may assume that $I = \{1, \dots, k\}$ for some k, $1 \leq k \leq n$, and $(\partial_j f)|_{\Bbbk^I} \not\equiv 0$ if and only if $i = k+1, \dots, k+l$. Then f can be expressed as

$$f = \sum_{j=1}^{l} x_{k+j} f_j(x_1, \dots, x_k) + \sum_{i \geq j \geq 1} x_{k+i} x_{k+j} f_{i,j}(x_1, \dots, x_n). \tag{140}$$

Let $h_j := \mathrm{In}_\nu(f_j)$, $j = 1, \dots, l$ and B be the $k \times l$ matrix with (i,j)-th entry $(\partial_i h_j)(a_1, \dots, a_k)$. We claim that $\mathrm{Rank}(B) < k$. Indeed, let $\nu_i := \nu(x_i)$, $i = 1, \dots, k$. For each $j = 1, \dots, l$, by assumption $(a_1, \dots, a_k)$ is a common zero of $\mathrm{In}_\nu((\partial_{k+j} f)|_{\Bbbk^I}) = h_j(x_1, \dots, x_k)$, so that $h_j(a_1 t^{\nu_1}, \dots, a_k t^{\nu_k}) = 0$ for all $t \in \Bbbk$. Note that

$$\begin{aligned} \frac{d}{dt}(h_j(a_1 t^{\nu_1}, \dots, a_k t^{\nu_k})) &= \sum_{i=1}^{k} \frac{\partial h_j}{\partial x_i}(a_1 t^{\nu_1}, \dots, a_k t^{\nu_k}) \nu_i a_i t^{\nu_i - 1} \\ &= \sum_{i=1}^{k} \frac{\partial h_j}{\partial x_i}(a_1, \dots, a_k) \nu_i a_i t^{\nu(h_j) - 1} \end{aligned}$$

Setting $t = 1$, it follows that $a'B = 0$, where $a' := (\nu_1 a_1, \dots, \nu_k a_k) \in \Bbbk^n$. Since ν is primitive, $a' \neq 0$, so that the map $\Bbbk^k \to \Bbbk^l$ given by multiplication by B on the right is *not* injective. Therefore, $\mathrm{Rank}(B) < k$, as claimed. Since $0 < [\Gamma_1, \dots, \Gamma_n]_0 < \infty$, corollary IX.13 implies that $l \geq k$, so that there is $b = (b_1, \dots, b_l) \neq 0 \in \Bbbk^l$ such that B times the transpose of b is zero. Let m be the number of non-zero coordinates of b. Without loss of generality we may assume that $b_j \neq 0$ if and only if $j = 1, \dots, m$. Then we have that

$$\sum_{j=1}^{m} b_j \frac{\partial h_j}{\partial x_i}(a_1, \dots, a_k) = 0 \tag{141}$$

for all $i = 1, \dots, k$. Fix positive integers $q, q_{k+m+1}, \dots, q_n$ and let ν' be the weighted order on $\Bbbk[x_1, \dots, x_n]$ such that

$$\nu'(x_i) = \begin{cases} \nu(x_i) & \text{if } i = 1, \dots, k, \\ q - \nu(f_{i-k}) & \text{if } i = k+1, \dots, k+m, \\ q_i & \text{if } i = k+m+1, \dots, n. \end{cases}$$

If $q \gg 1$ and $q_i \gg q$ for $i = k+m+1, \dots, n$, then identity (140) implies that $\mathrm{In}_{\nu'}(f) = \sum_{j=1}^{m} x_{k+j} h_j$. If $b'_1, \dots, b'_{n-k-m}$ are arbitrary elements in $\Bbbk^*$, it follows that $(a_1, \dots, a_k, b_1, \dots, b_m, b'_1, \dots, b'_{n-k-m})$ is a zero of $\mathrm{j}(\mathrm{In}_{\nu'}(f))$ on $(\Bbbk^*)^n$. Therefore, f is Newton degenerate, as required. □

Corollary XI.8. (Brzostowski and Oleksik [BO16, Theorem 3.1]). *Let $\Gamma_j := \mathrm{ND}(\partial_j f)$, $j = 1, \ldots, n$. Assume $f(0) = 0$ and f is Newton non-degenerate. Then the following are equivalent:*

(1) $\mu_0(f) < \infty$.
(2) $[\Gamma_1, \ldots, \Gamma_n]_0 < \infty$.
(3) $|\{j : \Gamma_j \cap \mathbb{R}^I \neq \emptyset\}| \geq |I|$ *for each* $I \subseteq [n]$.

PROOF. Combine corollary XI.5 and proposition XI.7. □

4.2. Inner Newton non-degeneracy. A *diagram* Γ in $\mathbb{R}^n$ is the Newton diagram of some subset of $\mathbb{R}^n_{\geq 0}$, and a *face* Δ of Γ is a compact face of $\Gamma + \mathbb{R}^n_{\geq 0}$. Δ is called an *inner face* if it is not contained in any proper coordinate subspace of $\mathbb{R}^n$. If $f = \sum_{\alpha \in \mathbb{Z}^n_{\geq 0}} c_\alpha x^\alpha$ is a power series in $(x_1, \ldots, x_n)$ and Δ is a subset of $\mathbb{R}^n$, we write $f_\Delta := \sum_{\alpha \in \Delta} c_\alpha x^\alpha$. We say that f is *inner Newton non-degenerate* if there is a *convenient* diagram Γ such that

(a) no point of $\mathrm{Supp}(f)$ "lies below" Γ, i.e. $\mathrm{Supp}(f) \subseteq \Gamma + \mathbb{R}^n_{\geq 0}$ and
(b) for every inner face Δ of Γ and for every nonempty subset I of $[n]$,

$$\Delta \cap \mathbb{R}^I \neq \emptyset \Rightarrow V(\mathrm{j}(f_\Delta)) \cap (\mathbb{k}^*)^I = \emptyset \tag{142}$$

The difference between Newton non-degeneracy and inner Newton non-degeneracy is most evident in the case of weighted homogeneous polynomials with isolated singularities, e.g. consider the polynomial $f := x_1 + (x_2 + x_3)^q$, where $q \geq 2$, from example XI.6. The Newton diagram of f is convenient and has only one inner face, namely the two-dimensional face Δ with vertices $(1, 0, 0), (0, q, 0), (0, 0, q)$. In particular, $f_\Delta = f$ and $\partial_1 f = 1$, which implies that f is inner non-degenerate. However, as we saw in example XI.6, f is Newton degenerate. In fact, C. T. C. Wall introduced inner Newton non-degeneracy in [Wal99] in order to find a condition which (in the case of power series with convenient Newton diagrams) is weaker than Newton non-degeneracy, but still wide enough to include all weighted homogeneous polynomials with isolated singularities. We now show that inner Newton non-degeneracy implies partial non-degeneracy at the origin. Given $\Delta \subseteq \mathbb{R}^n$, we write I_Δ for the smallest subset of $[n]$ such that $\Delta \subseteq \mathbb{R}^{I_\Delta}$.

Lemma XI.9. *Let Γ be a convenient diagram, $I \subseteq [n]$ and Δ be a face of $\Gamma \cap \mathbb{R}^I$. Pick an integral element $\nu \in (\mathbb{R}^I)^*$ centered at the origin such that $\Delta = \mathrm{In}_\nu(\Gamma \cap \mathbb{R}^I)$. Then there is an integral element $\nu' \in (\mathbb{R}^n)^*$ such that*

(1) ν' is centered at the origin,
(2) there is $\lambda > 0$ such that
  (a) $\langle \nu', \alpha \rangle = \lambda \langle \nu, \alpha \rangle$ for all $\alpha \in \mathbb{R}^{I_\Delta}$ (which means ν' is compatible with $\nu|_{\mathbb{R}^{I_\Delta}}$),
  (b) $\langle \nu', \alpha \rangle < \lambda \langle \nu, \alpha \rangle$ for all $\alpha \in \mathbb{R}^I_{\geq 0} \setminus \mathbb{R}^{I_\Delta}$ (here $\mathbb{R}^I_{\geq 0}$ denotes the set of all elements in $\mathbb{R}^I$ with nonnegative coordinates).
(3) $\Delta' := \mathrm{In}_{\nu'}(\Gamma)$ is an inner face of Γ.
(4) $\Delta = \Delta' \cap \mathbb{R}^{I_\Delta}$.

PROOF. We proceed by induction on $\delta := n - |I_\Delta|$. The lemma is obviously true if $\delta = 0$. Now assume $\delta = 1$. Then without loss of generality we may assume $I_\Delta = \{1, \ldots, n-1\}$. Note that I can be either I_Δ or $[n]$. For each pair of positive integers q, r, let $\nu_{q,r}$ be the element on $(\mathbb{R}^n)^*$ defined as follows:

$$\langle \nu_{q,r}, (\alpha_1, \ldots, \alpha_n) \rangle := q\langle \nu, (\alpha_1, \ldots, \alpha_{n-1}, 0) \rangle + r\alpha_n.$$

If $r \gg q$, then $\mathrm{In}_{\nu_{q,r}}(\Gamma) = \Delta$. On the other hand, since ν is centered at the origin and Γ is convenient, if $q \gg r$, then $\mathrm{In}_{\nu_{q,r}}(\Gamma)$ is a point on the x_n-axis. Therefore, we can find q', r' such that $\mathrm{In}_{\nu_{q',r'}}(\Gamma)$ contains Δ and also a point with positive n-th coordinate. We claim that the lemma holds with $\nu' := \nu_{q',r'}$. Indeed, if $I = I_\Delta$, then this is clear by the construction of $\nu_{q',r'}$. If $I = [n]$, then we also need to prove assertion (2b). But this follows from the observations that if α' is a point on $\mathrm{In}_{\nu_{q',r'}}(\Gamma)$ with positive n-th coordinate, then $\langle \nu, \alpha' \rangle > \min_\Gamma(\nu)$.

Now assume $\delta \geq 2$. Pick $\tilde{i} \notin I_\Delta$. The inductive hypothesis implies that there is an integral element $\tilde{\nu} \in (\mathbb{R}^{[n]\setminus\{\tilde{i}\}})^*$ such that assertions (1) to (4) of the lemma hold with Γ replaced by $\Gamma \cap \mathbb{R}^{[n]\setminus\{\tilde{i}\}}$, ν replaced by $\nu|_{(\mathbb{R}^{I\setminus\{\tilde{i}\}})^*}$ and ν' replaced by $\tilde{\nu}$. Extend $\tilde{\nu}$ to an element $\tilde{\nu}' \in (\mathbb{R}^n)^*$ by defining that for all $(\alpha_1, \ldots, \alpha_n) \in \mathbb{R}^n$,

$$\langle \tilde{\nu}', (\alpha_1, \ldots, \alpha_n) \rangle := \langle \tilde{\nu}, (\alpha_1, \ldots, \alpha_{\tilde{i}-1}, 0, \alpha_{\tilde{i}+1}, \ldots, \alpha_n) \rangle + \epsilon\alpha_{\tilde{i}},$$

where ϵ is a very small positive number. Then we can ensure that

(i) $\tilde{\Delta}' := \mathrm{In}_{\tilde{\nu}'}(\Gamma) = \mathrm{In}_{\tilde{\nu}}(\Gamma \cap \mathbb{R}^{[n]\setminus\{\tilde{i}\}})$,
(ii) there is $\tilde{\lambda} > 0$ such that
 (1) $\langle \tilde{\nu}', \alpha \rangle = \tilde{\lambda}\langle \nu, \alpha \rangle$ for all $\alpha \in \mathbb{R}^{I_\Delta}$,
 (2) $\langle \tilde{\nu}', \alpha \rangle < \tilde{\lambda}\langle \nu, \alpha \rangle$ for all $\alpha \in \mathbb{R}^I_{\geq 0} \setminus \mathbb{R}^{I_\Delta}$

Since $I_{\tilde{\Delta}'} = [n] \setminus \{\tilde{i}\}$, we can apply the $\delta = 1$ case of the lemma to obtain an integral element $\nu' \in (\mathbb{R}^n)^*$ such that assertions (1) to (4) of the lemma hold with I replaced by $[n]$ and ν replaced by $\tilde{\nu}'$. It is straightforward to check that the lemma holds with ν', which completes the proof. □

Proposition XI.10. ([Mon16]). *If $f \in \Bbbk[[x_1, \ldots, x_n]]$ is inner Newton non-degenerate, then it is partially non-degenerate at the origin.*

PROOF. Assume f is inner Newton non-degenerate with respect to a convenient diagram Γ. Pick $I \subseteq [n]$ and a weighted order ν on $\Bbbk[x_i : i \in I]$. Let $\Delta := \mathrm{In}_\nu(\Gamma \cap \mathbb{R}^I)$. Pick an integral element $\nu' \in (\mathbb{R}^n)^*$ which satisfies all assertions of lemma XI.9. In particular, $\Delta' := \mathrm{In}_{\nu'}(\Gamma)$ is an inner face of Γ and $\Delta = \Delta' \cap \mathbb{R}^{I_\Delta}$. Fix j, $1 \leq j \leq n$.

Claim XI.10.1. *One of the following holds:*

(i) either $\partial_j f_{\Delta'}$ is identically zero on $\Bbbk^{I_\Delta}$,
(ii) or $(\partial_j f_{\Delta'})|_{\Bbbk^{I_\Delta}} = \mathrm{In}_\nu((\partial_j f)|_{\Bbbk^I})$.

PROOF. Assume $\partial_j f_{\Delta'}$ is not identically zero on $\mathbb{k}^{I_\Delta}$. Then it is straightforward to check that

$$(\partial_j f_{\Delta'})|_{\mathbb{k}^{I_\Delta}} = (\mathrm{In}_{\nu'}(\partial_j f))|_{\mathbb{k}^{I_\Delta}} \tag{143}$$

Pick $\alpha \in \mathrm{Supp}((\partial_j f)|_{\mathbb{k}^I}) \setminus \mathrm{Supp}((\partial_j f_{\Delta'})|_{\mathbb{k}^{I_\Delta}})$. It suffices to show that $\alpha \notin \mathrm{Supp}(\mathrm{In}_\nu((\partial_j f)|_{\mathbb{k}^I}))$. Identity (143) implies that $\alpha \notin \mathrm{Supp}((\mathrm{In}_{\nu'}(\partial_j f))|_{\mathbb{k}^{I_\Delta}})$. If $\alpha \in \mathbb{R}^{I_\Delta}$, then the compatibility of ν' and $\nu|_{\mathbb{R}^{I_\Delta}}$ implies that $\alpha \notin \mathrm{Supp}(\mathrm{In}_\nu((\partial_j f)|_{\mathbb{k}^I}))$, as required. So assume $\alpha \in \mathbb{R}^I \setminus \mathbb{R}^{I_\Delta}$. Then lemma XI.9 implies that

$$\langle \nu, \alpha \rangle > \langle \nu', \alpha \rangle / \lambda$$

where λ is as in assertion (2) of lemma XI.9. Since $\langle \nu', \alpha \rangle \geq \nu'(\partial_j f) = \nu'((\partial_j f_{\Delta'})|_{\mathbb{k}^{I_\Delta}}) = \lambda \nu((\partial_j f_{\Delta'})|_{\mathbb{k}^{I_\Delta}})$, it follows that $\langle \nu, \alpha \rangle > \nu((\partial_j f)|_{\mathbb{k}^I})$, as required. □

Claim XI.10.1 implies that $V(\mathrm{In}_\nu((\partial_1 f)|_{\mathbb{k}^I}), \ldots, \mathrm{In}_\nu((\partial_n f)|_{\mathbb{k}^I})) \cap (\mathbb{k}^*)^I$ is contained in the product of $V((\partial_1 f_{\Delta'})|_{\mathbb{k}^{I_\Delta}}, \ldots, (\partial_n f_{\Delta'})|_{\mathbb{k}^{I_\Delta}}) \cap (\mathbb{k}^*)^{I_\Delta}$ with $(\mathbb{k}^*)^{I \setminus I_\Delta}$. The inner non-degeneracy of f with respect to Γ then implies that $V(\mathrm{In}_\nu(\partial_1 f|_{\mathbb{k}^I}), \ldots, \mathrm{In}_\nu(\partial_n f|_{\mathbb{k}^I})) \cap (\mathbb{k}^*)^I = \emptyset$, as required. □

Corollary XI.11. (Wall [Wal99, Lemma 1.2]). *If f is inner Newton non-degenerate and $f(0) = 0$, then $\mu_0(f) < \infty$.* □

If $p := \mathrm{char}(\mathbb{k})$ is non-zero, partial Newton non-degeneracy at the origin is strictly weaker than inner Newton non-degeneracy, e.g. $x^p + y^p + x^{p+1} + y^{p+1}$ is partially non-degenerate at the origin, but it is inner Newton *degenerate*. We do not know if this is true in zero characteristic (see section XI.6.2).

5. Newton number: Kushnirenko's formula for the generic Milnor number

Let Γ be a diagram in $\mathbb{R}^n$. We write $\bar{\Gamma}$ for the region bounded by the cone with base Γ and apex at the origin, and $V_k^-(\Gamma)$, $0 \leq k \leq n$, for the sum of k-dimensional Euclidean volumes of the intersections of $\bar{\Gamma}$ with the k-dimensional coordinate subspaces of $\mathbb{R}^n$ (in particular, $V_0(\Gamma)$ is defined to be 1). The *Newton number* of Γ is

$$\nu(\Gamma) := \begin{cases} \sum_{k=0}^{n} (-1)^{n-k} k! V_k^-(\Gamma) & \text{if } \Gamma \text{ is convenient,} \\ \sup\{\nu(\Gamma \cup \{me_1, \ldots, me_n\}) : m \geq 0\} & \text{otherwise,} \end{cases}$$

where e_j are the unit vectors along the (positive direction of the) axes of $\mathbb{R}^n$. Let $f \in \mathbb{k}[[x_1, \ldots, x_n]]$ such that $f(0) = 0$. A. Kushnirenko proved in [Kou76] that $\mu_0(f) \geq \nu(\mathrm{ND}(f))$, and $\mu_0(f) = \nu(\mathrm{ND}(f))$ if f is Newton non-degenerate and if either $\mathrm{char}(\mathbb{k}) = 0$ or $\mathrm{ND}(f)$ is convenient. C. T. C. Wall proved in [Wal99] that if $\mathrm{char}(\mathbb{k}) = 0$, then the equality $\mu_0(f) = \nu(\mathrm{ND}(f))$ continues to hold if f is inner Newton non-degenerate. In this section we prove these results and present some generalizations.

5.1. Preliminary results. Let $\mathcal{A}$ be a subset of $\mathbb{Z}^n_{\geq 0}$ not containing the origin. For each $m \geq 0$, let $\mathcal{A}_m := \mathcal{A} \cup \{(\alpha_1, \ldots, \alpha_n) : \sum_{j=1}^n \alpha_j \geq m\}$. For each $j = 1, \ldots, n$, let

$$\mathcal{A}'_{m,j} := \{\alpha - e_j : \alpha \in \mathcal{A}_m,\ \alpha - e_j \in \mathbb{Z}^n_{\geq 0}\})$$

where e_j are the unit vectors along the (positive direction of the) axes of $\mathbb{R}^n$. Note that $\partial_j \mathcal{A}_m \subseteq \mathcal{A}'_{m,j}$, and the inclusion is proper if $\operatorname{char}(\mathbb{k})$ is positive. In any event, theorem XI.3 and the monotonicity of intersection multiplicity (remark IX.3) imply that for each $m \geq 1$,

$$\mu_0(\mathcal{A}) \geq [\partial_1 \mathcal{A}, \ldots, \partial_n \mathcal{A}]_0 \geq [\partial_1 \mathcal{A}_m, \ldots, \partial_n \mathcal{A}_m]_0 \geq [\mathcal{A}'_{m,1}, \ldots, \mathcal{A}'_{m,n}]_0 \tag{144}$$

Lemma XI.12. *Let $\mathcal{A} \subseteq \mathbb{Z}^n_{\geq 0} \setminus \{0\}$. Define*

$$\mathcal{A}'_j := \{\alpha - e_j : \alpha \in \mathcal{A},\ \alpha - e_j \in \mathbb{Z}^n_{\geq 0}\},$$

where e_j are the j-th standard unit vector in $\mathbb{R}^n$, $j = 1, \ldots, n$. Assume each $\mathcal{A}'_j$ is convenient. *Then $\mathcal{A}$ is also convenient and $[\mathcal{A}'_1, \ldots, \mathcal{A}'_n]_0 = \nu(\mathrm{ND}(\mathcal{A}))$.*

Remark XI.13. Kushnirenko's theorem implies that the assumption "each $\mathcal{A}'_j$ is convenient" in lemma XI.12 is not needed for the equality $[\mathcal{A}'_1, \ldots, \mathcal{A}'_n]_0 = \nu(\mathrm{ND}(\mathcal{A}))$ to hold—see corollary XI.17 below.

Proof of Lemma XI.12. Note that $[\mathcal{A}'_1, \ldots, \mathcal{A}'_n]_0 = [g_1, \ldots, g_n]_0$ for all $g_1, \ldots, g_n \in \mathbb{k}[[x_1, \ldots, x_n]]$ such that

(i) $\mathrm{Supp}(g_j) = \mathcal{A}'_j$ for each j, and
(ii) $g_1, \ldots, g_n$ are non-degenerate at the origin.

We will show that there are $g_1, \ldots, g_n$ which satisfy both of the above properties and in addition satisfy $[g_1, \ldots, g_n]_0 = \nu(\mathrm{ND}(\mathcal{A}))$. Indeed, choose $g_1, \ldots, g_n$ which satisfy the above properties, and in addition satisfy

(iii) for each pair of subsets I, J of $[n]$ such that $|I| = |J|$, the restrictions $g_j|_{\mathbb{k}^I}$, $j \in J$ are $(\mathcal{A}'_j \cap \mathbb{R}^I : j \in J)$-non-degenerate at the origin.

Note that this is possible since each $\mathcal{A}'_j$ is convenient and since "generic" systems are non-degenerate at the origin (theorem IX.8). For each $I, J \subseteq [n]$ such that $|I| + |J| = n$, we write

$$[(g_i)_{i \in I}, (x_j)_{j \in J}]_0 := [g_{i_1}, \ldots, g_{i_k}, x_{j_1}, \ldots, x_{j_{n-k}}]_0$$

where $I = \{i_1, \ldots, i_k\}$ and $J = \{j_1, \ldots, j_{n-k}\}$. Since $\mathcal{A}'_j$ are convenient, corollary IX.13 and property (iii) imply that $[(g_i)_{i \in I}, (x_j)_{j \in [n] \setminus I}]_0$ is defined for each $I \subseteq [n]$. Consequently, corollary B.64 implies that

$$\begin{aligned}
[g_1, \ldots, g_n]_0 &= \sum_{I \subseteq [n]} (-1)^{n-|I|} [(x_i g_i)_{i \in I}, (x_j)_{j \in [n] \setminus I}]_0 \\
&= \sum_{I \subseteq [n]} (-1)^{n-|I|} [x_{i_1} g_{i_1}|_{\mathbb{k}^I}, \ldots, x_{i_{|I|}} g_{i_{|I|}}|_{\mathbb{k}^I}]_0,
\end{aligned}$$

where $i_1, \ldots, i_{|I|}$ are elements of I for each $I \subseteq [n]$. Fix $I \subseteq [n]$. Since the Newton diagram of the union of the supports of $x_{i_j} g_{i_j}|_{\mathbb{k}^I}$ is $\mathrm{ND}(\mathcal{A}) \cap \mathbb{R}^I$ (this is where the assumption $0 \notin \mathcal{A}$ is used!), property (iii) and proposition IX.30 imply that

$$[x_{i_1} g_{i_1}|_{\mathbb{k}^I}, \ldots, x_{i_{|I|}} g_{i_{|I|}}|_{\mathbb{k}^I}]_0 = V^-_{|I|}(\mathrm{ND}(\mathcal{A}) \cap \mathbb{R}^I).$$

The result now follows from the definition of $\nu(\cdot)$. □

Corollary XI.14. $[\mathcal{A}'_{m,1}, \ldots, \mathcal{A}'_{m,n}]_0 = \nu(\mathrm{ND}(\mathcal{A}_m))$ *for each* $m \geq 1$. □

Corollary XI.15. (Kushnirenko [Kou76, Theorem I, part (i)]). $\mu_0(f) \geq \nu(\mathrm{ND}(f))$ *for all* $f \in \mathbb{k}[[x_1, \ldots, x_n]]$*. In particular, if* $\nu(\mathrm{ND}(f)) = \infty$*, then* $\mu_0(f) = \infty$.

PROOF. Combine inequation (144) and corollary XI.14. □

5.2. Characteristic zero case. Continue to assume that $\mathcal{A}$ is a subset of $\mathbb{Z}^n_{\geq 0}$ not containing the origin.

THEOREM XI.16. *If* $\mathrm{char}(\mathbb{k}) = 0$*, then* $\mu_0(\mathcal{A}) = \nu(\mathrm{ND}(\mathcal{A}))$.

PROOF. If $\mathrm{char}(\mathbb{k}) = 0$, then theorem XI.3 implies that $\mu_0(\mathcal{A}) = [\partial_1 \mathcal{A}, \ldots, \partial_n \mathcal{A}]_0$. At first, consider the case that $[\partial_1 \mathcal{A}, \ldots, \partial_n \mathcal{A}]_0 < \infty$. For $m \gg 1$, proposition IV.27 implies that $[\partial_1 \mathcal{A}, \ldots, \partial_n \mathcal{A}]_0 = [\partial_1 \mathcal{A}_m, \ldots, \partial_n \mathcal{A}_m]_0$. Since in zero characteristic $\mathcal{A}'_{m,j} = \partial_j \mathcal{A}_m$, corollary XI.14 implies that $\mu_0(\mathcal{A}) = \nu(\mathrm{ND}(\mathcal{A}))$. On the other hand, if $[\partial_1 \mathcal{A}, \ldots, \partial_n \mathcal{A}]_0 = \infty$, then $\sup_m [\partial_1 \mathcal{A}_m, \ldots \partial_n \mathcal{A}_m]_0 = \infty$ (proposition IV.27) so that corollary XI.14 implies that $\nu(\mathrm{ND}(\mathcal{A})) = \infty$, as required. □

Corollary XI.17. *Lemma XI.12 holds even without the assumption that each* $\mathcal{A}'_j$ *is convenient.*

PROOF. In zero characteristic $\mu_0(\mathcal{A}) = [\mathcal{A}'_1, \ldots, \mathcal{A}'_n]_0$ (theorem XI.3); now use theorem XI.16. □

Corollary XI.18. (Cf. [Kou76, Characteristic zero case of Theorem I], [Wal99, Theorem 1.6], [BO16, Corollaries 3.10 and 3.11]). *Let* $f \in \mathbb{k}[[x_1, \ldots, x_n]]$ *be such that* $f(0) = 0$*. Assume* $\mathrm{char}(\mathbb{k}) = 0$*. Then* $\mu_0(f) = \nu(\mathrm{ND}(f))$ *whenever* f *is partially non-degenerate at the origin. In particular, if* f *is either Newton non-degenerate or inner Newton non-degenerate, then* $\mu_0(f) = \nu(\mathrm{ND}(f))$.

PROOF. Combine theorems XI.3 and XI.16 and propositions XI.7 and XI.10. □

5.3. The general case. We continue to use the notation of section XI.5.1. In particular, $\mathcal{A}$ is a subset of $\mathbb{Z}^n_{\geq 0}$ not containing the origin. Let $\mathcal{M}'_0(\mathcal{A})$ be as in theorem XI.3 the set of all $f \in \mathcal{L}_0(\mathcal{A})$ which are partially $\mathcal{A}$-non-degenerate at the origin. Proposition IV.27 and theorem XI.3, inequation (144) and corollary XI.14 imply the following result.

Proposition XI.19. $\mu_0(\mathcal{A}) = \nu(\mathrm{ND}(\mathcal{A}))$ *if and only if*

(1) either $\nu(\mathrm{ND}(\mathcal{A})) = \infty$*, or*
(2) (a) $\mathcal{M}'_0(\mathcal{A})$ *is nonempty, and*

(b) $[\partial_1 \mathcal{A}_m, \dots, \partial_n \mathcal{A}_m]_0 = \nu(\mathrm{ND}(\mathcal{A}_m))$ *for all* $m \gg 1$ *not divisible by* p. □

Remark XI.20. Corollaries IX.13 and XI.17 imply that the following are equivalent:

(1) $\nu(\mathrm{ND}(\mathcal{A})) = \infty$,
(2) $[\mathcal{A}'_1, \dots, \mathcal{A}'_n]_0 = \infty$, and
(3) there is a nonempty subset I of $[n]$ such that $|\{j : \mathcal{A}'_j \cap \mathbb{R}^I \neq \emptyset\}| < |I|$.

Let $p := \mathrm{char}(\Bbbk)$. Example XI.4 shows that if p is positive, then condition (2a) of proposition XI.19 is nontrivial. The example below shows that condition (2b) is nontrivial as well.

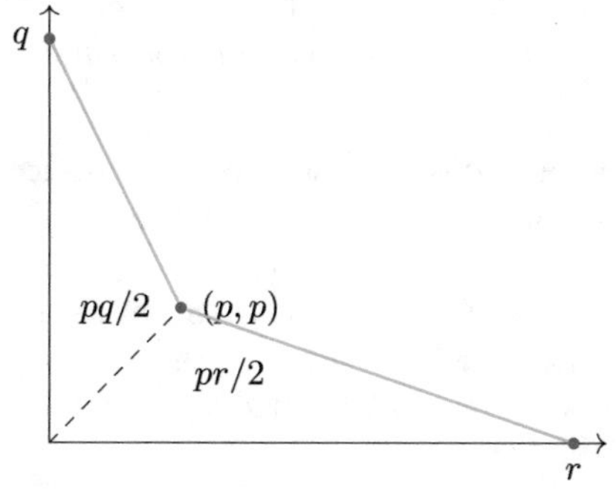

FIGURE 1. Subdivision of the area under the Newton diagram from example XI.21

Example XI.21. Assume p is positive. Let $f = x_1^r + x_1^p x_2^p + x_2^q$, where q, r are large positive integers not divisible by p. Let $\mathcal{A} := \mathrm{Supp}(f)$ and $\Gamma := \mathrm{ND}(f)$. It is straightforward to see from fig. 1 that $\nu(\Gamma) = 2p(q+r)/2 - (r+q) + 1 = (p-1)(q+r) + 1$. On the other hand, $\partial_1 f = r x_1^r$ and $\partial_2 f = q x_2^{q-1}$, so that f is partially non-degenerate at the origin. It follows that $\mathcal{M}'_0(\mathcal{A}) \neq \emptyset$ and $\mu_0(\mathcal{A}) = \mu_0(f) = [r x_1^{r-1}, q x_2^{q-1}]_0 = (r-1)(q-1)$. It follows that $\mu_0(\mathcal{A}) > \nu(\mathrm{ND}(\mathcal{A}))$ for sufficiently large q, r.

We now state a condition which guarantees that condition (2b) of proposition XI.19 holds. Given a subset $\mathcal{B}$ of $\mathbb{Z}^n_{\geq 0}$, let $J_{\mathcal{B}} := \{j \in [n] : \partial_j \mathcal{B} \neq \emptyset\} = \{j \in [n] :$ there is $(\beta_1, \dots, \beta_n) \in \mathcal{B}$ such that p does not divide $\beta_j\}$. The condition is the following:

(145)

for each $m \gg 1$ not divisible by p and for each face Δ of $\mathrm{ND}(\mathcal{A}_m)$, $J_{\mathcal{A}_m \cap \Delta} \neq \emptyset$, and the convex hulls of $\partial_j(\mathcal{A}_m \cap \Delta)$, $j \in J_{\mathcal{A}_m \cap \Delta}$, are dependent.

Proposition XI.22. *Assume (145) holds. Then condition (2b) of proposition XI.19 also holds. In particular, $\mu_0(\mathcal{A}) = \nu(\mathrm{ND}(\mathcal{A}))$ if either $\nu(\mathrm{ND}(\mathcal{A})) = \infty$ or $\mathcal{M}'_0(\mathcal{A})$ is nonempty.*

PROOF. Pick $m \gg 1$ not divisible by p. The arguments from the proof of lemma XI.12 show that it suffices to prove that $[x_1 g_1, \dots, x_n g_n]_0 = n! V_n^-(\mathrm{ND}(\mathcal{A}_m))$ for power series g_j such that $\mathrm{Supp}(g_j) = \partial_j \mathcal{A}_m$ and $g_1, \dots, g_n$ are non-degenerate at the origin. Condition (145) ensures that the Newton diagram of the union of the supports of $x_j g_j$ is $\mathrm{ND}(\mathcal{A}_m)$, and that the condition of assertion (2) of proposition IX.30 is also satisfied. Therefore, the result follows from proposition IX.30. □

Note that condition (145) is not necessary for (2b) of proposition XI.19 to hold—see section XI.6.3.

Corollary XI.23. ([Kou76, Positive characteristic case of Theorem I]). *Let $f \in \mathbb{k}[[x_1, \dots, x_n]]$ be such that $f(0) = 0$, $\mathrm{ND}(f)$ is convenient, and f is Newton non-degenerate. Then $\mu_0(f) = \nu(\mathrm{ND}(f))$.*

PROOF. Due to propositions XI.7 and XI.22, it suffices to show that $\mathcal{A} := \mathrm{Supp}(f)$ satisfies condition (145). Since $\mathcal{A}$ is convenient, $\mathrm{ND}(\mathcal{A}_m) = \mathrm{ND}(\mathcal{A})$ for $m \gg 1$. So pick a face Δ of $\mathrm{ND}(\mathcal{A})$. As in section XI.4.2, let f_Δ be the "portion" of f supported at Δ. Let $g_j := \partial_j f_\Delta$, $j = 1, \dots, n$. Since f is Newton non-degenerate, it follows that $g_1, \dots, g_n$ are BKK non-degenerate. Since $g_1, \dots, g_n$ have no common zero on $(\mathbb{k}^*)^n$, theorems VII.5, VII.7 and VII.33 imply that the Newton polytopes of g_j are dependent, as required. □

6. Open problems

6.1. Existence of non-degenerate polynomials. Let $\mathcal{A}$ be a subset of $\mathbb{Z}^n_{\geq 0}$ not containing the origin. The estimate of $\mu_0(\mathcal{A})$ from theorem XI.3 is exact if and only if the set $\mathcal{M}'_0(\mathcal{A})$ of power series which are partially $\mathcal{A}$-non-degenerate at the origin is nonempty. Theorem XI.3 shows that in characteristic zero $\mathcal{M}'_0(\mathcal{A})$ is always nonempty, and example XI.4 shows that in positive characteristic there are $\mathcal{A}$ such that $\mathcal{M}'_0(\mathcal{A})$ is empty. This motivates the following problem.

Problem XI.24. *In the case that $\mathrm{char}(\mathbb{k}) > 0$, characterize those $\mathcal{A}$ for which $\mathcal{M}'_0(\mathcal{A})$ is nonempty. Compute $\mu_0(\mathcal{A})$ for those $\mathcal{A}$ such that $\mathcal{M}'_0(\mathcal{A})$ is empty.*

Let $\mathcal{N}'^0_0(\mathcal{A})$ be the set of all power series f supported at $\mathcal{A}$ such that $\mathrm{ND}(f) = \mathrm{ND}(\mathcal{A})$, $\mathrm{ND}(\partial_j f) = \mathrm{ND}(\partial_j \mathcal{A})$ for each j, and f is Newton non-degenerate. Proposition XI.7 implies that $\mathcal{N}'^0_0(\mathcal{A}) \subseteq \mathcal{M}'_0(\mathcal{A})$ when $\mu_0(\mathcal{A}) < \infty$. The following is therefore a subproblem of problem XI.24 in that case.

Problem XI.25. *In the case that $\mathrm{char}(\mathbb{k}) > 0$, characterize those $\mathcal{A}$ for which $\mathcal{N}'^0_0(\mathcal{A})$ is nonempty.*

The proof of theorem XI.3 gives a sufficient condition for existence of Newton non-degenerate polynomials: let $\mathcal{B}$ be a finite subset of $\mathbb{Z}^n_{\geq 0}$ and B be the $n \times |\mathcal{B}|$ matrix whose columns are the elements of B. Let $\mathrm{Rank}_{\mathbb{k}}(\cdot)$ denote the rank of a matrix over $\Bbbk$. The condition we are interested in is the following:

$$\operatorname{Rank}_{\Bbbk}(B) = \dim(\operatorname{conv}(\mathcal{B})) + 1 \tag{146}$$

Lemma XI.26. *Assume (146) holds. Then the set of polynomials g supported at $\mathcal{B}$ such that $\partial_j g$, $j = 1, \ldots, n$, have no common zero on $(\Bbbk^*)^n$ contains a nonempty Zariski open subset of the space of all polynomials supported at $\mathcal{B}$.*

PROOF. This is in fact the main content of the proof of theorem XI.3 (the only place where the zero characteristic played a role in that proof was to ensure that (146) holds). □

Corollary XI.27. *If (146) holds with $\mathcal{B} = \Delta \cap \mathcal{A}$ for each face Δ of $\mathrm{ND}(\mathcal{A})$, then $\mathcal{N}_0'^0(\mathcal{A}) \neq \emptyset$.* □

Assertion (1) of theorem XI.3 implies that for $\mathcal{A}$ to admit polynomials which are partially non-degenerate at the origin, it is necessary that

$$[\partial_1 \mathcal{A}, \ldots, \partial_n \mathcal{A}]_0 \tag{147}$$

Corollary IX.13 implies that (147) is equivalent to the condition that $|\{j : \partial_j \mathcal{A} \cap \mathbb{R}^I \neq \emptyset\}| \geq |I|$ for each $I \subseteq [n]$. The following is an immediate corollary of proposition XI.7 and corollary XI.27.

Corollary XI.28. *Assume (147) holds and (146) holds with $\mathcal{B} = \Delta \cap \mathcal{A}$ for each face Δ of $\mathrm{ND}(\mathcal{A})$, then $\mathcal{M}_0'(\mathcal{A}) \neq \emptyset$.* □

Question XI.29. Is the condition from corollary XI.27 necessary for $\mathcal{N}_0'^0(\mathcal{A})$ to be nonempty?

6.2. Relation among non-degeneracy conditions. Given a power series f, let us write (N), (I), (P) to denote, respectively, the conditions that f is Newton non-degenerate, inner Newton non-degenerate and partially non-degenerate at the origin. If $p := \operatorname{char}(\Bbbk)$ is non-zero, then (P) does not imply (I), e.g. $x^p + y^p + x^{p+1} + y^{p+1}$ is partially non-degenerate at the origin, but it is not inner Newton degenerate. This observation together with the discussion from section XI.4 implies the relations depicted in fig. 2, where "$(N_{\mu_0<\infty})$" denotes the condition that f is Newton non-degenerate and $\mu_0(f) < \infty$.

Problem XI.30. *Determine if the question-marked implications from fig. 2 are valid.*

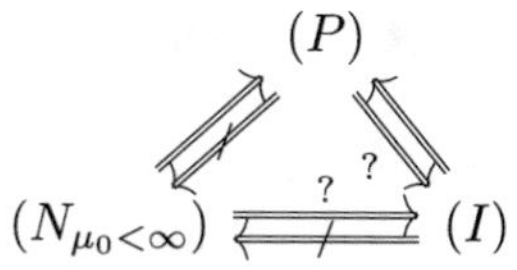

(A) zero characteristic

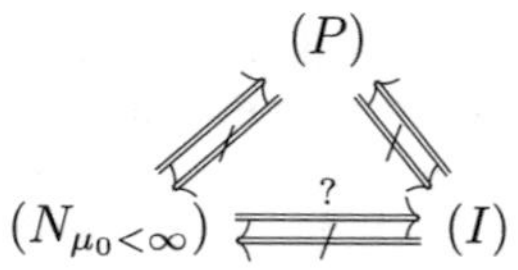

(B) positive characteristic

FIGURE 2. Relation among non-degeneracy conditions

We now show that in zero characteristic the implication $(P) \Rightarrow (I)$ does hold in dimension ≤ 3.

Proposition XI.31. *Pick $f \in \Bbbk[[x_1, \ldots, x_n]]$ such that $f(0) = 0$. Assume f is partially non-degenerate at the origin. If $n \leq 3$ and $\operatorname{char}(\Bbbk)$ is zero, then f is also inner Newton non-degenerate.*

PROOF. Since all the $\partial_j f$ can not be identically zero on any axis, $\mathrm{ND}(f)$ satisfies the following property:

(148) the distance from any axis to $\mathrm{ND}(f)$ can not be greater than 1.

This leads to the possibilities of fig. 3 in the case that $n = 2$. If $\mathrm{ND}(f)$ is convenient, take $\Gamma = \mathrm{ND}(f)$, otherwise, take Γ to be the union of $\mathrm{ND}(f)$ and edges (with appropriate slopes) from the end points of $\mathrm{ND}(f)$ to some points on the axes (e.g. the "dashed edges" in fig. 3). It is straightforward to check that f is inner Newton non-degenerate with respect to Γ.

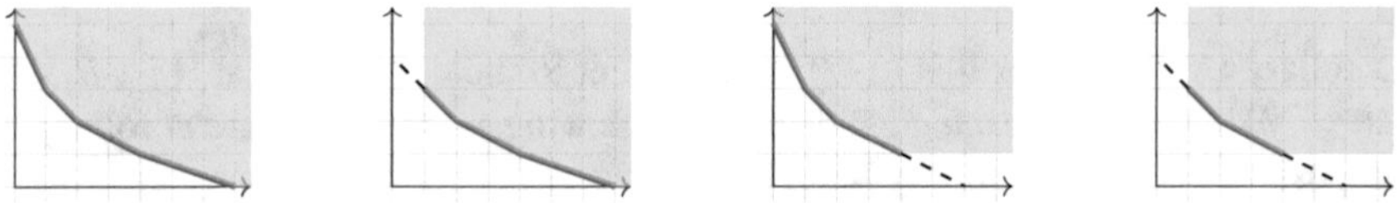

FIGURE 3. Four possibilities for $\mathrm{ND}(f)$ in dimension two

Now assume that $n = 3$. If $\mathrm{ND}(f)$ does not touch (at least) one of the three coordinate hyperplanes, then f would be divisible by some x_i. The partial non-degeneracy of f at the origin would then imply that $f = x_i g$ such that $g(0) \neq 0$, i.e. $\mathrm{ND}(f) = \{e_i\}$, where e_i is the i-th standard unit vector in $\mathbb{R}^3$. In that case f would be inner Newton non-degenerate with any convenient diagram with a vertex at e_i. Therefore, we may assume that $\mathrm{ND}(f)$ touches every coordinate hyperplane. Let Γ be the Newton diagram of $f + x_1^N + x_2^N + x_3^N$ for some $N \gg 1$. We claim that f is inner Newton non-degenerate with respect to Γ. Indeed, let Δ be an inner face of Γ and $I \subseteq \{1, 2, 3\}$ be such that $\Delta \cap \mathbb{R}^I \neq \emptyset$. We will check that the inner non-degeneracy condition (142) holds for Δ and I. Let $\Delta' := \Delta \cap \mathbb{R}^I$. If $\Delta' \cap \mathrm{ND}(f) = \emptyset$, then $\Delta' = \{Ne_i\}$ for some i. Assume without loss of generality that $i = 1$. Let $\alpha^{(2)}$ (respectively, $\alpha^{(3)}$) be the point on the intersection of $\mathrm{ND}(f)$ with the (x_1, x_2)-plane (respectively, (x_1, x_3)-plane) which is closest to x_1-axis. Since Δ is an inner face, it is straightforward to check that when N is sufficiently large, Δ must be the triangle with vertices $Ne_1, \alpha^{(2)}, \alpha^{(3)}$. Property (148) implies that either $\alpha^{(2)} = (k, 1, 0)$ for some $k \geq 0$ or $\alpha^{(3)} = (k, 0, 1)$ for some $k \geq 0$. Then $x_1^k \in \mathfrak{j}(f_\Delta)$, and therefore $V(\mathfrak{j}(f_\Delta))$ does not contain any point on x_1-axis other than possibly the origin, as required. It remains to consider the case that $\Delta'' := \Delta' \cap \mathrm{ND}(f) \neq \emptyset$. If $|I| = 1$ or $|I| = 3$, it is straightforward to check that violation of the inner non-degeneracy condition (142) implies violation of partial non-degeneracy of f. So we may assume without loss of generality $I = \{1, 2\}$.

Note that $\dim(\Delta'') \leq \dim(\Delta') \leq 1$. If $\dim(\Delta'') = 0$, then $f_\Delta = cx_1^{\alpha_1}x_2^{\alpha_2} + x_3 g$ for some polynomial g, $(\alpha_1, \alpha_2) \in \mathbb{Z}^2_{\geq 0} \setminus \{0\}$ and $c \neq 0$. It is then clear that either $\partial_1 f_\Delta$ or $\partial_2 f_\Delta$ does not vanish at any point on $(\Bbbk^*)^I$. So assume $\dim(\Delta'') = 1$. It then follows that $\Delta'' = \Delta'$, and if N is sufficiently large, then Δ is in fact a (two dimensional) face of $\mathrm{ND}(f)$ containing Δ', see fig. 4a.

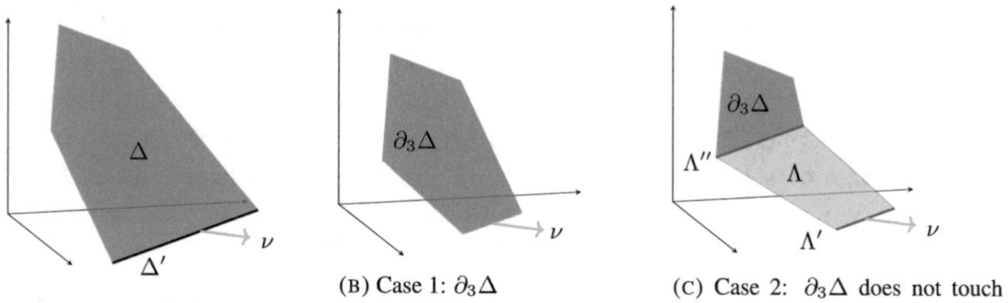

(A) Δ

(B) Case 1: $\partial_3\Delta$ touches (x_1, x_2)-plane

(C) Case 2: $\partial_3\Delta$ does not touch (x_1, x_2)-plane

FIGURE 4. Scenarios when $\dim(\Delta'') = 1$

If $\partial_3\Delta$ touches the (x_1, x_2)-plane, then applying the partial non-degeneracy condition (137) with $I = \{1, 2\}$ and ν equal to the "inner normal" to Δ' in $\mathbb{R}^I$ shows that $V(\mathfrak{j}(f_\Delta)) \cap (\Bbbk^*)^I = \emptyset$. So assume that $\partial_3\Delta$ does not touch (x_1, x_2)-plane. In this case we claim that $V(\partial_1 f_{\Delta'}, \partial_2 f_{\Delta'}) \cap (\Bbbk^*)^I = \emptyset$. Indeed, assume to the contrary that there is a common zero $a = (a_1, a_2, 0)$ of $\partial_1 f_{\Delta'}, \partial_2 f_{\Delta'}$ on $(\Bbbk^*)^I$. The partial non-degeneracy then implies that the intersection of $\partial_3\Gamma$ with the (x_1, x_2)-plane is nonempty, and if Λ' is the face of this intersection determined by ν, then $(\partial_3 f)_{\Lambda'}$ does *not* vanish at a. It is straightforward to check that there is a face Λ of $\partial_3\Gamma$ which contains Λ' and also intersects $\partial_3\Delta$. Let $\Lambda'' := \Lambda \cap \partial_3\Delta$. If $(\partial_3 f)_{\Lambda''}$ does not vanish after substituting $x_1 = a_1$ and $x_2 = a_2$, then one can find a zero of $(\partial_3 f)_\Lambda$ of the form (a_1, a_2, b) for some $b \in \Bbbk^*$, and this would contradict partial non-degeneracy condition (137) with $I = \{1, 2, 3\}$ and ν being an inner normal to Λ in $\mathbb{R}^3$. On the other hand, if $(\partial_3 f)_{\Lambda''}$ vanishes after substituting $x_1 = a_1$ and $x_2 = a_2$, then Λ'' must be an edge parallel to Δ' and the partial non-degeneracy condition (137) is violated with $I = \{1, 2, 3\}$ and ν being an inner normal to Λ'' in $\mathbb{R}^3$. This proves that $V(\partial_1 f_{\Delta'}, \partial_2 f_{\Delta'}) \cap (\Bbbk^*)^I = \emptyset$, as claimed. But then it is clear that $V(\mathfrak{j}(f_\Delta)) \cap (\Bbbk^*)^I$ is also empty, as required. □

6.3. Conditions for validity of Kushnirenko's formula in positive characteristic. Consider the case that $p := \mathrm{char}(\Bbbk) > 0$. The characterization of the conditions under which $\mu_0(\mathcal{A}) = \nu(\mathrm{ND}(\mathcal{A}))$ established in proposition XI.19, in particular condition (2b) of proposition XI.19, is *not* explicit. It would be interesting to find simple criteria under which this condition holds. The criterion from proposition XI.22 is sufficient, but not necessary, e.g. if $f = x + y^p$, then (145) fails (and f is not Newton non-degenerate as well), but $\mu_0(f) = \nu(\mathrm{ND}(f)) = 0$. Note that f is *inner Newton non-degenerate*. Y. Boubakri, G.-M. Greuel and T. Markwig claim in [BGM12, Theorem 3.5] that $\mu_0(f) = \nu(\mathrm{ND}(f))$ if f is inner Newton non-degenerate, i.e. Wall's result [Wal99, Theorem 1.6] extends to positive characteristics; however, their proof is wrong, as explained by J. Stevens in [Ste20, Example 2.5].

Problem XI.32. *Determine necessary and sufficient conditions for* $\mu_0(f) = \nu(\mathrm{ND}(f))$ *to hold when* $\mathrm{char}(\Bbbk) > 0$. *In particular, does it hold when* f *is inner Newton non-degenerate?*

6.4. Monotonicity of Newton number. Theorems XI.3 and XI.16 and the monotonicity of intersection multiplicity (remark IX.3) implies that the Newton number $\nu(\Gamma)$ of a diagram Γ grows monotonically with Γ. Even though the arguments in this implication involve nontrivial algebraic geometry, the monotonicity of Newton numbers is a purely "elementary" convex geometric statement. It is a problem of V. I. Arnold [Arn04, Problem 1982-16] to find an elementary proof of this statement. S. K. Lando wrote in the commentary of that problem that he gave an elementary proof for $n = 2$. For $n = 3$ an elementary proof was given by S. Brzostowski, T. Krasiński and J. Walewska [BKW19]. Since the monotonicity of the formula (88) for intersection multiplicity is obvious (see remark IX.3), one possible strategy would be to find an elementary proof of the following identity (which is an immediate consequence of theorems XI.3 and XI.16):

$$\nu(\Gamma) = \sum_{I \in \mathscr{T}_{A,1}} [\Gamma_1^I, \Gamma_{j_2}^I, \ldots, \Gamma_{j_{|I|}}^I]_0^* \times [\pi_{[n]\setminus I}(\Gamma_{j'_1}), \ldots, \pi_{[n]\setminus I}(\Gamma_{j'_{n-|I|}})]_0$$

where the right-hand side is as in theorem IX.1 with $\Gamma_j := \mathrm{ND}(\{\alpha - e_j : \alpha \in \Gamma + \mathbb{R}^n_{\geq 0},\ \alpha - e_j \in \mathbb{Z}^n_{\geq 0}\})$, $j = 1, \ldots, n$.

CHAPTER XII

Beyond this book

In this final chapter, we outline some of the natural directions of further study for a reader of this book, and point out a few interesting recent works which are accessible to someone equipped with the knowledge of algebraic and toric varieties developed in this book.

1. Toric varieties

An obvious direction for further pursuit is to study toric varieties. Standard introductions to the theory of toric varieties include [Oda88, Ful93, CLS11]. Unlike this book, in these expositions toric varieties are defined via *fans*, which reveals their combinatorial structure more cleanly. The combinatorics makes delightful appearances even in dimension two, e.g. continued fractions appear in resolutions of singularities [Ful93, Section 2.6] and Pick's formula appears in intersection theory [Ful93, Section 5.3]. A main attraction of toric varieties is their "computability," due to which, in the words of Fulton, "toric varieties have provided a remarkably fertile testing ground for general theories" of algebraic geometry. As a recent example, we mention the work [GM12] of D. Grigoriev and P. Milman where a study of toric varieties provides an evidence that *Nash blow ups*[1], which are in general somewhat poorly understood, might hold the key to a surprisingly efficient tool in

[1]The tangent space $T_a(X)$ at a nonsingular point a of a projective variety $X \subseteq \mathbb{P}^n$ of dimension d can be identified with a point on the *Grassmannian* $\mathrm{Gr}(d+1, n+1)$, which is the space of $(d+1)$-dimensional linear subspaces of $\mathbb{K}^{n+1}$. Let X' be the set of nonsingular points on X. The *Nash blow up* of X is the closure in $X \times \mathrm{Gr}(d+1, n+1)$ of the graph of the map $X' \to \mathrm{Gr}(d+1, n+1)$ given by $x \mapsto T_a(X)$.

P. Mondal, *How Many Zeroes?*, CMS/CAIMS Books in Mathematics 2,
https://doi.org/10.1007/978-3-030-75174-6_XII

the problem of resolution of singularities[2]. The article [GM12], which only uses elementary properties of "binomial varieties" (which are slight generalizations of toric varities), would be very much accessible for a reader of this book after some familiarity with the notion of normalization (which is covered e.g. in [CLS11, Section 1.3]).

2. Newton-Okounkov bodies

The Newton-Okounkov body is a recent very fruitful generalization of the Newton polytope of a (Laurent) polynomial. It was originally introduced by A. Okounkov, who associated in [Oko96, Oko03] convex bodies to ample divisors on a smooth variety. R. Lazarsfeld and M. Mustata [LM09], and independently, K. Kaveh and A. Khovanskii [KK12] made further generalizations of this construction. Since then, there have been numerous articles on these structures. The construction of Kaveh and Khovanskii in particular is very elementary and leads to simple proofs of nontrivial results from intersection theory and also convex geometry. The series of articles by Kaveh and Khovanskii on Newton-Okounkov bodies, in particular [KK10, KK12], would constitute an excellent reading material for a reader of this book.

3. Bézout problem

Another natural extension of the material of this book would be a deeper study of the Bézout problem of counting numbers of solutions of systems of polynomials. A. Khovanskii studied higher dimensional analogues of this problem. In particular, he computes in [Kho78] the Euler characteristics and in [Kho16] the number of irreducible components of generic complete intersections on the torus. P. Philippon and M. Sombra [PS08] studied a parametrized version of Bernstein's theorem over a nonsingular curve, and gave an answer in terms of an associated "mixed integral," which is a generalization of mixed volume to concave functions. In a sequel to this book, we describe an inductive algorithm to compute the precise number (counted with multiplicity) of solutions of any given system of n polynomials in n variables starting from the base estimate given by theorems X.4 and X.7.

4. Newton diagrams

Newton diagrams continue to be deeply studied in numerous algebraic and geometric problems. G. Rond and B. Schober [RS17] gave a simple proof of a new criterion for irreducibility of a polynomial with power series coefficients in terms

[2]John Nash asked in a private communication to Hironaka in early 1960s (as noted by Spivakovsky [Spi90]) whether singularities (in characteristic zero) can be resolved by a finite sequence of Nash blow ups. It remains an open problem for all dimensions ≥ 2. M. Spivakovsky [Spi90] showed that in dimension two a finite sequence of *normalized Nash blow ups*, i.e. Nash blow ups followed by normalizations, does suffice. Grigoriev and Milman [GM12] show that for toric varieties the Nash blow-up corresponds to a higher dimensional analogue of the Euclidean division of integers, and the resolution of singularities via Nash blow ups becomes a problem in combinatorics; moreover for toric surfaces the resolution of singularities via normalized Nash blow ups has a *polynomial* complexity which was very surprising since all general algorithms, which are essentially based on the original algorithm of H. Hironaka, have a much higher complexity.

of its Newton diagram. The *Łojasiewicz exponent*[3] of a complex hypersurface is an important invariant which is simple to define, but very hard to compute in many concrete situations. S. Brzostowski, T. Krasiński and G. Oleksik [BKO20] determined the Łojasiewicz exponent of a Newton non-degenerate surface singularity in $\mathbb{C}^3$ in terms of the Newton diagram. In general, many invariants of a hypersurface singularity can be computed from its Newton diagram if the singularity is non-degenerate in some sense. As we have seen in chapter XI, almost all singularities with a given diagram are non-degenerate. On the other hand, a given function is degenerate for most choices of coordinates, and for a degenerate function, it is usually very difficult, if possible at all, to find coordinates in which it is non-degenerate. However, sometimes a degenerate function becomes non-degenerate after adding a quadratic form in new variables, and invariants computed from the Newton diagram of the new function can shed light on the original singularity. Motivated by these considerations, V. I. Arnold asked whether this is always possible [Arn04, Problems 1975-3, 1976-8]. This problem is little understood—see the article [Ste20] by J. Stevens for an exposition.

5. Counting real zeroes

In parallel to studying the relation between the topological complexity (as measured e.g. by number of zeroes) of systems defined over $\mathbb{C}$ and their Newton polyhedra, V. I. Arnold and his students also explored the relation between the topological complexity of systems defined over $\mathbb{R}$ and the number of nonzero terms (or some other algebraic/combinatorial measure of the complexity) of the systems, initiating the theory of "Fewnomials". The canonical introduction to this subject is by A. Khovanskii [Kho91], who proved the conjecture of A. Kushnirenko that the number of real zeroes is bounded by the number of nonzero terms of systems of equations. Khovanskii's bound however is not optimal, and many authors have worked on improving it, e.g. T-Y. Li, J. M. Rojas and X. Wang [LRW03], F. Bihan and F. Sottile [BS07]. Sottile [Sot11] gave a beautiful introduction to these and other recent developments on the subject. A highly recommended read is Kushnirenko's letter[4] to Sottile where he describes the philosophy behind his conjectures, and early developments of the Fewnomial theory, including the contributions of K. Sevastyanov who tragically died young in an accident.

---

[3] If f is a polynomial or an analytic function near the origin on $\mathbb{C}^n$, its *Łojasiewicz exponent* at the origin is the smallest $\theta > 0$ such that $|\nabla f| \geq C|z|^\theta$ for some $C > 0$ and all z in a neighborhood of the origin.

[4] Dated February 26, 2008, available from Frank Sottile's website.

APPENDIX A

Commutative algebra results used in chapter III without a proof

The first major result we use without a proof is a special case of the "principal ideal theorem" of W. Krull, which is a fundamental tool in algebraic treatments of dimension. We use it frequently, starting from the proof that the dimension of the set of common zeroes of k polynomials in n-variables has dimension at least $n - k$ (theorem III.80).

THEOREM A.1. ([AM69, Corollary 11.17, Theorem 11.25]). *Let R be a finitely generated integral domain over a field k, f be a non-zero element of R and $\mathfrak{p}$ be an* isolated prime ideal[1] *of the principal ideal fR of R generated by f. If f is not a unit, then* $\operatorname{tr.deg.}_k(R/\mathfrak{p}) = \operatorname{tr.deg.}_k(R) - 1$ *(where* $\operatorname{tr.deg.}_k$ *is the transcendence degree over k).*

We use a few basic properties of *localization* and *completion* of Noetherian rings. Short introductions to these notions are provided in appendix B.7 and section III.14. The following result, also due to Krull, is used in our proof that near a nonsingular point a variety is a "locally complete intersection" (theorem III.105).

THEOREM A.2. ([AM69, Corollary 10.18]). *Let R be a Noetherian local ring with maximal ideal $\mathfrak{m}$. If $\mathfrak{q}$ is an ideal of R, then $\mathfrak{q} = \bigcap_{r \geq 0}(\mathfrak{q} + \mathfrak{m}^r)$.*

Theorem A.3 below is used in this book for the first time in the proof that the completion of the local ring at a point of a variety is isomorphic to the quotient of a power series ring (proposition III.117). We use theorems A.2 and A.3 only in the case that R is of the form $\{f/g : f, g \in k[x_1, \ldots, x_n],\ g(0) \neq 0\}$, where k is a field.

[1]This means $\mathfrak{p}$ is minimal among the prime ideals containing the ideal $\langle f \rangle$ of R generated by f. It follows from the results discussed in appendix B.10 that under the hypotheses of theorem A.1 the isolated prime ideals of $\langle f \rangle$ are precisely those (finitely many) prime ideals $\mathfrak{p}_1, \ldots, \mathfrak{p}_s$, determined by f uniquely up to reordering, which satisfy the following property: $\sqrt{\langle f \rangle} = \bigcap_j \mathfrak{p}_j$ and this presentation is *minimal* in the sense that $\sqrt{\langle f \rangle} \neq \bigcap_{j' \neq j} \mathfrak{p}_{j'}$ for any $j = 1, \ldots, s$.

P. Mondal, *How Many Zeroes?*, CMS/CAIMS Books in Mathematics 2,
https://doi.org/10.1007/978-3-030-75174-6_A

THEOREM A.3. (Exactness of completion [AM69, Proposition 10.12]). *Let $I \subset J$ be ideals of a Noetherian ring R and let $\bar{R} := R/I$. Let $\hat{R}$ (respectively $\hat{\bar{R}}$) be the completion of R (respectively $\bar{R}$) with respect to J (respectively $J\bar{R}$). Then $\hat{\bar{R}} \cong \hat{R}/I\hat{R}$.*

APPENDIX B

Miscellaneous commutative algebra

By a "ring" in this book, we mean a commutative ring with identity. In this chapter, we briefly recall several notions related to rings which are used in the book.

1. Integral domain, UFD, PID

A *zero-divisor* in a ring R is a non-zero element f such that $fg = 0$ for some non-zero $g \in R$. An *integral domain* is a ring without zero-divisors. An *irreducible element* or *prime* of an integral domain is a non-zero element which is a non-unit[1] and which can not be expressed as a product of two non-units. The notion of prime elements is modeled after properties of prime numbers; indeed, the prime elements in the ring of integers are precisely the prime numbers (and their negatives). A *unique factorization domain* (in short, *UFD*) is an integral domain such that every non-unit can be represented as a product of irreducible elements and this representation is unique up to multiplication by a unit or reordering of the irreducible factors. The ring of integers is a UFD, and so are polynomial rings (in a finite or infinite number of variables) over the integers or over a field. The ring of integers satisfies the additional property that each of its ideals is *principal*, i.e. generated by a single element. An integral domain with the latter property is called a *principal ideal domain* (in short, *PID*). The following is usually called the "fundamental theorem" or the "structure theorem" of finitely generated modules over a PID. Its proof can be found in any standard introductory abstract algebra text, e.g. [DF04].

THEOREM B.1. *Let M be a finitely generated module over a PID R. Then*

$$M \cong R^r \bigoplus R/\langle p_1^{\alpha_1} \rangle \bigoplus R/\langle p_2^{\alpha_2} \rangle \cdots \bigoplus R/\langle p_k^{\alpha_k} \rangle$$

for some $r, \alpha_1, \ldots, \alpha_k \geq 0$ and primes $p_1, \ldots, p_k$.

2. Prime and maximal ideals

Given elements $g_1, \ldots, g_k$ of a ring R, we write $\langle g_1, \ldots, g_k \rangle$ to denote the ideal generated by R. Recall that a *prime* ideal of R is a proper ideal $\mathfrak{p}$ such that $fg \in \mathfrak{p}$

[1] A unit of a ring is an element which has a multiplicative inverse.

P. Mondal, *How Many Zeroes?*, CMS/CAIMS Books in Mathematics 2,
https://doi.org/10.1007/978-3-030-75174-6_B

implies either f or g is in $\mathfrak{p}$. A *maximal* ideal of R is a proper ideal $\mathfrak{m}$ such that the only proper ideal of R containing $\mathfrak{m}$ is $\mathfrak{m}$ itself. It is straightforward to check that every maximal ideal is prime. The following are some other basic properties of prime and maximal ideals—see e.g. [AM69, Chapter 1] for their proof.

THEOREM B.2. *Let R be a non-zero ring (remember that by "ring," we mean a commutative ring with identity).*

(1) Given any non-unit $g \in R$, there is a maximal ideal of R containing g.
(2) If $\mathfrak{q}$ is an ideal of R contained in the union of finitely many prime ideals $\mathfrak{p}_1, \ldots, \mathfrak{p}_k$ of R, then $\mathfrak{q} \subseteq \mathfrak{p}_j$ for some j.
(3) If $\mathfrak{p}$ is a prime ideal of R containing the intersection of finitely many ideals $\mathfrak{q}_1, \ldots, \mathfrak{q}_k$, then $\mathfrak{p} \supseteq \mathfrak{q}_j$ for some j. If $\mathfrak{p} = \bigcap_{j=1}^k \mathfrak{q}_j$, then $\mathfrak{p} = \mathfrak{q}_j$ for some j.
(4) The intersection of all prime ideals of R is precisely its nilradical, *i.e. the ideal consisting of all* nilpotent[2] *elements of R.*

Assertion (2) of theorem B.2 is sometimes referred to as "prime avoidance"; see [Eis95, Lemma 3.3] for a stronger variation and geometric interpretation.

3. Noetherian rings, Hilbert's basis theorem, annihilators

A ring R is called *Noetherian*[3] if every nonempty set $\mathcal{I}$ of ideals in R has a maximal element with respect to $\subseteq$, i.e. there is an ideal $\mathfrak{q} \in \mathcal{I}$ such that the only element of $\mathcal{I}$ containing $\mathfrak{q}$ is $\mathfrak{q}$ itself. The following is a basic characterization of Noetherian rings (see e.g. [AM69, Proposition 6.2]).

Proposition B.3. *Given a ring R, the following are equivalent:*

(1) R is Noetherian.
(2) Every ideal of R is finitely generated *(i.e. generated by finitely many elements).*

A field is Noetherian, since the only proper ideal of a field is the zero ideal. It is straightforward to check that the ring of integers, or more generally, any PID is Noetherian. Hilbert's basis theorem (theorem B.4), which we prove next, implies that (quotients of) polynomial rings in finitely many variables over a field are also Noetherian. Most of the rings that appear in this book are Noetherian. The ring $R[x_1, x_2, \cdots]$ of polynomials in infinitely many variables x_j, $j \geq 1$, over a ring R is *not* Noetherian, since e.g. the sequence of ideals $\langle x_1, \ldots, x_n \rangle$, $n \geq 1$, does not have any maximal element.

THEOREM B.4. (Hilbert's basis theorem). *The ring of polynomials in finitely many variables over a Noetherian ring is also Noetherian.*

---

[2] $g \in R$ is called *nilpotent* if there is $n \geq 1$ such that $g^n = 0$.

[3] in honour of Emmy Noether, who was one of the founders of modern algebra. A very influential and highly respected mathematician, she was denied a permanent academic position for most of her career on account of being a woman.

PROOF. It suffices to show that $R[x]$ is Noetherian if R is a Noetherian ring and x is an indeterminate over R. Let $\mathfrak{q}$ be an ideal of $R[x]$. We will find a finite set of generators of R. Given $f = a_0 + a_1x + \cdots + a_dx^d \in R[x]$, $a_d \neq 0$, we call a_d the *leading coefficient* of f . Let $\mathfrak{r} \subseteq R$ be the set of leading coefficients of all polynomials in $\mathfrak{q}$. It is easy to check that $\mathfrak{r}$ is an ideal in R. Since R is Noetherian, there are polynomials $f_1, \ldots, f_r \in \mathfrak{q}$ whose leading coefficients generate $\mathfrak{r}$. Let $M := \max\{\deg(f_i) : i = 1, \ldots, r\}$. For each $m < M$, let $\mathfrak{r}_m$ be the ideal in R generated by the leading coefficients of all polynomials $f \in \mathfrak{q}$ such that $\deg(f) = m$. By Noetherianity of R, there is a finite set $\{f_{mj}\}_j$ of polynomials in $\mathfrak{q}$ of degree m whose leading coefficients generate $\mathfrak{r}_m$. We claim that $\mathfrak{q}$ is generated by all the f_i and all the f_{mj}. Indeed, let $f \in \mathfrak{q}$. If $d := \deg(f) \geq M$, then since the leading coefficient of f can be expressed as a sum of $b_i \in R$ times the leading coefficient of f_i, it follows that the degree of $f - \sum_i b_i x^{d-\deg(f_i)} f_i$ is smaller than d. Similarly, if $d < M$, then the degree of $f - \sum_j c_j f_{dj}$ is smaller than d for some $\{c_j\}_j \subseteq R$. In any event, the degree of f can be reduced by subtracting an element from the ideal $\mathfrak{q}'$ generated by the f_i and the f_{mj}. Repeating this process finitely many times yields that $f \in \mathfrak{q}'$, i.e. $\mathfrak{q} = \mathfrak{q}'$, which proves the theorem. □

Given $g \in R$, one writes $(0 : g)$ for the set of all $f \in R$ such that $fg = 0$; in other words $(0 : g)$ is the *annihilator* of g. It is clear that the set of zero-divisors is precisely the union of the annihilators of the non-zero elements of R.

Proposition B.5. *The set of zero-divisors in a Noetherian ring R is a union of prime ideals which are of the form $(0 : g)$ for some $g \in R$.*

PROOF. Let $\mathcal{I}$ be the set of all ideals of a Noetherian ring R of the form $(0 : g)$, where g varies over non-zero elements of R. By Noetherianity of R, each $\mathfrak{q} \in \mathcal{I}$ is contained in a maximal element $\mathfrak{p} \in \mathcal{I}$. We will show that $\mathfrak{p}$ is prime. Indeed, pick $g \in R$ such that $\mathfrak{p} = (0 : g)$. If $\mathfrak{p}$ is not prime, then there is $f_1, f_2 \in R \setminus \mathfrak{p}$ such that $f_1 f_2 \in \mathfrak{p}$, i.e. $f_1 f_2 g = 0$. But then $(0 : f_2 g)$ properly contains $\mathfrak{p}$. This contradiction (to the maximality of $\mathfrak{p}$) completes the proof. □

In corollary B.27 below, we prove a stronger version of proposition B.5.

4. (Algebraic) Field extensions

A *field extension* K/F is simply a pair consisting of a field K and a subfield F of K. The *degree* $[K : F]$ of the extension K/F is the dimension of K as a vector space over F; we say that K/F is a *finite* extension if $[K : F] < \infty$.

Example B.6. $\mathbb{C}$ is a finite extension of $\mathbb{R}$ with $[\mathbb{C} : \mathbb{R}] = 2$. $\mathbb{R}$ is *not* a finite extension of $\mathbb{Q}$ (since e.g. $\mathbb{Q}$ is countable, but $\mathbb{R}$ is not).

An element $\alpha \in K$ is *algebraic* over F if there is a non-zero polynomial $f \in F[t]$ (where t is an indeterminate) such that $f(\alpha) = 0 \in K$. A *transcendental* element over F is an element which is not algebraic over F. The extension K/F is called *algebraic* if all elements of K are algebraic over F.

Example B.7. All elements of F are trivially algebraic over F. All finite extensions of F are algebraic. Indeed, if $d := [K : F] < \infty$, then $1, \alpha, \alpha^2, \ldots, \alpha^d$ are linearly dependent over F for each $\alpha \in K$, which leads to an algebraic equation of α over F. If x is an indeterminate over F, then no element of $F(x) \setminus F$ is algebraic over F (where $F(x)$ is the field of rational functions in x). For $\mathbb{Q} \subseteq \mathbb{R}$, one has $\sqrt{2}$ is algebraic over $\mathbb{Q}$, but π is not. All elements of $\mathbb{C}$ are algebraic over $\mathbb{R}$.

K is said to be an *algebraic closure* of F if K is algebraic over F and if every polynomial $f \in F[x]$ has a root in K. We say that K is *algebraically closed* if K is an algebraic closure of itself. The existence of algebraic closures and algebraically closed fields are by now standard results of algebra (see e.g. [DF04, Section 13.4]). The *fundamental theorem of algebra*[4] states that $\mathbb{C}$ is algebraically closed, and in particular, $\mathbb{C}$ is the algebraic closure of $\mathbb{R}$. An algebraically closed field must be infinite, since given finitely many elements $\alpha_1, \ldots, \alpha_d$ of a field K, no α_j is a root of $1 + \prod_{j=1}^{d}(x - \alpha_j)$.

5. Hilbert's Nullstellensatz

Let K be a field. Given a subset X of K^n, we write $I(X)$ for the set of all polynomials in $K[x_1, \ldots, x_n]$ which are zero at all points of X; it is straightforward to check that $I(X)$ is an ideal of $K[x_1, \ldots, x_n]$. Conversely, if $\mathfrak{q}$ is an ideal of $K[x_1, \ldots, x_n]$, we write $V(\mathfrak{q})$ for the set of points on K^n on which each element of $\mathfrak{q}$ is zero; it is straightforward to check that $I(V(\mathfrak{q})) \supseteq \mathfrak{q}$ and $V(I(X)) \supseteq X$. The following result of David Hilbert describes the basic correspondence between $I(\cdot)$ and $V(\cdot)$ in the case that K is algebraically closed.

THEOREM B.8. (Hilbert's Nullstellensatz). *Assume K is algebraically closed. Then for each ideal $\mathfrak{q}$ of $K[x_1, \ldots, x_n]$, $I(V(\mathfrak{q}))$ is the radical $\sqrt{\mathfrak{q}}$ of $\mathfrak{q}$. In particular, the maximal ideals of $K[x_1, \ldots, x_n]$ are of the form $I(a)$ for $a \in K^n$.*

In this section, we prove the Nullstellensatz following an argument of Enrique Arrondo [Arr06]. We start with a simple version of the "normalization lemma" of Emmy Noether.

Lemma B.9. *Let K be an infinite field, and $f \in K[x_1, \ldots, x_n]$ be a polynomial with $d := \deg(f) \geq 1$. Then there are $\lambda_1, \ldots, \lambda_{n-1} \in K$ such that the coefficient of x_n^d in $f(x_1 + \lambda_1 x_n, \ldots, x_{n-1} + \lambda_{n-1} x_n, x_n)$ is non-zero.*

PROOF. Let f_d be the homogeneous component of f of degree d. The coefficient of x_n^d in $f(x_1 + \lambda_1 x_n, \ldots, x_{n-1} + \lambda_{n-1} x_n, x_n)$ is $f_d(\lambda_1, \ldots, \lambda_{n-1}, 1)$. Note that $f_d(x_1, \ldots, x_{n-1}, 1)$ is a non-zero polynomial in $(x_1, \ldots, x_{n-1})$. It is then straightforward to show, e.g. by induction on n, that there are $\lambda_1, \ldots, \lambda_{n-1} \in K$ such that $f_d(x_1, \ldots, x_{n-1}, 1) \neq 0$ (this is exercise III.11). □

THEOREM B.10. (Weak Nullstellensatz). *Let K be an algebraically closed field and $\mathfrak{q}$ be a proper ideal of $K[x_1, \ldots, x_n]$. Then $V(\mathfrak{q}) \neq \emptyset$.*

---

[4] See [ht] for a delightful collection of many proofs of the fundamental theorem of algebra.

PROOF. Without loss of generality we may assume $\mathfrak{q} \neq 0$. We proceed by induction on n. If $n = 1$, then $\mathfrak{q}$ must be a principal ideal generated by a nonconstant polynomial f. Since K is algebraically closed, f has a root $a \in K$. Then $V(\mathfrak{q})$ contains a and therefore is nonempty. Now assume $n > 1$. Due to lemma B.9 after a change of coordinates if necessary we may assume that $\mathfrak{q}$ contains a non-zero polynomial g of the form $g = g_0 + g_1 x_n + \cdots + g_{e-1}x_n^{e-1} + x_n^e$ where $e \geq 1$ and $g_0, \ldots, g_{e-1} \in K[x_1, \ldots, x_{n-1}]$. Let $\mathfrak{q}' := \mathfrak{q} \cap K[x_1, \ldots, x_{n-1}]$. By the inductive hypothesis there is $(a_1, \ldots, a_{n-1}) \in V(\mathfrak{q}')$.

Claim B.10.1. $\mathfrak{r} := \{f(a_1, \ldots, a_{n-1}, x_n) : f \in \mathfrak{q}\}$ *is a proper ideal of* $K[x_n]$.

PROOF. It is clear that $\mathfrak{r}$ is an ideal of $K[x_n]$. Assume to the contrary that $\mathfrak{r} = K[x_n]$. Then there is $f \in \mathfrak{q}$ such that $f(a_1, \ldots, a_{n-1}, x_n) = 1$. Write $f = f_0 + f_1 x_n + \cdots + f_d x_n^d$ with $f_0, \ldots, f_d \in K[x_1, \ldots, x_{n-1}]$. Then $f_0(a_1, \ldots, a_{n-1}) = 1$ and $f_1(a_1, \ldots, a_{n-1}) = \cdots = f_d(a_1, \ldots, a_{n-1}) = 0$. Let $R \in K[x_1, \ldots, x_{n-1}]$ be the *resultant* of f and g with respect to x_n, i.e. R is the determinant of the following $(d+e) \times (d+e)$ matrix:

$$\begin{pmatrix} f_0 & f_1 & \cdots & \cdots & \cdots & f_d & 0 & \cdots & 0 \\ 0 & f_0 & f_1 & \cdots & \cdots & \cdots & f_d & 0 & 0 \\ & & \ddots & & & & & \ddots & \\ 0 & \cdots & 0 & f_0 & f_1 & \cdots & \cdots & \cdots & f_d \\ g_0 & g_1 & \cdots & g_{e-1} & 1 & 0 & 0 & \cdots & 0 \\ 0 & g_0 & g_1 & \cdots & g_{e-1} & 1 & 0 & 0 & 0 \\ & & \ddots & & & & \ddots & & \\ & & & \ddots & & & & \ddots & \\ 0 & \cdots & \cdots & 0 & g_0 & g_1 & \cdots & g_{e-1} & 1 \end{pmatrix}$$

where the non-zero entries of the first e rows are translations of $(f_0, \ldots, f_d)$, and the next d rows are translations of $(g_0, \ldots, g_{e-1}, 1)$. If the first column of the above matrix is replaced by the first column plus x_n times the second column plus x_n^2 times the third column and so on, then the first colum turns into the column vector (which is the transpose of) $(f, x_n f, \cdots, x_n^{e-1} f, g, x_n g, \ldots, x_n^{d-1} g)$. Expanding the resulting determinant along the first column then shows that R is in the ideal generated by f and g. It follows that $R \in \mathfrak{q}'$. On the other hand, evaluating the entries of the above matrix at $(a_1, \ldots, a_{n-1})$ converts it into a lower triangular matrix whose diagonal entries are all 1; this implies that $R(a_1, \ldots, a_{n-1}) = 1$, which contradicts the fact that $(a_1, \ldots, a_{n-1}) \in V(\mathfrak{q}')$ and proves the claim. □

The preceding claim and the $n = 1$ case of the theorem (which we already proved) then show that there is $a_n \in K$ such that $f(a_1, \ldots, a_n) = 0$ for all $f \in \mathfrak{q}$, which completes the proof of the theorem. □

Now we prove the Nullstellensatz (theorem B.8) via a classical argument of Rabinowitsch[5]. Let $\mathfrak{q}$ be an ideal of $K[x_1, \ldots, x_n]$. The containment $\sqrt{\mathfrak{q}} \subseteq I(V(\mathfrak{q}))$

[5]Although Rabinowitsch's trick, which first appeared in [Rab29], is widely known in algebraic geometry, the man Rabinowitsch is not. The most widely accepted account is that he is the mathematical

is straightforward to verify. We will show that $\sqrt{\mathfrak{q}} \supseteq I(V(\mathfrak{q}))$. Fix a set of generators $f_1, \ldots, f_m$ of $\mathfrak{q}$. Given $f \in I(V(\mathfrak{q}))$, consider the ideal $\mathfrak{r}$ of $K[x_1, \ldots, x_{n+1}]$ generated by $f_1, \ldots, f_m, x_{n+1}f - 1$. Then $V(\mathfrak{r}) = \emptyset$, since whenever $f_1 = \cdots = f_m = 0$, then $f = 0$ and therefore $x_{n+1}f - 1 = -1$. Theorem B.10 then implies that $1 \in \mathfrak{r}$, i.e. there is an equation of the form $1 = \sum_i f_i h_i + (x_{n+1}f - 1)h$. Substituting $x_{n+1} = 1/y$ in the preceding equation and clearing the denominator via multiplying by an appropriate power of y yields an equation of the form $y^N = \sum_i f_i \tilde{h}_i + (f - y)\tilde{h}$ for some $\tilde{h}_1, \ldots, \tilde{h}_m, \tilde{h} \in K[x_1, \ldots, x_n, y]$. Substituting $y = f$ in the preceding equation then shows that $f^N \in \mathfrak{q}$, as required. □

6. Nakayama's lemma

In this section, we present a brief discussion of *Nakayama's lemma* following [AM69]. Let R be a ring, and $\mathfrak{j}$ be its *Jacobson ideal*, i.e. the intersection of all the maximal ideals of R.

Lemma B.11. *$1 + f$ is a unit in R for each $f \in \mathfrak{j}$.*

PROOF OF LEMMA B.11. Pick $f \in \mathfrak{j}$. If $1 + f$ is not a unit, then there is a maximal ideal $\mathfrak{m}$ of R that contains $1 + f$ (theorem B.2, assertion (1)). Since $\mathfrak{m}$ contains both f and $1 + f$, it contains 1, which is a contradiction. □

Lemma B.12. (Nakayama's lemma). *Let M be a finitely generated R-module and $\mathfrak{q}$ be an ideal of R contained in the Jacobson ideal of R. If $\mathfrak{q}M = M$, then $M = 0$.*

PROOF. Assume $M \neq 0$ and let $u_1, \ldots, u_k$ be a minimal set of generators of M. Since $u_k \in \mathfrak{q}M$, there is an equation of the form $u_k = \sum_{j=1}^{k} f_j u_j$, with each $f_j \in \mathfrak{q}$. It follows that $(1 - f_k)u_k = \sum_{j=1}^{k-1} u_j f_j$. Since $1 - f_k$ is a unit in R (lemma B.11), it follows that u_k is an R-linear combination of $u_1, \ldots, u_{k-1}$. This means that M can be generated by $u_1, \ldots, u_{k-1}$, a contradiction. □

Let R be a local ring with (the unique) maximal ideal $\mathfrak{m}$, and $\Bbbk = R/\mathfrak{m}$ be its residue field. Let M be a finitely generated R-module. Since $M/\mathfrak{m}M$ is annihilated by $\mathfrak{m}$, it is a finitely generated $R/\mathfrak{m}$-module, i.e. a finite-dimensional vector space over $\Bbbk$.

Corollary B.13. *Let $m_1, \ldots, m_k \in M$ be such that their images span $M/\mathfrak{m}M$ as a vector space over $\Bbbk$. Then $m_1, \ldots, m_k$ generate M as a module over R.*

PROOF. Let N be the submodule of M generated by $m_1, \ldots, m_k$. Since the image of N in $M/\mathfrak{m}M$ is all of $M/\mathfrak{m}M$, it follows that $M = N + \mathfrak{m}M$. But then $\mathfrak{m}(M/N) = (N + \mathfrak{m}M)/N = M/N$. Lemma B.12 then implies that $M = N$, as required. □

---

physicist George Yuri Rainich, who had published several articles under his original name "Rabinowitsch" before immigrating to USA and changing his name. There are however some doubt about this claim - see the comments to the *MathOverflow* answer [he].

7. Localization, local rings

If S is a multiplicative closed subset of a ring R, then the *localization* R_S of R with respect to S is the equivalence class of "fractions" $\{f/g : f \in R,\ g \in S\}$ under the equivalence relation that $f/g \sim f'/g'$ provided $(fg' - f'g)h = 0$ for some $h \in S$. It is easy to check that R_S is a commutative ring with respect to the usual rules of addition and multiplication of fractions. Two cases of localizations are especially relevant for our purpose:

- Case 1: $S = \{f^k\}_k$ for some $f \in R$. In this case, we denote R_S by R_f, and say that R_f is the *localization of R at f*.
- Case 2: $S = R \setminus \mathfrak{p}$ for some prime ideal $\mathfrak{p}$ of R. In this case, we denote R_S by $R_\mathfrak{p}$, and say that $R_\mathfrak{p}$ is the *localization of R at $\mathfrak{p}$*.

The following proposition compiles some of the basic properties of localizations. We refer to [AM69, Chapter 3] for a lucid discussion of these and other basic properties of localizations of commutative rings.

Proposition B.14. *R_S is the zero ring if and only if $0 \in S$. If R_S is not the zero ring, then (the equivalence class of) every element of S is invertible in R_S, and every ideal I of of R_S is generated by $R \cap I$. In particular, if R is Noetherian, then so is R_S.*

A *local ring* is a ring with a unique maximal ideal. Proposition B.14 implies that the localization $R_\mathfrak{p}$ of a ring R at a prime ideal $\mathfrak{p}$ is a local ring, and the maximal ideal of $R_\mathfrak{p}$ is generated by $\mathfrak{p}$.

Example B.15. Given $a = (a_1, \dots, a_n) \in \mathbb{k}^n$, where $\mathbb{k}$ is a field, the set S_a of all polynomials $f \in R := \mathbb{k}[x_1, \dots, x_n]$ such that $f(a) \neq 0$ is multiplicatively closed. In fact $\mathfrak{m}_a := R \setminus S_a = \{f \in R : f(a) = 0\}$ is a maximal ideal[6]. It follows that $R_{S_a} = R_{\mathfrak{m}_a}$ is a local ring and its maximal ideal is generated by $\mathfrak{m}_a$.

8. Discrete valuation rings

A *discrete valuation* on a field K is a surjective map ν from K onto $\mathbb{Z} \cup \{\infty\}$ such that $\nu(0) = \infty$, $\nu(fg) = \nu(f) + \nu(g)$ and $\nu(f+g) \geq \min\{\nu(f), \nu(g)\}$ for each $f, g \in K$. It is straightforward to check that the set of all $f \in K$ such that $\nu(f) \geq 0$ is a subring of K; it is called the *valuation ring* of ν.

Example B.16. A basic example of a discrete valuation is the *order* of rational functions in one variable over a field $\mathbb{k}$. Recall that the order of $f = \sum_j a_j t^j \in \mathbb{k}[t]$, where t is an indeterminate, is $\operatorname{ord}(f) := \min\{j : a_j \neq 0\}$, and the order can be extended to the field $\mathbb{k}(t)$ of rational functions by defining $\operatorname{ord}(f/g) := \operatorname{ord}(f) - \operatorname{ord}(g)$. The valuation ring of ord is precisely the localization of $\mathbb{k}[t]$ at the maximal ideal generated by t, i.e. the subring $\{f/g : f, g \in \mathbb{k}[t],\ g(0) \neq 0\}$ of $\mathbb{k}(t)$.

---

[6]It is straightforward to check that $\mathfrak{m}_a$ is an ideal. It is maximal since if $f \notin \mathfrak{m}_a$, then $f - f(a) \in \mathfrak{m}_a$ and $1 = \frac{1}{f(a)}(f - (f - f(a)))$.

A *discrete valuation ring* is an integral domain which is the valuation ring of a discrete valuation on its field of fractions. We now record some properties of discrete valuation rings: their verification is straightforward, and is left as an exercise.

Proposition B.17. *Let ν be a discrete valuation on a field K, and $R := \{f \in K : \nu(f) \geq 0\}$ be the valuation ring of ν.*

(1) The units of R are precisely the elements f in K with $\nu(f) = 0$.
(2) $\mathfrak{m} := \{f \in K : \nu(f) > 0\}$ is the unique maximal ideal of R; in particular, R is a local ring.
(3) If f is any element in $\mathfrak{m}$ such that $\nu(f) = 1$, then $\mathfrak{m}$ is the (principal) ideal of R generated by f. More generally, every proper ideal of R is a principal ideal generated by f^k for some $k \geq 1$.
(4) If R contains a field $\Bbbk$ which is isomorphic to $R/\mathfrak{m}$, then for each $g \in R$, R/gR is a vector space over $\Bbbk$ of dimension $\nu(g)$.
(5) R uniquely determines ν, i.e. if ν' is a discrete valuation on K such that R is also the valuation ring of ν', then $\nu' = \nu$. □

Proposition B.17 in particular implies that a discrete valuation ring R is a local ring whose maximal ideal is principal. A *parameter* of R is a generator of its maximal ideal.

9. Krull dimension

A *chain* of prime ideals of length $n \geq 0$ in a ring R is a finite sequence $\mathfrak{p}_0 \subsetneq \mathfrak{p}_1 \subsetneq \cdots \subsetneq \mathfrak{p}_n$ of prime ideals of R. The *Krull dimension* of R is the supremum of the lengths of all chains of prime ideals in R.

Example B.18. Since the only maximal ideal of a nontrivial field F is the zero ideal, the Krull dimension of F is zero. Each of the following rings has Krull dimension one: the ring of integers, the ring of polynomials in one variable over a field, a discrete valuation ring.

10. Primary decomposition

A proper ideal $\mathfrak{q}$ of a ring R is *primary* if $fg \in \mathfrak{q}$ implies either $f \in \mathfrak{q}$ or $g^k \in \mathfrak{q}$ for some $k \geq 1$.

Proposition B.19. *Let $\mathfrak{q}$ be an ideal of a ring R and $\sqrt{\mathfrak{q}}$ be the radical*[7] *of $\mathfrak{q}$.*

(1) If $\mathfrak{q}$ is primary, then $\sqrt{\mathfrak{q}}$ is prime.
(2) If $\sqrt{\mathfrak{q}}$ is maximal, *then $\mathfrak{q}$ is primary.*
(3) Write $\mathfrak{p} := \sqrt{\mathfrak{q}}$. Assume $\mathfrak{q}$ and $\mathfrak{q}'$ are primary ideals of R such that $\sqrt{\mathfrak{q}'} = \sqrt{\mathfrak{q}} = \mathfrak{p}$. Then $\mathfrak{q} \cap \mathfrak{q}'$ is also primary and $\sqrt{\mathfrak{q} \cap \mathfrak{q}'} = \mathfrak{p}$.

PROOF. The proof of the first assertion is straightforward from the definition of a primary ideal - it is left as an exercise. For the second assertion, assume $\sqrt{\mathfrak{q}}$ is maximal.

[7]The radical of $\mathfrak{q}$ is the ideal consisting of all $g \in R$ such that $g^k \in \mathfrak{q}$ for some $k \geq 1$.

Claim B.19.1. *If $g \notin \sqrt{\mathfrak{q}}$, then g is a unit in $R/\mathfrak{q}$.*

PROOF. Indeed, there is $f \in \sqrt{\mathfrak{q}}$ and $u, v \in R$ such that $gu + fv = 1$. If $k \geq 1$ is such that $f^k \in \mathfrak{q}$, then the relation $(gu + fv)^k = 1$ reduces to $gu' = 1$ in $R/\mathfrak{q}$ for an appropriate $u' \in R$. □

Pick $f, g \in R$ such that $fg \in \mathfrak{q}$. If $g \notin \sqrt{\mathfrak{q}}$, then claim B.19.1 implies that $f \in \mathfrak{q}$. Therefore, $\mathfrak{q}$ is primary. It remains to prove the third assertion. It is straightforward to check that $\sqrt{\mathfrak{q} \cap \mathfrak{q}'} = \sqrt{\mathfrak{q}} \cap \sqrt{\mathfrak{q}'} = \mathfrak{p}$. Now pick $fg \in \mathfrak{q} \cap \mathfrak{q}'$ such that $f \notin \mathfrak{q} \cap \mathfrak{q}'$. Then f is not in either $\mathfrak{q}$ or $\mathfrak{q}'$, and since both are primary with radical $\mathfrak{p}$, it follows that $g \in \mathfrak{p} = \sqrt{\mathfrak{q} \cap \mathfrak{q}'}$, so that $\mathfrak{q} \cap \mathfrak{q}'$ is also primary. □

In the examples below $\Bbbk$ denotes a field.

Example B.20. The ideal $\mathfrak{q}$ of $\Bbbk[x, y]$ generated by x^2 and xy is *not* primary, since $xy \in \mathfrak{q}$, but $x \notin \mathfrak{q}$ and $y \notin \sqrt{\mathfrak{q}} = \langle x \rangle$. In particular, assertion (2) of proposition B.19 does not hold if "maximal" is replaced by "prime."

Example B.21. The primary ideals of $\Bbbk[t]$ are the zero ideal and the ideals generated by f^n, $n \geq 1$, for irreducible polynomials f. More generally, the non-zero primary ideals of a PID are precisely the ideals generated by powers of irreducible elements.

Example B.22. The ideal $\mathfrak{q}$ of $\Bbbk[x, y]$ generated by x and y^2 is primary (proposition B.19, assertion (2)). The radical of $\mathfrak{q}$ is the maximal ideal $\mathfrak{m}$ generated by x, y. Note that $\mathfrak{m} \supsetneq \mathfrak{q} \supsetneq \mathfrak{m}^2$, so that $\mathfrak{q}$ is *not* a prime-power.

A *primary decomposition* of an ideal $\mathfrak{q}$ is an expression

$$\mathfrak{q} = \bigcap_{j=1}^{k} \mathfrak{q}_j \tag{149}$$

of $\mathfrak{q}$ as a finite intersection of primary ideals $\mathfrak{q}_j$. Given such a primary decomposition, excluding any redundant ideal from the right-hand side and then grouping the ideals with the same radical, it can be ensured that

- $\mathfrak{q}$ can *not* be expressed as an intersection of less than k of the $\mathfrak{q}_j$, and
- the radicals $\sqrt{\mathfrak{q}_j}$ of $\mathfrak{q}_j$ are distinct (due to assertion (3) of proposition B.19);

in that case, we say that (149) is a *minimal* primary decomposition, and that the prime ideals $\sqrt{\mathfrak{q}_j}$ are *associated with* $\mathfrak{q}$ (that $\sqrt{\mathfrak{q}_j}$ are prime follows from proposition B.19). Note that

$$\sqrt{\mathfrak{q}} = \bigcap_{j=1}^{k} \sqrt{\mathfrak{q}_j}$$

is a (not necessarily minimal) primary decomposition of the radical $\sqrt{\mathfrak{q}}$ of $\mathfrak{q}$; in particular, $\sqrt{\mathfrak{q}}$ is a finite intersection of *prime* ideals associated with $\mathfrak{q}$.

Example B.23. Let $\mathfrak{q} := \langle x^2, xy \rangle \subset \Bbbk[x, y]$. We saw in example B.20 that $\mathfrak{q}$ is not primary. Each of the following is a minimal primary decomposition of $\mathfrak{q}$ (since $\langle x \rangle$

is prime and the radical of the other ideal in the decomposition is maximal)

$$\mathfrak{q} = \langle x \rangle \cap \langle x^2, y \rangle = \langle x \rangle \cap \langle x^2, xy, y^k \rangle \ (k \geq 1).$$

In particular, minimal primary decompositions are in general *not* unique[8].

Example B.24. Let S is a multiplicatively closed subset of R and $\mathfrak{q}$ be an ideal of R such that $\mathfrak{q} \cap S = \emptyset$. If $\mathfrak{q}$ is primary, then it is straightforward to check that the ideal $\mathfrak{q}_S$ generated by $\mathfrak{q}$ in the localization R_S of R is also primary, and moreover, $\sqrt{\mathfrak{q}_S} = (\sqrt{\mathfrak{q}})_S$. It follows that if $\mathfrak{q} = \bigcap_{j=1}^k \mathfrak{q}_j$ is a minimal primary decomposition of $\mathfrak{q}$, then the following is a minimal primary decomposition of $\mathfrak{q}_S$:

$$\mathfrak{q}_S = \bigcap_i (\mathfrak{q}_{j_i})_S,$$

where $\{j_i\}$ is the subset of $\{1, \ldots, k\}$ consisting of those j such that $\mathfrak{q}_j \cap S = \emptyset$.

Proposition B.25. *Let $\mathfrak{q}$ be an ideal in R which has a primary decomposition. Then every prime ideal in R containing $\mathfrak{q}$ contains one of the prime ideals associated with $\mathfrak{q}$. The minimal ideals among those associated with $\mathfrak{q}$ are precisely the minimal elements in the set of all ideals in R containing $\mathfrak{q}$.*

PROOF. Follows immediately from assertion (3) of theorem B.2. □

The following is a fundamental property of Noetherian rings; it is a combination of [AM69, Theorem 4.5, Proposition 7.17].

THEOREM B.26. *Assume R is Noetherian. Then every proper ideal has a primary decomposition. In particular, every radical ideal*[9] *is a finite intersection of prime ideals. The prime ideals associated with an ideal $\mathfrak{q}$ are precisely those prime ideals of R which occur in the set of ideals $(\mathfrak{q} : f) := \{g \in R : fg \in \mathfrak{q}\}$ as f varies over R, and hence are uniquely determined by $\mathfrak{q}$.*

Corollary B.27. *The set of zero-divisors in a Noetherian ring R is the union of the prime ideals associated with the zero ideal. Every prime ideal of R contains a prime ideal associated with the zero ideal. The minimal ideals among the prime ideals associated with the zero ideal are precisely the minimal elements of the set of* minimal[10] *prime ideals of R. In particular, every element of a minimal prime ideal of R is a zero-divisor.*

PROOF. The first assertion follows from combining proposition B.5 and theorem B.26. The remaining assertions then follow from proposition B.25. □

---

[8]However, the *isolated* primary ideals (i.e. primary ideals whose radicals are minimal among the radicals of all primary ideals appearing in the decomposition) in a primary decomposition are in fact unique (e.g. the ideal $\langle x \rangle$ will appear in every primary decomposition of $\langle x^2, xy \rangle$)—see [AM69, Corollary 4.11].

[9]An ideal is *radical* if it equals its own radical.

[10]A *minimal* prime ideal $\mathfrak{p}$ of a ring R is a prime ideal $\mathfrak{p}$ such that the only prime ideal of R contained in $\mathfrak{p}$ is $\mathfrak{p}$ itself.

Corollary B.28. *Let f be a non zero-divisor in a ring R and $\mathfrak{p}$ be a minimal prime ideal of R. If R is Noetherian, then (the image of) f remains a non zero-divisor in $R/\mathfrak{p}$.*

PROOF. Corollary B.27 implies that $f \notin \mathfrak{p}$. Therefore, if $g \notin \mathfrak{p}$, then $fg \notin \mathfrak{p}$. Therefore, f is not a zero-divisor in $R/\mathfrak{p}$, as required. □

11. Length of modules

Let M be a module over a ring R. A *composition series* of M of length $n \geq 0$ is a sequence

$$M = M_0 \supsetneq M_1 \supsetneq \cdots \supsetneq M_n = 0 \tag{150}$$

of R-submodules which is "maximal," or equivalently, each quotient M_{i-1}/M_i, $1 \leq i \leq n$, is *simple*, (that is, has no non-zero proper submodule). Not every module has a composition series. In fact the following are equivalent ([AM69, Proposition 6.8]):

(1) M has a composition series,
(2) M satisfies *both* ascending and descending chain conditions[11].

If M has a composition series, then all composition series of M have the same length ([AM69, Proposition 6.7]). The *length* $l(M)$ of M is defined to be infinite if it has no composition series; otherwise $l(M)$ is the length of any composition series of M. If (150) is a composition series of M, then each M_{i-1}/M_i is isomorphic to $R/\mathfrak{m}$ for some maximal ideal $\mathfrak{m}$ of R.

Example B.29. $\mathbb{Z}$ does not have a composition series as a module over itself. Indeed, $\mathbb{Z}$-modules of $\mathbb{Z}$ are simply ideals of $\mathbb{Z}$. Therefore if $1 = n_0 < n_1 < \cdots$ is an infinite sequence of positive integers, then there is an infinite sequence $\langle n_0 \rangle \supsetneq \langle n_0 n_1 \rangle \supsetneq \langle n_0 n_1 n_2 \rangle \supsetneq \cdots$ of ideals of $\mathbb{Z}$; consequently $l(\mathbb{Z}) = \infty$. On the other hand, if $R = \mathbb{Z}/12\mathbb{Z}$, then both of the following are composition series of R (as a module over R or over $\mathbb{Z}$):

$$\mathbb{Z}/12\mathbb{Z} \supsetneq 2\mathbb{Z}/12\mathbb{Z} \supsetneq 4\mathbb{Z}/12\mathbb{Z} \supsetneq 0$$
$$\mathbb{Z}/12\mathbb{Z} \supsetneq 3\mathbb{Z}/12\mathbb{Z} \supsetneq 6\mathbb{Z}/12\mathbb{Z} \supsetneq 0.$$

It follows that $l(\mathbb{Z}/12\mathbb{Z}) = 3$. Note that for each of these composition series the successive quotients are isomorphic to either $\mathbb{Z}/2\mathbb{Z}$ or $\mathbb{Z}/3\mathbb{Z}$. In general, given primes p_j and nonnegative integers n_j, $j = 1, \ldots, k$, it is straightforward to check that the length of $\mathbb{Z}/(\prod_j p_j^{n_j})\mathbb{Z}$ as a module over $\mathbb{Z}$ is $\sum_j n_j$ and each successive quotient of each of its composition series is isomoprhic to $\mathbb{Z}/p_j\mathbb{Z}$ for some j.

Proposition B.30. *Let M be a module over a ring R with a composition-series $M = M_0 \supsetneq M_1 \supsetneq \cdots \supsetneq M_n = 0$. Fix i, $1 \leq i \leq n$.*

(1) Each M_{i-1}/M_i is isomorphic to $R/\mathfrak{m}$ for some maximal ideal $\mathfrak{m}$ of R.

---

[11] M satisfies *ascending* (respectively, *descending*) *chain condition* if for every chain $M_0 \subseteq M_1 \subseteq \cdots$ (respectively, $M_0 \supseteq M_1 \supseteq \cdots$) of submodules of M, there is k such that $M_j = M_k$ for all $j \geq k$.

(2) *Let* $\mathfrak{m}$ *be as in assertion (1). If there is a subfield* k *of* R *which maps isomorphically onto* $R/\mathfrak{m}$*, then* M_{i-1}/M_i *is a one-dimensional vector space over* k.

PROOF. Pick any non-zero element $\bar{m} \in M_{i-1}/M_i$. Since M_{i-1}/M_i is simple, the map $R \to M_{i-1}/M_i$ given by $f \mapsto f\bar{m}$ is a surjective map and its kernel must be a maximal ideal of R. This implies the first assertion. The second assertion follows immediately from the first. □

Corollary B.31. *Assume* R *is a local ring with maximal ideal* $\mathfrak{m}$*, and there is a field* $k \subseteq R$ *which maps isomorphically onto* $R/\mathfrak{m}$*. Then for any* R*-module* M*, the length of* M *as an* R*-module is equal to the length of* M *as a* k*-module, which in turn is equal to the dimension of* M *as a* k*-vector space.* □

Proposition B.32. *Let* $\mathfrak{p}$ *be a* minimal *prime ideal of a Noetherian ring* R*. Then the localization* $R_\mathfrak{p}$ *of* R *at* $\mathfrak{p}$ *has a finite length as an* $R_\mathfrak{p}$*-module.*

PROOF. Since $\mathfrak{p}$ is a minimal prime ideal of R, the ideal generated by $\mathfrak{p}$ is the unique prime ideal of $R_\mathfrak{p}$, i.e. the Krull dimension of $R_\mathfrak{p}$ is zero. Since $R_\mathfrak{p}$ is also Noetherian, it follows that it is *Artinian*[12] ([AM69, Theorem 8.5]) as well. This implies that the length $R_\mathfrak{p}$ as a module over itself is finite ([AM69, Proposition 6.8]), as required. □

12. (In)Separable field extensions

Let K/F be a finite extension of a field F and $\alpha \in K$ be algebraic over F. Pick $f \in F[t]$ such that f is *monic*[13], $f(\alpha) = 0$, and f has the smallest possible degree among all non-zero polynomials $g \in F[t]$ such that $g(\alpha) = 0$. It is straightforward to see, since $F[t]$ is a PID, that these properties uniquely determine f, and f is *irreducible* in $F[t]$; we say that f is the *minimal polynomial* of α over F. We say that α is *separable* over F if f has distinct roots in the algebraic closure $\bar{F}$ of F. Recall that a polynomial g in one variable (over any field) has distinct roots if and only if the "greatest common divisor" $\gcd(g, g')$ of g and its derivative g' is a nonzero constant. Since $\gcd(f, f')$ is an element of $F[t]$ and since f is irreducible in $F[t]$, this immediately implies the following:

Proposition B.33. α *is separable over* F *if and only if the derivative of its minimal polynomial is a non-zero polynomial in* $F[t]$.

We say that K/F is a *separable* extension if every element of K is separable (and in particular, algebraic) over F.

Example B.34. If $p := \mathrm{char}(F) = 0$, then the derivative of any nonconstant polynomial over F is a non-zero polynomial. It follows that every element which is algebraic over F is also separable, i.e. every algebraic extension of F is separable. Now assume $p > 0$. Let x, y be indeterminates over F. Then $y^p - x$ is irreducible

[12] A ring R is *Artinian* if it satisfies the descending chain condition as a module over itself, i.e. if for every descending chain of ideals $I_0 \supseteq I_1 \supseteq \cdots$, there is k such that $I_j = I_k$ for all $j \geq k$.

[13] A polynomial in a single variable t is monic if the coefficient of its highest degree term is 1.

in $F(x)[y]$, so that $R := F(x)[y]/\langle y^p - x\rangle$ is an integral domain. Let K be the field of fractions of R. Then K/F is a algebraic, but *not* separable (since the image of y in K not separable over $F(x)$).

We now prove the *primitive element theorem* in the case that the base field is infinite. It is also true for finite fields (see e.g. [DF04, Chapter 14, Proposition 17]), but we do not use that case in this book. The following proof is taken from [Ful89, Problem 6-31].

THEOREM B.35. *Let K/F be a finite separable extension of fields. Assume $|F| = \infty$. Then there is $\alpha \in K$ such that $K = F(\alpha)$. Moreover, given $\alpha_1, \dots, \alpha_n \in K$ such that $K = F(\alpha_1, \dots, \alpha_n)$, one can ensure that $\alpha = \sum_j \lambda_j \alpha_j$ for some $\lambda_1, \dots, \lambda_n \in F$.*

PROOF. A straightforward induction on n shows that it suffices to prove the second assertion of the theorem for $n = 2$. So assume $K = F(\alpha_1, \alpha_2)$. Let $f_i \in F[t]$ be the minimal polynomial of α_i. Since K/F is separable, over the algebraic closure $\bar{K}$ of K there are factorizations of the form $f_1 = \prod_{j=1}^{d_j}(t - \alpha_{1,j})$, where $\alpha_1 = \alpha_{1,1} \neq \alpha_{1,j}$ for any $j > 1$. Since F is infinite, there is $\lambda \in F$ such that $\lambda\alpha_1 + \alpha_2 \neq \lambda\alpha_{1,j_1} + \alpha_2'$ for all $j_1 > 1$ and all roots α_2' of $f_2(t)$ in $\bar{K}$. Let $\alpha := \lambda\alpha_1 + \alpha_2$ and $f(t) := f_2(\alpha - \lambda t) \in F'(t)$, where $F' := F(\alpha)$. Then it is straightforward to check that $f(\alpha_1) = 0$, and for each $j_1 > 1$, $f(\alpha_{1,j_1}) \neq 0$. It follows that the greatest common divisor of f and f_1 in $F'(t)$ is $t - \alpha_1$, which means the ideal generated by f and f_1 in $F'(t)$ is the same as the ideal generated by $t - \alpha_1$. In particular, $t - \alpha_1 = g(t)f(t) + g_1(t)f_1(t)$ for some $g, g_1 \in F'(t)$. But then $\alpha_1 = -(g(0)f(0) + g_1(0)f_1(0)) \in F'$. It follows that $\alpha_2 := \alpha - \lambda\alpha_1 \in F'$ as well, so that $F' = K$, as required. □

Remark B.36. The arguments of the proof of theorem B.35 does not use the separability of α_2. Therefore, the following generalization of theorem B.35 holds: "Let $K = F(\alpha_1, \dots, \alpha_n)$ be a finite extension of F such that $\alpha_1, \dots, \alpha_{n-1}$ are separable over F. If $|F| = \infty$, then $K = F(\sum_j \lambda_j \alpha_j)$ for some $\lambda_1, \dots, \lambda_n \in F$."

A field F of characteristic p is called *perfect* if either $p = 0$ or if for every $\alpha \in F$, there is $\beta \in F$ such that $\beta^p = \alpha$. Following [Sha94, Appendix 5, Proposition 1], we now prove a result of F. K. Schmidt for infinite perfect fields (note that it is also true in the case when the (perfect) field is finite, see e.g. [ZS75a, Chapter II, Theorem 31]).

Corollary B.37. *Let F be a perfect field and K/F is a finitely generated field extension. If $|F| = \infty$, then there are $\alpha_1, \dots, \alpha_{d+1} \in K$ such that $\alpha_1, \dots, \alpha_d$ are algebraically independent over F, α_{d+1} is separable over $F(\alpha_1, \dots, \alpha_d)$, and $K = F(\alpha_1, \dots, \alpha_{d+1})$.*

PROOF. Pick $\beta_1, \dots, \beta_n \in K$ such that $K = F(\beta_1, \dots, \beta_n)$. Let d be the maximal number of the β_j which are algebraically independent over F. Without loss of generality we may assume that $\beta_1, \dots, \beta_d$ are algebraically independent over F.

Claim B.37.1. *For each $j = d, \dots, n$, reordering $\beta_1, \dots, \beta_j$ if necessary, we may ensure that $\beta_1, \dots, \beta_d$ are algebraically independent over F and $F(\beta_1, \dots, \beta_j) = F(\beta_1, \dots, \beta_d, \gamma_j)$ for some $\gamma_j \in K$.*

PROOF. The claim is clearly true for $j = d$. We proceed by induction and assume it is true for j, $d \leq j \leq n-1$. By definition of d, β_{j+1} is algebraic over $F(\beta_1, \dots, \beta_d)$. Let f be a non-zero irreducible polynomial in $F[t_1, \dots, t_d, t_{j+1}]$ (where the t_i are indeterminates) such that $f(\beta_1, \dots, \beta_d, \beta_{j+1}) = 0$. We claim that there is i such that $\partial f / \partial t_i$ is a non-zero polynomial. Indeed, otherwise $p := \text{char}(F)$ is non-zero and each t_i occurs in f in powers which are multiples of p. Then f is of the form $f = \sum b_{i_1,\dots,i_d,i_{j+1}} t_1^{pi_1} \cdots t_d^{pi_d} t_{j+1}^{pi_{j+1}}$. Choose $a_{i_1,\dots,i_d,i_{j+1}} \in F$ such that $a^p_{i_1,\dots,i_d,i_{j+1}} = b_{i_1,\dots,i_d,i_{j+1}}$. Then $f = (\sum a_{i_1,\dots,i_d,i_{j+1}} t_1^{i_1} \cdots t_d^{i_d} t_{j+1}^{i_{j+1}})^p$, which contradicts the irreducibility of f. Reorder $\beta_1, \dots, \beta_d, \beta_{j+1}$ if necessary so that $i = j+1$. Then $\beta_1, \dots, \beta_d$ still remain algebraically independent over F, and β_{j+1} is *separable* over $F(\beta_1, \dots, \beta_d)$ (proposition B.33). It then follows from the inductive hypothesis and remark B.36 that $F(\beta_1, \dots, \beta_{j+1}) = F(\beta_1, \dots, \beta_d, \gamma_j, \beta_{j+1}) = F(\beta_1, \dots, \beta_d, \gamma_{j+1})$ for some $\gamma_{j+1} \in K$, as required. □

Let γ_n be as in claim B.37.1. The arguments of the proof of claim B.37.1 show that reordering $\beta_1, \dots, \beta_d, \gamma_n$ if necessary, we may ensure that γ_n is separable over $F(\beta_1, \dots, \beta_d)$, which proves the corollary. □

Given an algebraic extension K/F, it is a standard result from algebra (see e.g. [DF04, Section 14.9]) that the set $\overline{F}_{sep}$ of all elements in K which are separable over F is a field; we say that $\overline{F}_{sep}$ is the *separable closure* of F in K. The *separable degree* (respectively *inseparable degree*) of K/F is $[\overline{F}_{sep} : F]$ (respectively $[K : \overline{F}_{sep}]$). The following result compiles a few basic properties regarding these notions (see e.g. [DF04, Section 14.9]):

Proposition B.38. *Assume $p := \text{char}(K)$ is non-zero*[14]*. Then the extension $K/\overline{F}_{sep}$ is* purely inseparable, *i.e. for each $\alpha \in K$, there is a non-zero integer m such that the minimal polynomial of α over $\overline{F}_{sep}$ is of the form $t^{p^m} - \alpha$. Moreover, the degree of K/F is the product of its separable degree and the inseparable degree, i.e.*

$$[K : F] = [K : \overline{F}_{sep}][\overline{F}_{sep} : F] \tag{151}$$

Example B.39. Assume $p := \text{char}(F) > 0$. Choose positive integers k, q, where q is relatively prime to p. If x, y are indeterminates, it follows as in example B.34 that $R := F(x)[y]/\langle y^{qp^k} - x \rangle$ is an integral domain and the quotient field K of R is not a separable extension over $F(x)$. Note that (the image of) y^{p^k} in K is separable over $F(x)$, since its minimal polynomial in $F(x)[t]$ is $t^q - x$, which is separable. It is not hard to see $\overline{F(x)}_{sep}$ is the field $F(x)(y^{p^k})$ generated over $F(x)$ by y^{p^k}. It follows

[14]In zero characteristic $\overline{F}_{sep} = K$ (example B.34), so that the conclusions of proposition B.38 are trivially true.

that $[\overline{F(x)}_{sep} : F(x)] = q$, $[K : \overline{F(x)}_{sep}] = p^k$, so that $[K : \overline{F(x)}_{sep}][\overline{F(x)}_{sep} : F(x)] = qp^k = [K : F(x)]$, as implied by (151).

13. Rings of formal power series over a field

A *(formal) power series* over a field $\mathbb{k}$ in variables $(x_1, \ldots, x_n)$ is a formal expansion of the form $f = \sum_{\alpha \in \mathbb{Z}^n_{\geq 0}} c_\alpha x^\alpha$, where $c_\alpha \in \mathbb{k}$, and x^α is a shorthand for the monomial $x_1^{\alpha_1} \cdots x_n^{\alpha_n}$. The *power series ring* $\mathbb{k}[[x_1, \ldots, x_n]]$ is the set of all such power series; it has the structure of a ring induced by the usual multiplication and product of power series. In this section, we write $\hat{R}$ for $\mathbb{k}[[x_1, \ldots, x_n]]$. It is straightforward to see that $\hat{R}$ is an algebra over $\mathbb{k}$, and also an *integral domain*.

Proposition B.40. *$\hat{R}$ is a* local ring, *i.e. it has a unique maximal ideal. The maximal ideal $\hat{\mathfrak{m}}$ of $\hat{R}$ consists of all power series with zero constant term. Every element in $\hat{R} \setminus \hat{m}$ is a unit in $\hat{R}$.*

PROOF. That $\hat{\mathfrak{m}}$ is a maximal ideal follows from the isomorphism $\hat{R}/\hat{R} \cong k$. If $f \in \hat{R} \setminus \hat{\mathfrak{m}}$, then $f = c(1+g)$ for some $c \in \mathbb{k} \setminus \{0\}$ and $g \in \hat{\mathfrak{m}}$. It is straightforward to check that

$$\tilde{f} := \frac{1}{c}(1 + \sum_{d \geq 1} (-1)^d g^d) \tag{152}$$

is a well-defined element in $\hat{R}$ and $\tilde{f} f = 1$. Therefore, f is a unit. It follows that $\hat{\mathfrak{m}}$ is the unique maximal ideal of $\hat{R}$. □

We now describe all the $\mathbb{k}$-algebra automorphisms[15] of $\hat{R}$. Given $f = \sum_\alpha c_\alpha x^\alpha \in \hat{R}$, Recall that the *homogeneous component of f of degree d* is $f_d := \sum_{|\alpha| = d} c_\alpha x^\alpha$, where $|\alpha| := \sum_j \alpha_j$; note that f_d is a *polynomial* for each $d \geq 0$. The *order* $\operatorname{ord}(f)$ of a power series f is the smallest m such that $c_\alpha \neq 0$ for some α with $|\alpha| = m$.

THEOREM B.41. *Let $f_1, \ldots, f_n \in \hat{m}mm$.*

(1) The map $x_j \mapsto f_j$, $j = 1, \ldots, n$, induces a well-defined $\mathbb{k}$-algebra homomorphism from $\Phi : \hat{R} \to \hat{R}$. All $\mathbb{k}$-algebra homomorphisms from $\hat{R}$ to itself is of this form.

(2) Φ is an isomorphism if and only if the linear parts of the f_j are linearly independent over $\mathbb{k}$.

PROOF. Assertion (1) is straightforward. The ($\Leftarrow$) implication of assertion (2) is also straightforward in the case that all f_j are linear. Moreover, the ($\Rightarrow$) implication of assertion (2) also follows from the linear case of the ($\Leftarrow$) implication. Now we prove the general case of the ($\Leftarrow$) implication of assertion (2). Due to the linear case, after an automorphism of $\hat{R}$ if necessary we may assume that the linear part of each f_j is precisely x_j, $j = 1, \ldots, n$. It is then straightforward to check that for any $f \in \hat{R}$,

[15]A $\mathbb{k}$-algebra automorphism of $\hat{R}$ is a $\mathbb{k}$-algebra isomorphism from $\hat{R}$ to itself.

(153) $\operatorname{ord}(\Phi(f)) = \operatorname{ord}(f)$. Moreover, if $m = \operatorname{ord}(f)$, then the homogeneous component of degree m of $\Phi(f)$ is the same as that of f.

Property (153) immediately implies that Φ is injective. To see that Φ is surjective, pick $f \in \hat{R}$. Let $m_0 := \operatorname{ord}(f)$ and g_0 be the degree-m_0 homogeneous component of f. Property (153) implies that the order m_1 of $f_1 := f - \Phi(g_0)$ is greater than m_0; let g_1 be the degree-m_1 homogeneous component of f_1 and set $f_2 := f_1 - \Phi(g_1)$. Continuing this way, we get a power series $g = g_0 + g_1 + \cdots \in \hat{R}$ such that $\Phi(g) = f$, as required. □

Corollary B.42. *If $f_1, \ldots, f_r \in \hat{\mathfrak{m}}$ has linearly independent (over $\Bbbk$) linear parts, then they generate a prime ideal in $\hat{R}$.* □

Let $R := \{g/f : f, g \in \Bbbk[x_1, \ldots, x_n],\ f(0) \neq 0\}$; in other words, R is the *localization* of the polynomial ring $k[x_1, \ldots, x_n]$ at the ideal generated by polynomials which vanish at the origin (example B.15). There is a natural map $R \to \hat{R}$ which is identity on polynomials, and for each polynomial f such that $f(0) \neq 0$, maps $1/f$ to a power series as in (152). The following is straightforward to verify:

Proposition B.43. *The map $R \to \hat{R}$ is injective.*

From now on, we will treat R as a subring of $\hat{R}$. Recall that R is also a local ring and $\mathfrak{m} := \hat{\mathfrak{m}} \cap R$ is the unique maximal ideal of R (example B.15).

Proposition B.44. $(\hat{\mathfrak{m}})^d \cap R = \mathfrak{m}^d$ *for each $d \geq 0$.*

PROOF. It is clear that $(\hat{\mathfrak{m}})^d \cap R \supseteq \mathfrak{m}^d$ for each d. for the opposite inclusion, assume $g/f = h \in \hat{\mathfrak{m}}^d$ for some $f, g \in \Bbbk[x_1, \ldots, x_n]$ such that $f(0) \neq 0$. But then the identity $g = fh \in \hat{R}$ implies that all monomials in g must have order $\geq d$. This means $g \in \mathfrak{m}^d$, as required. □

In the next section, we use "monomial orders" on $\hat{R}$ to derive some of its basic properties in a simple way, e.g. we show that $\hat{R}$ is Noetherian (corollary B.54) and that the dimensions of the quotients of $\hat{R}$ are "finitely determined" (theorem B.56).

14. Monomial orders on rings of formal power series

14.1. Monomial orders on $\mathbb{Z}^n_{\geq 0}$. A *monomial order* on $\mathbb{Z}^n_{\geq 0}$ is a binary relation $\preceq$ on $\mathbb{Z}^n_{\geq 0}$ such that

(a) $\preceq$ is a *total order*[16],
(b) $\preceq$ is compatible with the addition on $\mathbb{Z}^n_{\geq 0}$, i.e. if $\alpha \preceq \beta$, then $\alpha + \gamma \preceq \beta + \gamma$ for all $\gamma \in \mathbb{Z}^n_{\geq 0}$, and
(c) $0 \preceq \alpha$ for each $\alpha \in \mathbb{Z}^n_{\geq 0}$.

---

[16]A *total order* on a set S is a binary relation $\preceq$ on S which is reflexive (i.e. $x \preceq x$ for all $x \in S$), transitive (i.e. if $x \preceq y$ and $y \preceq z$ then $x \preceq z$), antisymmetric (i.e. if $x \preceq y$ and $y \preceq x$ then $x = y$), and totally comparable (i.e. either $x \preceq y$ or $y \preceq x$ for each $x, y \in S$).

We show below in corollary B.47 that every monomial order $\preceq$ is also a *well order* on $\mathbb{Z}^n_{\geq 0}$, i.e. for every nonempty subset S of $\mathbb{Z}^n_{\geq 0}$, there is a unique $\alpha \in S$ such that $\alpha \preceq \alpha'$ for all $\alpha' \in S$.

Example B.45. The *lexicographic order* $\preceq_{lex}$ on $\mathbb{Z}^n_{\geq 0}$ is defined as follows: if $\alpha, \beta \in \mathbb{Z}^n_{\geq 0}$, then $\alpha \preceq_{lex} \beta$ if either $\alpha = \beta$, or $\alpha \neq \beta$ and the first non-zero coordinate from the left of $\alpha - \beta$ is negative. Replacing "left" to "right" in the preceding definition leads to *reverse lexicographic order* $\preceq_{rlex}$. It is straightforward to check that $\preceq_{lex}$ and $\preceq_{rlex}$ are monomial orders.

A *corner point* of a subset S of $\mathbb{Z}^n_{\geq 0}$ is an element $\alpha \in S$ such that there is no $\alpha' \in S$, $\alpha' \neq \alpha$, such that $\alpha = \alpha' + \beta$ for some $\beta \in \mathbb{Z}^n_{\geq 0}$.

Lemma B.46. *Let S be a nonempty subset of $\mathbb{Z}^n_{\geq 0}$. The set $\mathcal{C}_S$ of corner points of S is finite and nonempty. Moreover, $S + \mathbb{Z}^n_{\geq 0} = \mathcal{C}_S + \mathbb{Z}^n_{\geq 0}$.*

PROOF. For any $\alpha = (\alpha_1, \ldots, \alpha_n) \in S$, define $S_{\leq \alpha} := \{\beta \in S : \alpha - \beta \in \mathbb{Z}^n_{\geq 0}\}$. Since $S_{\leq \alpha}$ is a finite set, it has a corner point β. It is clear that β is also a corner point of S and $\alpha \in \beta + \mathbb{Z}^n_{\geq 0}$. This proves the second assertion, and in addition shows that $\mathcal{C}_S$ is nonempty. It remains to show that $\mathcal{C}_S$ is finite. Assume to the contrary that it is infinite. Let $\alpha^0 = (\alpha^0_1, \ldots, \alpha^0_n)$ be an arbitrary element of S. For each $\alpha = (\alpha_1, \ldots, \alpha_n) \in \mathcal{C}_S \setminus \{\alpha^0\}$, there is j such that $\alpha_j < \alpha^0_j$. Fix j_1 such that there are infinitely many $\alpha \in \mathcal{C}_S$ with $\alpha_{j_1} < \alpha^0_{j_1}$. Since there are finitely many choices for α_{j_1}, it follows that there is $a_{j_1} < \alpha^0_{j_1}$ such that $\mathcal{C}^1_S := \{\alpha \in \mathcal{C}_S : \alpha_{j_1} = a_{j_1}\}$ is infinite. Now fix $\alpha^1 \in \mathcal{C}^1_S$. Replacing α^0 by α^1 and $\mathcal{C}_S$ by $\mathcal{C}^1_S$ and running the above procedure yields $j_2 \neq j_1$ and $a_{j_2} < \alpha^1_{j_2}$ such that $\mathcal{C}^2_S := \{\alpha \in \mathcal{C}^1_S : \alpha_{j_2} = a_{j_2}\}$ is infinite. Continuing this process, we will end up with an infinite set $\mathcal{C}^n_S$. But this is absurd, since $\mathcal{C}^n_S$ will consist of a single element $(a_1, \ldots, a_n)$ by construction. This contradiction implies that $\mathcal{C}_S$ was finite to begin with, which completes the proof. □

Corollary B.47. *Every monomial order on $\mathbb{Z}^n_{\geq 0}$ is also a well order on $\mathbb{Z}^n_{\geq 0}$.*

PROOF. Let $\preceq$ be a monomial order on $\mathbb{Z}^n_{\geq 0}$ and S be a nonempty subset of $\mathbb{Z}^n_{\geq 0}$. Let $\mathcal{C}_S$ be the set of corner points of S. Since $\mathcal{C}_S$ is finite and nonempty, it has a unique minimal element β_0 with respect to $\preceq$. For every $\alpha \in S$, lemma B.46 implies that there is $\beta \in \mathcal{C}_S$ such that $\alpha - \beta \in \mathbb{Z}^n_{\geq 0}$, so that properties (b) and (c) of monomial orders imply that $\beta \preceq \alpha$. It follows that β_0 is the minimal element of S with respect to $\preceq$. □

Corollary B.48. *If $S \subseteq \mathbb{Z}^n_{\geq 0}$, then the convex hull of $S + \mathbb{R}^n_{\geq 0}$ is a (convex) polyhedron.*

PROOF. Lemma B.46 implies that $S + \mathbb{R}^n_{\geq 0} = \mathcal{C}_S + \mathbb{R}^n_{\geq 0}$, where $\mathcal{C}_S$ is the *finite* set of corner points of S. It is straightforward to see that the convex hull of the Minkowski addition of any finite set with $\mathbb{R}^n_{\geq 0}$ is a convex polyhedron. □

Example B.49. If $S = \{(0,0)\} \cup \{(-n, 1) : n \geq 0\} \subseteq \mathbb{Z}^2$, then the convex hull of $S + \mathbb{R}^2_{\geq 0}$ is the upper half-plane excluding the negative x-axis $\{(a, 0) : a < 0\}$. In particular, corollary B.48 may not hold if $\mathbb{Z}^n_{\geq 0}$ is replaced by $\mathbb{Z}^n$.

We say that a monomial order $\preceq$ on $\mathbb{Z}^n_{\geq 0}$ has *finite depth* if for every $\alpha \in \mathbb{Z}^n_{\geq 0}$, the set $[\mathbb{Z}^n_{\geq 0}]_{\preceq \alpha} := \{\beta \in \mathbb{Z}^n_{\geq 0} : \beta \preceq \alpha\}$ is finite.

Example B.50. The lexicographic order $\preceq_{lex}$ from example B.45 does *not* have finite depth. The *graded lexicographic order* $\preceq_{grlex}$ on $\mathbb{Z}^n_{\geq 0}$ is defined as follows: if $\alpha = (\alpha_1, \ldots, \alpha_n)$ and $\beta = (\beta_1, \ldots, \beta_n) \in \mathbb{Z}^n_{\geq 0}$, then $\alpha \preceq_{grlex} \beta$ if either $\sum_j \alpha_j < \sum_j \beta_j$, or if $\sum_j \alpha_j = \sum_j \beta_j$ and $\alpha \preceq_{lex} \beta$. It is straightforward to check that $\preceq_{grlex}$ is a monomial order on $\mathbb{Z}^n_{\geq 0}$ of finite depth.

14.1. Monomial orders on rings of formal power series over a field. Following appendix B.13, we write $\hat{R}$ for the ring $\Bbbk[[x_1, \ldots, x_n]]$ of formal power series in indeterminates $(x_1, \ldots, x_n)$ over a field $\Bbbk$ and $\hat{m}$ for the (unique) maximal ring of $\hat{R}$. A *monomial order* on $\hat{R}$ is simply a monomial order $\preceq$ on $\mathbb{Z}^n_{\geq 0}$, which induces an ordering on the set of monomials in $(x_1, \ldots, x_n)$ by the relation: $x^\alpha \preceq x^\beta$ if and only if $\alpha \preceq \beta$. In this section, we use monomial orders to deduce some of the basic properties of $\hat{R}$. Fix a monomial order $\preceq$ on $\hat{R}$, or equivalently, on $\mathbb{Z}^n_{\geq 0}$. For each nonempty subset S of $\mathbb{Z}^n_{\geq 0}$, we write $\mathrm{In}_\preceq(S)$ for the minimal element of S with respect to $\preceq$. For $f = \sum_\alpha c_\alpha x^\alpha \in \hat{R}$, the *support* $\mathrm{Supp}(f)$ of f is the set $\{\alpha : c_\alpha \neq 0\} \subseteq \mathbb{Z}^n_{\geq 0}$; if $f \neq 0$ and $\alpha := \mathrm{In}_\preceq(\mathrm{Supp}(f))$, we say that α is the *initial exponent* of f denoted by $\exp_\preceq(f)$ and $c_\alpha x^\alpha$ is the *initial form* $\mathrm{In}_\preceq(f)$ of f. For each subset Q of $\hat{R}$, we write $\exp_\preceq(Q) := \{\exp_\preceq(f) : f \in Q,\ f \neq 0\} \subseteq \mathbb{Z}^n_{\geq 0}$ for the set of initial exponents of non-zero elements in Q.

Example B.51. Consider $f_1 := x_1^2 + x_1^3 - x_2$ and $f_2 := x_1^3 + x_1x_2 + x_2^2 \in \Bbbk[[x_1, x_2]]$. Then $\mathrm{In}_{\preceq_{lex}}(f_1) = x_2$ and $\mathrm{In}_{\preceq_{lex}}(f_2) = x_2^2$, so that $\exp_{\preceq_{lex}}(f_1) = (0, 1)$ and $\exp_{\preceq_{lex}}(f_2) = (0, 2)$. Substituting $x_2 = x_1^2 + x_1^3 - f_1$ in the expression of f_2 shows that $g := 2x_1^3 + x_1^4 + (x_1^2 + x_1^3)^2$ is in the ideal I of $\Bbbk[[x_1, x_2]]$ generated by f_1, f_2. If $\mathrm{char}(\Bbbk) \neq 2$, then $\mathrm{In}_{\preceq_{lex}}(g) = 2x_1^3$, so that $\exp_{\preceq_{lex}}(g) = (3, 0)$, and it is not hard to see that $\exp_{\preceq_{lex}}(I)$ is in fact the set of all elements on $\mathbb{Z}^2_{\geq 0}$ on and "above" the line joining $(0, 1)$ and $(3, 0)$ (see fig. 1c).

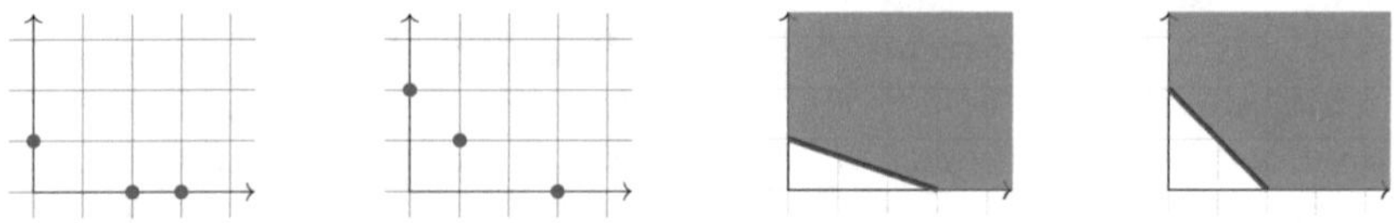

(A) $\mathrm{Supp}(f_1)$ (B) $\mathrm{Supp}(f_2)$ (C) $\exp_{\preceq_{lex}}(\langle f_1, f_2\rangle)$ (D) $\exp_{\preceq_{rlex}}(\langle f_1, f_2\rangle)$

FIGURE 1. Supports of $f_1 := x_1^2 + x_1^3 - x_2$ and $f_2 := x_1^3 + x_1x_2 + x_2^2$, and exponents of $\langle f_1, f_2\rangle$ when $\mathrm{char}(\Bbbk) \neq 2$.

Example B.52. To run the same computations as in example B.51 with $\preceq_{rlex}$, note that $\mathrm{In}_{\preceq_{rlex}}(f_1) = x_1^2$ and $\mathrm{In}_{\preceq_{rlex}}(f_2) = x_1^3$, so that $\exp_{\preceq_{rlex}}(f_1) = (2, 0)$ and $\exp_{\preceq_{rlex}}(f_2) = (3, 0)$. Now $h_1 := f_2 - x_1(f_1 - f_2) = 2x_1x_2 + x_1^2x_2 + x_2^2 + x_1x_2^2$ and $h_2 := x_1h_1 - x_2(2f_1 - f_2) = 2x_2^2 + 2x_1x_2^2 + x_1^2x_2^2 + x_2^3$, so that if $\mathrm{char}(\Bbbk) \neq 2$, then

$$\begin{aligned} \operatorname{In}_{\preceq_{rlex}}(h_1) &= 2x_1x_2, & \exp_{\preceq_{rlex}}(h_1) &= (1,1) \\ \operatorname{In}_{\preceq_{rlex}}(f_1) &= 2x_2^2, & \exp_{\preceq_{rlex}}(h_2) &= (0,2) \end{aligned}$$

It is then not hard to see that $\exp_{\preceq_{rlex}}(I)$ is the set of all elements on $\mathbb{Z}^2_{\geq 0}$ on and above the line joining $(0,2)$ and $(2,0)$ (see fig. 1d).

Using lemma B.46 it is straightforward to check that every ideal of $\hat{R}$ which is generated by monomials is finitely generated. We now prove the more general fact that every ideal of $\hat{R}$ is finitely generated.

THEOREM B.53. *Let $\preceq$ be a monomial order of finite depth on $\hat{R}$. Let I be an ideal of $\hat{R}$ and $\mathcal{C}_I = \{\alpha_1, \ldots, \alpha_s\}$ be the set of corner points of $\exp_{\preceq}(I)$. For each $i = 1, \ldots, s$, pick $f_i \in I$ such that $\operatorname{In}_{\preceq}(f_i) = x^{\alpha_i}$.*

(1) Each $g \in \hat{R}$ can be expressed as $g = \sum_i f_i h_i + g'$ for some $h_1, \ldots, h_s, g' \in \hat{R}$ such that either $g' = 0$, or $\exp_{\preceq}(g) \preceq \exp_{\preceq}(g')$ and $\exp_{\preceq}(g') \notin \exp_{\preceq}(I)$.
(2) I is generated by $f_1, \ldots, f_s$.
(3) If $g \in \hat{R} \setminus \{0\}$ is such that $\operatorname{Supp}(g) \subset \mathbb{Z}^n_{\geq 0} \setminus \exp_{\preceq}(I)$, then $g \notin I$. In particular, $\dim_{\Bbbk}(\hat{R}/I) = |\mathbb{Z}^n_{\geq 0} \setminus \exp_{\preceq}(I)|$ and if $\dim_{\Bbbk}(\hat{R}/I) < \infty$, then $\{x^\alpha : \alpha \in \mathbb{Z}^n_{\geq 0} \setminus \exp_{\preceq}(I)\}$ form a $\Bbbk$-basis of $\hat{R}/I$.

PROOF. At first, we prove the first assertion. Pick $g \in \hat{R}$. If $\alpha := \exp_{\preceq}(g) \notin \exp_{\preceq}(I)$, there is nothing to do. Otherwise pick the smallest i_1, $1 \leq i_1 \leq s$, such that $\alpha = \alpha_{i_1} + \beta_1$ for some $\beta_1 \in \mathbb{Z}^n_{\geq 0}$. Then $\operatorname{In}_{\preceq}(g) = c_1 x^{\alpha_{i_1} + \beta_1}$ for some $c_1 \in \Bbbk$. Write $g_0 := g$ and $g_1 := g - c_1 x^{\beta_1} f_{i_1}$. Continuing with g_1 and repeating this procedure, yields a sequence of elements $(g_k)_{k \geq 0}$ such that $\exp_{\preceq}(g_k) \precneqq \exp_{\preceq}(g_{k+1})$ for each k. Either this sequence is infinite, in which case the finite depth of $\preceq$ ensures that f is a $\hat{R}$-linear combination of the f_j, or it stops at a stage k, in which case $\exp_{\preceq}(g_k) \notin \exp_{\preceq}(I)$. This implies the first assertion. The second assertion follows from the first. The third assertion follows from the first assertion and finite depth of $\preceq$. □

Corollary B.54. *$\hat{R}$ is Noetherian.* □

Remark B.55. The requirement that $\preceq$ in theorem theorem B.53 have finite depth is only a technical convenience which makes the proof shorter. All the assertions of theorem B.53 are true without this requirement. For example, in examples B.51 and B.52, we have applied two different choices of $\preceq$ (none of which is of finite depth) to the same ideal I, and have seen that $|\mathbb{Z}^2_{\geq 0} \setminus \exp_{\preceq}(I)|$ remains the same even though $\exp_{\preceq}(I)$ were different. In particular, $\dim_{\Bbbk}(\Bbbk[[x_1, x_2]]/I) = 3$, and taking the monomials with exponents in $\mathbb{Z}^2_{\geq 0} \setminus \exp_{\preceq}(I)$, we get two bases of $\Bbbk[[x_1, x_2]]/I$, namely $(1, x_1, x_1^2)$ and $(1, x_1, x_2)$.

Given formal power series $f_1, \ldots, f_s \in \hat{R}$, we now show that the dimension of $\hat{R}/\langle f_1, \ldots, f_s \rangle$ as a vector space over $\Bbbk$ can be *finitely determined*, i.e. it can be determined by *polynomials* $g_1, \dots, g_s$ provided the g_j are "sufficiently close" to to f_j, where the "closeness" of elements of R will be measured by monomial orders of finite depth. Let $\preceq$ be a monomial order on $\hat{R}$. For each $f = \sum_\alpha c_\alpha x^\alpha \in \hat{R}$ and each $\beta \in \mathbb{Z}^n_{\geq 0}$ write $[f]_{\preceq\beta} := \sum_{\alpha \preceq \beta} c_\alpha x^\alpha$.

THEOREM B.56. *Assume $\preceq$ has finite depth. Let I be an ideal of $\hat{R}$ generated by $f_1, \dots, f_s$. For each $\beta \in \mathbb{Z}^n_{\geq 0}$, let $[I]_{\preceq\beta}$ be the ideal of $\hat{R}$ generated by $[f_j]_{\preceq\beta}$, $j = 1, \dots, s$.*

(1) If $\dim_{\mathbb{k}}(\hat{R}/I) < \infty$, then there is $\beta \in \mathbb{Z}^n_{\geq 0}$ such that $\exp_{\preceq}(I) = \exp_{\preceq}([I]_{\preceq\beta'})$ and $\dim_{\mathbb{k}}(\hat{R}/I) = \dim_{\mathbb{k}}(\hat{R}/[I]_{\preceq\beta'})$ for all $\beta' \succeq \beta$.

(2) If $\dim_{\mathbb{k}}(\hat{R}/I) = \infty$, then for each $N \geq 0$, there is $\beta \in \mathbb{Z}^n_{\geq 0}$ such that $\dim_{\mathbb{k}}(\hat{R}/[I]_{\preceq\beta'}) \geq N$ for all $\beta' \succeq \beta$.

PROOF. At first assume $\dim_{\mathbb{k}}(\hat{R}/I) < \infty$. Due to theorem B.53 to prove assertion item (1) it suffices to show that $\exp_{\preceq}([I]_{\preceq\beta'}) = \exp_{\preceq}(I)$ if β' is sufficiently "high" with respect to $\preceq$. Theorem B.53 implies that $\mathbb{Z}^n_{\geq 0} \setminus \exp_{\preceq}(I)$ is finite. For every finite subset S of $\mathbb{Z}^n_{\geq 0}$, write $\mathrm{ld}_{\preceq}(S)$ for the maximal element of S with respect to $\preceq$. Let $\beta_1 := \mathrm{ld}_{\preceq}(\mathbb{Z}^n_{\geq 0} \setminus \exp_{\preceq}(I))$. Then for each $\beta' \succeq \beta_1$ and each $g_1, \dots, g_s \in \hat{R}$, we have

$$\sum_j g_j [f_j]_{\preceq\beta'} = \sum_j g_j([f_j]_{\preceq\beta'} - f_j) + \sum_j g_j f_j$$

Let $h := \sum_j g_j [f_j]_{\preceq\beta'}$, $h_1 := \sum_j g_j([f_j]_{\preceq\beta'} - f_j)$ and $h_2 := \sum_j g_j f_j$. Since $\exp_{\preceq}(h_2) \in \exp_{\preceq}(I)$ and $\exp_{\preceq}(h_1) \succneqq \mathrm{ld}_{\preceq}(\mathbb{Z}^n_{\geq 0} \setminus \exp_{\preceq}(I))$, it is straightforward to see that if $h = h_1 + h_2 \neq 0$, then $\exp_{\preceq}(h) \in \exp_{\preceq}(I)$. It follows that $\exp_{\preceq}([I]_{\preceq\beta'}) \subseteq \exp_{\preceq}(I)$. Now let $\mathcal{C}_I$ be the set of corner points of $\exp_{\preceq}(I)$ and $\beta_2 := \mathrm{ld}_{\preceq}(\exp_{\preceq}(\mathcal{C}_I))$. Fix $\alpha \in \mathcal{C}_I$. There is $g_1, \dots, g_s \in \hat{R}$ such that $\exp(\sum_j g_j f_j) = \alpha$. Pick $\beta' \succeq \beta_2$, and define h, h_1, h_2 as above. Then $\exp_{\preceq}(h_2) \precneqq \exp_{\preceq}(h_1)$, so that $\exp_{\preceq}(h) = \exp_{\preceq}(h_2) = \alpha$. Therefore, $\mathcal{C}_I \subset \exp_{\preceq}([I]_{\preceq\beta'})$ and consequently lemma B.46 implies that $\exp_{\preceq}(I) \subset \exp_{\preceq}([I]_{\preceq\beta'})$. It follows that $\exp_{\preceq}(I) = \exp_{\preceq}([I]_{\preceq\beta'})$ whenever $\beta' \succeq \beta := \mathrm{ld}_{\preceq}\{\beta_1, \beta_2\}$, which proves the first assertion. For the second assertion, fix $N \geq 0$. Take an arbitrary finite subset S of $\mathbb{Z}^n_{\geq 0} \setminus \exp_{\preceq}(I)$ such that $|S| \geq N$ and let $\beta := \mathrm{ld}_{\preceq}(S)$. The same argument as in the beginning of the proof suggests that for each $\beta' \succeq \beta$ and each $g_1, \dots, g_s \in \hat{R}$, $\exp_{\preceq}(\sum_j g_j [f_j]_{\preceq\beta'}) \notin S$, so that theorem B.53 implies that $\dim_{\mathbb{k}}(\hat{R}/[I]_{\preceq\beta'}) \geq |S| \geq N$, as required. □

15. Primitive elements of $\mathbb{Z}^n$

The results we prove in this section follow almost directly from the fundamental theorem of finitely generated modules over a PID (theorem B.1); here, we give more elementary proofs. An element in $\mathbb{Z}^n$ is *primitive* if it is non-zero and the greatest common divisor of its non-zero coordinates is one. Every member of a basis of $\mathbb{Z}^n$ is primitive. The first result below shows that the converse is also true. In this section, we use $\langle \cdot, \cdot \rangle$ to denote the standard coupling of elements of $\mathbb{Z}^n$ and $(\mathbb{Z}^n)^*$.

Lemma B.57. *Let n be a positive integer and G be a subgroup of $\mathbb{Z}^n$.*

(1) If α is a primitive element in $\mathbb{Z}^n$, then there is a basis of $\mathbb{Z}^n$ containing α.

(2) $G \cong \mathbb{Z}^m$ for some $m \leq n$, and there is a basis $(\alpha_1, \ldots, \alpha_n)$ of $\mathbb{Z}^n$ and positive integers $k_1, \ldots, k_m$ such that $(k_1\alpha_1, \ldots, k_m\alpha_m)$ is a basis of G.

PROOF. For the first assertion, let $\alpha_1 := \alpha$. Since α_1 is primitive, there is $\beta_1 \in (\mathbb{Z}^n)^*$ such that $\langle \beta_1, \alpha_1 \rangle = 1$. Let $H_1 := \beta_1^\perp := \{\gamma \in \mathbb{Z}^n : \langle \beta_1, \gamma \rangle = 0\}$. If $H_1 \neq 0$, pick a primitive element $\alpha_2 \in H_1$ and $\beta_2 \in (\mathbb{Z}^n)^*$ such that $\langle \beta_2, \alpha_2 \rangle = 1$, and set $H_2 := H_1 \cap \beta_2^\perp$. Continue in this way up to the n-th step. It is straightforward to see that $\beta_1, \ldots, \beta_n$ are linearly independent (over $\mathbb{R}$), so that $H_n = 0$.

Claim B.57.1. *$(\alpha_1, \ldots, \alpha_n)$ is a basis of $\mathbb{Z}^n$.*

PROOF. Indeed, given any $\delta \in \mathbb{Z}^n$, let $d_1 := \langle \beta_1, \delta \rangle$. Then $\delta_1 := \delta - d_1\alpha_1 \in H_1$. Let $d_2 := \langle \beta_2, \delta_1 \rangle$. Then $\delta_2 := \delta_1 - d_2\alpha_2 \in H_2$. In this way, we get that $\delta_n = \delta - \sum_j d_j\alpha_j \in H_n = 0$, as required. □

The above claim proves the first assertion. For the second assertion, we may assume $G \neq \{0\}$. For each non-zero $\alpha \in G$, let d_α be the greatest common divisor of the non-zero coordinates of α. Pick $\alpha \in G$ with the smallest possible d_α. Due to the first assertion, we may assume without loss of generality that $\alpha = d_\alpha(1, 0, \ldots, 0)$. The minimality of d_α implies that for all $\beta = (\beta_1, \ldots, \beta_n) \in G$, the first coordinate β_1 of β is divisible by d_α. It follows that $G = \mathbb{Z}\alpha + G'$, where $G' := G \cap (\{0\} \times \mathbb{Z}^{n-1})$. The second assertion holds for G' by induction on n, which in turn implies that it holds for G. □

Corollary B.58. *Let $\phi : \mathbb{Z}^n \to \mathbb{Z}^m$ be a homomorphism of abelian groups and r be the rank (over $\mathbb{Q}$) of the matrix of ϕ. Then the matrix of ϕ with respect to appropriate bases of $\mathbb{Z}^n$ and $\mathbb{Z}^m$ is of the form*

$$[\phi] = \left[\begin{array}{c|c} D & 0 \\ \hline 0 & 0 \end{array}\right]$$

where D is an $r \times r$-diagonal matrix whose diagonal entries are positive integers.

PROOF. Let e_i^m and e_j^n denote respectively the i-th standard unit element in $\mathbb{Z}^m$ and the j-th standard unit element in $\mathbb{Z}^n$. Let $\alpha_1, \ldots, \alpha_m$ be the rows of the matrix of ϕ with respect to the standard bases of $\mathbb{Z}^n$ and $\mathbb{Z}^m$. Let G be the subgroup of $\mathbb{Z}^n$ generated by $\alpha_1, \ldots, \alpha_m$. Lemma B.57 implies that after a change of basis of $\mathbb{Z}^m$, we may assume that G is generated by $k_1e_1^n, \ldots, k_re_r^n$, where each k_j is a positive integer. Pick $\gamma_1 = (\gamma_{1,1}, \ldots, \gamma_{1,m}) \in \mathbb{Z}^m$ such that $\sum_j \gamma_{1,j}\alpha_j = k_1e_1^n$. Then γ_1 must be a primitive element of $\mathbb{Z}^m$, so after a change of basis in $\mathbb{Z}^m$, we may assume that $\gamma_1 = e_1^m$, which in turn implies that $\alpha_1 = k_1e_1^n$. Note that for each $j = 2, \ldots, m$, α_j is of the form $d_jk_1e_1^n + \alpha_j'$, with α_j' in the subgroup G' of $\mathbb{Z}^n$ generated by $k_2e_2^n, \ldots, k_re_r^n$. Therefore, after a change of basis of $\mathbb{Z}^m$ the form $e_j^m \mapsto e_j^m + d_jk_1e_1^m$, $j = 2, \ldots, m$, the matrix of ϕ is of the form

$$\left[\begin{array}{c|c} k_1 & 0 \\ \hline 0 & M \end{array}\right]$$

for some $(m-1)\times(n-1)$ matrix M. Now apply induction (say, on n) to the homomorphism from $\mathbb{Z}^{n-1}\to\mathbb{Z}^{m-1}$ induced by M. □

16. Symmetric multiadditive functions on a commutative semigroup

Throughout this section $(\mathcal{K},+)$ is a commutative semigroup and n is a positive integer. Let ρ be a mapping from $\mathcal{K}^n$ to $\mathbb{R}$. We say that ρ is *symmetric* if $\rho(f_1,\ldots,f_n)=\rho(f_{\sigma_1},\ldots,f_{\sigma_n})$ for each permutation σ of $(1,2,\ldots,n)$. We say that ρ is *multiadditive* if

$$\begin{aligned}\rho(h_1,\ldots,h_{j-1},qf+rg,h_{j+1}\ldots,h_n)&=q\rho(h_1,\ldots,h_{j-1},f,h_{j+1}\ldots,h_n)\\&\quad+r\rho(h_1,\ldots,h_{j-1},g,h_{j+1}\ldots,h_n)\end{aligned}\tag{154}$$

for each $q,r\in\mathbb{Z}_{\geq 0}$, $j\in\{1,\ldots,n\}$ and $f,g,h_1,\ldots,h_{j-1},h_{j+1},\ldots,h_n\in\mathcal{K}$. Throughout this section, we use ρ to denote a symmetric multiadditive function from $\mathcal{K}^n$ to $\mathbb{R}$.

16.1. Existence from polynomial functions.

For nonnegative integers $j_1,\ldots,j_k,n$ such that $n=j_1+\cdots+j_k$, we write $\binom{n}{j_1,\ldots,j_k}$ for the *multinomial coefficient* $n!/(j_1!\cdots j_k!)$. Recall that $\binom{n}{j_1,\ldots,j_k}$ is precisely the coefficient of $x_1^{j_1}\cdots x_k^{j_k}$ in $(x_1+\cdots+x_k)^n$.

Lemma B.59. *Let $\nu:\mathcal{K}\to\mathbb{R}$ be a function which satisfies the following property: for each $s\geq 1$ and $f_1,\ldots,f_s\in\mathcal{K}$, there are $\nu_\alpha(f_1,\ldots,f_s)\in\mathbb{R}$ for all $\alpha\in\mathcal{E}_s:=\{(\alpha_1,\ldots,\alpha_s)\in(\mathbb{Z}_{\geq 0}^s:\alpha_1+\cdots+\alpha_s=n\}$ such that for all $\lambda_1,\ldots,\lambda_s\in\mathbb{Z}_{\geq 0}$,*

$$\nu(\lambda_1f_1+\cdots+\lambda_sf_s)=\sum_{\alpha\in\mathcal{E}_s}\nu_\alpha(f_1,\ldots,f_s)\lambda_1^{\alpha_1}\cdots\lambda_s^{\alpha_s}\tag{155}$$

Then $\rho(f_1,\ldots,f_n):=\frac{1}{n!}\nu_{(1,\ldots,1)}(f_1,\ldots,f_n)$ is a symmetric multiadditive function from $\mathcal{K}^n$ to $\mathbb{R}$ such that $\rho(f,\ldots,f)=\nu(f)$.

PROOF. Fix $f\in\mathcal{K}$. Applying (155) with $s=1$ shows that $\nu(\lambda f)=\nu_n(f)\lambda^n$ for each $\lambda\in\mathbb{Z}_{\geq 0}$. Setting $\lambda=1$, we have $\nu_n(f)=\nu(f)$, and therefore $\nu(\lambda f)=\lambda^n\nu(f)$ for all $\lambda\in\mathbb{Z}_{\geq 0}$. It follows that

$$\nu(\lambda_1f+\cdots+\lambda_nf)=(\lambda_1+\cdots+\lambda_n)^n\nu(f)=\sum_\alpha\binom{n}{\alpha_1,\ldots,\alpha_n}\lambda_1^{\alpha_1}\cdots\lambda_n^{\alpha_n}\nu(f)$$

equating the coefficients of $\lambda_1\cdots\lambda_n$ of the middle and the rightmost polynomial in $(\lambda_1,\ldots,\lambda_n)$ yields that

$$\rho(f,\ldots,f)=\frac{1}{n!}\binom{n}{1,\ldots,1}\nu(f)=\nu(f)$$

It is clear that ρ is symmetric in its arguments. For multiadditivity, write elements of $\mathbb{Z}^{n+1}$ as $(\alpha, \beta, \bar{\gamma}) := (\alpha, \beta, \gamma_1, \ldots, \gamma_{n-1})$ and note that

$$\nu(\lambda f + \mu g + \tau_1 h_1 + \cdots + \tau_{n-1} h_n) = \sum_{(\alpha,\beta,\bar{\gamma}) \in \mathcal{E}_{n+1}} \nu_{(\alpha,\beta,\bar{\gamma})}(f, g, h_1, \ldots, h_{n-1}) \lambda^\alpha \mu^\beta \tau_1^{\gamma_1} \cdots \tau_{n-1}^{\gamma_{n-1}}$$

so that

$$\nu(\lambda(f + g) + \tau_1 h_1 + \cdots + \tau_{n-1} h_n) = \sum_{(\alpha,\beta,\bar{\gamma}) \in \mathcal{E}_{n+1}} \nu_{(\alpha,\beta,\bar{\gamma})}(f, g, h_1, \ldots, h_{n-1}) \lambda^{\alpha+\beta} \tau_1^{\gamma_1} \cdots \tau_{n-1}^{\gamma_{n-1}}$$

It follows that

$$\nu_{(\delta,\bar{\gamma})}(f + g, h_1, \ldots, h_{n-1}) = \sum_{\alpha+\beta=\delta} \nu_{(\alpha,\beta,\bar{\gamma})}(f, g, h_1, \ldots, h_{n-1})$$

In particular,

$$\begin{aligned}
\rho(f + g, h_1, \ldots, h_{n-1}) &= \frac{1}{n!}\nu_{(1,\ldots,1)}(f + g, h_1, \ldots, h_{n-1}) \\
&= \frac{1}{n!}(\nu_{(1,0,1,\ldots,1)}(f, g, h_1, \ldots, h_{n-1}) \\
&\quad + \nu_{(0,1,\ldots,1)}(f, g, h_1, \ldots, h_{n-1})) \\
&= \rho(f, h_1, \ldots, h_{n-1}) + \rho(g, h_1, \ldots, h_{n-1})
\end{aligned}$$

which implies that ρ is multiadditive, and completes the proof. □

16.2. Identities from homogeneous polynomials. The properties of symmetric multiadditive functions imply that its summands can be "expanded like polynomials," i.e. given $f_1, \ldots, f_N \in \mathcal{K}$ and $\lambda_{ij} \in \mathbb{Z}_{\geq 0}$,

$$\rho(\sum_{j=1}^N \lambda_{1j} f_j, \ldots, \sum_{j=1}^N \lambda_{nj} f_j) = \sum_{1 \leq j_1, \ldots, j_n \leq N} (\prod_{i=1}^n \lambda_{i,j_i}) \rho(f_{j_1}, \ldots, f_{j_n})$$

This is an analogue of the polynomial identity

$$\prod_{i=1}^n (\sum_{j=1}^N \lambda_{ij} x_j) = \sum_{1 \leq j_1, \ldots, j_n \leq N} \prod_{i=1}^n \lambda_{i,j_i} \prod_{i=1}^n x_{j_i}$$

where $x_1, \ldots, x_N$ are indeterminates. This observation immediately implies the principle that symmetric multiadditive functions "respect" polynomial identities:

Proposition B.60. *Let $x_1, \ldots, x_N$ be indeterminates and $f_1, \ldots, f_N \in \mathcal{K}$. Assume the following identity holds in $\mathbb{R}[x_1, \ldots, x_N]$:*

$$\sum_i r_i \prod_{j=1}^N (\sum_{k=1}^N \lambda_{ijk} x_k)^{\alpha_{ij}} = 0$$

where $r_i \in \mathbb{R}$ and $\lambda_{ijk}, \alpha_{ij} \in \mathbb{Z}_{\geq 0}$, and $\sum_{j=1}^{N} \alpha_{ij} = n$. Then

$$\sum_i r_i \rho(\sum_{k=1}^{N} \lambda_{i1k} f_k, \dots, \sum_{k=1}^{N} \lambda_{i1k} f_k, \dots, \sum_{k=1}^{N} \lambda_{iNk} f_k, \dots, \sum_{k=1}^{N} \lambda_{iNk} f_k) = 0$$

where $\sum_{k=1}^{N} \lambda_{ijk} f_k$ is repeated α_{ij}-times for each i, j.

□

We now use this observation to show that a symmetric multiadditive function is uniquely determined by its diagonal part.

Lemma B.61. *Let $x_1, \dots, x_n$ be indeterminates and $I \subseteq [n]$. Let $I := \{i_1, \dots, i_k\}$, where $k = |I|$. Write*

$$s_I := (x_{i_1} + \cdots + x_{i_k})^n$$
$$r_I := \sum_{\substack{j_1 + \cdots + j_k = n \\ j_l \geq 1,\ l=1,\dots,k}} \binom{n}{j_1, \dots, j_k} x_{i_1}^{j_1} \cdots x_{i_k}^{j_k}$$

Then

$$r_I = s_I - \sum_{\substack{J \subset I \\ |J| = k-1}} s_J + \sum_{\substack{J \subset I \\ |J| = k-2}} s_J + \cdots + (-1)^{k-1} \sum_{\substack{J \subset I \\ |J| = 1}} s_J \tag{156}$$

In particular,

$$n! x_1 \cdots x_n = \sum_{\substack{I \subseteq [n] \\ I \neq \emptyset}} (-1)^{n-|I|} (\sum_{i \in I} x_i)^n \tag{157}$$

PROOF. It suffices to prove identity (156), since identity (157) follows from (156) by setting $I = [n]$. Straightforward algebra shows that

$$s_I = \sum_{j_1 + \cdots + j_k = n} \binom{n}{j_1, \dots, j_k} x_{i_1}^{j_1} \cdots x_{i_k}^{j_k} = r_I + \sum_{J \subsetneq I} r_J = r_I + \sum_{\substack{J \subsetneq I \\ |J| = k-1}} r_J + \sum_{\substack{J \subsetneq I \\ |J| \leq k-2}} r_J$$

$$= r_I + \sum_{\substack{J \subsetneq I \\ |J| = k-1}} (s_J - \sum_{J' \subsetneq J} r_{J'}) + \sum_{\substack{J \subsetneq I \\ |J| \leq k-2}} r_J$$

$$= r_I + \sum_{\substack{J \subsetneq I \\ |J| = k-1}} s_J + \sum_{\substack{J \subsetneq I \\ |J| \leq k-2}} (1 - k + |J|) r_J = r_I + \sum_{\substack{J \subsetneq I \\ |J| = k-1}} s_J + \sum_{l=1}^{k-2} \sum_{\substack{J \subset I \\ |J| = l}} (1 - k + l) r_J$$

$$= r_I + \sum_{\substack{J \subsetneq I \\ |J| = k-1}} s_J - \sum_{\substack{J \subset I \\ |J| = k-2}} (s_J - \sum_{J' \subsetneq J} r_{J'}) + \sum_{l=1}^{k-3} \sum_{\substack{J \subset I \\ |J| = l}} (1 - k + l) r_J$$

$$= r_I + \sum_{\substack{J\subsetneq I\\|J|=k-1}} s_J - \sum_{\substack{J\subset I\\|J|=k-2}} s_J + \sum_{l=1}^{k-3}\sum_{\substack{J\subset I\\|J|=l}} (1-(k-l)+\binom{k-l}{2})r_J$$

In this way, at every step writing $r_J = s_J - \sum_{J'\subsetneq J} r_{J'}$ and rearranging terms, and observing that $r_J = s_J$ whenever $|J| = 1$, we will have

$$s_I = r_I + \sum_{\substack{J\subsetneq I\\|J|=k-1}} s_J - \sum_{\substack{J\subset I\\|J|=k-2}} s_J + \cdots + (-1)^{k-3}\sum_{\substack{J\subset I\\|J|=2}} s_J$$
$$+ \sum_{\substack{J\subset I\\|J|=1}} (1-(k-1)+\binom{k-1}{2} + \cdots + (-1)^{k-2}\binom{k-1}{k-2})s_J$$

Since $\sum_{j=0}^{k-2}(-1)^j\binom{k-1}{j} = (1-1)^{k-1} - (-1)^{k-1} = -(-1)^{k-1}$, the lemma follows. □

Corollary B.62. (See e.g. [hl]) *ρ is uniquely determined by its diagonal part, i.e. the map from $\mathcal{K}$ to $\mathbb{R}$ which sends $f \mapsto \rho(f,\ldots,f)$. More precisely,*

$$\rho(f_1,\ldots,f_n) = \frac{1}{n!}\sum_{\substack{I\subseteq[n]\\I\neq\emptyset}} (-1)^{n-|I|}\rho(\sum_{i\in I} f_i,\ldots,\sum_{i\in I} f_i) \tag{158}$$

PROOF. Combine proposition B.60 and identity (157). □

Lemma B.63. *Let $x_1,\ldots,x_k,y_1,\ldots,y_k$ be indeterminates. Then*

$$x_1\cdots x_k = \sum_{I\subseteq[k]} (-1)^{k-|I|}\prod_{i\in I}(x_i+y_i)\prod_{j\in[k]\setminus I} y_j \tag{159}$$

PROOF. We prove this by induction on k. For $k=1$ it boils down to the identity $x_1 = (x_1+y_1) - y_1$. In the general case, write

$$x_1\cdots x_k = \prod_{i=1}^k (x_i+y_i) - \sum_{j=1}^k (\prod_{i=1}^{j-1} x_i) y_j (\prod_{i=j+1}^k (x_i+y_i))$$

Applying induction to the expression $\prod_{i=1}^{j-1} x_i$ yields that

$$x_1\cdots x_k = \prod_{i=1}^k (x_i+y_i) - \sum_{j=1}^k (\sum_{I\subseteq[j-1]} (-1)^{j-1-|I|}\prod_{i\in I}(x_i+y_i)\prod_{i'\in[j-1]\setminus I} y_{i'}) y_j (\prod_{i=j+1}^k (x_i+y_i))$$
$$= \prod_{i=1}^k (x_i+y_i) - \sum_{j=1}^k \sum_{\substack{[k]\supseteq I\supseteq[k]\setminus[j]\\ I\not\ni j}} (-1)^{k-|I|-1}\prod_{i\in I}(x_i+y_i)\prod_{i'\in[k]\setminus I} y_{i'}$$
$$= \sum_{I\subseteq[k]} (-1)^{k-|I|}\prod_{i\in I}(x_i+y_i)\prod_{i'\in[k]\setminus I} y_{i'}$$

This completes the proof. □

For the next result, we assume $\rho : \mathcal{K}^n \to \mathbb{R}$ is a symmetric multiadditive "rational" map, i.e. ρ may not be defined everywhere on $\mathcal{K}^n$, but if $\rho(f_1, \ldots, f_n)$ is defined, then $\rho(f_{\sigma_1}, \ldots, f_{\sigma_n})$ are defined for all permutations σ of $(1, \ldots, n)$, and all of them take the same value in $\mathbb{R}$; and identity (154) holds whenever ρ is defined on at least two of the three elements of $\mathcal{K}^n$ that appear on (154). Given $h_1, \ldots, h_n \in \mathcal{K}$ and $I, J \subseteq [n]$ such that $|I|+|J| = n$, we write $\rho((h_i)_{i\in I}, (h_j)_{j\in J})$ for $\rho(h_{i_1}, \ldots, h_{i_k}, h_{j_1}, \ldots, h_{j_{n-k}})$ (provided it is defined), where $I = \{i_1, \ldots, i_k\}$ and $J = \{j_1, \ldots, j_{n-k}\}$. The following is an immediate consequence of the $k = n$ case of identity (159) and proposition B.60.

Corollary B.64. *Let $f_1, \ldots, f_n, g_1, \ldots, g_n \in \mathcal{K}$ be such that $\rho((f_i)_{i\in I}, (g_j)_{j\in[n]\setminus I})$ is defined for each $I \subseteq [n]$. Then*

$$\rho(f_1, \ldots, f_n) = \sum_{I\subseteq[n]} (-1)^{n-|I|} \rho((f_i + g_i)_{i\in I}, (g_j)_{j\in[n]\setminus I}) \tag{160}$$

□

APPENDIX C

Some results related to schemes

1. Macaulay's Unmixedness Theorem

In this section, we prove the *unmixedness theorem* of F. S. Macaulay for polynomial rings $R := k[x_1, \ldots, x_n]$ over an arbitrary field k following [ZS75b, Section VII.8]. This theorem is a crucial ingredient of intersection theory on nonsingular varieties, and was used in a fundamental way (via lemma IV.29) in many places in this book. The statement of this theorem requires the notion of *dimension*[1] $\dim(\mathfrak{q})$ of an ideal $\mathfrak{q}$ of R. If $\mathfrak{q}$ is prime, then $\dim(\mathfrak{q})$ is the transcendence degree (of the field of fractions) of $R/\mathfrak{q}$ over k. In general $\dim(\mathfrak{q})$ is the maximum of the dimensions of the prime ideals *associated with* $\mathfrak{q}$. Since the minimal elements of the set of prime ideals of R containing $\mathfrak{q}$ are precisely the minimal elements of the set of ideals associated with $\mathfrak{q}$ (proposition B.25), it follows that $\dim(\mathfrak{q})$ is the maximum of the dimensions of the *minimal* prime ideals containing $\mathfrak{q}$. One of the most fundamental properties of dimensions is given by Krull's principal ideal theorem (theorem A.1)—the following result is its straightforward consequence.

Proposition C.1. *Let k be an arbitrary field, and $\mathfrak{q}$ be a* proper *ideal of $k[x_1, \ldots, x_n]$ generated by $f_1, \ldots, f_m$, $m \leq n$. Then* $\dim(\mathfrak{q}) \geq n - m$. *If in addition* $\dim(\mathfrak{q}) = n - m$, *then* $\dim(\langle f_1, \ldots, f_j \rangle) = n - j$, *for each* $j = 1, \ldots, m$. ☐

The following is a special case of Macaulay's unmixedness theorem (theorem C.3).

Proposition C.2. *Let $g = g_1^{q_1} \cdots g_r^{q_r} \in R := k[x_1, \ldots, x_n]$, where g_j are irreducible (non-constant) polynomials and q_j are positive integers. Then $h \in R$ is a zero-divisor in $R/\langle g \rangle$ if and only if h is divisible by some g_j.*

PROOF. This immediately follows from the fact that R is a unique factorization domain. ☐

[1] It is in fact the same as the dimension of the algebraic variety determined by $\mathfrak{q}$ as defined in section III.11.

P. Mondal, *How Many Zeroes?*, CMS/CAIMS Books in Mathematics 2,
https://doi.org/10.1007/978-3-030-75174-6_C

THEOREM C.3. (Macaulay's unmixedness[2] theorem [ZS75b, Theorem VII.26]). *Let k be an arbitrary field, and $f_1, \dots, f_m \in k[x_1, \dots, x_n]$, $m \leq n$, be such that* $\dim(\langle f_1, \dots, f_m \rangle) = n - m$. *Then for each $j = 1, \dots, m$, f_j is a non zero-divisor in $k[x_1, \dots, x_n]/\langle f_1, \dots, f_{j-1} \rangle$.*

PROOF. Write $R := k[x_1, \dots, x_n]$. We proceed by induction on m. The case of $m = 1$ is obvious, and the case of $m = 2$ follows from proposition C.2. So assume $m \geq 3$. If f_m is a zero-divisor in $R/\langle f_1, \dots, f_{m-1} \rangle$, then it is contained in a prime ideal $\mathfrak{p}$ associated with $\mathfrak{q} := \langle f_1, \dots, f_{m-1} \rangle$. □

Claim C.3.1. $\dim(\mathfrak{p}) < n - m + 1$.

PROOF. By assumption $\mathfrak{p}$ is *not* a minimal prime ideal containing $\mathfrak{q}$, i.e. $\mathfrak{p} \supsetneq \mathfrak{p}' \supseteq \mathfrak{q}$ for some prime ideal $\mathfrak{p}'$ of R with $\dim(\mathfrak{p}') = n - m + 1$. The claim follows from applying Krull's principal ideal theorem (theorem A.1) to $R/\mathfrak{p}'$ (or from theorem III.78). □

Let $d := \dim(\mathfrak{p})$. Then there are $x_1, \dots, x_d$ which are algebraically independent over k in $R/\mathfrak{p}$. This means $\mathbb{k}[x_1, \dots, x_d] \cap \mathfrak{p} = 0$. Let $\tilde{R} := k(x_1, \dots, x_d)[x_{d+1}, \dots, x_n]$. Given an ideal $\mathfrak{r}$ of R, we write $\tilde{\mathfrak{r}}$ for the ideal of $\tilde{R}$ generated by $\mathfrak{r}$.

Claim C.3.2. *Let $\mathfrak{r}$ be a prime ideal of R.*

(1) *If $\mathfrak{r} \cap k[x_1, \dots, x_d] = \{0\}$, then* $\dim(\tilde{\mathfrak{r}}) = \dim(\mathfrak{r}) - d$.
(2) *The prime ideals associated to $\tilde{\mathfrak{q}}$ are precisely the ideals $\tilde{\mathfrak{r}}$ corresponding to those prime ideals $\mathfrak{r}$ associated with $\mathfrak{q}$ such that $\mathfrak{r} \cap k[x_1, \dots, x_d] = \{0\}$.*

PROOF. The first assertion follows immediately from choosing a transcendence basis (over k) of $R/\mathfrak{r}$ containing $x_1, \dots, x_d$. The second assertion follows from example B.24. □

Since $\dim(\mathfrak{q}) = n-(m-1)$ (proposition C.1), the first assertion of claim C.3.2 implies that $\tilde{\mathfrak{q}}$ is a proper ideal of $\tilde{R}$ with dimension $\leq n - d - (m - 1)$. Since $\tilde{\mathfrak{q}}$ is generated by $m - 1$ elements, proposition C.1 then implies that $\dim(\tilde{\mathfrak{q}})$ is precisely $n - d - (m - 1)$. On the other hand, applying the first assertion of claim C.3.2 with $\mathfrak{r} = \mathfrak{p}$ shows that $\tilde{\mathfrak{p}}$ is a zero-dimensional ideal of $\tilde{R}$ which is also associated with $\tilde{\mathfrak{q}}$. But then claim C.3.3 below gives a contradiction and proves theorem C.3.

Claim C.3.3. *Let $\mathfrak{b} \subset \mathfrak{a}$ be ideals in $R := k[x_1, \dots, x_r]$ such that*

(1) *$\mathfrak{a}$ is prime,*
(2) $\dim(\mathfrak{a}) = 0$,
(3) *$\mathfrak{b}$ is generated by $m - 1$ elements, and*
(4) $\dim(\mathfrak{b}) = r - (m - 1)$.

If $r \geq m$, then $\mathfrak{a}$ is not *associated with $\mathfrak{b}$.*

---

[2]An ideal is *unmixed* if all its associated prime ideals have the same dimension. Theorem C.3 is called the "unmixedness theorem" due to this equivalent formulation: "Let $\mathfrak{q}$ be an ideal of $k[x_1, \dots, x_n]$ generated by m polynomials. If $\dim(\mathfrak{q}) = n - m$, then $\mathfrak{q}$ is unmixed." The equivalence of this statement with theorem C.3 follows in straightforward manner from corollary B.27 and Krull's principal ideal theorem (theorem A.1).

To complete the proof of theorem C.3 it remains to prove claim C.3.3. Pick a set $g_1, \dots, g_{m-1}$ of generators of $\mathfrak{b}$. Let $\mathfrak{c}$ be the ideal of R generated by $g_1, \dots, g_{m-2}$ (recall that $m \geq 3$). Proposition C.1 implies that $\dim(\mathfrak{c}) = r - (m-2)$. If $h \in R$ is not in any of the minimal prime ideal associated with $\mathfrak{c}$, the induction hypothesis then applies to $g_1, \dots, g_{m-2}, h$ and implies that h is *not* a zero-divisor in $R/\mathfrak{c}$. Theorem B.26 and proposition C.1 then imply that all prime ideals associated to $\mathfrak{c}$ have dimension $r - m + 2$; enumerate these ideals as $\mathfrak{p}_1, \dots, \mathfrak{p}_s$. Let $\mathfrak{p}_{s+1}, \dots, \mathfrak{p}_{s'}$ be the *minimal* (i.e. $(r - m + 1)$-dimensional) prime ideals associated to $\mathfrak{b}$. Since $r \geq m$, each $\mathfrak{p}_j$ has dimension greater than zero. We will use the following result to construct a special type of polynomial which does not belong to any $\mathfrak{p}_j$.

LEMMA C.3.4. *Given finitely many positive dimensional prime ideals* $\mathfrak{p}_1, \dots, \mathfrak{p}_{s'}$ *in* $R := k[x_1, \dots, x_r]$, *there is* e, $1 \leq e \leq s'$, *and a polynomial* $h(x_1, \dots, x_{e-1})$ *such that the image of* $y_e := x_e + h(x_1, \dots, x_{e-1})$ *is transcendental over* k *in* $R/\mathfrak{p}_j$ *for each* j.

PROOF. Pick the smallest integer i_1 such that x_{i_1} is transcendental over k in $R/\mathfrak{p}_j$ for some j. Reorder the $\mathfrak{p}_j$ in a way that x_{i_1} is transcendental over k in $R/\mathfrak{p}_1, \dots, R/\mathfrak{p}_{s_1}$ and algebraic over k in $R/\mathfrak{p}_j$ for $j > s_1$. If $s_1 < s'$, then pick the smallest integer $i_2 > i_1$ such that x_{i_2} is transcendental over k in $R/\mathfrak{p}_j$ for some $j > s_1$. Then reorder $\mathfrak{p}_{s_1+1}, \dots, \mathfrak{p}_{s'}$ in a way that x_{i_2} is transcendental over k in $R/\mathfrak{p}_{s_1+1}, \dots, R/\mathfrak{p}_{s_2}$ and algebraic over k in $R/\mathfrak{p}_j$ for $j > s_2$. Since each $\mathfrak{p}_j$ is positive dimensional, we can continue in this way until there is an integer t such that $s_t = s'$. In particular,

(a) x_{i_t} is transcendental over k in $R/\mathfrak{p}_j$ for each $j = s_{t-1} + 1, \dots, s_t = s'$.

Now fix j, $s_{t-2} + 1 \leq j \leq s_{t-1}$. We claim that there is at most one integer q such that $x_{i_t} + x_{i_{t-1}}^q$ is algebraic over k in $R/\mathfrak{p}_j$. Indeed, otherwise there would be $q_1 \neq q_2$ such that $x_{i_{t-1}}^{q_1} - x_{i_{t-1}}^{q_2}$ is algebraic over k in $R/\mathfrak{p}_j$; but this is impossible since by construction $x_{i_{t-1}}$ is transcendental over k in $R/\mathfrak{p}_j$. Therefore

(b) there is a positive integer q_{t-1} such that $x_{i_t} + x_{i_{t-1}}^{q_{t-1}}$ is transcendental over k in $R/\mathfrak{p}_j$ for each $j = s_{t-2} + 1, \dots, s_{t-1}$.

Continuing in this way one can choose integers $q_{t-1}, q_{t-2}, \dots, q_1$ such that for each $t' = 1, \dots, t-1$,

(c) $x_{i_t} + x_{i_{t-1}}^{q_{t-1}} + \cdots + x_{i_{t-t'}}^{q_{t-t'}}$ is transcendental over k in $R/\mathfrak{p}_j$ for each $j = s_{t-t'-1} + 1, \dots, s_{t-t'}$.

Since for each $t' = 1, \dots, t$ and each $i < i'_t$, x_i is algebraic over k in $R/\mathfrak{p}_j$ for each $j > s_{t'-1}$, it follows that $x_{i_t} + x_{i_{t-1}}^{q_{t-1}} + \cdots + x_{i_1}^{q_1}$ satisfies the conclusion of the lemma with $e := i_t$. □

We go back to the proof of claim C.3.3. Let y_e be the polynomial from lemma C.3.4. Note that $R = k[x_1, \dots, x_{e-1}, y_e, x_{e+1}, \dots, x_r]$. Since $\dim(\mathfrak{a}) = 0$, the image of y_e in $R/\mathfrak{a}$ is algebraic over k. Let $f := \phi(y_e) \in k[y_e]$ be the polynomial with the minimum degree in y_e which is zero in $R/\mathfrak{a}$. Note that

(i) $f \in \mathfrak{a}$, and

(ii) f is *not* in any $\mathfrak{p}_j$, since the image of y_e is transcendental over k in $R/\mathfrak{p}_j$.

Claim C.3.5. *g_{m-1} is not a zero-divisor in $R/\langle \mathfrak{c}, f\rangle$.*

PROOF. Note that $R/\langle f\rangle \cong k(\bar{y}_e)[x_1,\ldots,x_{e-1},x_{e+1},\ldots,x_r]$, where $\bar{y}_e$ is the image of y_e in $R/\mathfrak{a}$. Let $\bar{\mathfrak{b}}$ be the ideal of $R/\langle f\rangle$ generated by $\mathfrak{b}$. Krull's principal ideal theorem (theorem A.1) and observation (ii) above imply that $\dim(\bar{\mathfrak{b}}) = \dim(\mathfrak{b}) - 1 = r - (m-1) - 1 = r - 1 - (m-1)$. Since $\bar{\mathfrak{b}}$ is generated by $m-1$ elements (namely the images $\bar{g}_i$ of g_i, $1 \le i \le m-1$) in the polynomial ring $R/\langle f\rangle$ over $k(\bar{y}_e)$, the induction hypothesis implies that $\bar{g}_{m-1}$ is not a zero-divisor in $(R/\langle f\rangle)/\langle \bar{g}_1,\ldots,\bar{g}_{m-2}\rangle \cong R/\langle f,\mathfrak{c}\rangle$, as required. □

Returning to the proof of claim C.3.3, assume to the contrary that $\mathfrak{a}$ is associated with $\mathfrak{b}$. Then there is $g \in R$ such that $\mathfrak{a} = (\mathfrak{b} : g)$ (theorem B.26); in particular $gf \in \mathfrak{b} = \langle \mathfrak{c}, g_{m-1}\rangle$. Therefore, there is $h \in R$ such that $gf - g_{m-1}h \in \mathfrak{c}$, i.e. $g_{m-1}h \in \langle \mathfrak{c}, f\rangle$. Claim C.3.5 then implies that $h \in \langle \mathfrak{c}, f\rangle$, i.e. $h - af \in \mathfrak{c}$ for some $a \in R$. It follows that $gf - g_{m-1}af = (g - g_{m-1}a)f \in \mathfrak{c}$. Since f is not in any prime ideal associated with $\mathfrak{c}$ (observation (ii)), f is *not* a zero divisor in $R/\mathfrak{c}$ (theorem B.26 and corollary B.27). It follows that $g - g_{m-1}a \in \mathfrak{c}$, so that $g \in \langle \mathfrak{c}, g_{m-1}\rangle = \mathfrak{b}$. But then $(\mathfrak{b} : g) = R \not\subseteq \mathfrak{a}$. This contradiction finishes the proof of claim C.3.3 and theorem C.3. □

2. Properties of order at a point on a possibly non-reduced curve

In section IV.3.2, we defined the notion of "order at a point on a possibly non-reduced curve," and stated some of its properties without proof. In this section, we prove these results, namely proposition IV.21 and theorem IV.24, in respectively proposition C.4 and theorem C.8. The proof uses somewhat more involved commutative algebra than the rest of the book. Recall that for a point a on a possibly non-reduced curve C and $f \in \mathcal{O}_{a,C}$, the order $\operatorname{ord}_a(f)$ of f at a is the dimension of $\mathcal{O}_{a,C}/f\mathcal{O}_{a,C}$ as a vector space over $\mathbb{k}$.

Proposition C.4. (proposition IV.21) *Let a be a point on a possibly non-reduced curve C and $f \in \mathcal{O}_{a,C}$. Let $C' := \operatorname{Supp}(C)$. Recall that $\mathcal{O}_{a,C'}$ is a quotient of $\mathcal{O}_{a,C}$.*

(1) If the image of f is a non zero-divisor in $\mathcal{O}_{a,C'}$, then $\operatorname{ord}_a(f) < \infty$. In particular, if f is a non zero-divisor in $\mathcal{O}_{a,C}$, then $\operatorname{ord}_a(f) < \infty$.
(2) If f is a non zero-divisor in $\mathcal{O}_{a,C}$, then $\operatorname{ord}_a(f) < \infty$.
(3) $\operatorname{ord}_a(f) = 0$ if and only if f is invertible in $\mathcal{O}_{a,C}$.
(4) If f is a non zero-divisor in $\mathcal{O}_{a,C}$ and $g \in \mathcal{O}_{a,C}$, then $\operatorname{ord}_a(fg) = \operatorname{ord}_a(f) + \operatorname{ord}_a(g)$.

PROOF. If the image of f is a non zero-divisor in $\mathcal{O}_{a,C'}$, then a version of Krull's principal ideal theorem [AM69, Corollary 11.18] implies that the quotient $\mathcal{O}_{a,C}/f\mathcal{O}_{a,C}$ of $\mathcal{O}_{a,C}$ by the ideal generated by f is a Noetherian local ring of Krull dimension zero. Therefore, it is also Artinian [AM69, Theorem 8.5] and a finite-dimensional vector space over $\mathbb{k}$ [AM69, Exercise 8.3], which proves the first assertion. The second assertion follows from the first, and the third assertion is straightforward to prove. For the last assertion, without loss of generality we may assume that $\mathrm{ord}_a(g) < \infty$. Let $(f_1, \ldots, f_l)$ be a basis of $\mathcal{O}_{a,C}/f\mathcal{O}_{a,C}$ and $(g_1, \ldots, g_m)$ be a basis of $\mathcal{O}_{a,C}/g\mathcal{O}_{a,C}$ over $\mathbb{k}$. Let $h \in \mathcal{O}_{a,C}$. Write K_f and K_g for respectively the $\mathbb{k}$-linear span of the f_i and of the g_j. Then $\mathcal{O}_{a,C} = K_f + f\mathcal{O}_{a,C} = K_f + fK_g + fg\mathcal{O}_{a,C}$, which implies that $f_1, \ldots, f_l, fg_1, \ldots, fg_m$ spans $\mathcal{O}_{a,C}$ modulo the ideal generated by fg. We claim that they are also linearly independent over $\mathbb{k}$. Indeed, pick $c_i, d_j \in \mathbb{k}$ such that $\sum_i c_i f_i + \sum_j d_j f g_j = fgh$, $h \in \mathcal{O}_{a,C}$. Then $\sum_i c_i f_i \in f\mathcal{O}_{a,C}$, so that $c_1 = \cdots = c_l = 0$. It follows that $f \sum_j d_j g_j = fgh$. Since f is a non zero-divisor, $\sum_j d_j g_j = gh \in g\mathcal{O}_{a,C}$. It follows that $d_1 = \cdots = d_m = 0$ as well. □

The proof of theorem IV.24 will be long. We start with a few auxiliary results.

Lemma C.5. *Let R be a ring containing a field k, and $t \in R$ be such that t is transcendental over k, and R contains the ring $k[[t]]$ of power series in t over k. Assume R/tR is a finite-dimensional vector space over k generated by (the images in R/tR of) $f_1, \ldots, f_m \in R$. Then $f_1, \ldots, f_m$ generate R as a module over $k[[t]]$.*

PROOF. Given $g \in R$, it can be successively expressed as $g = \sum_j f_j h_{0,j} + tg_1 = \sum_j f_j(h_{0,j} + th_{1,j}) + t^2 g_2$, and so on, resulting in an expression of the form $g = \sum_j f_j h_j$ with $h_j \in \mathbb{k}[[t]]$. □

Proposition C.6. *Let a be a point on an possibly non-reduced affine curve C. Assume $C' := \mathrm{Supp}(C)$ is irreducible and nonsingular at a. Let t be a regular function on C such that $t|_{C'}$ is a parameter of $\mathcal{O}_{a,C'}$. Assume t is not a zero-divisor in $\mathcal{O}_{a,C}$. Then*

(1) $\mathrm{ord}_a(t) = \mu_{C'}(C)$, *where $\mu_{C'}(C)$ is the multiplicity of C' in C (defined in section IV.3.2 in the paragraph preceding theorem IV.24),*
(2) $\mathrm{ord}_a(f) = \mathrm{ord}_a(f|_{C'})\,\mathrm{ord}_a(t)$ *for each regular function f on C.*

PROOF. We start with the first assertion. Without loss of generality we may assume that C is the closed subscheme of $\mathbb{k}^{n+1}$ determined by an ideal J in $S := \mathbb{k}[x_0, \ldots, x_n]$, and a corresponds to the point $x_0 = \cdots = x_n = 0$, and t is the restriction of x_0. Then $\hat{\mathcal{O}}_{a,C} \cong \hat{S}/J\hat{S}$ where $\hat{S} := \mathbb{k}[[x_0, \ldots, x_n]]$ (theorems A.3 and III.118). Let $r := \dim_{\mathbb{k}}(\hat{S}/(x_0\hat{S} + J\hat{S})) = \dim_{\mathbb{k}}(\hat{\mathcal{O}}_{a,C}/t\hat{\mathcal{O}}_{a,C})$. Assertion (2) of proposition C.4 and exercise III.120 imply that $r = \dim_{\mathbb{k}}(\mathcal{O}_{a,C}/t\mathcal{O}_{a,C}) = \mathrm{ord}_a(t)$. In particular, $r < \infty$, so that lemma C.5 implies that $\hat{S}/J\hat{S}$ is a finitely generated $\mathbb{k}[[x_0]]$-module. Since t is a non zero-divisor in $\mathcal{O}_{a,C}$, it follows that x_0 is a non zero-divisor in $\hat{\mathcal{O}}_{a,C} = \hat{S}/J\hat{S}$ (this is due to the general version of exactness of completions, see e.g. [AM69, Exercise 10.4.]). The fundamental theorem of finitely generated modules over a PID (theorem B.1) then implies that $\hat{S}/J\hat{S} \cong \mathbb{k}[[x_0]]^s$ as a module over $\mathbb{k}[[x_0]]$ for some $s \geq 0$. But then $s = \dim_{\mathbb{k}}(\hat{S}/(x_0\hat{S} + J\hat{S})) = r$. Since $t|_{C'}$ is a parameter of $\mathcal{O}_{a,C'}$, for each $i = 1, \ldots, n$, there is $u_i \in \mathcal{O}_{a,C}$ and a positive integer m_i such that $x_i - u_i x_0^{m_i}$ is nilpotent in $\mathcal{O}_{a,C}$. Pick a representative $\psi_i \in \hat{S}$ of u_i and set $y_i := x_i - x_0^{m_i}\psi_i$. Pick $\phi_1, \ldots, \phi_r \in \hat{S}$ such that $\hat{S}/J\hat{S} = \bigoplus_{i=1}^r \phi_i \mathbb{k}[[x_0]]$ as a $\mathbb{k}[[x_0]]$-module. Since the linear parts of $(x_0, y_1, \ldots, y_n)$ are linearly independent, theorem B.41 implies that $\hat{S} = \Bbbk[[x_0, y_1, \dots, y_n]]$. Moreover, since each y_j is nilpotent modulo $J\hat{S}$, each ϕ_i can be expressed as a $\Bbbk[[x_0]]$-linear combination of finitely many monomials in $(y_1, \dots, y_n)$. Let y^{α_j}, $j = 1, \dots, N$, be a minimal collection of such monomials such that every other monomial in $(y_1, \dots, y_n)$ is their $\Bbbk((x_0))$-linear combination modulo $J\hat{S}$. After a Gauss-Jordan elimination process and re-orderings of the α_j if necessary we may assume that $\phi_i = y^{\alpha_i} + \sum_{j=r+1}^{N} \phi_{i,j}(x_0) y^{\alpha_j}$, $i = 1, \dots, r$, where $\phi_{i,j}(x_0) \in \Bbbk((x_0))$. Now express y^{α_1} as a $\Bbbk[[x_0]]$-linear combination of the ϕ_i modulo $J\hat{S}$:

$$y^{\alpha_1} = \sum_{i=1}^{r} \rho_i(x)\phi_i = \sum_{i=1}^{r} \rho_i(x) y^{\alpha_i} + \sum_{i=1}^{r} \sum_{j=r+1}^{N} \rho_i(x)\phi_{i,j}(x) y^{\alpha_j}$$

The minimality assumption on the y^{α_i} implies that $\rho_1(x) = 1$, and $\rho_i(x)\phi_{i,j}(x) = 0$ if either $i > 1$ or $j > 1$. It follows that $y^{\alpha_1} = \phi_1$. The same arguments inductively show that $\phi_i = y^{\alpha_i}$ for each $i = 1, \dots, r$. In particular, $N = r$.

Claim C.6.1. *$\mathcal{O}_{C',C}$ (which was defined in section IV.2.5) can be identified with a $\Bbbk$-subalgebra of $\mathcal{O}_{a,C}[1/t]$.*

PROOF. Elements of $\mathcal{O}_{C',C}$ are of the form f/g where $f, g \in \mathcal{O}_{a,C}$ such that $g|_{C'} \not\equiv 0$. Let $m := \operatorname{ord}_a(g|_{C'})$. Lemma IV.13 implies that there is invertible $u \in \mathcal{O}_{a,C}$ such that $g = ut^m - h$ where h is a nilpotent in $\mathcal{O}_{a,C}$. Pick k such that $h^{k+1} = 0$. Write $f' := f/u$ and $h' := h/u$. Then $f/g = f'/(t^m - h') = t^{-m(k+1)} f' \sum_{j=0}^{k} t^{mj} h'^{k-j}$ which is an element of $\mathcal{O}_{a,C}[1/t]$. Since t is a non zero-divisor in $\mathcal{O}_{a,C}$, it follows that the map $\mathcal{O}_{C',C} \to \mathcal{O}_{a,C}[1/t]$ is injective. □

Claim C.6.1 and proposition III.115 imply that there are $\Bbbk$-subalgebra homomorphisms

$$\mathcal{O}_{C',C} \hookrightarrow \hat{\mathcal{O}}_{a,C}[1/t] \cong \hat{S}_{x_0}/J\hat{S}_{x_0} \tag{161}$$

where the first map is injective and $\hat{S}_{x_0} := \hat{S}[1/x_0] = \Bbbk((x_0))[[y_1, \dots, y_n]]$. Given any finite collection of elements $\beta_1, \dots, \beta_s \in \mathbb{Z}^n_{\geq 0}$, let $\hat{J}$ be the ideal of $\hat{S}_{x_0}/J\hat{S}_{x_0}$ generated by $y^{\beta_1}, \dots, y^{\beta_s}$ and $M := \hat{J} \cap \mathcal{O}_{C',C}$.

Claim C.6.2. *$y^{\beta_1}, \dots, y^{\beta_s}$ generate M as an ideal of $\mathcal{O}_{C',C}$.*

PROOF. Pick $h \in M$. Due to (161) there is $k \geq 0$ such that ht^k is represented by an element in $\mathcal{O}_{a,C}$ and hx_0^k is represented by a $\Bbbk[[x_0]]$-linear combination of the $y^{\beta_1}, \dots, y^{\beta_s}$ in $\hat{S}/J\hat{S}$. Write M' for the ideal of $\mathcal{O}_{a,C}$ generated by $y^{\beta_1}, \dots, y^{\beta_s}$ and N' for the ideal of $\mathcal{O}_{a,C}$ generated by M' and ht^k. Let $L' := N'/M'$ be the quotient of N'/M' as a module over $\mathcal{O}_{a,C}$. Let $\hat{L}', \hat{M}', \hat{N}'$ be the completion (with respect to the maximal ideal of $\mathcal{O}_{a,C}$) of respectively L', M', N'. Theorem A.3 implies that $\hat{L}' \cong \hat{N}'/\hat{M}'$, and then [AM69, Proposition 10.13] implies that $\hat{L}' \cong (\hat{\mathcal{O}}_{a,C} \otimes_{\mathcal{O}_{a,C}} N')/(\hat{\mathcal{O}}_{a,C} \otimes_{\mathcal{O}_{a,C}} M') = 0$. A theorem of Krull [AM69, Theorem

10.17] then implies that $L' = 0$, so that $ht^k \in M'$. Since t is invertible in $\mathcal{O}_{C',C}$, it follows that $h \in M'$, as required. $\square$

Recall that the only condition satisfied by $\alpha_1, \ldots, \alpha_r$ is that

(i) $\{y^{\alpha_i} : i = 1, \ldots, r\}$ is a minimal collection of monomials in $(y_1, \ldots, y_n)$ such that every other monomial is a $\Bbbk((x_0))$-linear combination modulo $J\hat{S}$.

While choosing such $\{\alpha_i\}$, we can start from monomials with *maximum* possible degree and then sequentially adjoin monomials with smaller degrees to ensure that for each $i \geq 2$,

(ii) $y^{\alpha_i + e_j}$ is a $\Bbbk((x_0))$-linear combination modulo $J\hat{S}$ of $y^{\alpha_1}, \ldots, y^{\alpha_{i-1}}$ for each $j = 1, \ldots, n$ (where e_j is the j-th standard unit vector in $\mathbb{Z}^n$).

For each $i = 0, \ldots, r$, let $\hat{J}_i$ be the ideal of $\hat{S}_{x_0}/J\hat{S}_{x_0}$ generated by $y^{\alpha_1}, \ldots, y^{\alpha_{r-i}}$ and $M_i := \hat{J}_i \cap \mathcal{O}_{C',C}$. Fix i, $0 \leq i \leq r-1$. Claim C.6.2 implies that M_i/M_{i+1} is generated by $y^{\alpha_{r-i}}$ as an $\mathcal{O}_{C',C}$-module. Since $y^{\alpha_{r-i}} \in M_i \setminus M_{i+1}$, it follows that $M_i/M_{i+1} \neq 0$. On the other hand, if h is a nilpotent element in $\mathcal{O}_{C',C}$, then as an element of $\hat{S}_{x_0}/\hat{J}\hat{S}_{x_0}$, h is in the ideal generated by $y_1, \ldots, y_n$, and the choice of the α_j ensures that $hM_i \subseteq M_{i+1}$. Combining these observations, we see that as an $\mathcal{O}_{C',C}$-module, $M_i/M_{i+1} \cong \mathcal{O}_{C',C}/\mathfrak{n}$, where $\mathfrak{n}$ is the (maximal) ideal of nilpotent elements of $\mathcal{O}_{C',C}$. It follows that $\mathcal{O}_{C',C} = M_0 \supset M_1 \supset \cdots \supset M_r = 0$ is a composition series of $\mathcal{O}_{C',C}$, so that $\mu_{C'}(C) = r$, which proves the first assertion of proposition C.6.

For the second assertion, we may assume without loss of generality that $q := \operatorname{ord}_a(f|_{C'}) < \infty$. Then there is $u \in \mathcal{O}_{a,C}$ such that $u|_{C'}$ is invertible and $(t^q - uf)|_{C'} = 0$. Lemma IV.13 implies that u is invertible in $\mathcal{O}_{a,C}$. Note that $h := t^q - uf$ is nilpotent in $\mathcal{O}_{a,C}$. By replacing C by a smaller neighborhood[3] of a in C if necessary we may assume h is nilpotent in S/J. Let z be a new indeterminate which we think of as the last coordinate of $\Bbbk^{n+1} \times \Bbbk = \Bbbk^{n+2}$, and let D be the closed subscheme of $\Bbbk^{n+2}$ determined by the ideal K in $T := \Bbbk[x_0, \ldots, x_n, z]$ generated by J and $z^q - h$. In T/K, we have $uf = t^q - z^q = \prod_{i=1}^{q}(t - \zeta_i z)$, where the ζ_i are the q-th roots of unity in $\Bbbk$. Note that $D' := \operatorname{Supp}(D)$ is isomorphic to C'; in particular a can be naturally identified with a point on D, which by an abuse of notation, we also denote by a. Since z is nilpotent in T/K, it follows that $(t - \zeta_i z)|_{D'} = t|_{D'}$ is a parameter of $\mathcal{O}_{a,D'}$ for each i. Since $T/K = \sum_{i=0}^{q-1} z^i S/J \cong (S/J)^q$ as a module over S/J, it follows that

(i) t is not a zero-divisor in $\mathcal{O}_{a,D}$, which implies that $t - \zeta_i z$ is not a zero-divisor in $\mathcal{O}_{a,D}$ for any i (since it is easy to check that if g_1 is a zero-divisor and g_2 is nilpotent, then $g_1 + g_2$ is a zero-divisor), and
(ii) $\operatorname{ord}_a(g|_D) = q \operatorname{ord}_a(g|_C)$ for each for each $g \in S/J$.

---

[3] Since C is a *possibly non-reduced* curve, by a "neighborhood of a in C," we mean an open subscheme of C containing a.

These observations together with the first assertion and proposition C.4 imply that $q\operatorname{ord}_a(f|_C) = \operatorname{ord}_a((uf)|_D) = \sum_{i=0}^{q-1}\operatorname{ord}_a((t-\zeta_i z)|_D) = q\operatorname{ord}_a(t|_D) = q^2\operatorname{ord}_a(t|_C)$, so that $\operatorname{ord}_a(f|_C) = q\operatorname{ord}_a(t|_C)$, as required. □

Lemma C.7. *Let D be a possibly non-reduced affine curve over $\mathbb{k}$ defined by an ideal I of $\mathbb{k}[x_1,\ldots,x_n]$. If $f \in R := \mathbb{k}[x_1,\ldots,x_n]/I$ is such that $r := \dim_{\mathbb{k}}(R/fR) < \infty$. Then $r = \sum_{a\in D}\operatorname{ord}_a(f)$.*

PROOF. Since $r < \infty$, there are finitely many zeroes of f on D. Denote them by $a_1,\ldots,a_k$; let $\mathfrak{m}_j$ be the maximal ideal of a_j in R and $\iota_j : R \to R_{\mathfrak{m}_j}$ be the natural map. □

Claim C.7.1. $fR = \bigcap_j \iota_j^*(fR_{\mathfrak{m}_j})$.

PROOF. (À LA MUMFORD [Mum95, Proposition 1.11]). We only need to show the "$\supset$" inclusion. Let $h \in \iota_j^*(fR_{\mathfrak{m}_j})$ for each j. Then for each maximal ideal $\mathfrak{m}$ of R, there exists $u \notin \mathfrak{m}$ such that $uh \in fR$. It follows that the ideal $(fR : h) := \{u \in R : uh \in fR\}$ of R is not contained in any maximal ideal of R. The Nullstellensatz then implies that $1 \in (fR : h)$, as required. □

Claim C.7.2. *If $j \neq j'$, then $\iota_j^*(fR_{\mathfrak{m}_j}) + \iota_{j'}^*(fR_{\mathfrak{m}_{j'}}) = R$.*

PROOF. Since $R_{\mathfrak{m}_j}/fR_{\mathfrak{m}_j}$ and $R_{\mathfrak{m}_{j'}}/fR_{\mathfrak{m}_{j'}}$ are Artinian local rings, their maximal ideals are nilpotent [AM69, Proposition 8.6]. Therefore, there exists q such that $(\mathfrak{m}_j)^q R_{\mathfrak{m}_j} \subseteq fR_{\mathfrak{m}_j}$ and $(\mathfrak{m}_{j'})^q R_{\mathfrak{m}_{j'}} \subseteq fR_{\mathfrak{m}_{j'}}$. Since $1 \in (\mathfrak{m}_j)^q + (\mathfrak{m}_{j'})^q$ [why?], the claim follows.

Claims C.7.1 and C.7.2 and the Chinese remainder theorem [AM69, Proposition 1.10] imply that

$$r = \dim_{\mathbb{k}}(R/fR) = \sum_{j=1}^{k}\dim_{\mathbb{k}}(R/\iota_j^*(fR_{\mathfrak{m}_j})) \tag{162}$$

Claim C.7.3. $R/\iota_j^*(fR_{\mathfrak{m}_j}) \cong R_{\mathfrak{m}_j}/fR_{\mathfrak{m}_j}$ *for each j.*

PROOF. It is straightforward to see that the natural map $R/\iota_j^*(fR_{\mathfrak{m}_j}) \to R_{\mathfrak{m}_j}/fR_{\mathfrak{m}_j}$ is injective. For surjectivity, note that if $h = g_1/g_2$ is an element of $(\mathfrak{m}_j)^q R_{\mathfrak{m}_j}$ for some $q \geq 0$, where $g_1 \in R$ and $g_2 \in R\setminus\mathfrak{m}_j$, then $h - cg_1 \in (\mathfrak{m}_j)^{q+1}R_{\mathfrak{m}_j}$, where $c \in \mathbb{k}$ is the image of g_2^{-1} in $\mathbb{k} \cong R_{\mathfrak{m}_j}/\mathfrak{m}_j R_{\mathfrak{m}_j}$. It follows by an induction on q that for each $h \in R_{\mathfrak{m}_j}$ and $q \geq 1$, there exists $h' \in R$ such that $h - h' \in (\mathfrak{m}_j)^{q+1}R_{\mathfrak{m}_j}$. Choosing q such that $(\mathfrak{m}_j)^{q+1}R_{\mathfrak{m}_j} \subseteq fR_{\mathfrak{m}_j}$ (which is possible due to the arguments in the proof of claim C.7.2) yields the required result. □

The result follows from (162) and claim C.7.3. □

THEOREM C.8. (theorem IV.24). *Let a be a point on a possibly non-reduced curve C. Let $C_1,\ldots,C_s$ be the irreducible components of* $\operatorname{Supp}(C)$ *containing a and $\pi_i : \tilde{C}_i \to C_i$ be the desingularizations of C_i. If f is a non zero-divisor in $\mathcal{O}_{a,C}$,* *then*

(163)

$$\operatorname{ord}_a(f) = \sum_i \mu_{C_i}(C) \operatorname{ord}_a(f|_{C_i}) = \sum_i \mu_{C_i}(C) \sum_{\tilde{a} \in \pi_i^{-1}(a)} \operatorname{ord}_{\tilde{a}}(\pi_i^*(f|_{C_i}))$$

PROOF. Write $C' := \operatorname{Supp}(C)$. Without loss of generality we may assume that

(i) C is affine, i.e. the closed subscheme of $\mathbb{k}^n$ determined by an ideal $\mathfrak{a}$ of $\mathbb{k}[x_1, \ldots, x_n]$; in particular $\mathbb{k}[C'] = \mathbb{k}[x_1, \ldots, x_n]/\sqrt{\mathfrak{a}}$;
(ii) $C' \setminus \{a\}$ is nonsingular;
(iii) f is in the maximal ideal of $\mathcal{O}_{a,C}$;
(iv) f restricts to a regular function on C', and
(v) the restriction of f to every irreducible component of C' is non-constant.

Let $\bar{C}'$ be the (unique) compactification of C' such that $\bar{C}' \setminus \{a\}$ is nonsingular (given a closed embedding of C' into an affine space, $\bar{C}'$ can be explicitly constructed by taking the closure of C' in an projective completion of the affine space and then resolving the singularities at infinity of the closure of C'). The restriction $f' := f|_{C'}$ of f to C' induces a morphism $\bar{C}' \to \mathbb{P}^1$ (corollary III.109); we use f' to denote this morphism as well. Note that $f'(a) = 0 \in \mathbb{k} \subset \mathbb{P}^1$ (since f is in the maximal ideal of $\mathcal{O}_{a,C}$). It follows that

(vi) There is a finite set S of $\mathbb{k} \setminus \{0\}$ such that $f'^{-1}(\mathbb{k} \setminus S)$ is an *affine* curve[4] (exercise III.97).

Let $D' := f'^{-1}(\mathbb{k} \setminus S)$ and $p' := \prod_{s \in S}(f' - s) \in \mathbb{k}[C'] \cap \mathbb{k}[D']$. Then $D' \supseteq C' \setminus V(p')$. Let $q' \in \mathbb{k}[D']$ such that $q'(a) \neq 0$ and $D' \setminus V(q') \subseteq C'$. There is $N \geq 0$ such that $p'^N q' \in \mathbb{k}[C']$ (since $\mathbb{k}[D'] \subseteq \mathbb{k}[C']_{p'}$). Let $g' := p'^{N+1} q' \in \mathbb{k}[f']$. It is straightforward to check that g' is regular on $C' \cup D'$, $g'(a) \neq 0$, $D' \setminus V(g') = C' \setminus V(g')$, and $\mathbb{k}[D']_{g'} = \mathbb{k}[C']_{g'}$. In other words, if $g \in \mathbb{k}[x_1, \ldots, x_n]$ is such that $g' = g|_{C'}$, and we write X for the open subscheme $C \setminus V(g)$ of C, and X' for the support of X, then there is a sequence of morphisms as in fig. 1a.

$$\begin{array}{ccccc} X & & & & \\ \big\downarrow{\scriptstyle \text{Supp}} & & & & \\ X' & \xrightarrow{\cong} & D' \setminus V(g') & \hookrightarrow & D' \end{array}$$

(A)

$$\begin{array}{ccccc} X & \xrightarrow{\cong} & D \setminus V(g) & \hookrightarrow & D \\ \big\downarrow{\scriptstyle \text{Supp}} & & \big\downarrow{\scriptstyle \text{Supp}} & & \big\downarrow{\scriptstyle \text{Supp}} \\ X' & \xrightarrow{\cong} & D' \setminus V(g') & \hookrightarrow & D' \end{array}$$

(B)

FIGURE 1. Compactification of $C \setminus V(g)$.

Up to an embedded isomorphism, we can treat X as an "affine scheme," i.e. a closed subscheme of an affine space (example IV.10). We now construct an affine scheme D with support D' such that the diagram in fig. 1b commutes, where the

[4]In fact $f'^{-1}(\mathbb{k})$ is also affine; we did not use this fact since all the proofs we know of it use the *Riemann-Roch* theorem, which we did not cover in this book.

"$\cong$" on the top left denotes an embedded isomorphism. Indeed, choose $\mathbb{k}$-algebra generators $f_1 = 1/g, f_2 = g, f_3, \ldots, f_k$ of $A := (\mathbb{k}[x_1, \ldots, x_n]/\mathfrak{a})_g$ such that

(vii) for each $j = 2, \ldots, k$, the image $\bar{f}_j$ in $\mathbb{k}[X']$ of f_j is in the image of the natural map $\mathbb{k}[D'] \to \mathbb{k}[X']$, and
(viii) $\mathbb{k}[D'] = \mathbb{k}[\bar{f}_2, \ldots, \bar{f}_k]$.

Let $B := \mathbb{k}[f_2, \ldots, f_k] \subset A$, and D be the "closed subscheme of $\mathbb{k}^{k-1}$ with coordinate ring B," i.e. D is the closed subscheme of $\mathbb{k}^{k-1}$ corresponding to the kernel of the surjective map $\mathbb{k}[x_2, \ldots, x_k] \to B$ which maps each x_j to f_j. Then it is straightforward to check that the diagram in fig. 1b commutes. For each $i = 1, \ldots, s$, we write D_i for the irreducible component of D containing C_i. Let $p := \prod_{s \in S}(f - s) \in \mathbb{k}[f]$, so that $p' = p|_{C'}$. Then p' is invertible in $\mathbb{k}[D']$, which implies that p is invertible in B.

Claim C.8.1. *B (respectively $\mathbb{k}[D']$) is a finitely generated module over $\mathbb{k}[f]_p$ (respectively $\mathbb{k}[f']_{p'}$).*

PROOF. Theorem IV.19 implies that each $\mathbb{k}[D_i]$ is a finitely generated $\mathbb{k}[f']_{p'}$-module. Since the natural map $\mathbb{k}[D'] \to \prod_i \mathbb{k}[D_i]$ is injective, $\mathbb{k}[D']$ is isomorphic to a $\mathbb{k}[f']_{p'}$-submodule of $\prod_i \mathbb{k}[D_i]$, and therefore also a finitely generated module over $\mathbb{k}[f']_{p'}$. Pick a finite collection $g'_1, \ldots, g'_{k'}$ of $\mathbb{k}[f']_{p'}$-module generators of $\mathbb{k}[D']$. Pick $g_i \in B$ such that $g'_i = g_i|_{D'}$, $i = 1, \ldots, k'$. Let $h_1, \ldots, h_l$ be generators of the ideal $\mathfrak{n}$ of nilpotent elements of B. There is $m \geq 0$ such that $\mathfrak{n}^{m+1} = 0$. We claim that B is generated as a $\mathbb{k}[f]_p$-module by $g_i h_1^{\alpha_1} \cdots h_m^{\alpha_m}$, $1 \leq i \leq k'$, $\alpha \in \mathbb{Z}^l_{\geq 0}$, $\alpha_1 + \cdots + \alpha_l \leq m$. Indeed, let $u \in B$. Then there are $\phi_1, \ldots, \phi_{k'} \in \mathbb{k}[f]_p$ such that $u_1 := u - \sum_{j=1}^{k'} \phi_j g_j \in \mathfrak{n}$. Then $u_1 = \sum_{j=1}^{l} u_{1,j} h_j$. Expressing the $u_{1,j}$ as $\mathbb{k}[f]_p$-linear combinations of the g_j modulo $\mathfrak{n}$ and continuing as above gives the claim. □

The fundamental theorem of finitely generated modules over a PID (theorem B.1) implies that as a $\mathbb{k}[f]_p$-module B has a decompositions of the form:

$$B \cong (\mathbb{k}[f]_p)^r \bigoplus \left(\bigoplus_j \mathbb{k}[f]_p / \langle \phi_j(f) \rangle \right)$$

such that each ϕ_j is a polynomial in an indeterminate t. For each $c \in \mathbb{k}$, it follows that $f - c$ is a zero-divisor in B if and only if $c \notin S$ (since $f - c$ is invertible in $\mathbb{k}[f]_p$) and $t - c$ divides some $\phi_j(t)$ in $\mathbb{k}[t]$; in particular, there are only finitely many such $c \in \mathbb{k}$. It is then straightforward to see that $r = \dim_{\mathbb{k}}(B/\langle f - c \rangle)$ for all $c \in \mathbb{k}$ such that $f - c$ is a non zero-divisor in B. Lemma C.7 then implies that

$$r = \sum_{\substack{b \in D \\ f(b) = c}} \operatorname{ord}_b(f - c)$$

for all $c \in \mathbb{k}$ such that $f - c$ is a non zero-divisor in $\mathbb{k}[D]$. Now pick an arbitrary point $a^* \in C \setminus \{a\}$. Applying the above construction with a^* and $C^* := C \setminus \{a\}$

respectively in place of a and C yields an affine possibly non-reduced curve D^* such that $\mathrm{Supp}(D^*)$ containing an open neighborhood of a^*, and the same arguments show that there is an integer r^* such that

$$r^* = \sum_{\substack{b \in D^* \\ f(b)=c}} \mathrm{ord}_b((f-c)|_{D^*})$$

for all $c \in \Bbbk$ such that $f - c$ is a non zero-divisor in the "coordinate ring" of D^*. Since D and D^* are "birational"[5] and since our construction guarantees that f is a non zero-divisor in coordinate rings of both D and D^*, it follows that $r = r^*$ and

$$\sum_{\substack{b \in D \\ f(b)=0}} \mathrm{ord}_b(f) = \sum_{\substack{b \in D^* \\ f(b)=0}} \mathrm{ord}_b(f|_{D^*}) \tag{164}$$

Let $a_1 = a, a_2, \ldots, a_k$ be the points of $f^{-1}(0)$ on D. For each $j > 1$, the construction of D^* shows that $a_j \in D^*$ and a_j has a neighborhood on D which is isomorphic to D^*. Let $S^* := (f|_{\mathrm{Supp}(D^*)})^{-1}(0) \setminus \{a_2, \ldots, , a_k\} \subset D^*$. Identity (164) implies that

$$\mathrm{ord}_a(f) = \sum_{b \in S^*} \mathrm{ord}_b(f|_{D^*})$$

It follows from the construction of D^* that $\mathrm{Supp}(D^*)$ is nonsingular at every point on S. For each $b \in S^*$, let $D^*_{i_b}$ be the (unique) irreducible component of D^* containing b. Proposition C.6 then implies that

$$\mathrm{ord}_a(f) = \sum_{b \in S^*} \mu_{D^*_{i_b}}(D^*)\, \mathrm{ord}_b(f|_{D^*_{i_b}})$$

Since there $D^*_{i_b}$ and C_i have isomorphic nonempty Zariski open subsets such that the corresponding open subschemes of D^* and C are isomorphic, it follows that $\mu_{D^*_{i_b}}(D^*) = \mu_{C_{i_b}}(C)$, and since the desingularization $\tilde{C}_i$ of C_i is isomorphic to $D^*_{i_b}$ near b, it follows that $\mathrm{ord}_b(f|_{D^*_{i_b}}) = \mathrm{ord}_b(\pi^*_{i_b}(f|_{C_{i_b}}))$, which proves the theorem. □

[5]i.e. there are open subschmes U, U^* respectively of D, D^* such that $\mathrm{Supp}(U)$ (respectively, $\mathrm{Supp}(U^*)$) intersects each irreducible component of D (respectively, D^*), and there is an embedded isomorphism between U and U'.

APPENDIX D

Notation

$:=$	is defined as
$\cong$	isomorphic
$\|\| \cdot \|\|$	Euclidean length
$\coprod$	disjoint union
$\langle \nu, \cdot \rangle$, $\nu \in (\mathbb{R}^n)^*$	the function induced by ν on $\mathbb{R}^n$
$\langle \beta, \cdot \rangle$, $\beta \in \mathbb{R}^n$	dot product with β
$\eta^\perp$	$\{\alpha \in \mathbb{R}^n : \langle \eta, \alpha \rangle = 0\}$
$\langle f, g, \ldots, \rangle$	ideal generated by $f, g, \ldots$
$\Bbbk[[x]]$	ring of formal power series in x with coefficients in $\Bbbk$
$\Bbbk((x))$	field of Laurent series in x with coefficients in $\Bbbk$
$\Bbbk[X]$	ring of rational functions on the algebraic variety X defined over $\Bbbk$
$[n]$, $n \in \mathbb{Z}$	$\{1, 2, \ldots, n\}$
$[T]$, $T : (\Bbbk^*)^n \to (\Bbbk^*)^m$	$n \times m$ matrix of exponents of coordinates of T
x^α, $\alpha = (\alpha_1, \ldots, \alpha_n)$	$x_1^{\alpha_1} \cdots x_n^{\alpha_n}$
$\mathbb{1}_n$	identity matrix of size $n \times n$
$\mathrm{aff}(\cdot)$	affine hull
$\mathrm{cone}(S)$	cone generated by S
$\mathrm{conv}(\cdot)$	convex hull
$\partial_i(\cdot)$	$\partial(\cdot)/\partial x_i$
$\mathrm{fund}(H)$	volume of any fundamental lattice parallelotope of H

P. Mondal, *How Many Zeroes?*, CMS/CAIMS Books in Mathematics 2,
https://doi.org/10.1007/978-3-030-75174-6_D

$\gcd(a, b, \cdots)$	greatest common positive divisor of the nonzero elements from $a, b, \cdots$
$\mathrm{In}_\nu(\cdot)$	initial form of a (Laurent) polynomial or minimizing face of a polyhedron
$\mathfrak{j}(\cdot)$	ideal generated by the partial derivatives of a polynomial
$\mathrm{lcm}(a, b, \cdots)$	lowest common positive multiple of the nonzero elements from $a, b, \cdots$
ld	leading form of a (Laurent) polynomial or maximizing face of a polyhedron
$\max_{\mathcal{P}}(\nu)$	$\max\{\langle \nu, \alpha \rangle : \alpha \in \mathcal{P}\}$
$\min_{\mathcal{P}}(\nu)$	$\min\{\langle \nu, \alpha \rangle : \alpha \in \mathcal{P}\}$
MV	mixed volume
MV'_ν	normalized mixed volume
$\mathfrak{n}(\cdot)$	nilradical
$\mathrm{ND}(\cdot)$	Newton diagram
$\mathrm{NP}(\cdot)$	Newton polytope
$\mathcal{O}_{Z,X}$	local ring of the variety X at its subvariety Z
$\mathbb{Q}_{>0}$	$\{q \in \mathbb{Q} : q > 0\}$
$\mathbb{Q}_{\geq 0}$	$\{q \in \mathbb{Q} : q \geq 0\}$
$\mathbb{R}_{>0}$	$\{r \in \mathbb{R} : r > 0\}$
$\mathbb{R}_{\geq 0}$	$\{r \in \mathbb{R} : r \geq 0\}$
$\mathrm{relint}(\cdot)$	relative interior
$\mathrm{Supp}(f)$	support of f
$\mathrm{tr.\,deg.}_k(\cdot)$	transcendence degree over (a field) k
$V(f, g, \ldots)$	the set of zeroes of (or depending on the context, the closed subscheme defined by) $f, g, \ldots$
Vol'_H	normalized lattice vlume on an affine subspace H
Vol'_ν	Vol'_H, where $H := \{\alpha \in \mathbb{R}^n : \langle \nu, \alpha \rangle = 0\}$
$\mathbb{Z}_{>0}$	$\{0, 1, 2, \ldots, \}$
$\mathbb{Z}_{\geq 0}$	$\{1, 2, \ldots, \}$

Bibliography

[AGZV85] V.I. Arnold, S.M. Gusein-Zade, A.N. Varchenko, *Singularities of Differentiable Maps*, vol. 1 (Birkhäuser, 1985)

[AM69] M.F. Atiyah, I.G. Macdonald, *Introduction to Commutative Algebra* (Addison-Wesley Publishing Co., Reading, Mass.-London-Don Mills, Ont., 1969)

[AM75] S.S. Abhyankar, T.T. Moh, Embeddings of the line in the plane. J. Reine Angew. Math. **276**, 148–166 (1975)

[And07] K. Andersen, *The Geometry of an Art. The History of the Mathematical Theory of Perspective from Alberti to Monge* (Springer, New York, NY, 2007)

[Arn04] V.I. Arnold (ed.), *Arnold's Problems*. Translated and Revised Edition of the 2000 Russian Original (Springer, Berlin; PHASIS, Moscow, 2004)

[Arr06] E. Arrondo, Another elementary proof of the Nullstellensatz. Am. Math. Mon. **113**(2), 169–171 (2006)

[AY83] L.A. Ajzenberg, A.P. Yuzhakov, *Integral Representations and Residues in Multidimensional Complex Analysis*. Transl. from the Russian by H.H. McFaden, ed. by Lev J. Leifman (1983)

[Ber75] D.N. Bernstein, The number of roots of a system of equations. Funkcional. Anal. i Priložen. **9**(3), 1–4 (1975)

[BGM12] Y. Boubakri, G.-M. Greuel, T. Markwig, Invariants of hypersurface singularities in positive characteristic. Rev. Mat. Complut. **25**(1), 61–85 (2012)

[BK86] E. Brieskorn, H. Knörrer, *Plane Algebraic Curves* (Birkhäuser Verlag, Basel, 1986). Translated from the German by John Stillwell

[BKO20] S. Brzostowski, T. Krasiński, G. Oleksik, The łojasiewicz exponent of non-degenerate surface singularities (2020), https://arxiv.org/abs/2010.06071

[BKW19] S. Brzostowski, T. Krasiński, J. Walewska, Arnold's problem on monotonicity of the Newton number for surface singularities. J. Math. Soc. Japan **71**(4), 1257–1268 (2019)

[BO16] S. Brzostowski, G. Oleksik, On combinatorial criteria for non-degenerate singularities. Kodai Math. J. **39**(2), 455–468 (2016)

[BS07] F. Bihan, F. Sottile, New fewnomial upper bounds from Gale dual polynomial systems. Mosc. Math. J. **7**(3), 387–407 (2007)

[BZ88] Yu.D. Burago, V.A. Zalgaller, *Geometric Inequalities*. Transl. from the Russian by A.B. Sossinsky. Transl. from the Russian by A.B. Sossinsky. Berlin etc.: Springer, transl. from the Russian by A.B. Sossinsky edition (1988)

[CLS11] D.A. Cox, J.B. Little, H.K. Schenck, *Toric Varieties*, Graduate Studies in Mathematics, vol. 124 (American Mathematical Society, Providence, RI, 2011)

[Dam99] J. Damon, A global weighted version of Bezout's theorem, in *The Arnoldfest (Toronto, ON, 1997)*, Fields Institute Communications, vol. 24 (American Mathematical Society,

Providence, RI, 1999), pp. 115–129

[DF04] D.S. Dummit, R.M. Foote, *Abstract Algebra*, 3rd edn. (Wiley, Chichester, 2004)

[dh] damiano (https://mathoverflow.net/users/4344/damiano). Algebraic geometry examples. MathOverflow. https://mathoverflow.net/q/34258 (version: 2010-08-02)

[Eis95] D. Eisenbud, *Commutative Algebra*, Graduate Texts in Mathematics, vol. 150 (Springer, New York, 1995). With a view toward algebraic geometry

[Est12] A. Esterov, Multiplicities of degenerations of matrices and mixed volumes of Cayley polyhedra. J. Singul. **6**, 27–36 (2012)

[FG87] J.V. Field, J.J. Gray, *The Geometrical Work of Girard Desargues* (1987)

[Ful89] W. Fulton, *Algebraic Curves. An Introduction to Algebraic Geometry*. Notes written with collab. of R. Weiss. new ed. Redwood City, CA etc.: Addison-Wesley Publishing Company, Inc., new ed. edition (1989)

[Ful93] W. Fulton, *Introduction to Toric Varieties*, Annals of Mathematics Studies, vol. 131 (Princeton University Press, Princeton, NJ, 1993). The William H. Roever Lectures in Geometry

[GK03] R. Goldman, R. Krasauskas (eds.), *Topics in Algebraic Geometry and Geometric Modeling. Proceedings of the Workshop on Algebraic Geometry and Geometric Modeling*, July 29–August 2, 2002, Vilnius, Lithuania (American Mathematical Society (AMS), Providence, RI, 2003)

[GKZ94] I.M. Gel′fand, M.M. Kapranov, A.V. Zelevinsky, *Discriminants, Resultants, and Multidimensional Determinants*. Mathematics: Theory & Applications (Birkhäuser Boston Inc., Boston, MA, 1994)

[GM12] D. Grigoriev, P.D. Milman, Nash resolution for binomial varieties as Euclidean division. A priori termination bound, polynomial complexity in essential dimension 2. Adv. Math. **231**(6), 3389–3428 (2012)

[Gra11] J. Gray, *Worlds out of Nothing. A Course in the History of Geometry in the 19th Century*, 2nd corrected edn. (Springer, London, 2011)

[Gri96] P.A. Griffiths, *Introduction to Algebraic Curves*. Transl. from the Chinese by Kuniko Weltin. 2nd edn., vol. 76 (American Mathematical Society, Providence, RI, 1996)

[hba] R. Borcherds (https://mathoverflow.net/users/51/richard-borcherds). Algebraic geometry examples. MathOverflow, https://mathoverflow.net/q/34110 (version: 2018-06-24)

[hbb] A. Bayer (https://mathoverflow.net/users/7437/arend-bayer). Algebraic geometry examples. MathOverflow, https://mathoverflow.net/q/34175 (version: 2010-08-02)

[he] G. Elencwajg (https://mathoverflow.net/users/450/georges-elencwajg). Pseudonyms of famous mathematicians. MathOverflow, https://mathoverflow.net/q/45195 (version: 2015-10-31)

[hh] Y. Huang (https://mathoverflow.net/users/1657/yuhao-huang). Algebraic geometry examples. MathOverflow, https://mathoverflow.net/q/36627 (version: 2010-08-25)

[HJS13] M.I. Herrero, G. Jeronimo, J. Sabia, Affine solution sets of sparse polynomial systems. J. Symb. Comput. **51**, 34–54 (2013)

[HJS19] M.I. Herrero, G. Jeronimo, J. Sabia, On the multiplicity of isolated roots of sparse polynomial systems. Discrete Comput. Geom. **62**(4), 788–812 (2019)

[hl] T. Leinster (https://mathoverflow.net/users/586/tom-leinster). Do the elementary properties of mixed volume characterize it uniquely? MathOverflow. https://mathoverflow.net/q/71980 (version: 2011-08-03)

[HS95] B. Huber, B. Sturmfels, A polyhedral method for solving sparse polynomial systems. Math. Comp. **64**(212), 1541–1555 (1995)

[HS97] B. Huber, B. Sturmfels, Bernstein's theorem in affine space. Discrete Comput. Geom. **17**(2), 137–141 (1997)

[ht] Anweshi (https://mathoverflow.net/users/2938/anweshi). Ways to prove the fundamental theorem of algebra. MathOverflow, https://mathoverflow.net/q/10535 (version: 2010-01-04)

[Kho78] A.G. Khovanskii, Newton polyhedra, and the genus of complete intersections. Funktsional. Anal. i Priložhen. **12**(1), 51–61 (1978)

[Kho16] A.G. Khovanskii, Newton polytopes and irreducible components of complete intersections. Izv. Math. **80**(1), 263–284 (2016)

[Kho20] A.G. Khovanskii, Newton polyhedra and good compactification theorem. Arnold Math. J. (2020)

[Kho91] A.G. Khovanskij. *Fewnomials*. Transl. from the Russian by Smilka Zdravkovska, vol. 88. (American Mathematical Society, Providence, RI, 1991)

[KK10] K. Kaveh, A.G. Khovanskii, Mixed volume and an extension of intersection theory of divisors. Mosc. Math. J. **10**(2), 343–375, 479 (2010)

[KK12] K. Kaveh, A.G. Khovanskii, Newton-Okounkov bodies, semigroups of integral points, graded algebras and intersection theory. Ann. Math. (2) **176**(2), 925–978 (2012)

[Kol07] J. Kollár, *Lectures on Resolution of Singularities*, vol. 166 (Princeton University Press, Princeton, NJ, 2007)

[Kou76] A.G. Kouchnirenko, Polyèdres de Newton et nombres de Milnor. Invent. Math. **32**(1), 1–31 (1976)

[LM09] R. Lazarsfeld, M. Mustaţă, Convex bodies associated to linear series. Ann. Sci. Éc. Norm. Supér. (4) **42**(5), 783–835 (2009)

[LRW03] T.-Y. Li, J. Maurice Rojas, X. Wang, Counting real connected components of trinomial curve intersections and *m*-nomial hypersurfaces. Discrete Comput. Geom. **30**(3), 379–414 (2003)

[LW96] T.Y. Li, X. Wang, The BKK root count in $\mathbf{C}^n$. Math. Comp. **65**(216), 1477–1484 (1996)

[Mil68] J.W. Milnor, Singular points of complex hypersurfaces. *Annals of Mathematics Studies*. No. 61 (Princeton University Press and the University of Tokyo Press, Princeton, N.J., 1968), 122 p

[Min41] F. Minding, Ueber die bestimmung des grades einer durch elimination hervorgehenden gleichung. J. Reine Angew. Math. **1841**(22), 178–183 (1841)

[Mon10] P. Mondal, Towards a Bezout-type theory of affine varieties, http://hdl.handle.net/1807/24371, March 2010. Ph.D. Thesis

[Mon16] P. Mondal, Intersection multiplicity, Milnor number and Bernstein's theorem, http://arxiv.org/abs/1607.04860 (2016)

[MS87] A. Morgan, A. Sommese, A homotopy for solving general polynomial systems that respects m-homogeneous structures. Appl. Math. Comput. **24**, 101–113 (1987)

[Mum95] D. Mumford, *Algebraic Geometry. I*. Classics in Mathematics (Springer, Berlin, 1995). Complex projective varieties, Reprint of the 1976 edition

[Oda88] T. Oda, *Convex Bodies and Algebraic Geometry. An Introduction to the Theory of Toric Varieties*, vol. 15 (Springer, Berlin, 1988)

[Oka79] M. Oka, On the bifurcation of the multiplicity and topology of the Newton boundary. J. Math. Soc. Japan **31**, 435–450 (1979)

[Oka97] M. Oka, *Non-degenerate Complete Intersection Singularity* (Hermann, Paris, 1997)

[Oko96] A. Okounkov, Brunn-Minkowski inequality for multiplicities. Invent. Math. **125**(3), 405–411 (1996)

[Oko03] A. Okounkov, Why would multiplicities be log-concave? in *The Orbit Method in Geometry and Physics*. In honor of A. A. Kirillov. Papers from the International Conference, Marseille, France, December 4–8, 2000 (Birkhäuser, Boston, MA, 2003), pp. 329–347

[PS08] P. Philippon, M. Sombra, A refinement of the Bernštein-Kušnirenko estimate. Adv. Math. **218**(5), 1370–1418 (2008)

[Rab29] J.L. Rabinowitsch, Zum Hilbertschen Nullstellensatz. Math. Ann. **102**, 520 (1929)

[Rei88] M. Reid, *Undergraduate Algebraic Geometry*, vol. 12 (Cambridge University Press, Cambridge (UK), 1988)

[Roj94] J.M. Rojas, A convex geometric approach to counting the roots of a polynomial system. Theoret. Comput. Sci. **133**(1), 105–140 (1994). Selected papers of the Workshop on Continuous Algorithms and Complexity (Barcelona, 1993)

[Roj99] J.M. Rojas, Toric intersection theory for affine root counting. J. Pure Appl. Algebra **136**(1), 67–100 (1999)

[Ros05] B.A. Rosenfeld, The analytic principle of continuity. Am. Math. Mon. **112**(8), 743–748 (2005)

[RS17] G. Rond, B. Schober, An irreducibility criterion for power series. Proc. Am. Math. Soc. **145**(11), 4731–4739 (2017)

[rshs] roy smith (https://mathoverflow.net/users/9449/roy-smith). Algebraic geometry examples. MathOverflow, https://mathoverflow.net/q/44301 (version: 2010-10-31)

[RW96] J.M. Rojas, X. Wang, Counting affine roots of polynomial systems via pointed Newton polytopes. J. Complexity **12**(2), 116–133 (1996)

[Sch98] A. Schrijver, *Theory of Linear and Integer Programming*, Repr. edition (Wiley, Chichester, 1998)

[Sha94] I.R. Shafarevich, *Basic Algebraic Geometry. 1*, 2nd edn. (Springer, Berlin, 1994). Varieties in projective space, Translated from the 1988 Russian edition and with notes by Miles Reid

[Sot11] F. Sottile. *Real solutions to equations from geometry*, vol. 57. (American Mathematical Society (AMS), Providence, RI, 2011)

[Spi90] M. Spivakovsky, Sandwiched singularities and desingularization of surfaces by normalized Nash transformations. Ann. Math. (2) **131**(3), 411–491 (1990)

[Ste20] J. Stevens, Conjectures on stably newton degenerate singularities. Arnold Math. J. **7**(3), 441–465 (2021). https://arxiv.org/abs/1406.0328

[Wal99] C.T.C. Wall, Newton polytopes and non-degeneracy. J. Reine Angew. Math. **509**, 1–19 (1999)

[Wat79] W.C. Waterhouse, Gauss on infinity. Hist. Math. **6**, 430–436 (1979)

[Zie95] G.M. Ziegler, *Lectures on Polytopes*, Graduate Texts in Mathematics, vol. 152. (Springer, New York, 1995)

[ZS75a] O. Zariski, P. Samuel, *Commutative Algebra. Vol. 1*. With the cooperation of I. S. Cohen. 2nd edn., vol. 28 (Springer, New York, NY, 1975)

[ZS75b] O. Zariski, P. Samuel, *Commutative Algebra. Vol. II* (Springer, New York, 1975). Reprint of the 1960 edition, Graduate Texts in Mathematics, vol. 29

Index

P. Mondal, *How Many Zeroes?*, CMS/CAIMS Books in Mathematics 2,
https://doi.org/10.1007/978-3-030-75174-6

Printed in the United States
by Baker & Taylor Publisher Services